Factories of the Future

Tullio Tolio · Giacomo Copani
Walter Terkaj

Editors

Factories of the Future

The Italian Flagship Initiative

Editors
Tullio Tolio
Director of the Italian Flagship Project
"Factories of the Future", Direttore del
Progetto Bandiera "La Fabbrica del Futuro"
CNR - National Research Council of Italy
Rome, Italy

and

Dipartimento di Meccanica
Politecnico di Milano
Milan, Italy

Giacomo Copani
CNR-STIIMA, Istituto di Sistemi e
Tecnologie Industriali Intelligenti per il
Manifatturiero Avanzato
Milan, Italy

Walter Terkaj
CNR-STIIMA, Istituto di Sistemi e
Tecnologie Industriali Intelligenti per il
Manifatturiero Avanzato
Milan, Italy

https://doi.org/10.1007/978-3-319-94358-9

Preface

Manufacturing plays a key role both in advanced economies and developing countries because of the large contribution to the overall employment, value added, gross domestic product (GDP) and social welfare. Manufacturing is also a pillar for the tertiary sector, since manufacturing activities generate the need for services and in turn manufacturing provides products and technologies for the operation of the service sector. Furthermore, manufacturing is fundamental to guarantee national independence and security and to design the future of our societies. Continuously evolving grand challenges compel the manufacturing sector to innovate its processes, technologies and business models to continue sustaining the national economies and progress.

This book presents the philosophy and the findings of the Italian Flagship Project *Factories of the Future* (*La fabbrica del futuro 2012–2018*). This flagship project was a major national research program promoted by the Italian Ministry of University, Innovation and Research (MIUR) and coordinated by the National Research Council of Italy (CNR) to innovate the manufacturing sector and address global challenges. Starting from an analysis of research policies, Chap. 1 outlines the main ongoing research programs and initiatives both at international and Italian level. Among these initiatives, the Italian Flagship Project *Factories of the Future* is presented in details. The roadmap for research and innovation implemented by the flagship project is based on five research priorities that can be seen as different views of the same factory of the future: *Evolutionary and Reconfigurable Factory, Sustainable Factory, Factory for the People, Factory for Customised and Personalised Products, Advanced-Performance Factory*.

On the basis of the five research priorities, the flagship project funded 18 research projects and 14 demonstrators. The findings of the specific scientific and technological research projects and demonstrators are reported in Chaps. 2–19.

Chapter 20 proposes seven future *missions* resulting from the flagship project that can be set ahead for the manufacturing industry. Missions are of vital importance to guarantee the evolution of our societies by means of new systemic solutions. Missions will also foster the important role of manufacturing as a backbone for the employment and wealth of our national and European economies. Indeed missions such as *Circular Economy*, *Rapid and Sustainable Industrialisation*, *Robotic Assistant*, *Factories for Personalised Medicine*, *Internet of Actions*, *Factories close to the People*, and *Turning Ideas into Products* will have a relevant societal impact and at the same time will require to address significant scientific and technological challenges which will be particularly important in view of the next strategic initiatives at national and European level, including *Horizon Europe*. Moreover, the demonstration and exploitation of results related to missions require proper research infrastructures. Therefore, Chap. 21 analyzes and gives examples of different types of pilot plant together with a discussion about funding mechanisms needed to support industrial research and make pilot plants sustainable.

Milan, Italy

Tullio Tolio
Giacomo Copani
Walter Terkaj

Acknowledgements

The Director of the Italian Flagship Project *Factories of the Future* (*La fabbrica del futuro*) gratefully thanks Prof. Francesco Jovane for his visionary approach to manufacturing research that triggered the launch of the flagship project Factories of the Future in the context of the National Research Plan (PNR 2011–2013). Many thanks also to Prof. Quirico Semeraro and Prof. Vincenzo Nicolò for the scientific supervision and guidance of the activities of the two main streams of the flagship project. A special appreciation goes to Dott.ssa Federica Rossi, Vice-director of the flagship project. Finally, warm thanks to the present and past members of the Implementation Support Group (ISG) of the Flagship Project *Factories of the Future*: Walter Terkaj, Giacomo Copani, Eleonora Schiariti, Emanuela Alfieri, Daniele Dalmiglio, Davide Ceresa, and Anna Valente.

Contents

Part I
Introduction

Chapter 1
The Italian Flagship Project: Factories of the Future

Walter Terkaj and Tullio Tolio

Abstract This chapter deals with the central role of manufacturing in developed and developing countries, assessing how relevant it is from economic and social perspectives. The current international and Italian manufacturing contexts are analysed by highlighting the main criticalities and the impact of relevant global megatrends. Then, the main ongoing industrial research initiatives are presented both at international and Italian level. Based on the elaboration of current context and research initiatives, the Italian Flagship Project *Factories of the Future* defined five research priorities for the future of the manufacturing industry. Based on these priorities, the flagship project funded a total of 18 small-sized research projects after a competition based on calls for proposals. The results of the funded research projects are analysed in terms of scientific and industrial results, while providing references for more detailed descriptions in the specific chapters.

1.1 The Importance of Manufacturing Industry

Manufacturing industry plays a central role in the economy of developed countries for the generation of wealth, jobs and a growing quality of life. Beyond being a sector capable of directly producing wealth and employment, manufacturing industry is a fundamental pillar of the whole economy. In 2016 the value added of manufacturing represented 15.6% and 14.3% of the Gross Domestic Product (GDP) at world and European Union level, respectively. A relevant number of persons are directly

W. Terkaj (✉)
CNR-STIIMA, Istituto di Sistemi e Tecnologie Industriali Intelligenti per il Manifatturiero Avanzato, Milan, Italy
e-mail: walter.terkaj@stiima.cnr.it

T. Tolio
Director of the Italian Flagship Project "Factories of the Future", Direttore del Progetto Bandiera "La Fabbrica del Futuro", CNR - National Research Council of Italy, Rome, Italy

T. Tolio
Dipartimento di Meccanica, Politecnico di Milano, Milan, Italy

T. Tolio et al. (eds.), *Factories of the Future,*
https://doi.org/10.1007/978-3-319-94358-9_1

3

employed in industries, in particular 22.5% and 24% at world and European Union level, respectively.[1] During the last ten years the share of manufacturing value added with respect to the GDP has been declining in developed areas like USA, European Union, Japan, whereas the share of service value added has been increasing. However, it must be stressed that services and manufacturing are strictly interwoven. Indeed, manufacturing industry produces the goods needed to support the delivery of several services and, more important, it generates the need of acquiring new services, thus increasing the demand in the market. The interactions between manufacturing and services can be found along whole industrial value chain [1]:

- Upstream services in the value chain, e.g. product design, innovation activities, research and development. The acceleration of production and information technologies innovation requires more and more specific and advanced scientific and technical support.
- Core services in the value chain, e.g. services strictly related to production activities like supply management, process engineering, production engineering, maintenance services.
- Downstream services in the value chain, e.g. marketing, distribution, pre- and after-sales services to generate further value added. The concept of servitization [2] is included in these services; about 4% of manufacturing gross output was due to secondary services, i.e. services sold together with the product [1].
- Transversal services, e.g. Information and Communication Technologies (ICT) related services, management and strategy consulting to support an global enhancement of company competitiveness.

In Europe the average content of services in manufactured goods accounts for about 40% of their total value. The most relevant services are related to distribution (15%), transport and communication (8%), and business [1]. Moreover, it is important to highlight the role of manufacturing as a job multiplier, i.e. each direct job in manufacturing leads to additional jobs in service activities. In USA it has been estimated that the manufacturing multiplier is equal to 1.58 on average [3], but it can be higher depending on the region and how technology intensive is the manufacturing sector. In the European Union one out of four jobs in the private sector is in manufacturing industry, and at least another one out of four is in services that depend on industry as a client or supplier [4].

The world economy is constantly evolving through scenarios of global change and development that have an impact on the lives of people, companies and communities, thus generally influencing the society and the economy [5]. The analysis of socio-economic megatrends is important to understand and anticipate what a sustainable manufacturing industry will have to cope with in the future [6]. Among the others, it is possible to identify five megatrends that will deeply affect the structure of industry: demographic change, new emerging markets, scarcity of resources, climate change and acceleration of technology process [7].

---

[1] Source: World Bank Open Data, https://data.worldbank.org/.

The global population is expected to grow from 7.55 billion 2017 to 8.55 billion in 2030 and 9.77 billion in 2050, with a growth rate much higher in developing countries.[2] Together with the expected increase of the overall population, also the population ageing is a relevant process at global level. By 2030, over 22% of the population in high-income countries will consist of people with age 65 and above, thus determining a significant rise of the old-age dependency rate. These demographic changes will pave the way for the need of always more customized products (niche products), as well as services, to cope with new specific needs in terms of comfort, health and well-being of individuals or communities. In addition, it will be necessary to find a right balance between the need to let over-65 people prolong their working life and the need to offer job opportunities for the young generations while improving, at the same time, the worker well-being in terms of satisfaction, safety and inclusivity. Moreover, it is foreseen that almost 60% of the global population will be living in cities by 2030 pushing the development of the so-called megacities.[3] Hence, new production models implementing *urban manufacturing* strategies (factories as a good neighbour of cities) needs to be settled to permit the workers to combine the work with their personal lives [8].

The extraordinary population growth in the developing countries will push also the growth of their economies, even though at lower rates. This will lead to a significant increase in the middle class population. If the middle class population was about three billion people in 2015 (half of them living in Asia), then it is expected to be over five billion by 2030 [9], thus creating new emerging markets. In this scenario, the manufacturing industry will have to cope with continuously mutating market conditions being able to manage complex global networks of enterprises. Indeed, the globalisation process has already created new markets but also new competitors, therefore the need of innovation becomes stronger and stronger. Hence, factors such as a strong industrial tradition, manufacturing culture, consolidated design skills, the presence of research institutions and technology transfer centres will play a key role to enhance competitive advantages.

An overall increase in natural resource consumption is foreseen because of the socio-economic growth in developing countries. In particular, the world energy consumption is expected to increase by 28% between 2015 and 2040. The increase will be higher in non-OECD countries (+41%) than in OECD countries (+9%) [10]. Also the consumption of water, food and several raw materials are going to increase in the next few years. Therefore, manufacturing industry needs to adopt the circular economy paradigm [11] to considerably reduce waste through re-use, remanufacturing and recycling. At the same time, production systems should become always more efficient in the use of raw materials.

A more efficient use of the resources is also driven by the need to cope with climate change that has started to affect the overall planet. Hence, economies need to become more resilient to change, efficient in the use of resources and rely on high levels of eco-innovation to remain competitive.

---

[2] Source: United Nations, DESA/Population Division, https://esa.un.org/unpd/wpp/DataQuery/.

[3] Source: World Bank Open Data, Population estimates and projections, https://data.worldbank.org/.

Finally, the continuous acceleration of the technological progress, the integration of advanced technologies and cross-fertilisation of different technologies will be increasingly important to enhance innovation and create new markets. In addition, greater flexibility and reconfigurability will be essential requirements for the future manufacturing industry to cope with the turbulence of markets and the unpredictable changes in demand.

1.2 Italian Manufacturing Industry

In 2016 Italy was the seventh manufacturing country in the world in terms of value added (2.2% of the total), even though the position is declining since it was at the fifth place in 2007 (4.9% of the total). Indeed, the manufacturing sector of countries like China, India, and Rep. of Korea is growing at a higher rate. China has already overtaken the USA manufacturing sector as the first in the world, representing 26.2% of the global manufacturing value added.

In Europe, Italy is the second manufacturing country behind Germany, in spite of representing just the fourth Gross Domestic Product (GDP)[4] behind Germany, France and United Kingdom.

Manufacturing plays a key role in the Italian economy since its total output was equal to 897 billion euro (28.6% of the total) and the value added equal to 245 billion euro (16.3% of the total) in 2016.[5] The detailed contribution of the various manufacturing sectors is shown in Fig. 1.1. The production of machinery and equipment is the top sector. Moreover, Italy is among the largest producers in all the sectors and in the textile sector holds the first position, as shown in Table 1.1.

In 2016 almost 3.9 million Italian people were directly employed in the manufacturing industry (15.5% of the total), earning 20.6% of the total compensation of employees. However, the overall unemployment rate at 11.7% in 2016 was still higher than the pre-crisis level and youth unemployment was particularly significant; moreover, a significant regional heterogeneity can be noticed [12].

In spite of the still relevant position of its manufacturing sector, the Italian economy has been slowly declining and is characterized by historical problems related to productivity [13] that is not converging in the euro area [14]. Indeed, the real GDP per hour has been flat in Italy, whereas it has been growing in other countries of the euro area [12]. After the recessions in 2008–2009 and again in 2011–2013, the Italian manufacturing sector has been slowly recovering competitiveness, even if at a lower rate than the other main European countries. The second recession of the years 2011–2013 was mainly caused by a decreasing or stagnant internal demand [15] and a positive role was played by the export of Italian manufacturing companies, that keeps on growing (+1.1% in 2016) and represents 3% of the world export, even though the world trade is slowing down [15]. However, a growth based on export is

---

[4]Source: World Bank Open Data, https://data.worldbank.org/.

[5]Source: eurostat, http://ec.europa.eu/eurostat/data/database.

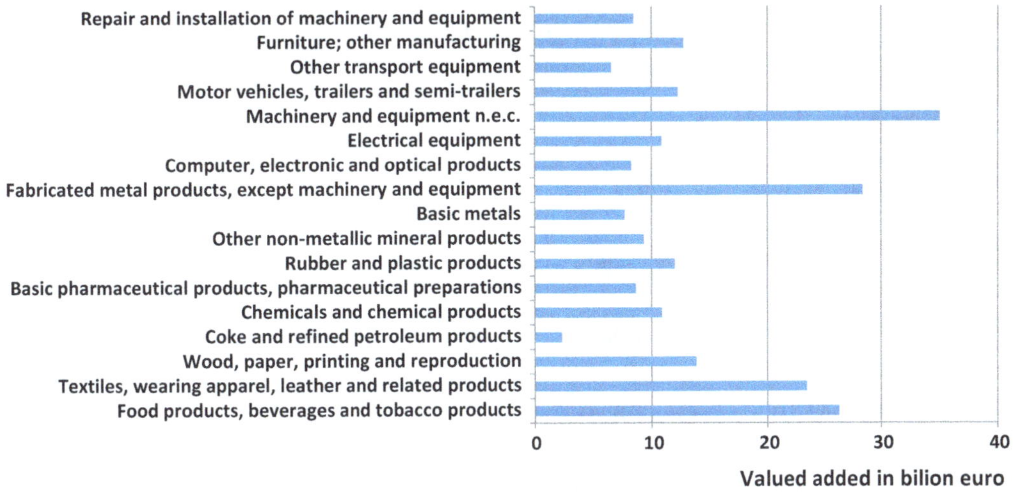

Fig. 1.1 Value added of the Italian manufacturing sectors (*Source* Eurostat 2015, NACE Rev.2 A*64 classification)

hardly sustainable in the long run and exposes the economy to external shocks (like it happened in 2008). Moreover, it must be noted that the Italian post-crisis recovery of competitiveness has been achieved also thanks to a general policy of wage moderation, in particular real public wages have decreased since the start of the recession mainly because of missed inflation adjustments [12]. Even if wage moderation and the enhancement of labour market flexibility may lead to short-term increase of competitiveness, these policies have counterproductive consequences in the mid-long run, because on average labour and total factor productivity are depressed [14].

Given the current scenario and boundary conditions, investments in industrial research and innovation represent one of the few viable options to improve the competitiveness of Italian manufacturing companies and economy in general. In particular, a higher share of ICT investments may lead to a long-run increase in labour productivity [14]. Therefore, the following sections will present the main international (Sect. 1.3) and Italian (Sect. 1.4) research initiatives on manufacturing.

1.3 International Research Initiatives on Manufacturing

The Technology Platform Manu*future*[6] [16] was launched in the first decade of the new millennium to develop and implement the European research and innovation strategy. The Strategic Research Agenda published in 2006 [17] proposed a change of paradigm by fostering the industrial transformation to high-added-value products, processes and services, while keeping a relevant share in the future world manufacturing output and protecting the employment in a knowledge-driven economy [18]. The following key pillars of the future manufacturing were identified:

---

[6]http://www.manufuture.org/.

Table 1.1 Position of Italian manufacturing sectors in Europe

Manufacturing sector	Largest value added in Europe	Position of Italy in Europe
Food products, beverages and tobacco products	France	5
Textiles, wearing apparel, leather and related products	Italy	1
Wood, paper, printing and reproduction	Germany	3
Coke and refined petroleum products	Germany	5
Chemicals and chemical products	Germany	5
Basic pharmaceutical products, pharmaceutical preparations	Switzerland	5
Rubber and plastic products	Germany	3
Other non-metallic mineral products	Germany	2
Basic metals	Germany	2
Fabricated metal products, except machinery and equipment	Germany	2
Computer, electronic and optical products	Germany	5
Electrical equipment	Germany	2
Machinery and equipment n.e.c.	Germany	2
Motor vehicles, trailers and semi-trailers	Germany	3
Other transport equipment	France	5
Furniture; other manufacturing	Germany	2
Repair and installation of machinery and equipment	France	4

Source Eurostat 2016, NACE Rev.2 A*64 classification

- New added-value products and product/services
- Advanced industrial engineering
- New business models
- Infrastructure and education
- Emerging manufacturing sciences and technologies

The financial crisis of 2007–2008 showed even more the importance of manufacturing in the European economy. A *recovery plan* for the European economy was published in 2008 [19], leading to the formation of Public Private Partnerships (PPP) between the European Commission and private companies to support the investments in strategic areas and activities:

- Factories of the Future (FoF)[7]
- Energy-efficient Buildings (EeB)
- Sustainable Process Industry (SPIRE)
- European Green Vehicles Initiative (EGVI)

After the initial PPPs proved the viability of the approach, other PPPs were launched in the following years:

- Robotics
- Photonics
- Advanced 5G networks for the Future Internet (5G)
- High Performance Computing (HPC)

Each PPP is linked to an association representing the private side that interfaces with the public sector (i.e. the European Union). EFFRA[8] (European Association for the Factories of the Future) is the private association of FoF PPP and is composed of companies, trade associations, universities and research institutes. During the years EFFRA produced strategic roadmaps continuously updating the research priorities based on the socio-economic context and the technological progress. The first strategic multi-annual roadmap [20] identified four industrial needs and R&D challenges:

- Sustainable manufacturing
- High performance manufacturing
- ICT-enabled intelligent manufacturing
- Exploiting new materials through manufacturing

In 2013 EFFRA prepared a new multi-annual roadmap [21] providing input for the definition of open calls in the Horizon 2020 Program.[9] The following research and innovation priorities were proposed:

- *Advanced manufacturing processes.* The products of the future are expected to be more complex (3D, nano-micro-meso-macro-scale, smart), therefore innovative manufacturing processes need to be developed to provide complex and enhanced functionalities in a cost effective way.
- *Adaptive and smart manufacturing systems.* The manufacturing industry of the future should also respond and adapt in an agile manner to the mutating market and factory demands, developing innovative manufacturing equipment at component and system level, including mechatronics, control and monitoring systems while exploiting intelligent robots and machines that can work with human operators in a safe, autonomous and reliable manner.

---

[7] https://www.effra.eu/factories-future.

[8] https://www.effra.eu/.

[9] https://ec.europa.eu/programmes/horizon2020/.

- *Digital, virtual and resource-efficient factories.* Industrial plants (assets, inventories, production and assembly lines) need to be designed, monitored and maintained thorough new paradigms based on integrated and scalable digital factory models with multi-level semantic access to all the factory resources (i.e. assets, machines, workers and objects).
- *Collaborative and mobile enterprises.* A strong collaborative network and a highly dynamic supply chain are becoming more and more key factors for manufacturing companies. Innovation efforts to make collaborative enterprises mobile enable to quickly take decisions along the value chain. Innovative and user friendly mobile manufacturing applications help to take the decisions independently from the location of the enterprise or the decision-maker.
- *Human-centred manufacturing.* Future factories can increase flexibility, agility, and competitiveness by enhancing the role played by workers that continuously develop their skills and competencies. New technologies help to transfer skills to new generations of workers, while assisting ageing, disabled and multi-cultural workers with better information.
- *Customer-focused manufacturing.* Customers have been demonstrating to be able to influence product development, therefore factories of the future will need to follow a user-centred paradigm to make an impact on the market. Customers will be involved in the manufacturing value chain, from product and process design to innovative services by collecting explicit or tacit customer requirements.

Taking in consideration the wave of Industry 4.0, EFFRA produced recommendations for the work programme 18-19-20 of the FoF PPP under Horizon 2020 [22] by identifying five key priorities:

- Agile Value Networks: lot-size one and distributed manufacturing.
- The Human Factor: human competences in synergy with technological assets.
- Excellence in Manufacturing: advanced manufacturing processes and services for zero-defect processes and products.
- Interoperable digital manufacturing platforms: connecting manufacturing services.
- Sustainable Value Networks: manufacturing in a circular economy.

Several platforms, networks and clusters have been created in the European countries aiming at fostering economic and industrial innovation. For instance, the *Platform Industrie 4.0* in Germany,[10] *Usine du Future*[11] and *Alliance Industrie du Futur*[12] in France, *Catapult* network[13] and its *High Value Manufacturing (HVM)* division[14] in UK, *Piano Impresa 4.0*[15] and the Technology Cluster *Intelligent Factories*[16] in

---

[10] https://www.plattform-i40.de.

[11] http://industriedufutur.fim.net/.

[12] http://www.industrie-dufutur.org/.

[13] https://catapult.org.uk/.

[14] https://hvm.catapult.org.uk/.

[15] http://www.sviluppoeconomico.gov.it/index.php/it/industria40.

[16] http://www.fabbricaintelligente.it/english/.

Italy (see Sect. 1.4.2). Furthermore, the German, Italian, and French governments launched a *Trilateral Cooperation for Smart Manufacturing* in 2017 to promote digitising manufacturing. This cooperation deals with three topics: (1) Standardisation and reference architectures; (2) small and medium-sized enterprises (SMEs) engagement and testbeds; (3) Policy support.

As it happened in Europe, also the USA acknowledged the central role of manufacturing after the 2008–2009 crisis. The American Recovery and Reinvestment Act of 2009 [23] included two billion dollars dedicated to grants for the manufacturing of advanced batteries and components, a temporary expansion of availability of industrial development bonds to facilities manufacturing intangible property (i.e. patents, know-how, copyrights, formula and similar categories), and credit for investment in advanced energy facilities. The report on *Ensuring American Leadership in Advanced Manufacturing* [24] pointed out that the United States was losing the global leadership in manufacturing while other countries were advancing their industries and R&D. Advanced manufacturing is essential to national security and has the potential to create and retain high-quality jobs in the United States. In particular, the report advised the Federal Government to launch an Advanced Manufacturing Initiative (AMI) providing coordinated federal support to academia and industry, public-private partnerships, development and dissemination of design methodologies, shared facilities and infrastructure to help SMEs. This recommendation led to the creation of *Manufacturing USA*[17] (also known as the National Network for Manufacturing Innovation program) after the *Revitalize American Manufacturing and Innovation Act of 2014* was approved [25]. *Manufacturing USA* takes care of coordinating federal resources and programs and has already established 14 manufacturing innovation institutes based on public-private partnerships.

1.4   Italian Research Initiatives on Manufacturing

Similarly to what happened at international level (Sect. 1.3), a set of complementary actions have been established also at Italian level to involve and enhance manufacturing excellences in the international competitive scenario, paying particular attention to advanced and high value added manufacturing. Among these actions, the National Research Program (PNR, Programma Nazionale della Ricerca) 2011–2013 [26] identified two key initiatives: Flagship Projects (Sect. 1.4.1) and Technological Clusters (Sect. 1.4.2). More recently in 2017, the Ministry of Economic Development in Italy (MISE) designed the *National Plan Enterprise 4.0* (Sect. 1.4.3).

---

[17]https://www.manufacturingusa.com/.

1.4.1 Flagship Projects

Flagship projects are research programmes approved by the Interministerial Committee for Economic Planning (CIPE) and funded by the Italian Ministry of Education, Universities and Research (MIUR) to address strategic themes that represent a national priority. These themes were selected after an assessment based on the contributions provided by public research bodies. PNR 2011–13 identified 14 flagship projects and, among them, seven were coordinated by the National Research Council of Italy (CNR), including the project *Factories of the Future* (*La Fabbrica del Futuro – Piattaforma manifatturiera nazionale*) that will be presented in Sect. 1.5. Indeed, according to PNR, CNR has a multi-disciplinary mission supporting MIUR, fostering the Italian presence in international projects, and facilitating the integration of various actors (e.g. research institutes, universities, local institutions, governmental institutions, companies and industrial consortia) involved in research activities by favouring the technology transfer. The total budget allocated to flagship projects was equal to 1.77 billion euro.

1.4.2 Technological Clusters

Technological Clusters[18] are networks of public and private partners working at national level to foster industrial research, education and training, and technology transfer. The main goal is to gather and better exploit resources to answer the specific needs of the regions and markets.

Each cluster is dedicated to a focused technological and application area playing a relevant role for the national population. Clusters are supposed to coordinate proposals and strategies to increase innovation and industrial competitiveness by means of the following actions:

- Positioning the national research and industrial system in the international scenario.
- Collecting and sharing best practices and competences while fostering networking and collaborations between industry and research.
- Supporting the cooperation between national, regional and local policy makers in the research and innovation field.
- Optimising the exploitation of research and innovation programs that are coherent with the national and European initiatives.
- Working to create the best environment to attract investments and competencies.
- Promoting *Made in Italy* excellence.

In 2012 the MIUR Ministry, after assessing various proposals, approved the institution of eight technological clusters: Aerospace (Aerospazio), Agrifood, Green Chemistry (Chimica verde), Intelligent Factories (Fabbrica Intelligente), Transport and mobility systems of land and ocean surface (Mezzi e sistemi per la mobilità

18http://www.miur.gov.it/cluster.

di superficie terrestre e marina), Advanced Life Science (Scienze della Vita), Smart Living Technologies (Tecnologie per gli ambienti di vita), Smart Communities Technologies (Tecnologie per le Smart Communities). Four technological clusters were added in 2016 to fully cover the 12 research priority areas defined in the PNR for years 2015–2020: Cultural Heritage Technologies (Tecnologie per il Patrimonio Culturale), Design, creativity and Made in Italy (Design, creatività e Made in Italy), Maritime economy (Economia del Mare), Energy (Energia).

Among the technological clusters, the Cluster Intelligent Factories[19] (CFI) is focused on the manufacturing domain and developed a Roadmap for research and innovation including the following strategic action lines [7]:

- Strategies for personalised production.
- Strategies, methods and tools for industrial Sustainability.
- Factories for Humans.
- High-efficiency production systems.
- Innovative production processes.
- Evolutive and adaptive production systems.
- Strategies and management for next generation production systems.

1.4.3   *National Plan Enterprise 4.0*

The *National Plan Enterprise 4.0* (Piano Nazionale Impresa 4.0)[20] was developed by MISE to provide the manufacturing companies with incentives for investments in innovation and competitiveness, mainly focused on digitising industry (so-called *Industry 4.0*). The plan includes the following main initiatives:

- Super-and hyper-depreciation as a way to support the implementation of advanced technologies in Italian manufacturing companies.
- Support for companies requesting bank loans to invest in capital goods, machinery, plant, and digital technologies.
- Tax credit for research and development.
- Special rate of taxation for revenues related to the use of intellectual property rights (e.g. industrial patents).
- Support for innovative start-ups and SMEs.
- Supports for large investments in the industrial, tourism and environmental protection sectors.
- Co-funding of projects related to industrial research and experimental development of new products, processes, services or the relevant improvement of already existing ones.
- Support for the training of employees in the areas of sales and marketing, computer science and techniques, production technologies.

---

[19]http://www.fabbricaintelligente.it/english/.

[20]http://www.sviluppoeconomico.gov.it/index.php/it/industria40.

Among the ongoing initiatives coordinated by MISE it is possible to mention the Lighthouse Plants, i.e. greenfield or brownfield production plants that are designed to make extensive use of Industry 4.0 technologies. A lighthouse plant is characterized by a significant innovation project (10 million euro or more) that continuously provides results to be implemented into the production plant. Therefore, the lighthouse plant is designed to evolve through the years while becoming a national and international reference to show best practices for technological development. The Cluster Intelligent Factories supports MISE to identify and propose lighthouse plants and later to coordinate the awareness and dissemination of their results. Currently, four Lighthouse plants have already been approved by MISE and two of them started their activities (cf. Chap. 21 for a detailed description of the Lighthouse plants initiative [27]).

1.5 Flagship Project Factories of the Future

The Italian Flagship Project *Factories of the Future*[21] (*La Fabbrica del Futuro – Piattaforma manifatturiera nazionale*) is one of the 14 Flagship Projects (Sect. 1.4.1) that started in January 2012 and lasted till December 2018 with a total funding of 10 million euro. CNR was the coordinator of this flagship project and played a key role both as research body and also as facilitator and integrator of the various actors involved in the activities: research institutes, universities, government institutions, manufacturing companies and industrial consortia.

The flagship project defined five strategic macro-objectives (Sect. 1.5.1) for factories of the future to be pursued thanks to the development, enhancement, and application of key enabling technologies (Sect. 1.5.2). The strategic macro-objectives took inspiration from the research priorities identified at international level (Sect. 1.3), while considering the evolution of the global industrial contexts (Sect. 1.1) and, above all, the peculiarities of the Italian manufacturing context (Sect. 1.2).

The flagship project was organised into two subprojects to create a stable national community characterized by scientific excellence of research. *Subproject 1*[22] aimed at establishing a multi-disciplinary cooperation among research organisations operating in specific scientific domains to exploit synergies for the development of integrated solutions. The goal of *Subproject 2*[23] was to strengthen the systemic cooperation among national research centres and universities with complementary competences.

Each subproject published a call for proposals based on the strategic macro-objectives to increase the competitiveness of the Italian manufacturing industry, paying particular attention to *Made in Italy* products in the global context. Section 1.5.3 presents the calls for proposals and the evaluation process that led to the selection

---

[21] http://www.fabbricadelfuturo-fdf.it/?lang=en.

[22] http://www.fabbricadelfuturo-fdf.it/projects/subproject-1/?lang=en.

[23] http://www.fabbricadelfuturo-fdf.it/projects/subproject-2/?lang=en.

of 18 small-sized research projects. The key results of the funded research projects together with the main dissemination activities of the flagship project are presented in Sect. 1.5.4.

1.5.1 Strategic Macro-objectives

Five strategic macro-objectives were defined for factories of the future:

- Evolutionary and Reconfigurable Factory (Sect. 1.5.1.1)
- Sustainable Factory (Sect. 1.5.1.2)
- Factory for the People (Sect. 1.5.1.3)
- Factory for Customised and Personalised Products (Sect. 1.5.1.4)
- Advanced-Performance Factory (Sect. 1.5.1.5)

These objectives are not disjoint and must be intended as complementary perspectives concurring together to build the holistic concept of *Factories of the Future*.

1.5.1.1 Evolutionary and Reconfigurable Factory

A relevant characteristic of high-tech *Made in Italy* products (e.g. machinery, medical equipment, mechanical products, textile and wearing apparel) is the continuous reduction of product life cycles, because of the fast innovation of materials, ICT, artificial intelligence, mechatronics, and the fast evolving needs of the client.

Complex and variable market demands combined with the technological evolution of products and processes lead to the need of factories that are able to react and evolve themselves by exploiting flexibility [28], reconfigurability [29], changeability [30–32], and scalability [33] to stay competitive in dynamic production contexts [34].

Enabling technologies (Sect. 1.5.2) such as high automation, ICT tools, digital twins, and an integrated and efficient logistics can provide factories with the ability to change processes and configurations in a fast and cost effective way.

Hence, production systems must be endowed with operational flexibility and a high reconfigurability to cope with the co-evolution of product-process-manufacturing systems [35]. Production systems will be required to evolve and reconfigure themselves at various factory levels, from the global logistics network to the single production resource. The factory evolution will consists of changes and reconfiguration of production resources and system layout, and changes in planning and production management policies.

Specific research and innovation topics for *evolutionary and reconfigurable factories* will cover:

- Development of new methodologies and tools to model and design flexible and reconfigurable production systems [29], control systems, automation systems [36], machines and fixtures [37].

- Development of methodologies to support the integrated design of products-processes-systems in evolution [35].
- Design of optimisation approaches to reduce set-up and ramp-up times in production systems [38].

1.5.1.2 Sustainable Factory

The concept of *green products* has a strong impact on the global scenario and involves also *Made in Italy* products. According to the Life Cycle Assessment (LCA) approach, *green products* must be characterized by a limited environmental impact during their whole life cycle, including the production phase [39, 40]. Therefore, sustainable production requires factories to guarantee limited energy consumption of industrial plants, systems, and processes, while producing limited industrial waste and consuming a reduced amount of natural resources [41].

Factories will have to be compliant with stricter and stricter energy consumption and emissions regulations, considering both the consumption of the workstations and the lighting and conditioning systems of the building. Factories will be able to reduce their environmental impact also by exploiting clean energy sources, cogeneration, industrial symbiosis [42] as well as re-using any available source. In addition, factories will have to be sustainable also from a societal perspective, integrating worker skills and contributing to the growth of the local economies [43]. Furthermore, besides being sustainable in the production, there is the need of a new generation of factories able to manage the final stages of the product life cycle by implementing product de-manufacturing, re-manufacturing, reuse, recycling and recovery, thus generating new opportunities and resources (de-production factories) [44]. Both production and de-production factories must be part of a network that is sustainable as a whole in terms of supply chain management and overall business model [45].

Research and innovation lines for *sustainable factories* include:

- new materials and production technologies exploiting renewable and green sources of energy and wastes of production processes [46];
- ICT tools and digital twins supporting the integrated control and management of factories, considering energy and environmental aspects of both productions systems and buildings [47];
- methodologies and tools for the modeling, design and management of processes, machines, systems and factories characterized by an efficient consumption of resources [41];
- methodologies and tools to model and design new products endowed with many lives since their conception;
- human-robot interaction and artificial intelligence to disassemble products after their use [48];
- new business models for an efficient exploitation of production resources, through the offer of targeted services [49];

- methodologies and tools for the modeling, design and management of factories for de-manufacturing able to regain the functions of the products and their components [50].

1.5.1.3 Factory for the People

Factories of the future must be designed and managed taking in due consideration societal and demographic changes such as the increase in the retirement age of workers and the global aging of the population (see Sect. 1.1). Technologies, machine tools and workplaces will be designed not only for young employees but also for workers with relevant accumulated knowledge [51]. The complexity and evolution of the working environment will require continuous training of the workforce [52].

Indeed, manufacturing history show that culture, know-how and skills of the workers play a key role in the success of manufacturing companies. Therefore, decisions regarding the factory design and location cannot be taken while considering only short-term cost minimisation, but it is necessary to fully evaluate the socio-economic impacts of phenomena like industrial de-localization [43].

The continuous improvement of robotics and automation technologies will make human-machine interaction even more relevant in the future of manufacturing. New forms of interactions must be investigated to better exploit human-machine cooperation in a shared and safe manufacturing environment [53].

Safety and ergonomics have a strong impact on productivity and profitability, therefore they should be addressed in a proactive way and not only in reaction to regulations [54].

New factories will provide an environment where people can face difficult production contexts characterized by products with short life cycles and high variability, thus requiring a quick adaptation of the production systems and the generation of new knowledge. Operators must be trained in a multidisciplinary way to flexibly manage the planning and execution of complex production plants. Furthermore, the high rate of technological obsolescence requires more and more attention to the ease of use of production resources, placing people in a central position within the factory environment. Specific research and innovation topics to develop *factories for the people* will include:

- study of socio-economic aspects to assess the impact and exploitation of knowledge and technology, considering standardisation and ethical issues;
- development of technologies that can improve working conditions thanks to ergonomic studies, reduction of risks related to dangerous processes by means of higher automation and remote control [55], more effective training using augmented and virtual reality [56], reduction of noise emissions [57] and air pollution, and telework;
- interactive human-intelligent machines cooperation in the factory environment to better exploit human intuition and skills in changing working conditions [48];

- development of adaptive and reactive human-machine interfaces (voice recognition, gesture recognition, autonomous moving machines) to better support an effective collaboration.

1.5.1.4 Factory for Customised and Personalised Products

The offer of personalised products and services that are difficult to replicate allows competing with high value-added products in the global market. This represents an important strategic opportunity for the Italian manufacturing industry that is traditionally focused on meeting the customer requirements by exploiting process and product know-how together with an attitude for innovation. This is particularly relevant in sectors such as textiles, wearing apparel, footwear, glasses and accessories, luxury goods, and furniture.

A full personalisation based on the specific customer needs (e.g. biometric characteristics, non-standard size and shape) represents an evolution of the *mass customization* concept offering products in pre-defined variants [58]. This asks for shifting the focus from high production volumes, process capability, component standardisation and modularity to factories able to offer *one-of-a-kind* products [59]. The production of personalised products asks for factories implementing fast innovation cycles thanks to modern technologies and approaches that can further increase flexibility, efficiency and ability to offer highly personalised products in a very short time [60].

Customer-driven factories can be designed thanks to a close cooperation with end users. Indeed, the collection and analysis of customer preferences through innovative technologies and the testing of product prototypes can help to improve the customer experience along all the phases of the product life cycle [61].

Factories for customised and personalised products will have to address the following research and innovation lines.

- ICT tools to support product personalisation (e.g. augmented and virtual reality [62], design and simulation of human-product interaction);
- new business models to optimise production and logistics supporting the *one-of-a-kind* paradigm;
- new approaches for the design, management and cooperation of supply chains and single manufacturing companies aimed at the production of personalised products [63];
- new tools and services based on innovative monitoring and maintenance techniques to support the use of products along their life cycle [64].

1.5.1.5 Advanced-Performance Factory

The production of customised and evolving products in a sustainable and human-oriented way poses serious challenges for the factory performance. High-performance factories will meet the demand by minimising all the inefficiencies associated with internal and external logistics, management of inter-operational buffers, transformation processes and their parameters, management and maintenance policies, software and hardware tools, quality inspection and control techniques. Both production systems and production processes can concur to increase the factory performance if monitoring data are properly collected [65]. Necessary enabling technologies include advanced sensors, innovative mechatronic components, ICT platforms [66], and digital twins [67–69].

The elaboration of data collected from the field with innovative techniques including data fusion and artificial intelligence, will enable advanced-performance factories to autonomously identify the causes of anomalies, failures and disturbances, implement adaptive strategies (e.g. predictive maintenance) modify operating modes so that the factory can constantly operate in conditions of high efficiency and zero defects [47].

Increasingly efficient transformation processes will reduce cycle times of the transformation operations, thus improving also the factory service level. New transformation processes will be needed to produce new products making use of innovative materials.

Specific research and innovation topics to develop *advanced-performance factories* will include:

- new high-performance transformation and transportation processes and systems;
- digital twins of processes, machines and systems [70] together with model predictive control and multicriteria optimization.
- models and platforms for the collection and fusion of shop floor data aimed at improving the technical efficiency of the production systems also by means of artificial intelligence [71];
- methodologies to support the design and modeling of quality control systems, management policies [47], and maintenance policies;
- new hardware and software solutions, data fusion, digital twins and artificial intelligence to continuously monitor and optimize manufacturing systems performance.

1.5.2 Enabling Technologies

Enabling technologies are technologies with a high content of knowledge and capital that are associated with intense research and development activities, involving highly skilled employment [72]. These technologies are characterized by rapid and integrated innovation cycles and *enable* innovation in a wide range of applications involving products, services, processes, and systems. Enabling technologies are of strategic importance at the systemic, multidisciplinary and trans-sectoral levels, because they incorporate skills deriving from different scientific-technological areas to induce structural changes and disruptive solutions with respect to the state of the art.

The list of enabling technologies has evolved during the last ten years depending on the technological progress. The list proposed by EFFRA in 2013 [21] included:

- advanced manufacturing processes;
- information and communication technologies;
- mechatronics for advanced manufacturing systems;
- modelling, simulation and forecasting methods and tools;
- manufacturing strategies;
- knowledge-workers.

Enabling technologies represent the basis for the innovation of *Made in Italy* production, providing solutions for a large number of applications and sectors that will offer new products and services. Herein, the following *enabling technologies* have been identified as relevant for factories of the future:

- Artificial Intelligence, digital twins, and digital factory technologies for intelligent factories;
- production technologies;
- de-manufacturing and material recovery technologies;
- factory reconfiguration technologies;
- control technologies of production resources and systems;
- resource management and maintenance technologies;
- technologies for monitoring and quality control;
- human-machine interaction technologies.

The wide adoption of ICT in manufacturing industry can help to improve the overall efficiency, adaptability and sustainability of the production systems. The interoperability among software tools and digital twins [67, 69] is crucial for sharing and transferring data along all the phases of the factory life cycle. For instance, integrated and interoperable solutions could include software tools for:

- Product Life-cycle Management (PLM) and platforms for Life-Cycle Assessment (LCA), at product level [40];
- wireless sensors and solutions for remote resources monitoring [65], CAD (Computer Aided Design), CAE (Computer Aided Engineering), Computer Aided Process Planning (CAPP), Computer Aided Manufacturing (CAM), at process level;
- evaluating the production system performance [73], multi-level simulation of reconfigurable factories, collaborative factory design in Virtual Reality (VR) [66] and Augmented Reality (AR) environments, ontologies for the conceptual modeling of the factory and its elements [74], at production system level.

Modern ICT offers new solutions (e.g. Cyber Physical Systems—CPS [75], Internet of Things—IoT [76], and Big Data Analytics [77]) with high potential impact on manufacturing. However, new digital technologies are associated with relevant challenges and risks for manufacturing companies because it is necessary to acquire or outsource advanced services, cyber security is under threat [78], and reference technical standard are still under development.

Factories of the future will have to adopt production technologies that enable the efficient use of resources [41] and are based on *clean* processes. Therefore, it is necessary to search for new processes characterized by low energy consumption, exploitation of renewable resources, increased efficiency and reduced emissions. New modular and flexible technologies will be needed for the production of non-standard products, even in small batches. Examples of enabling production technologies are high-speed machining, high performance tools, modular and reconfigurable handling technologies, non-conventional manufacturing processes (e.g. water jet, plasma, laser, ultrasonic machining—USM), micro-machining, micro-assembly and micro-factories [79].

Sustainable manufacturing involves both the production process and the management of the life cycle and reuse of materials and components [80]. Production technologies will have to minimise energy, materials' use, as well as the production of waste. The contribution of robotic disassembly technologies, advanced automation and human-robot cooperation will be fundamental to enable efficient re-use and re-manufacturing applications [48]. Finally, a key role is played by advanced technologies for the shredding and separation of materials to recover and recycle materials with commercial value.

New system and machine architectures enabling fast hardware and automation systems will provide competitive advantages to the factories of the future. Methodologies and tools to support the reconfiguration of machines and production systems will have to model uncertain information about the production context.

The design of control systems distributed over a network of heterogeneous devices is enabled by the introduction of advanced fieldbus communication techniques and intelligent devices endowed with microprocessors and programmable hardware. Traditional modelling techniques (e.g. based on IEC 61131 standard and programmable logic controllers - PLCs) are inadequate for distributed systems, since they can hardly meet the requirements of reusability, reconfigurability and flexibility for the development of control applications. Research on distributed and reconfigurable controls will rely on technical standards such as IEC 61499.

Production planning, scheduling and maintenance planning will have a strong impact on the performance of the factories of the future that are coping with changes in the market (e.g. demand) and the production environment (e.g. reconfiguration of the production system, availability of resources, etc.). An effective management of production resources can be supported by techniques such as reactive and robust production planning [81, 82], preventive and predictive maintenance [83], integrated production and maintenance planning, condition based maintenance (CBM), self-learning and self-organization algorithms for self-repair of systems.

Zero-defect production will enable factories of the future to increase their efficiency thanks to proactive improvement processes and intelligent measurement systems. The acquisition of data from the field requires designing accurate and low-cost sensor networks, whereas the elaboration of monitoring data asks for multi-resolution and multi-scale algorithms. Factory operations will be assisted by integrated methods that are able to process monitoring data and evaluate the system performance depending on possible adaptive reconfigurations.

Table 1.2 Calls for proposals

Call for proposals	N. submitted proposals	N. funded projects	Total cost [million euro]	Duration	Participants
Subproject 1	22	9	5.8	2 years	CNR institutes
Subproject 2	21	9	3	1 year and 4 months	CNR institutes and universities
Prototypes	15	14	1.4	4 months	CNR institutes and universities

Factories of the future will need to cope with the continuous increase of factory automation. Efficient and reactive technologies for human/machine interactions in advanced production environments will guarantee employment levels and ergonomics [53]. Innovative industrial robots will perform a wide range of tasks in spite of significant knowledge gaps. Advanced graphical interfaces will enable the use of increasingly complex software tools. Virtual and digital environment will be used to support training and enhance human skills [56].

1.5.3 Calls for Proposals and Research Projects

Each call for proposals of the two subprojects included four macro-objectives selected among the ones defined in Sect. 1.5.1 to demonstrate how enabling technologies (Sect. 1.5.2) can be developed and applied to innovate manufacturing processes. The project proposals had to follow a template consisting of project description, partnership, and impact. An additional call for proposals was published to enhance the result of the previously funded research projects through the development of hardware/software prototypes.[24]

Table 1.2 reports the summary of the three calls for proposals, pointing out the total cost of the research projects (including co-funding), the number of submitted and funded proposals, the duration of the projects and the admissible participants. Manufacturing companies could not be funded, but each proposal had to include an Industrial Interest Group to prove that manufacturing companies are interested in the proposed research topics.

Table 1.3 and Table 1.4 report the topics included in the call of Subproject 1 and Subproject 2, respectively.

The submission and evaluation of the project proposals were managed through a third-party informative system provided by CINECA[25] (Italian consortium of universities and research centres) to guarantee robustness and maintain anonymity along the

---

[24] http://www.fabbricadelfuturo-fdf.it/projects/prototypes/?lang=en.

[25] https://www.cineca.it/en.

Table 1.3 Topics of the Subproject 1 call for proposals

Macro-objective	Call topic
Advanced-Performance Factory	Mechatronic devices for high-performance factories
	Innovative technologies to realise components with advanced functional properties
Evolutionary and Reconfigurable Factory	Evolving and reconfigurable control
Factory for the People	Design and development of robotics systems cooperating with human operators
Sustainable Factory	Sustainable and interoperable factories
	De-manufacturing factories

Table 1.4 Topics of the Subproject 2 call for proposals

Macro-objective	Call topic
Factory for Customised and Personalised Products	Methodologies for the joint design of customised and personalised products, processes and productions systems
Evolutionary and Reconfigurable Factory	Optimisation of co-evolution of production systems
Factory for the People	Human-machine interaction technologies
Sustainable Factory	Technologies and methodologies for sustainable factories

Table 1.5 Reviewers and reviews

Call	N. national reviewers	N. international reviewers	N. reviews
Subproject 1	19	8	66
Subproject 2	17	9	63

process. The proposals were evaluated by a set of independent national and international reviewers selected from a pool of 167 experts registered in the MIUR database. The reviewers for each proposal were automatically selected by an algorithm based on matching between the call topic and the expertise of the reviewer, both of them identified by ERC (European Research Council)[26] keywords. Three reviewers (two national and one international) were assigned to each proposal and each reviewer evaluated a maximum of three proposals (Table 1.5).

The project proposals were evaluated by the reviewers according to the following criteria:

- Technical-scientific quality (max 5 points, threshold 4 points)
- Project organization and planning of activities (max 5 points, threshold 3 points)
- Scientific and industrial impacts (max 5 points, threshold 3 points)

---

[26] https://erc.europa.eu/.

Table 1.6 Subproject 1 research projects

Acronym	Research project title	Macro-objective
GECKO [84]	Generic Evolutionary Control Knowledge-based mOdule	Evolutionary and Reconfigurable Factory
Zero Waste PCBs [85]	Integrated Technological Solutions for Zero Waste Recycling of Printed Circuit Boards (PCBs)	Sustainable Factory
FACTOTHUMS [86]	FACTOry Technologies for HUMans Safety	Factory for the People
SNAPP [87]	Surface Nano-structured Coating for Improved Performance of Axial Piston Pumps	Advanced-Performance Factory
NanoTWICE [88]	composite Nanofibres for Treatment of air and Water by an Industrial Conception of Electrospinning	Advanced-Performance Factory
PLUS [89]	Plastic Lab-on-chips for the optical manipUlation of Single-cells	Advanced-Performance Factory
MaCISte [90]	Mature CIGS-based solar cells technology	Advanced-Performance Factory
MECAGEOPOLY [91]	Mechano-chemistry: an innovative process in the industrial production of poly-sialate and poly-silanoxosialate geopolymeric binders used in building construction	Advanced-Performance Factory
SILK.IT [92]	Silk Italian Technology for Industrial Biomanufacturing	Advanced-Performance Factory

Finally, the Executive Committee (consisting of Director, vice-Director, Subproject 1 coordinator and Subproject 2 coordinator) of the flagship project approved the ranking of the proposals based on the evaluation of the reviewers and published the list of funded projects. Table 1.6, Table 1.7, and Table 1.8 report the list of funded projects after the Subproject 1, Subproject 2, and Prototypes calls, respectively.

1.5.4   Results of the Flagship Project

A total of 21 CNR institutes and six universities participated in the 18 funded research projects. In addition, 55 private companies joined the various Industrial Interest

Table 1.7 Subproject 2 research projects

Acronym	Research project title	Macro-objective
IMET2AL [93]	genomIc Model prEdictive ConTrol Tools for evolutionAry pLants	Evolutionary and Reconfigurable Factory
Pro2Evo [94]	Product and Process Co-Evolution Management via Modular Pallet configuration	Evolutionary and Reconfigurable Factory
WEEE Reflex [95]	Highly Evolvable E-waste Recycling Technologies and Systems	Evolutionary and Reconfigurable Factory
PROBIOPOL [96]	Innovative and Sustainable Production of Biopolymers	Sustainable Factory
Xdrone [97]	Haptic teleoperation of UAV equipped with X-ray spectrometer for detection and identification of radio-active materials in industrial plants.	Factory for the People
Made4Foot [98]	Innovative Methodologies, ADvanced processes and systEms for Fashion and Wellbeing in Footwear	Factory for Customised and Personalised Products
Fab@Hospital [99]	Hospital Factory for Manufacturing Customized, Patient Specific 3D Anatomo-Functional Model and Prostheses	Factory for Customised and Personalised Products
POLYPHAB [100]	POLYmer nanostructuring by two-PHoton Absorption	Factory for Customised and Personalised Products
CHINA [101]	Customized Heat exchanger with Improved Nano-coated surface for earth moving machines Applications	Factory for Customised and Personalised Products

Groups, thus effectively creating a large community working on topics for factories of the future.

The main scientific results of the research projects are summarised in Table 1.9 in terms of publications on international journals, proceedings of international conferences, and chapters of international books. The citations of the journal articles demonstrate how the results of the research have already achieved a scientific impact, even though the publication date is still recent. In addition, two patents were successfully published.

The flagship project as a whole was disseminated during more than 20 national and international events related to manufacturing industry and research. In particular, synergies were established with the Italian Cluster Intelligent Factories (Sect. 1.4.2) and with European initiatives such as Manu*future* and EFFRA (Sect. 1.3).

Table 1.8 Prototype projects

Acronym	Prototype title	Related research project
AUTOSPIN	Automated electrospinning plant for industrial manufacturing of functional composite nanofibres	NanoTWICE
APPOS	Axial Piston Pump Prototype Assembled with Oleophobic Surfaces Components	SNAPP
AWESOME	Advanced and WEarable SOlutions for human Machine interaction with Enhanced capabilities	FACTOTHUMS
CD-NET	Tool for Customer-driven Supply Networks configuration	Made4Foot
F@H for 3D plates	Fab@Hospital for bone plate fabrication and patient anatomy reconstruction using rapid prototyping technologies	Fab@Hospital
IC+	Imaging Citometry in Plastic Ultra-mobile Systems	PLUS
PCB-ID	In-line automated device for the identification of components and the characterization of materials and value in waste PCBs	Zero Waste PCBs
Pro2ReFix	Product and Process Co-Evolution Management via Reconfigurable Fixtures	Pro2Evo
ProBioType	Prototyping ProBioPol results	PROBIOPOL
ProBioType II	Prototyping ProBioPol Results II (UV/TiO2 Photocatalytic Reactor)	PROBIOPOL
Rolling CIGS	Roll-to-roll deposition of flexible CIGS-based solar cells	MaCISTe
ShredIT	Self-Optimizing Shredding Station for Demanufacturing Plants	Zero Waste PCBs
THESIS	Thermally Improved Heat Exchangers prototypes with Superhydrophobic Internal Surfaces: new assembly procedures and materials	CHINA
WEEE ReFlex CPS	Cyber-Physical System (CPS) for reconfigurable e-waste recycling processes	Weee ReFlex
X-Drone2	Improved Haptic-guided UAV for detection and identification of radioactive materials	XDrone

Table 1.9 Scientific results of the research projects

Research project	Int'l journal articles	Citations of int'l journal articles[a]	Int'l conference articles	Int'l book chapters	Presentations at conferences and workshops
IMET2AL	2	9	1	1	1
Pro2Evo	5	78	7	2	4
GECKO	3	34	20	1	3
WEEE Reflex	1	0	8	1	8
PROBIOPOL	2	14	11	1	11
Zero Waste PCBs	0	0	5	1	7
FACTOTHUMS	7	28	38	1	2
Xdrone	1	1	2	1	5
Made4Foot	2	7	7	1	6
Fab@Hospital	10	36	6	2	6
POLYPHAB	3	8	4	1	1
CHINA	5	33	5	1	2
SNAPP	5	35	8	1	6
NanoTWICE	6	78	15	1	12
PLUS	15	398	12	1	15
MaCISte	6	30	5	1	2
MECAGEOPOLY	1	6	3	1	4
SILK.IT	11	102	2	1	9
Total	85	897	159	20	104

[a]Retrieved on 13th July 2018

The flagship project participated in the BI-MU[27] 2016 exhibition in Milan. BIMU is the largest Italian fair of machine tools and other capital goods related to manufacturing and is the second largest exhibition in this field in Europe. The Flagship Project *Factories of the Future* presented the results to the selected public attending the exhibition by organising a stand (Figs. 1.2 and 1.3) of 170 m^2 dedicated to the 14 prototypes (Table 1.8) resulting from the activities of various research projects (Fig. 1.4).

Four conferences of the flagship project were organised in 2012, 2013, 2016 and 2018. During these conferences the calls for proposals and/or the results of the research projects were presented.

The Final Event of the Flagship Project was conceived as a one-week national research road-show (26–30 November, 2018) visiting laboratories of institutes and universities that are active on advanced manufacturing research. Forty young researchers participated in the final event to work on new ideas for factories of the future in a creative and multi-disciplinary environment.[28]

---

[27]http://www.bimu.it/en/.
[28]http://eventofinale.fabbricadelfuturo-fdf.it/

Fig. 1.2 *Factories of the Future* stand (left side) at BI-MU 2016 exhibition

Fig. 1.3 *Factories of the Future* stand (right side) at BI-MU 2016 exhibition

The flagship project promoted an international collaboration with *Automotive Partnership Canada* after a memorandum of understanding between CNR and the Natural Sciences and Engineering Research Council of Canada (NSERC). In this

Fig. 1.4 Examples of the Prototypes at the *Factories of the Future* stand: **a** ShredIT, **b** X-Drone2 (left) and IC+ (right), **c** ProBioType II (left) and Pro2ReFix (right), **d** ProBioType

framework a concurrent call for joint research projects[29] was launched in the area of manufacturing research, with the Canadian side focusing on automotive manufacturing. The collaboration resulted in one joint research project [86].

Finally, this very book represents a contribution to the dissemination of the whole flagship project. The following chapters will present in details the scientific and industrial results of the 18 research projects (see Tables 1.6 and 1.7) [84–101]. Grounding on the experience and results of the flagship project, the final two chapters of this book present an outlook on future manufacturing research by proposing *missions* aimed at fostering growth and innovation [102] and discussing research infrastructures and funding mechanisms [27].

Acknowledgements This work has been partially funded by the Italian Ministry of Education, Universities and Research (MIUR) under the Flagship Project "Factories of the Future—Italy" (Progetto Bandiera "La Fabbrica del Futuro").
The authors would like to thank Giacomo Copani and Margherita De Vivo for their contributions to this chapter.

---

[29]http://www.apc-pac.ca/Apply-Demande/CanIta-CanIta_eng.asp.

References

1. Stehrer R, Baker P, Foster-McGregor N, Koenen J, Leitner S, Schricker J, Strobel T, Vieweg H-G, Vermeulen J, Yagafarova A (2014) Study on the relation between industry and services in terms of productivity and value creation. Final report. ECSIP consortium: wiiw, Ifo and Ecorys
2. Baines TS, Lightfoot HW, Benedettini O, Kay JM (2009) The servitization of manufacturing: a review of literature and reflection on future challenges. J Manuf Technol Manag 20(5):547–567
3. The Manufacturing Institute (2009) Facts about modern manufacturing, 8th edn.
4. European Commission (2010) Communication from the Commission to the European Parliament, the Council, the European Economic and Social Committee and the Committee of the Regions: An Integrated Industrial Policy for the Globalisation Era Putting Competitiveness and Sustainability at Centre Stage. Document 52010DC0614. https://eur-lex.europa.eu/legal-content/EN/TXT/?qid=1531930246089&uri=CELEX:52010DC0614
5. Jovane F, Yoshikawa H, Alting L, Boër CR, Westkamper E, Williams D, Tseng M, Seliger G, Paci AM (2008) The incoming global technological and industrial revolution towards competitive sustainable manufacturing. CIRP Ann 57(2):641–659. https://doi.org/10.1016/J.CIRP.2008.09.010
6. Westkämper E (2014) Global "Megatrend's" grand societal challenges. In: Towards the re-industrialization of Europe: a concept for manufacturing for 2030. Springer, Berlin, Heidelberg. https://doi.org/10.1007/978-3-642-38502-5_4
7. Associazione Cluster Fabbrica Intelligente (2015) Roadmap per la ricerca e l'innovazione—research and innovation roadmap. http://www.fabbricaintelligente.it/english/roadmap/
8. Kumar M, Graham G, Hennelly P, Srai J (2016) How will smart city production systems transform supply chain design: a product-level investigation. Int J Prod Res 54(23):7181–7192
9. Kharas H (2017) The unprecedented expansion of the global middle class: an update. Brookings India. http://hdl.handle.net/11540/7251
10. U.S. Energy Information Administration (2017) International Energy Outlook 2017. https://www.eia.gov/outlooks/ieo/pdf/0484(2017).pdf
11. Ghisellini P, Cialani C, Ulgiati S (2016) A review on circular economy: the expected transition to a balanced interplay of environmental and economic systems. J Clean Prod 114:11–32
12. Marino F, Nunziata L (2017) The labor market in Italy, 2000–2016. IZA World of Labor, 407
13. Bagnai A (2016) Italy's decline and the balance-of-payments constraint: a multicountry analysis. Int Rev Appl Econ 30(1):1–26
14. Bagnai A, Mongeau Ospina CA (2017) Monetary integration vs. real disintegration: single currency and productivity divergence in the euro area. J Econ Policy Reform
15. ISTAT (2017) Rapporto sulla competitività dei settori produttivi. Edizione 2017
16. Jovane F, Westkämper E (2008) The European strategic initiative ManuFuture. In: The ManuFuture road: towards competitive and sustainable high-adding-value manufacturing. Springer, Berlin, Heidelberg, pp 53–87. https://doi.org/10.1007/978-3-540-77012-1_4
17. Manufuture (2006) Strategic research agenda assuring the future of manufacturing in Europe. http://www.manufuture.org/documents/manufuture-documents/
18. Jovane F, Westkämper E, Williams D (2008) The ManuFuture road: towards competitive and sustainable high-adding-value manufacturing. Springer Science & Business Media
19. Commission of the European Communities (2008) A European economic recovery plan. http://ec.europa.eu/economy_finance/publications/pages/publication13504_en.pdf
20. European Commission (2010) Factories of the future PPP—strategic multi-annual roadmap. Prepared by the Ad-hoc Industrial Advisory Group
21. European Commission (2013) Factories of the future: multi-annual roadmap for the contractual PPP under Horizon 2020. Prepared by EFFRA

22. EFFRA (2016) Factories 4.0 and beyond—recommendations for the work programme 18-19-20 of the FoF PPP under Horizon 2020
23. U.S. Congress (2009) H.R.1—American Recovery and Reinvestment Act of 2009. https://www.congress.gov/bill/111th-congress/house-bill/1/text
24. President's Council of Advisors on Science and Technology, President's Innovation and Technology Advisory Committee (2011) Ensuring American leadership in advanced manufacturing. https://eric.ed.gov/?id=ED529992
25. U.S. Congress (2009) TITLE VII—Revitalize American Manufacturing and Innovation Act of 2014. https://www.gpo.gov/fdsys/pkg/BILLS-113hr83enr/pdf/BILLS-113hr83enr.pdf
26. Ministero dell'Istruzione, dell'Università e della Ricerca (2011) Programma Nazionale della Ricerca 2011–2013. http://www.miur.it/Documenti/ricerca/pnr_2011_2013/PNR_2011-2013_23_MAR_2011_web.pdf
27. Tolio T, Copani G, Terkaj W (2019) Key research priorities for factories of the future—part II: pilot plants and funding mechanisms. In: Tolio T, Copani G, Terkaj W (eds) Factories of the future. Springer
28. Terkaj W, Tolio T, Valente A (2009) Focused flexibility in production systems. In ElMaraghy HA (ed) Changeable and reconfigurable manufacturing systems. Springer. https://doi.org/10.1007/978-1-84882-067-8_3, pp 47–66
29. Koren Y, Heisel U, Jovane F, Moriwaki T, Pritschow G, Ulsoy G, Van Brussel H (1999) Reconfigurable manufacturing systems. CIRP Ann 48(2):527–540. https://doi.org/10.1016/S0007-8506(07)63232-6
30. Wiendahl H-P, ElMaraghy HA, Nyhuis P, Zäh MF, Wiendahl H-H, Duffie N, Brieke M (2007) Changeable manufacturing—classification. Des Oper CIRP Ann 56(2):783–809. https://doi.org/10.1016/J.CIRP.2007.10.003
31. Andersen A-L, Larsen JK, Brunoe TD, Nielsen K, Ketelsen C (2018) Critical enablers of changeable and reconfigurable manufacturing and their industrial implementation. J Manuf Technol Manag 29(6):983–1002. https://doi.org/10.1108/JMTM-04-2017-0073
32. Andersen A-L, Brunoe TD, Nielsen K, Bejlegaard M (2018) Evaluating the investment feasibility and industrial implementation of changeable and reconfigurable manufacturing concepts. J Manuf Technol Manag 29(3):449–477. https://doi.org/10.1108/JMTM-03-2017-0039
33. Putnik G, Sluga A, ElMaraghy H, Teti R, Koren Y, Tolio T, Hon B (2013) Scalability in manufacturing systems design and operation: state-of-the-art and future developments roadmap. CIRP Ann 62(2):751–774. https://doi.org/10.1016/J.CIRP.2013.05.002
34. Terkaj W, Tolio T, Valente A (2009) Designing manufacturing flexibility in dynamic production contexts. In Tolio T (ed) Design of flexible production systems. Springer, pp 1–18
35. Tolio T, Ceglarek D, ElMaraghy HA, Fischer A, Hu SJ, Laperrière L, Newman ST, Váncza J (2010) SPECIES—co-evolution of products, processes and production systems. CIRP Ann 59(2):672–693. https://doi.org/10.1016/J.CIRP.2010.05.008
36. Jovane F, Koren Y, Boër CR (2003) Present and future of flexible automation: towards new paradigms. CIRP Ann 52(2):543–560. https://doi.org/10.1016/S0007-8506(07)60203-0
37. Bejlegaard M, ElMaraghy W, Brunoe TD, Andersen A-L, Nielsen K (2018) Methodology for reconfigurable fixture architecture design. CIRP J Manuf Sci Technol. https://doi.org/10.1016/J.CIRPJ.2018.05.001 (in press)
38. Matta A, Tomasella M, Valente A (2008) Impact of ramp-up on the optimal capacity-related reconfiguration policy. Int J Flex Manuf Syst 19(3):173–194
39. Westkämper E, Alting L, Arndt G (2000) Life cycle management and assessment: approaches and visions towards sustainable manufacturing (keynote paper) CIRP Ann 49(2):501–526. https://doi.org/10.1016/S0007-8506(07)63453-2
40. Hauschild M, Jeswiet J, Alting L (2005) From life cycle assessment to sustainable production: status and perspectives. CIRP Ann 54(2):1–21. https://doi.org/10.1016/S0007-8506(07)60017-1
41. Duflou JR, Sutherland JW, Dornfeld D, Herrmann C, Jeswiet J, Kara S, Hauschild M, Kellens K (2012) Towards energy and resource efficient manufacturing: a processes and systems approach. CIRP Ann 61(2):587–609. https://doi.org/10.1016/J.CIRP.2012.05.002

42. Chertow MR (2007) "Uncovering" industrial symbiosis. J Ind Ecol 11(1):11–30
43. Sutherland JW, Richter JS, Hutchins MJ, Dornfeld D, Dzombak R, Mangold J, Robinson S, Hauschild MZ, Bonou A, Schönsleben P, Friemann F (2016) The role of manufacturing in affecting the social dimension of sustainability. CIRP Ann 65(2):689–712. https://doi.org/10.1016/J.CIRP.2016.05.003
44. Tolio T, Bernard A, Colledani M, Kara S, Seliger G, Duflou J, Battaia O, Takata S (2017) Design, management and control of demanufacturing and remanufacturing systems. CIRP Ann 66(2):585–609. https://doi.org/10.1016/J.CIRP.2017.05.001
45. Wiendahl H-P, Lutz S (2002) Production in networks. CIRP Ann 51(2):573–586. https://doi.org/10.1016/S0007-8506(07)61701-6
46. Umeda Y, Takata S, Kimura F, Tomiyama T, Sutherland JW, Kara S, Herrmann C, Duflou JR (2012) Toward integrated product and process life cycle planning—an environmental perspective. CIRP Ann 61(2):681–702. https://doi.org/10.1016/J.CIRP.2012.05.004
47. Colledani M, Tolio T, Fischer A, Iung B, Lanza G, Schmitt R, Váncza J (2014) Design and management of manufacturing systems for production quality. CIRP Ann 63(2):773–796. https://doi.org/10.1016/J.CIRP.2014.05.002
48. Bley H, Reinhart G, Seliger G, Bernardi M, Korne T (2004) Appropriate human involvement in assembly and disassembly. CIRP Ann 53(2):487–509
49. Copani G, Behnam S (2018) Remanufacturing with upgrade PSS for new sustainable business models. CIRP J Manuf Sci Technol (in press)
50. Colledani M, Copani G, Tolio T (2014) De-manufacturing systems. Procedia CIRP 17:14–19
51. Wallen ES, Mulloy KB (2006) Computer-based training for safety: comparing methods with older and younger workers. J Saf Res 37(5):461–467
52. Abele E, Chryssolouris G, Sihn W, Metternich J, ElMaraghy H, Seliger G, Sivard G, ElMaraghy W, Hummel V, Tisch M, Seifermann S (2017) Learning factories for future oriented research and education in manufacturing. CIRP Ann 66(2):803–826. https://doi.org/10.1016/J.CIRP.2017.05.005
53. Krüger J, Lien TK, Verl A (2009) Cooperation of human and machines in assembly lines. CIRP Ann 58(2):628–646. https://doi.org/10.1016/J.CIRP.2009.09.009
54. Shikdar AA, Sawaqed NM (2003) Worker productivity, and occupational health and safety issues in selected industries. Comput Ind Eng 45(4):563–572
55. Krüger J, Wang L, Verl A, Bauernhansl T, Carpanzano E, Makris S, Fleischer J, Reinhart G, Franke J, Pellegrinelli S (2017) Innovative control of assembly systems and lines. CIRP Ann 66(2):707–730. https://doi.org/10.1016/J.CIRP.2017.05.010
56. Leu MC, ElMaraghy HA, Nee AYC, Ong SK, Lanzetta M, Putz M, Zhu W, Bernard A (2013) CAD model based virtual assembly simulation, planning and training. CIRP Ann 62(2):799–822. https://doi.org/10.1016/J.CIRP.2013.05.005
57. Casas WJP, Cordeiro EP, Mello TC, Zannin PHT (2014) Noise mapping as a tool for controlling industrial noise pollution. J Sci Ind Res 73:262–266
58. Tseng MM, Piller F (2003) The customer centric enterprise—advances in mass customization and personalization. Springer, Berlin, Heidelberg
59. ElMaraghy H, Schuh G, ElMaraghy W, Piller F, Schönsleben P, Tseng M, Bernard A (2013) Product variety management. CIRP Ann 62(2):629–652. https://doi.org/10.1016/J.CIRP.2013.05.007
60. Jovane F, Koren Y, Boër CR (2003) Present and future of flexible automation: towards new paradigms. CIRP Ann 52(2):543–560. https://doi.org/10.1016/S0007-8506(07)60203-0
61. Tseng MM, Kjellberg T, Lu SC-Y (2003) Design in the new e-commerce era. CIRP Ann 52(2):509–519. https://doi.org/10.1016/S0007-8506(07)60201-7
62. Lu S-Y, Shpitalni M, Gadh R (1999) Virtual and augmented reality technologies for product realization. CIRP Ann 48(2):471–495. https://doi.org/10.1016/S0007-8506(07)63229-6
63. Macchion L, Fornasiero R, Vinelli A (2017) Supply chain configurations: a model to evaluate performance in customised productions. Int J Prod Res 55(5):1386–1399

64. Takata S, Kirnura F, van Houten FJAM, Westkamper E, Shpitalni M, Ceglarek D, Lee J (2004) Maintenance: changing role in life cycle management. CIRP Ann 53(2):643–655. https://doi.org/10.1016/S0007-8506(07)60033-X
65. Teti R, Jemielniak K, O'Donnell G, Dornfeld D (2010) Advanced monitoring of machining operations. CIRP Ann 59(2):717–739. https://doi.org/10.1016/J.CIRP.2010.05.010
66. Kádár B, Terkaj W, Sacco M (2013) Semantic Virtual Factory supporting interoperable modelling and evaluation of production systems. CIRP Ann Manuf Technol 62(1):443–446
67. Rosen R, von Wichert G, Lo G, Bettenhausen KD (2015) About the importance of autonomy and digital twins for the future of manufacturing. IFAC-PapersOnLine 48:567–572. https://doi.org/10.1016/j.ifacol.2015.06.141
68. Boschert S, Rosen R (2016) Digital twin—the simulation aspect. In: Hehenberger P, Bradley D (eds) Mechatronic futures. Springer, Cham, pp 59–74
69. Negri E, Fumagalli L, Macchi M (2017) A review of the roles of digital twin in CPS-based production systems. Procedia Manuf 11:939–948. https://doi.org/10.1016/j.promfg.2017.07.198
70. Modoni GE, Caldarola EG, Sacco M, Terkaj W (2018) Synchronizing physical and digital factory: benefits and technical challenges. In: Proceedings of 12th CIRP conference on intelligent computation in manufacturing engineering
71. Gao R, Wang L, Teti R, Dornfeld D, Kumara S, Mori M, Helu M (2015) Cloud-enabled prognosis for manufacturing. CIRP Ann 64(2):749–772. https://doi.org/10.1016/J.CIRP.2015.05.011
72. European Commission (2009) Preparing for our future: developing a common strategy for key enabling technologies in the EU. COM(2009) 512 final
73. Colledani M, Magnanini MC, Tolio T (2018) Impact of opportunistic maintenance on manufacturing system performance. CIRP Ann 67(1):499–502. https://doi.org/10.1016/J.CIRP.2018.04.078
74. Terkaj W, Tolio T, Urgo M (2015) A virtual factory approach for in situ simulation to support production and maintenance planning. CIRP Ann Manuf Technol 64(1):451–454
75. Monostori L, Kádár B, Bauernhansl T, Kondoh S, Kumara S, Reinhart G, Sauer O, Schuh G, Sihn W, Ueda K (2016) Cyber-physical systems in manufacturing. CIRP Ann 65(2):621–641. https://doi.org/10.1016/J.CIRP.2016.06.005
76. Xu LD, He W, Li S (2014) Internet of things in industries: a survey. IEEE Trans Ind Inform 10(4):2233–2243
77. Tsai CW, Lai CF, Chao HC, Vasilakos AV (2016) Big data analytics. In: Furht B, Villanustre F (eds) Big data technologies and applications. Springer, Cham
78. Conti M, Dehghantanha A, Franke K, Watson S (2018) Internet of Things security and forensics: challenges and opportunities. Future Generation Computer Systems 78(2):544–546
79. Fassi I, Shipley D (2017) Micro-manufacturing technologies and their applications. Springer, Cham
80. Seliger G (2007) Sustainability in manufacturing. Springer, Berlin
81. Tolio T, Urgo M, Váncza J (2011) Robust production control against propagation of disruptions. CIRP Ann 60(1):489–492
82. Sobaszek Ł, Gola A, Świć A (2018) Predictive scheduling as a part of intelligent job scheduling system. In: Burduk A, Mazurkiewicz D (eds) Intelligent systems in production engineering and maintenance – ISPEM 2017. ISPEM 2017. Advances in Intelligent Systems and Computing, vol 637. Springer, Cham
83. Roy R, Stark R, Tracht K, Takata S, Mori M (2016) Continuous maintenance and the future—foundations and technological challenges. CIRP Ann 65(2):667–688. https://doi.org/10.1016/J.CIRP.2016.06.006
84. Ballarino A, Brusaferri A, Cesta A, Chizzoli G, CibrarioBertolotti I, Durante L, Orlandini A, Rasconi R, Spinelli S, Valenzano A (2019) Knowledge based modules for adaptive distributed control systems. In: Tolio T, Copani G, Terkaj W (eds) Factories of the future. Springer
85. Copani G, Colledani M, Brusaferri A, Pievatolo A, Amendola E, Avella M, Fabrizio M (2019) Integrated technological solutions for zero waste recycling of Printed Circuit Boards (PCBs). In: Tolio T, Copani G, Terkaj W (eds) Factories of the future. Springer

86. Pecora A, Maiolo L, Minotti A, Ruggeri M, Dariz L, Giussani M, Iannacci N, Roveda L, Pedrocchi N, Vicentini F (2019) Systemic approach for the definition of a safer human-robot interaction. In: Tolio T, Copani G, Terkaj W (eds) Factories of the future. Springer

87. Bonanno A, Raimondo MR, Zapperi S (2019) Surface nano-structured coating for improved performance of axial piston Pumps. In: Tolio T, Copani G, Terkaj W (eds) Factories of the future. Springer

88. Bonura L, Bianchi G, Sanchez Ramirez DO, Carletto RA, Varesano A, Vineis C, Tonetti C, Mazzuchetti G, Lanzarone E, Ortelli S, Blosi M (2019) Monitoring systems of an electro-spinning plant for the production of composite nanofibres. In: Tolio T, Copani G, Terkaj W (eds) Factories of the future. Springer

89. Martínez Vázquez R, Trotta G, Volpe A, Paturzo M, Modica F, Bianco V, Coppola S, Ancona A, Ferraro P, Fassi I (2019) Plastic lab-on-chip for the optical manipulation of single cells. In: Tolio T, Copani G, Terkaj W (eds) Factories of the future. Springer

90. Gilioli E, Albonetti C, Bissoli F, Bronzoni M, Ciccarelli P, Rampino S, Verucchi R (2019) CIGS-based flexible solar cells. In: Tolio T, Copani G, Terkaj W (eds) Factories of the future. Springer

91. Ciccioli P, Capitani D, Gualtieri S, Soragni E, Belardi G, Plescia P, Contini G (2019) Mechano-chemistry of rock materials for the industrial production of new geopolymeric cements. In: Tolio T, Copani G, Terkaj W (eds) Factories of the future. Springer

92. Benfenati V, Toffanin S, Chieco C, Sagnella A, DiVirgilio N, Posati T, Varchi G, Natali M, Ruani G, Muccini M, Zamboni R (2019) Silk Italian technology for industrial biomanufacturing. In: Tolio T, Copani G, Terkaj W (eds) Factories of the future. Springer

93. Cataldo A, Cibrario Bertolotti I, Scattolini R (2019) Model predictive control tools for evolutionary plants. In: Tolio T, Copani G, Terkaj W (eds) Factories of the future. Springer

94. Urgo M, Terkaj W, Giannini F, Pellegrinelli S, Borgo S (2019) Exploiting modular pallet flexibility for product and process co-evolution through zero-point clamping systems. In: Tolio T, Copani G, Terkaj W (eds) Factories of the future. Springer

95. Copani G, Picone N, Colledani M, Pepe M, Tasora A (2019) Highly evolvable E-waste recycling technologies and systems. In: Tolio T, Copani G, Terkaj W (eds) Factories of the future. Springer

96. Ortelli S, Costa AL, Torri C, Samorì C, Galletti P, Vineis C, Varesano A, Bonura L, Bianchi G (2019) Innovative and sustainable production of biopolymers. In: Tolio T, Copani G, Terkaj W (eds) Factories of the future. Springer

97. Aleotti J, Micconi G, Caselli S, Benassi G, Zambelli N, Bettelli M, Calestani D, Zappettini A (2019) Haptic teleoperation of UAV equipped with gamma-ray spectrometer for detection and identification of radio-active materials in industrial plants. In: Tolio T, Copani G, Terkaj W (eds) Factories of the future. Springer

98. Macchion L, Marchiori I, Vinelli A, Fornasiero R (2019) Proposing a tool for supply chain configuration: an application to customised production. In: Tolio T, Copani G, Terkaj W (eds) Factories of the future. Springer

99. Lanzarone E, Marconi S, Conti M, Auricchio F, Fassi I, Modica F, Pagano C, Pourabdollahian G (2019) Hospital factory for manufacturing customised, patient-specific 3D anatomo-functional models and prostheses. In: Tolio T, Copani G, Terkaj W (eds) Factories of the future. Springer

100. Zandrini T, Suriano R, DeMarco C, Osellame R, Turri S, Bragheri F (2019) Polymer nanostructuring by two-photon absorption. In: Tolio T, Copani G, Terkaj W (eds) Factories of the future. Springer

101. Bonanno A, Raimondo MR, Pinelli M (2019) Use of nanostructured coating to improve heat exchanger efficiency. In: Tolio T, Copani G, Terkaj W (eds) Factories of the future. Springer

102. Tolio T, Copani G, Terkaj W (2019) Key research priorities for factories of the future—part I: missions. In: Tolio T, Copani G, Terkaj W (eds) Factories of the future. Springer

Part II
Evolutionary and Reconfigurable Factory

Chapter 2
Model Predictive Control Tools for Evolutionary Plants

Andrea Cataldo, Ivan Cibrario Bertolotti and Riccardo Scattolini

Abstract The analysis and design of control system configurations for automated production systems is generally a challenging problem, in particular given the increasing number of automation devices and the amount of information to be managed. This problem becomes even more complex when the production system is characterized by a fast evolutionary behaviour in terms of tasks to be executed, production volumes, changing priorities, and available resources. Thus, the control solution needs to be optimized on the basis of key performance indicators like flow production, service level, job tardiness, peak of the absorbed electrical power and the total energy consumed by the plant. This paper proposes a prototype control platform based on Model Predictive Control (MPC) that is able to impress to the production system the desired functional behaviour. The platform is structured according to a two-level control architecture. At the lower layer, distributed MPC algorithms control the pieces of equipment in the production system. At the higher layer an MPC coordinator manages the lower level controllers, by taking full advantage of the most recent advances in hybrid control theory, dynamic programming, mixed-integer optimization, and game theory. The MPC-based control platform will be presented and then applied to the case of a pilot production plant.

A. Cataldo (✉)
CNR-STIIMA, Istituto di Sistemi e Tecnologie Industriali Intelligenti per il Manifatturiero Avanzato, Milan, Italy
e-mail: andrea.cataldo@stiima.cnr.it

I. C. Bertolotti
CNR-IEIIT, Istituto di Elettronica e di Ingegneria dell'Informazione e delle Telecomunicazioni, Turin, Italy

R. Scattolini
Dipartimento di Elettronica Informazione e Bioingegneria, Politecnico di Milano, Milan, Italy

© The Author(s) 2019
T. Tolio et al. (eds.), *Factories of the Future,*
https://doi.org/10.1007/978-3-319-94358-9_2

2.1 Scientific and Industrial Motivations

The configuration of the control system represents a key phase during the design and management of large scale complex production systems characterized by strongly interacting and possibly spatially distributed subsystems (such as power networks, transport networks, data traffic networks, and irrigation networks). The increasing amount of information to be managed and the need of flexibility and reconfigurability in industrial production contexts have a direct impact on control systems to deal with typical manufacturing problems such as routing, scheduling, and planning. Therefore, also the control systems must be flexible and scalable to cope with the selection of different control policies, plug-and-play operations, changes in the demand, modified conditions in the factory environment (e.g. addition or removal of sensors and actuators), self-reconfiguration after the malfunctioning of parts of the system.

Model Predictive Control (MPC) [1, 2] is widely used to control continuous industrial processes, such as chemical and petrochemical plants or pulp industry. However, its application in the discrete manufacturing industry is still in its infancy, although great advantages could be achieved in the design of the overall production system architecture in terms of scalability, adaptivity to changing environments, robustness to communication limitations and faults, asynchronous communication, reconfigurability with respect to the addition, replacement or removal of subsystems. These properties bring to the analysis and identification of most suitable distributed and hierarchical Model Predictive Control methodologies to be applied to flexible and time variant production systems, characterized by a very fast evolution concerning the whole product and process life cycle, from the ramp-up to the end-of-life.

This work proposes a two-layer control architecture based on MPC solutions to support the design and management of control system configurations for manufacturing plants. The lower layer of the architecture is dedicated to control the single piece of equipment and to consider the possible dynamic mutual influences and overall constraints, for example on the maximum admissible energy consumption. The higher layer is in charge of coordinating the whole system guaranteeing the fulfilment of operational and production constrains. The proposed control architecture has been implemented into a prototype control platform.

This chapter is organized as follows. The state of the art related to MPC is discussed in Sect. 2.2. The proposed solution is introduced in Sect. 2.3 and then further detailed in Sect. 2.4. The industrial case and the experiments can be found in Sect. 2.5, whereas the conclusions are drawn in Sect. 2.6.

2.2 State of the Art

Model Predictive Control (MPC) is the most popular and widely used advanced control technique in view of its ability to cope with linear and nonlinear models and to consider state, input, and output constraints directly in the design phase. In addition, the parameters that characterize the controller working function can be easily tuned, starting from empirical models of the system to be controlled, models obtained with simple experiments on the plant like step or impulse responses. For these reasons, there are nowadays thousands of applications of MPC (e.g. [3]) and strong theoretical results have been developed [4].

MPC algorithms are usually designed and implemented according to a centralized approach, where all the information collected by the sensors are sent to a unique central station that computes the values of the commands to be transmitted to the plant actuators. This centralized structure guarantees the optimality of the control action and is suitable for many industrial plants. However, the implementation of a centralized structure becomes too demanding in terms of computational burden and transmission of information because of the growing complexity of the systems and of the increasing amount of information to be managed. For these reasons, and also to deal with large scale complex systems, many distributed versions of MPC, also called DMPC (Distributed MPC), have been proposed in recent years. The survey papers [5, 6] and the book [7] thoroughly describe the most recent methods. In DMPC, local MPC regulators are used to control parts of the system and exchange information according to a flat structure (i.e. single-level structure), in order to coordinate their actions.

Another way to handle the control of large scale complex systems is to resort to multilayer hierarchical control structures, instead of a flat one. At the lowest layer, a local MPC controller is designed for each subsystem, while at the higher layer an MPC controller (also named supervisor) coordinates and assigns high-level tasks to groups of subsystems [5]. The hierarchical control structure can be effective and provides a common framework to solve routing, scheduling, and planning problems.

Standard real-time protocols (e.g. Modbus [8]) are typically chosen to implement inter-module communication, as well as communication with the plant environment, in industrial applications. As outlined in [9] and further asserted in [10, 11], choosing the Modbus protocol offers significant advantages. For instance, it guarantees the resiliency of traditional fieldbus solutions. At the same time, it enables the improved bandwidth, open connectivity, and standardization of Ethernet-based networks.

About platforms and tools available for the MPC design and implementation, there are different industrial solutions implementing proprietary MPC algorithms, so they do not allow modifying the structure of the control algorithm but only set the related parameters to characterize the control behaviour. On the contrary, some tools (e.g. HYSDEL [12], YALMIP [13]) allow to design and implement MPC algorithms, starting from the system modelling through logic propositions, translating them into linear expressions Mixed Logical Dynamical (MLD) model [14], and then running the control algorithm.

2.3 Problem Statement and Proposed Approach

The reference production system consists of production resources (e.g. machine tools), storage spaces (e.g. buffer slots), transport systems (e.g. conveyors), and pallets that are used to move the work-in-progress parts. Two relevant industrial problems related to system control are investigated: *Production scheduling* and *Pallet routing*.

The *Production scheduling problem* is related to the design of a control algorithm for the production scheduling and buffer management of a multiple-line production plant. It is assumed that the machines can operate at different speeds corresponding to different energy demands. The controller must be tuned to optimally move the pallets from a source node to one of the available machines and to decide the processing speed through the minimization of a suitable cost function.

The *Pallet routing problem* deals with controlling the movement of the pallets along the network of machines and transportation system to optimize an objective function, while coping with problems such as traffic, starvation, bottleneck and deadlock.

The proposed approach exploits distributed and hierarchical MPC algorithms (see Sect. 2.2) to design a flexible and scalable *genomic MPC-based Control Platform*. This platform can assist and support industrial production system designers to distribute and integrate specific advanced model predictive control solutions into the production system architecture, in order to impress to the whole automated production system an adaptable behaviour. The term *genomic control* comes from the analogy with the DNA model, in which the combination and the specific integration of *control kernels* (nucleotides) into the production system architecture (helical structure) defines the automated production system as a whole. Control kernels will be hosted within a virtual execution environment to guarantee their isolation and platform independency, while preserving their real-time execution characteristics.

Among open-source virtualization products, Xen [15] can be mentioned as a relevant Virtual Machine Monitor (VMM) for the x86, x86_64, and IA64 architectures. Even though Xen has not been designed with real-time execution in mind, the research community is actively working to overcome this limit [16, 17]. More specifically, the RT-Xen project [18] aims at developing a fully real-time, Xen-based hypervisor. However, the existing open-source virtualization products are characterized by two main shortcomings:

- Due to their focus on data centres and server farms, most existing virtualization products were not specifically designed for real-time execution. For this reason, detailed real-time performance data are usually not publicly available and it is unclear if, and to what extent, any real-time software hosted in a Virtual Machine (VM) instead of a physical processing node will still satisfy its timing requirements.
- The scientific literature (e.g. [19]) has shown that the isolation of different applications hosted on the same processing node might be less than perfect. This is also true for VMs and especially for what concerns undue timing interferences, which are of paramount interest for real-time execution.

For these reasons, in the design of the execution environment, special attention is paid to analyse the configuration parameters of the virtualization software, as well as their effect on the real-time characteristics of the system. Secondly, it is necessary to carry out a preliminary evaluation of the degree of isolation provided by the virtualization software with respect to concurrent execution of multiple VMs on the same processing node.

Each control kernel is equipped with a specific standardized software communication interface to exchange data with the external environment (i.e. hardware devices in a production system). The same interface also takes care of the internal communication between control kernels. Following the analysis reported in Sect. 2.2, Modbus has been chosen for internal communication between MPC modules.

The careful deployment of control kernels in a virtual execution and communication environment further improves flexibility because it makes control kernels mostly independent on the physical characteristics of the system without sacrificing their real-time performance.

The obtained control solutions are optimized on the basis of key performance indicators like flow production, peak of absorbed electrical power and total energy consumed by the plant, so that such control algorithms are able to imprint to the production system the desired functional properties. This is especially useful for contemporary automated production systems that are often characterized by a fast and evolutionary behaviour.

2.4   MPC-Based Control Platform

As anticipated, the proposed MPC-based control platform is structured in a two-layer software control structure.

The *lower layer* manages a set of atomic control kernels (i.e. MPC modules) that are deputed to implement specific MPC functionalities. Such control kernels are selectively distributed and then integrated into the industrial production system hardware devices (e.g. hardware controllers, sensors and actuator drivers) by the automation plant designer to control individual pieces of equipment in the production system. In particular, an innovative MPC algorithm for nonlinear systems has been developed. The common approach to deal with nonlinear systems is to consider simpler models obtained through linearization procedures, so that the resulting optimization problems to be solved on-line are quite simple and compatible with computational limitations, although the neglected nonlinearities can lead to poor performance. Alternatively, purely nonlinear MPC algorithms require a heavy computational effort and pose difficult optimization problems. An MPC algorithm has been developed to overcome these limitations, by considering the model of the system linearized along the planned trajectories. This method is an evolution of the approach presented in [20] with enhanced properties with respect to the original one.

The *higher layer* supports the design, evaluation and implementation of the production system automation architecture and the related control functionalities. In particular, this software layer implements an MPC plant coordinator taking full advantage of the most recent advances in hybrid control theory. The software implementation is based on a workbench suitable to support the automation engineer to formalize and implement the flexibility and reconfigurability concepts of an evolutionary production system by converting automatically them into a specific selection and combination of control kernels. Furthermore, innovative solutions based on MPC for manufacturing systems described by Mixed Logical Dynamical (MLD) models have been developed. MLD enables the description of the dynamics characterizing event-driven systems and gives a mathematical formulation of the logical and operational constraints to be considered by using Boolean variables.

The control platform has been implemented in C++ programming language and named Dynamic Control Platform for Industrial Plants (DCPIP). The Object-Oriented programming paradigm [21] has been chosen to support modularity and guarantee easy maintainability. Moreover, the internal data architecture is dynamic, i.e. the platform dynamically builds the necessary data structure, according to the production system structure, to cope with the specific plant to be controlled. Indeed, the platform contains a block library where a certain number of specific and predefined operating machine control modules (the so-called *kernels*) are stored.

DCPIP offers specific structural and functional characteristics to deal with the control kernels and with the higher software level pertaining to the coordinator control algorithms.

The MPC controller was developed in MATLAB, making use also of the YALMIP, HYSDEL and CPLEX software tools. The MPC algorithm is managed by the DCPIP control environment by launching the following computation phases:

- Building of the optimization problem (model of system and cost function).
- The state of the system to be controlled is acquired (Input).
- Solving of the optimization problem.
- Application of the calculated control action.

The DCPIP can be interfaced with either a real plant or a virtual plant that can be modelled as discrete event simulator. No change is required in the DCPIP if the simulator accurately represents the real plant from the point of view of the Input/Output variables.

The control kernels represent software modules that are dynamically generated during the DCPIP start-up to provide specific control and communication functionalities. Since the Object-Oriented paradigm is used for the DCPIP implementation, then the generic kernel can be described in terms of a class object that are instantiated according to the specific plant to be managed. The control kernels needed to design the DCPIP are:

- *Task_Manager*
- *Line_Supervisor / Plant_j_Line_Supervisor*
- *Machine / Machine_j*
- *Controller / Controller_j*
- *Interface_vs_ext / Interface_vs_Txt_file / TCP_IP*

                                45

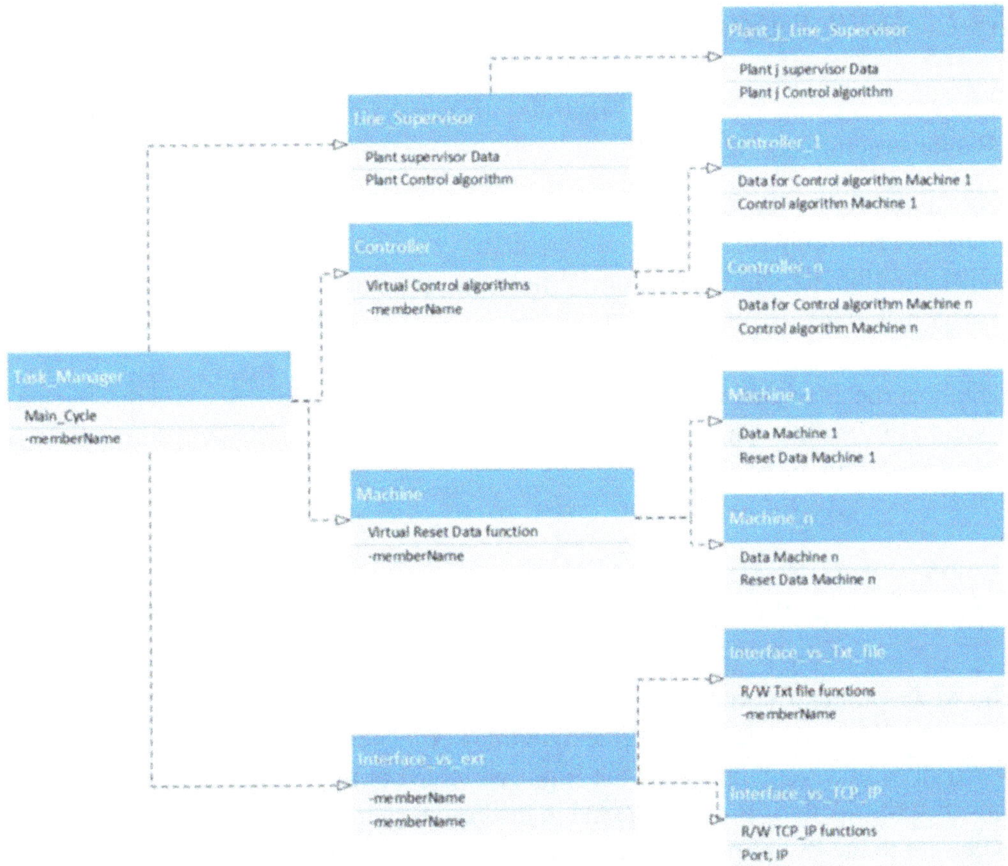

Fig. 2.1 Kernels architecture

A hierarchy of the different kernel classes is shown in Fig. 2.1 and further details are presented in the following subsections.

The control software platform has to communicate the Input/Output process and monitoring variables not only via text files but also via fieldbus (i.e. TCP). This is very important in order to distribute the control platform on different devices (controllers and PCs) connected to each other in a network. This allows to allocate the control platform, the optimization process, the simulator of the plant and the plant controllers all on different devices and to use different software technologies to run such software components (i.e. Virtual Machines).

A kernel hierarchic architecture (including plant supervisor, machine controllers, and communication software modules) requires the implementation of a cross-communication between all the software components, in order to be able both to exchange data with the plant simulator/real plant and to coordinate the different control kernels according to specific supervisor algorithms run into specific control kernel themselves.

2.4.1 Main_program

This kernel contains the software instructions needed to dynamically start the whole CP data structure configuration:

- Instantiation of the interface between the control architecture and the external process environment.
- Instantiation of the pointer to the kernel *Line_Supervisor*. This software module is responsible for the management of the data structure of the configured resources in the Plant or Simulator.
- Start of the *Task_Manager* main cycle. This software program schedules the activation of the different control algorithms for each configured resource in the Plant or Simulation model.

2.4.2 Task_Manager

The *Task_Manager* kernel contains all the variable and methods for the execution of each kernel. In particular, it starts the dynamic kernel generation, performs the control platform Main Cycle that scans the Input variable coming from the PLC/Simulator, runs the control algorithms contained in the kernels and updates the Output variables to be sent to the PLC/Simulator. It must be highlighted that the Input acquisition and the Output updating are carried out by means of the communication functionalities implemented in dedicated kernels.

2.4.3 Line_Supervisor / Plant_j_Line_Supervisor

The *Line_Supervisor* kernel implements the control algorithm of the plant supervisor controller. In particular, it communicates with each machine control kernel (horizontal or cross communication) besides with the PLC/Simulator by means of the I/O data exchanged via text files or TCP protocol.

This class takes advantage of the object-oriented paradigm because the classes derived via inheritance from *Line_Supervisor* (*Plant_j_Line_Supervisor*) can implement specific control algorithms for the considered industry sector (Petro-chemical, Pulp&Paper, Wood, Steel, Automotive, etc.) by means of on virtual functions.

2.4.4 Machine / Machine_j

The *Machine* kernel contains variables declaration and relative methods to perform the data structure management of the generic machine that must be controlled. The

Machine_j kernels represents the specific machine control algorithm implementation. This means that the machine control strategies must be implemented in such kernels.

2.4.5 Controller / Controller_j

The *Controller* kernel contains the control algorithm of the corresponding machine to be controlled. Each machine is characterized by its own control functionalities, therefore the *Controller* class has some virtual functions that are specialized in the derived class controllers *Controller_j*.

2.4.6 Interface_vs_ext / Interface_vs_Txt_file/TCPIP

The *Interface_vs_ext* kernel implements the communication of the *Machine_j* controller and *Line_Supervisor* controller kernels with the external environment, e.g. all the Input/Output data exchanged between the kernels and the PLC/Simulator. This class has methods to execute the Read/Write functions on the Input/Output data, according to the different communication methods (e.g. text file and TCP protocol). Furthermore, this class is virtual because there are different communication channels and further communication protocols could be defined and added to the control platform structure. The *Interface_vs_Txt_file* and *Interface_vs_TCPIP* kernels implement the specific communication methods, being classes derived from the parent class *Interface_vs_ext*. The Read/Write functions are specialized according to the relative communication channel (Text files or TCPIP protocol).

2.5 Testing and Validation of Results

This section presents how the control platform (Sect. 2.5.2) was customized for a reference industrial case (Sect. 2.5.1) to run a set of experiments (Sect. 2.5.3).

2.5.1 Industrial Case

The reference industrial case is a de-manufacturing pilot plant (Fig. 2.2) implemented in the lab of CNR-STIIMA (ex CNR-ITIA) [22–24]. The system was designed for testing and repairing of printed circuit boards (PCBs) and it consists of four cells and a transport line based on transport modules, in particular:

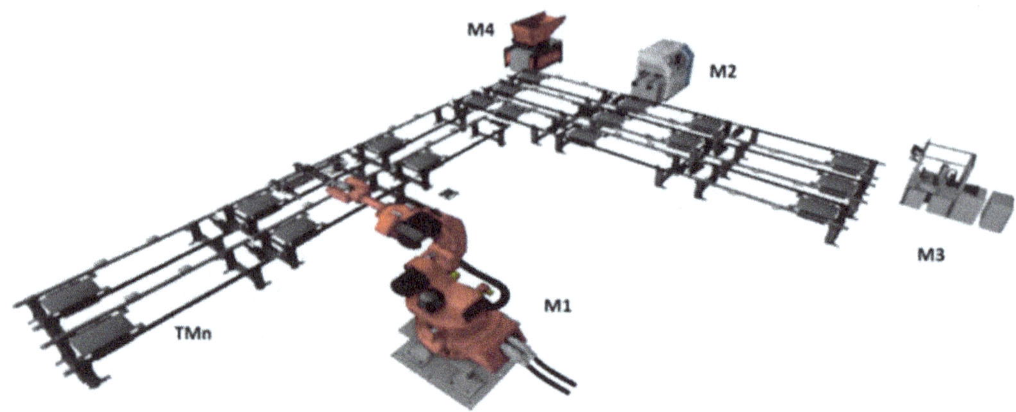

Fig. 2.2 De-manufacturing pilot plant

- *Cell M1* consists of a robot cell able to load/unload the PCBs on a pallet that is then placed on the transport line to be moved towards other machines.
- *Cell M2* consists of a testing machine where PCBs are tested in order to identify the specific faults.
- *Cell M3* consists of a reworking machine where PCBs are repaired by replacing failed components.
- *Cell M4* that is able to unload the PCB from the pallet. Then the PCB is sent to the recycling area of the plant.
- 15 *Transport Modules TMi* (with *i* ranging from 1 to 15), connected together to compose a modular and flexible transport line.

The typical sequence of operations performed by the de-manufacturing plant consists of the following steps:

- The board is loaded on a pallet by M1.
- The transport line moves the pallet to M2 where the board is tested and possible failures are identified to decide whether the board is sent to M3 or M4.
- If a repair is needed, then the pallet with the board is sent first to M3 and back to M2 to test it again.
- If the board is properly working, then it is sent back to M1 where it is unloaded from the pallet and stored in the warehouse. Otherwise the board is sent to M4.
- After the board is removed, the pallet is ready to load a new board and start a new cycle.

The Transport Modules enable to move the pallet and stop it in specific positions based on the transport line topology, on the configuration of each module in terms of mechanical structure, on the automation system instrumentation (i.e. actuators and sensors), and on specific low level control system functionalities. Thus, the pallet can be moved from a specific position to another one called *Buffer Zone* ($BZ_{i,j}$), or in general *Node N_k*, where $BZ_{i,j}$ represents the *j*-th stop position on the *i*-th transport module [25]. If these Buffer Zones are considered from a mathematical point of view, then the plant can be seen as a graph whose nodes represents all the possible

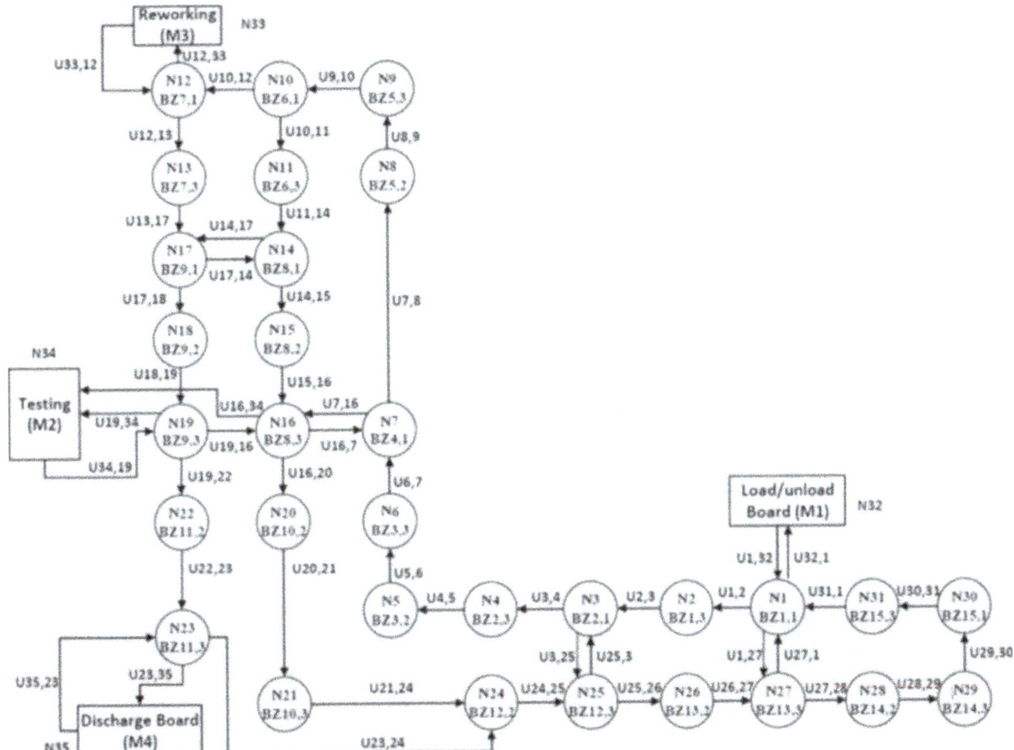

Fig. 2.3 Pilot plant graph representation

positions the pallets can take and the arcs define the possible movements between those positions. Each machine can be seen as a special node that must be visited by the pallet for a minimum amount of time (i.e. the processing time). The resulting reachability graph is depicted in Fig. 2.3. This kind of graph can also be automatically generated thanks to artificial intelligence (AI) techniques as discussed in [26].

2.5.2   De-manufacturing Pilot Plant Control Platform Implementation

The extended automation system architecture involving the control platform (Sect. 2.4) and the Plant/Simulator of the industrial case were designed according to the functional structure depicted in Fig. 2.4. A specific MPC algorithm has been developed [27] for the de-manufacturing pallet transport line [28].

The following software components of the control platform are hosted on a PC:

- *DCPIP* (see Sect. 2.4). Such control system acquires via text file the Input from the Hard Disk of the PC, elaborates it and writes the corresponding Output via text file on the same Hard Disk.

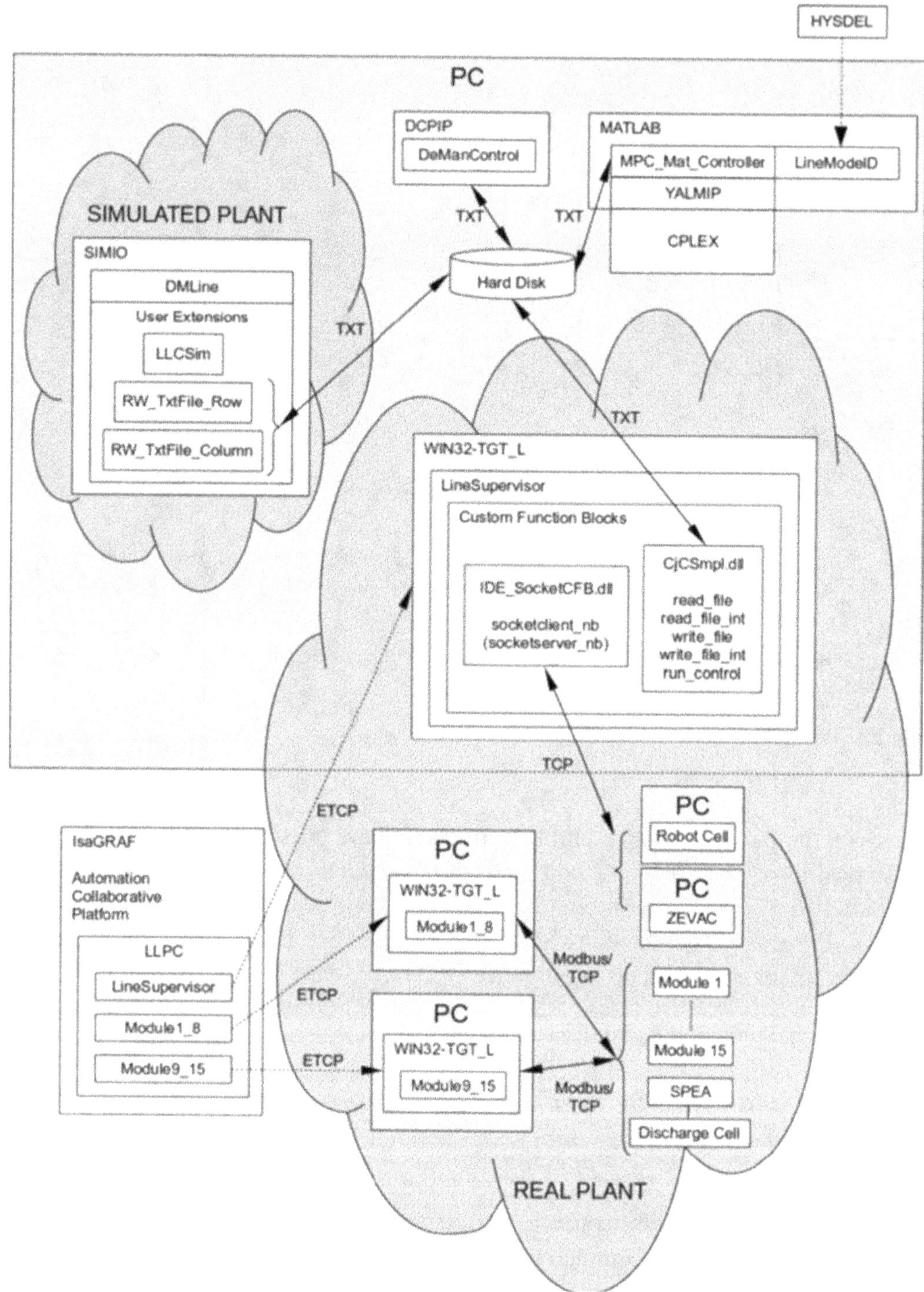

Fig. 2.4 Extended automation system architecture

- *MATLAB* where the Model Predictive Controller is implemented and run (see Sect. 2.4).
- *Simulated Plant*, implemented using SIMIO commercial tool. The simulator represents the real plant from the point of view of the Input/Output variables and performs the plant dynamics, by reading via text file the control actions that are sent by the control environment and calculates the relative process variables.
- *Plant Target WIN32-TGT-L* consists of the low-level controller of the plant that communicates with the software control environment the same way as the *Simulated Plant*.

The software configuration of the real plant consists of:

- Soft PLC (i.e. PLC algorithms running on PC and implemented in ISaGRAF platform) used to implement the plant *low level* control functionalities.
- PC, used as controllers of specific plant devices (Robot cell M1 and reworking machine M3).

For what concerns communication, a CAN-based Modbus adaptation layer (called Modbus CAN) has been designed and implemented. The protocol was successfully analysed with the help of a model checker, to prove several properties of interest related to correctness and communication error tolerance.

In order to evaluate and improve VM-based execution in a realistic scenario, close to typical industrial automation applications, the measurements have been carried out on a setup based on Modbus TCP communications between a master node and a slave node. In this setup, the master node coincides with the execution environment under test and is based upon a Tecnint Leonardo PC BOX industrial embedded PC. The slave node has been built using a standard personal PC that mimics the behaviour of a typical Modbus slave. Performance evaluation was based on round-trip time measurements of Modbus TCP frame exchanges. As a result, the Xen credit scheduler was modified to improve the real-time behaviour of virtual machines dedicated to real-time execution.

2.5.3 Experiments

Several experiments have been carried out to assess the performance of the proposed approach. The obtained experimental results show a satisfactory behaviour in terms of the number of pallets that can be loaded on the transportation line and of the average time required to move the pallets from the initial node to their final destination. Notably, the computational load required to solve on line the optimization problem required by the solution of MPC is fully acceptable for the considered case.

The results obtained addressing the *Production scheduling* and *Pallet routing* problems (Sect. 2.3), extensively described in [27, 29], show that the developed algorithms, based on the MLD representation of the systems considered in the two cases, are highly flexible, so that they can be easily adapted to different problems.

Table 2.1 Task priority assignment for Modbus TCP protocol stacks

	Master	Slave
Ethernet receive task	+4	+3
LWIP main task	+3	+2
Modbus TCP receive helper	+2	n/a
Application	+1	+1
Idle RTOS task	0	0

Moreover, it is easy to obtain different behaviours of the controlled manufacturing system by properly tuning easy-to-understand parameters of the algorithm, such as the cost function and the constraints defining the on-line optimization problem to be solved in MPC. The MPC-MLD approach also enables to cope with dynamic changes of the production environment, such as minimum production and maximum absorbed power requirements in the *Production scheduling problem*, or local malfunctioning of the transportation line in the *Pallet routing problem*. Finally, the important issue related to the computational load required to solve on-line MILP problems has not resulted to be a real bottleneck both in the considered simulation environment and in the real application of these methodologies to a complex pilot plant (see [27]).

For what concerns virtual execution and communication, the experiments have been focused on developing and analysing an appropriate priority assignment scheme for the various tasks involved in the Modbus master and slave protocol stacks. The main goal of the analysis has been to determine a priority assignment scheme that minimizes communication jitter, based on previous experience on jitter analysis gained on similar applications [30]. As it can be seen in Table 2.1, jitter reduction has been achieved by favouring application-level message processing with respect to the lower layers of the protocol stacks.

Afterwards, the performance of Modbus TCP communication has been evaluated and compared with a plain data exchange through a direct TCP connection. The experimental results presented in [30] show that the real-time characteristics of Modbus TCP communication are affected by the underlying TCP segment acknowledge mechanism. More specifically, this mechanism induces a significant communication jitter that, as a result, may hinder the applicability of this method in demanding real-time systems. It is also worth noting that the dependency between segment acknowledgment and jitter is not straightforward and depends on multiple factors, e.g. the above-mentioned priority assignment, the protocol stack architecture, and the traffic initiator (resulting in an asymmetry between the timing behaviour of master requests and slave responses). Further experiments carried out on the same testbed have also provided additional insights on the overhead of the Modbus protocol stack layer with respect to a streamlined, custom TCP-based protocol. This information can be profitably used, for instance, in the design of other similarly distributed embedded systems.

The implemented MPC takes into account the graph-based representation of the plant (Fig. 2.3). Figure 2.5 shows how the control platform works if two pallets are concurrently moved along the plant that is managed as if it consists of multiple feeder

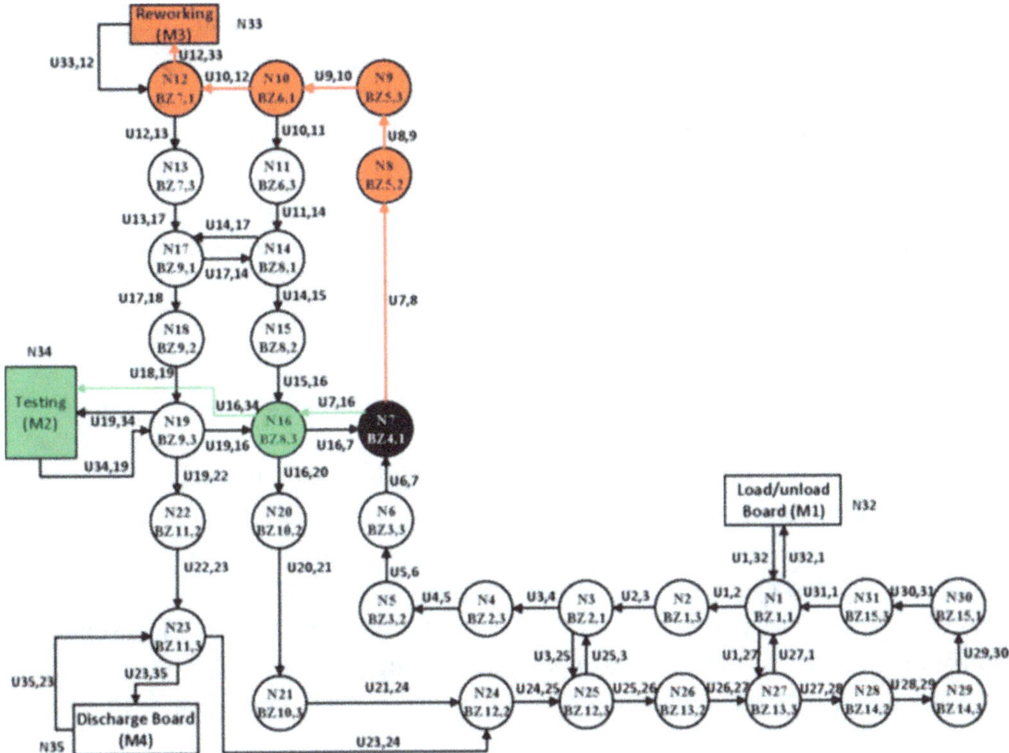

Fig. 2.5 Pilot plant test path definition

lines. One line feeds the reworking machine M3 and is composed of four buffer nodes (N_8, N_9, N_{11}, N_{12}); the other line feeds the testing machine M2 and is composed of one buffer node (N_{16}). Therefore, the approach proposed in Sect. 2.3 can be applied in this part of the pilot plant that is equivalent to a generic plant with two operating machines working with different processing time and two feeder lines with different length.

2.6 Conclusions and Future Research

Motivated by the advanced capabilities of MPC in the process industry and the increasing demands of high performing controllers, this work has applied MPC also to the discrete manufacturing industry by developing a new optimization-based control. We focused on the buffer management and the production scheduling of a multiple-line production plant, aiming to provide optimal scheduling strategies with respect to production requirements. Experimental results show that the algorithm can be highly adapted to obtain different behaviours, by means of simple and easy-to-understand parameters of the cost function. Moreover, such algorithm allows to dynamically change the minimum production and the maximum energy available and to choose, if present, which possible constraint should be violated if necessary [29]. All these features can be hardly achieved with standard controllers or with simple scheduling. The activities related to real-time communication led to a better understanding of the TCP acknowledgment mechanism in the context of Modbus TCP communication [30], as well as the definition of a CAN transport layer for Modbus [31].

The research activity related to the design of hierarchical control systems for the manufacturing industry has been described in a number of journal and conference papers. Specifically, the *Production scheduling problem*, together with the methodology adopted for its solution, has been the object of the paper [29], and related results have been reported in [32]. The results obtained in the *Pallet routing problem* have been extensively described in [27, 28]. Finally, an innovative methodology to compute the energy consumption of manufacturing systems has been proposed in [33].

Concerning future developments of the activity, the computational effort is obviously one of the main obstacles to the diffusion of Model Predictive Control in manufacturing systems, and many improvements are still required to design efficient control algorithms. Distributed optimization methods must be developed, in particular for MLD models, in order to be able to control larger and larger systems and to provide new and efficient solutions. Another important topic of future research concerns the possibility to customize the algorithms and the creation of libraries of models to reduce the development times.

Acknowledgements This work has been funded by the Italian Ministry of Education, Universities and Research (MIUR) under the Flagship Project "Factories of the Future—Italy" (Progetto Bandiera "La Fabbrica del Futuro") [34], Sottoprogetto 2, research project "Genomic Model Predictive Control Tools for Evolutionary Plants" (IMET2AL).

References

1. Camacho EF, Bordons Alba C (2007) Model predictive control. Springer
2. Maciejowski JM (2002) Predictive control with constraints. Pearson Education Limited, Prentice Hall
3. Qin SJ, Badgwell TA (2003) A survey of industrial model predictive control technology. Control Eng Pract 11:733–764
4. Rawlings JB, Mayne D (2009) Model predictive control: theory and design. Nob Hill Publishing, Madison, WI
5. Scattolini R (2009) Architectures for distributed and hierarchical model predictive control—a review. J Process Control 19:723–731
6. Christofides P et al (2013) Distributed model predictive control: a tutorial review and future research directions. Comput Chem Eng 51:21–41
7. Maestre JM, Negenborn RR (2013) Distributed model predictive control made easy. Springer
8. Modbus-IDA (2006). Modbus application protocol specification V1.1b. http://www.modbus-ida.org/
9. Ballarino A et al (2014) On the performance of an automation solution based on IEC 61499 and OPC UA. In: Proceedings of the 17th IEEE conference on emerging technologies and factory automation (ETFA), pp 1–6
10. McConahey J (2011) Using Modbus for process control and automation, part 1. Appl Autom A12–A14

11. McConahey J (2012) Using Modbus for process control and automation, part 2. Appl Autom A12–A14
12. Torrisi FD, Bemporad A (2004) HYSDEL-a tool for generating computational hybrid models for analysis and synthesis problems. IEEE Trans Control Syst Technol 12(2):235–249
13. Lofberg J (2004) YALMIP: a Toolbox for Modeling and Optimization in MATLAB. In: 2004 IEEE international conference on robotics and automation, Taipei, pp 284–289
14. Bemporad A, Morari M (1999) Control of systems integrating logic, dynamics, and constraints. Automatica 407–427
15. Barham P et al (2003) Xen and the art of virtualization. In: Proceedings of the 19th ACM symposium on operating system principles, pp 164–177
16. Ongaro D, Cox AL, Rixner S (2008) Scheduling I/O in virtual machine monitors. In: Proceedings of the 4th ACM SIGPLAN/SIGOPS international conference on virtual execution environments, pp 1–10
17. Yu P et al (2010) Real-time enhancement for Xen hypervisor. In: Proceedings of the 8th IEEE/IFIP international conference on embedded and ubiquitous computing (EUC), pp 23–30
18. Xi S et al (2011) Rt-Xen: towards real-time hypervisor scheduling in Xen. In: Proceedings of the international conference on embedded software (EMSOFT), pp 39–48
19. Cereia M, Cibrario Bertolotti I (2009) Virtual machines for distributed real-time systems. Comput Stand Interfaces 31:30–39
20. Falcone P et al (2008) Linear time-varying model predictive control and its application to active steering systems: stability analysis and experimental validation. Int J Robust Nonlinear Control 18:862–875
21. Ghezzi C, Jazayeri M, Mandrioli D (1991) Fundamentals of software engineering. Prentice Hall
22. Colledani M, Copani G, Tolio T (2014) Management in manufacturing. In: Proceedings of the 47th CIRP conference on manufacturing systems
23. Copani G et al (2012) Integrated de-manufacturing systems as new approach to end-of-life management of mechatronic devices. In: 10th global conference on sustainable manufacturing, Istanbul, Turkey
24. Tolio T, Copani G, Terkaj W (2019) key research priorities for factories of the future—part II: pilot plants and funding mechanisms. In: Tolio T, Copani G, Terkaj W (eds) Factories of the future. Springer
25. Cataldo A, Scattolini R (2014) Logic control design and discrete event simulation model implementation for a de-manufacturing plant. Automazione-plus on-line J
26. Terkaj W, Urgo M, Andolfatto D (2017) Answer set programming for modeling and reasoning on modular and reconfigurable transportation systems. In: Proceedings of the 2017 federated conference on computer science and information systems
27. Cataldo A, Scattolini R (2016) Dynamic pallet routing in a manufacturing transport line with model predictive control. IEEE Trans Control Syst Technol 24:1812–1819
28. Cataldo A, Perizzato A, Scattolini R (2014) Modeling and model predictive control of a de-manufacturing plant. In: IEEE multi-conference on systems and control
29. Cataldo A, Perizzato A, Scattolini R (2015) Production scheduling of parallel machines with model predictive control. Control Eng Pract 28–40
30. Hu T, Cibrario Bertolotti I (2015) Overhead and ACK-induced jitter in Modbus TCP communication. In: Proceedings of the 1st IEEE international forum on research and technologies for society and industry (RTSI)
31. Cena G et al (2014) Design, verification, and performance of a Modbus–CAN adaptation layer. In: Proceedings of the 10th IEEE international workshop on factory communication systems (WFCS), pp 1–10

32. Cataldo A, Perizzato A, Scattolini R (2014) Management of a production cell lubrication system with model predictive control. In: IFIP international conference advances in production management systems, pp 131–138
33. Cataldo A, Scattolini R, Tolio T (2015) An energy consumption evaluation methodology for a manufacturing plant. CIRP J Manuf Sci Technol 11:53–61
34. Terkaj W, Tolio T (2019) The Italian flagship project: factories of the future. In: Tolio T, Copani G, Terkaj W (eds) Factories of the future. Springer

Chapter 3
Exploiting Modular Pallet Flexibility for Product and Process Co-evolution Through Zero-Point Clamping Systems

Marcello Urgo, Walter Terkaj, Franca Giannini, Stefania Pellegrinelli and Stefano Borgo

Abstract Flexibility and reconfigurability of production systems are typically exploited to cope with the changing production demand in terms of volume and variety. This work addresses the problem of enhancing the current flexibility of Flexible Manufacturing Systems (FMSs) by designing pallet configurations with zero-point modular fixtures. This class of equipment provides the ability of rapidly reconfiguring the pallets to match the production requirements, thus providing a strategic option to quickly manage the joint evolution of products and processes. An approach consisting of methods and tools is presented to overcome the main obstacles related to the use of zero-point clamping technologies in modern FMSs. The proposed approach ranges from the design of the pallet configuration to the pallet verification during the manufacturing executing phase. The feasibility of the overall approach has been demonstrated through the development of a prototype.

3.1 Scientific and Industrial Motivations

Manufacturing systems need fast reconfigurations to cope with changing customer needs (e.g. new models, model variants, and materials), new technologies, and also unexpected external and internal events. This problem has been identified as the need to tackle the co-evolution, i.e., the joint evolution of products, processes and the production systems [1]. The proper management of the co-evolution asks for production

M. Urgo (✉)
Dipartimento di Meccanica, Politecnico di Milano, Milan, Italy
e-mail: marcello.urgo@polimi.it

W. Terkaj · S. Pellegrinelli
CNR-STIIMA, Istituto di Sistemi e Tecnologie Industriali Intelligenti per il Manifatturiero Avanzato, Milan, Italy

F. Giannini
CNR-IMATI, Istituto di Matematica Applicata e Tecnologie Informatiche, Genoa, Italy

S. Borgo
CNR-ISTC, Istituto di Scienze e Tecnologie della Cognizione, Trento, Italy

© The Author(s) 2019
T. Tolio et al. (eds.), *Factories of the Future*,
https://doi.org/10.1007/978-3-319-94358-9_3

systems that are able to implement modifications of their usage or configuration [2]. A production system that is able to evolve can be associated with different *basic flexibility levels* (i.e. 1—*Flexibility*, 2—*Reconfigurability* and 3—*Changeability*), depending on the magnitude of actions that are needed to change its capability [3].

A Flexible Manufacturing System (FMS) is a production system consisting of CNC machining centres connected by automated transport systems, moving pallets with clamped workpieces, under the supervision of a centralised control system [4]. An FMS is typically endowed with both *Flexibility* and *Reconfigurability*. *Flexibility* consists in the ability to execute a wide range of operations thanks to general purpose machining centres that are equipped with tool magazines; furthermore, different routings, process plans and part mixes can be realized, while respecting the overall production capacity of the system. *Reconfigurability* in an FMS is given by the possibility of making changes to its equipment, such as: (1) addition/replacement of tools to execute new machining operations; (2) addition/reallocation of fixtures composing the pallets to process new part types or increase the throughput of existing part types; (3) acquisition of further machining centres to increase the overall throughput of the system.

The level of flexibility of FMSs has significantly increased and machining centres are typically programmable and equipped with automatic systems to quickly change tools and parts inside the working area defining their capability. However, important limitations to their flexibility still exist and the attention has now shifted from the machine/robot to the interface between the system and the workpieces. Moreover, a market demand characterized by small lot sizes and a high number of product variants entails a consistent increase in the number and type of fixtures and pallets needed by an FMS. In this case, an FMS can provide actual *Flexibility* only if the whole set of usable fixtures and pallets is concurrently available, so that the production plans can be optimised without further resource constraints. However, this solution leads to high investment costs in production resources and is not adopted by manufacturing companies. Therefore, changes in the configuration of the pallets usually require significant reconfiguration times (both for hardware modifications and validation of the setup), thus hindering the *Flexibility* of an FMS, since it is needed to move to the higher level of *Reconfigurability*. This limitation can be tackled only if it is possible to reduce the overall pallet reconfiguration time.

The adoption of zero-point clamping systems [5] is a solution with high potential for the reduction of pallet reconfiguration time. A zero-point clamping system is designed by assembling standard baseplates with fixtures and provides a reference *zero point* without needing to realign the modular fixture and the pallet, therefore rapid and safe changes of the fixtures can be operated. If different baseplates are available, then the pallet can be effectively reconfigured in a short time. Moreover, if the pallet consists of a tombstone with multiple housings, then the pallet configuration depends on both the type and position of the baseplates. However, the higher flexibility offered by zero-point clamping systems comes with a higher complexity both at system design and system management level.

The use of modular fixtures and reconfigurable pallets requires a different way of dealing with pallet configuration and process planning. The modularity and reconfigurability of the pallets give the opportunity of frequently changing the type, position and number of parts and, therefore, the associated process plan must be updated accordingly. This can be achieved through a modular structure of the machining process (i.e. part program) that enables to shift some decisions from the process plan definition (planning level) to the process plan execution (shop floor level). Thus, the flexibility offered by modular process plans must be paired with methods that are able to generate and validate the overall process plan that is fitting the actual pallet configuration.

In addition, even in modern FMSs, loading/unloading activities still remain mostly manual and, thus, prone to errors with the risk of potentially destructive crashes during high speed machining. As the variety of tasks to be executed by workers increases, the risk of errors becomes higher, hence, inspection devices and tools are needed to control the actual pallet configuration (e.g. state of the clamping, position, orientation and geometric dimensions of the parts).

Finally, the possibility of frequent and quick reconfiguration of pallets gives additional options for production planning, through more complex loading and management policies.

After presenting an overview of the state of the art (Sect. 3.2), this work proposes an approach to cope with design and management problems associated with the use of zero-point clamping systems in an FMS (Sect. 3.3). Given the different methodologies and tools integrated in the approach (Sect. 3.4), there is a need of structuring the involved heterogeneous data. Therefore, a reference data model is defined to enable the sharing of data and knowledge as well as to enhance the interoperability of the software tools. The proposed approach has been tested thanks to a hardware and software prototype (Sect. 3.5) that was realized to represent a realistic industrial case. Finally, conclusions are drawn in Sect. 3.6.

3.2 State of the Art

This section gives an overview of the state of the art for the scientific and industrial areas described in the previous section, i.e. (1) selection of modular fixtures, (2) design of pallet configurations and process planning, (3) inspection of objects in industrial environments, (4) production planning and (5) data modelling for pallets and fixtures.

Modern machine tools and systems are generally provided with reconfigurability options and flexibility degrees designed to meet production requirements. However, the focus is usually on machine tools, handling devices and tools, while the fixtures for clamping the parts have to be manufactured or significantly altered for each specific part. Hence, several researchers focused on fixture devices to enable quick reconfiguration and adaptation to customer demands. Ghandhi and Thompson [6] proposed an automated approach for the design of modular fixtures for flexible manufacturing systems, Perremans [7] used a feature-based description of modular fixturing elements, and Hunter et al. [8] presented an approach to formalize the fixture design process under consideration of functional requirements and their associated constraints. Other researches have addressed the configuration of flexible and modular fixtures. For instance, Wu et al. [9] presented an automated fixture configuration design approach taking as input fixturing surfaces and points to select and arrange proper modular fixture components while satisfying assembly constraints. Nevertheless, most of the approaches in the literature are focused on a static set of parts without the capability of managing both the initial design and the reconfiguration steps in a single approach. The evolving requirements can be tackled if the design of single pieces of fixture is integrated with the definition of pallet configurations by specifying the number and position of the workpieces on the pallet. The goal of pallet configuration is to guarantee the machinability of the workpieces on the pallet as well as their accessibility for loading, unloading and re-clamping. The machinability depends on the number and type of the machine tool controlled axes, as well as on the setup (i.e. orientation) and pattern (i.e. location) of the workpieces. Setup planning and pallet configuration have been simultaneously solved by exploiting a unique mathematical model considering one part type per pallet and one workpiece setup per pallet face [10]. Such limitations regarding the pallet configuration can be properly addressed only by jointly tackling also the process planning in a modular way, e.g. by defining the machining process as a Network Part Program (NPP) [11] that is a process plan (and then a part program) defined in terms of a Directed Acyclic Graph (DAG). An approach for pallet configuration based on NPP to deal with the simultaneous machining of different workpieces on the same pallet has been recently proposed [12].

Object inspection is largely employed in the manufacturing industry to guarantee that the produced parts are satisfying the required quality. Contact measurement systems (e.g. coordinate measuring machines or CMM) are the most commonly employed in the industrial practice [13]. However, they are sub-optimal due to several limitations in terms of low reconfigurability and customizability levels, high costs and the time required for the inspection. On the other hand, commercially available contactless inspection systems generally require manual intervention for the registration process of the scanned data to the ideal shape and are thus not usable to automatically detect deviations through pallet inspection as required in FMS scenarios [13, 14]. Various three-dimensional (3D) scanning technologies are available to provide a digital description of the 3D scanned object in terms of 3D meshes. Research has been carried out about mesh analysis and segmentation according to different perspectives [15, 16]. Moreover, different methodologies have been proposed for shape comparison enabling the identification of similar subparts [17].

The assignment of the machining operations to the available resources in an FMS is typically addressed by machine/resource loading approaches [18]. The FMS loading problem has been formulated in various variants, taking into consideration the different objectives, e.g. minimisation of the movements across the system, balancing of the machine workload, and maximisation of the job priorities [19]. The throughput of an FMS (but also a larger variety of its performance indicators) depends on the capacity of the machines, but also on the number of available fixtures; nevertheless, only a limited subset of the literature explicitly considers availability of pallets and fixtures [20].

Several works have addressed the problem of modelling pallets and fixtures in the context of manufacturing systems. Bugtai and Young [21] proposed an object-oriented approach to model the manufacturing information related to fixture elements and processes including a specific extension related to the fixturing. The FIXON ontology [22] applied the Description Logic (DL) language to fixtures and related components to support seamless information exchange. FixOnt [23] was proposed as domain ontology to classify workholding fixtures and fixture components while supporting automated fixture design in reconfigurable manufacturing systems. FixOnt reused and adapted the FIXON ontology [22], including several extensions to address specific requirements of the reconfigurable vise-type fixture not included in FIXON. Pellegrinelli et al. [24] proposed an ontology-based data model that integrates various knowledge domains relevant to the problem of design and management of an FMS. The ontology proposed in this paper is loosely based on the earlier work by Terkaj et al. [25] and integrates the FixOnt ontology [23], the ifcOWL ontology [26] and an ontology version of the STEP-NC standard [27].

3.3 Problem Formalization

This section presents a formalization of the addressed industrial problem by representing the workflow of activities related to the whole lifecycle of a flexible production system, as shown in Fig. 3.1 adopting the IDEF0 formalism. Based on the requirements to be met and the set of selectable resources, the activity *Design System* (*A1* in Fig. 3.1) takes care of designing the whole flexible production system in terms of hardware configuration, process plans, specific fixtures and inspection systems. The activity *Construct, Install & Maintain System* (*A2* in Fig. 3.1) implements the decisions taken during the (re)design phase and continuously updates the state of the system based on monitoring information. The activity *Execute & Monitor System* (*A3* in Fig. 3.1) receives as input the orders and the current system configuration and plans the manufacturing execution tasks that must be properly monitored. Herein, the attention will be focused on the activities *Design System* and *Execute & Monitor System* because they are relevant for the design and management of modular reconfigurable pallets in an FMS.

The activity *Design System* can be further detailed as shown in Fig. 3.2.

The activity *Design System Configuration* (*A11* in Fig. 3.2) consists of selecting the production resources and planning the layout of the system. Beyond the typical production resources, the design of the pallet inspection system is also included because it plays a crucial role to cope with the additional flexibility given by the adoption of zero-point clamping fixtures. Indeed, pallets are typically configured and verified before their utilization in traditional FMS systems. Once they are introduced in the FMS, the verification activities are limited and in charge of the workers that

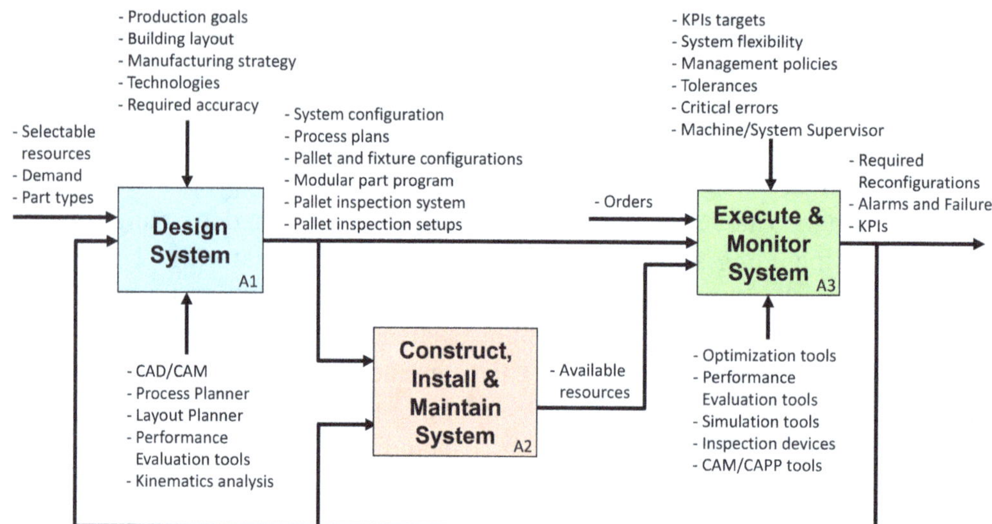

Fig. 3.1 IDEF0 diagram representing Design and Manage Flexible Production System activity

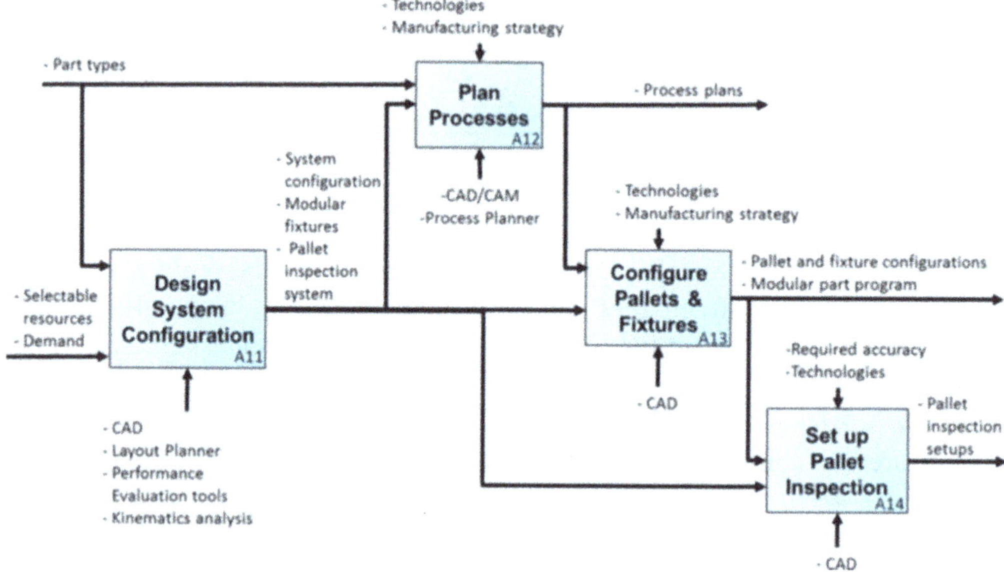

Fig. 3.2 IDEF0 diagram representing Design System activity

have to manage a restricted range of possible situations while loading and unloading the parts. On the contrary, when the reconfiguration of the pallet occurs more often and can be quickly executed, the risk of errors in the loading of the right parts on the right fixture increases and it is necessary to provide a fast and reliable verification method. Possible errors can be related to loading the wrong part type variant, or loading the right part type in the wrong process phase. The pallet inspection system will be reasonably installed in the load/unload station.

The activity *Plan Processes* (*A12* in Fig. 3.2) defines the process plans, i.e. the set of machining operations needed to transform the raw parts/materials into finished products. The operations are defined based on the available technologies and production resources (e.g. machine tools, types of fixtures), while considering manufacturing goals and constraints.

The activity *Configure Pallets & Fixtures* (*A13* in Fig. 3.2) aims at providing a solution for the allocation of parts (and consequently fixtures) on pallets to guarantee the feasibility of the machining. Therefore, it takes into consideration the characteristics of the parts and the associated process plan, the definition of the setups together with the process parameters. The fixture/pallet configuration activity is constrained by the available machine tools in terms of the number of axes, the dimension and the characteristics of the working envelope, the performance in terms of maximum spindle speed, available cutting power, quality characteristics like repeatability and accuracy, energy consumption, etc. When considering the additional degrees of freedom and, hence, of flexibility provided by the adoption of zero-clamping fixtures, the fixture/pallet configuration phase has to take into account the available modular fixtures and provide a reduced set of possible configurations in terms of different ways to assemble the available baseplates onto the pallet tombstone and the feasibility of the machining operations in the different pallet configurations. Hence, it provides a set of possible pallet reconfiguration options exploitable during the execution phase.

Once the possible fixture/pallet configurations are generated, the activity *Set up Pallet Inspection* (*A14* in Fig. 3.2) delivers the reference pallet inspection setups (e.g. master geometries) that will be used to validate the actual configured pallet during manufacturing execution.

The activity *Execute & Monitor System* (Fig. 3.3) starts with the *Load & Optimize* (*A31* in Fig. 3.3), i.e. the assignment of the operations to the production resources based on management policies taking into consideration and exploiting the benefits provided by zero-point clamping systems. Basically, management policies are responsible of selecting the best pallet configurations among the possible ones and define how the machines have to process the parts to optimize the use of the FMS. Management policies consider the part types to be produced together with their evolution within the considered planning horizon, the possible pallet configurations, the availability of pallets and fixtures in terms of tombstones and baseplates as well as the machinability of the operations in the different pallet configurations and the availability of the tools. The assessment of the quality of the loading policies can be done considering the dynamic behaviour of the system via performance evaluation methods that are used to calculate the typical performance of the system, e.g., the throughput, the utilization of the resources, the number of pallets in the queues, etc.

The activity *Make Pallet Inspection* (*A32* in Fig. 3.3) deals with methods and tools to verify the actual configuration of the pallets as well as the correct loading of parts by obtaining a 3D model of the real pallet (slave) to be checked against the ideal pallet configuration (master). The pallet check phase provides warnings or alarms when the final configuration of the pallet does not match the expected layout.

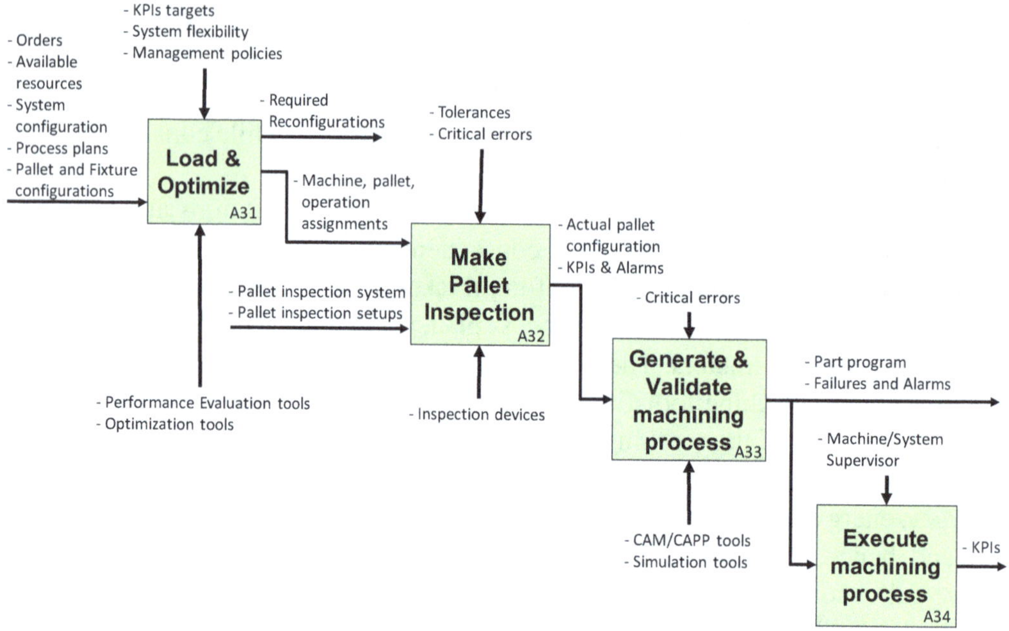

Fig. 3.3 IDEF0 diagram representing Execute & Monitor System activity

If the pallet check is successful, then the digital representation of the pallet is used to feed the activity *Generate & Validate machining process* (*A33* in Fig. 3.3) that tackles the generation of the part program of the complete pallet (i.e. the sequence of commands to be executed by a CNC machining centre). If the part program for the current pallet configuration has not yet been validated, then it is possible to run a process simulation to identify possible errors, e.g. collisions between a cutting tool and the fixtures, before finally moving to the activity *Execute machining process* (*A34* in Fig. 3.3).

3.4 Pallet Design and Management Approach

This section presents the proposed approach to support the design and management of modular reconfigurable pallets. The approach is based on the work by Urgo et al. [28] and consists of a set of methods and tools supporting most of the activities described in the previous section. Herein, the activities *Construct, Install & Maintain System* (*A2*) and *Execute machining process* (*A34*) are considered as out of scope. Furthermore, the assumption is made that the following activities have already been carried out: *Design System Configuration* (*A11*), *Plan Processes* (*A12*). Therefore, it is already known which are the selected production resources (i.e. machine tools, transporters, physical pallets, modular fixtures, etc.) and the set of machining operations to execute for each part type.

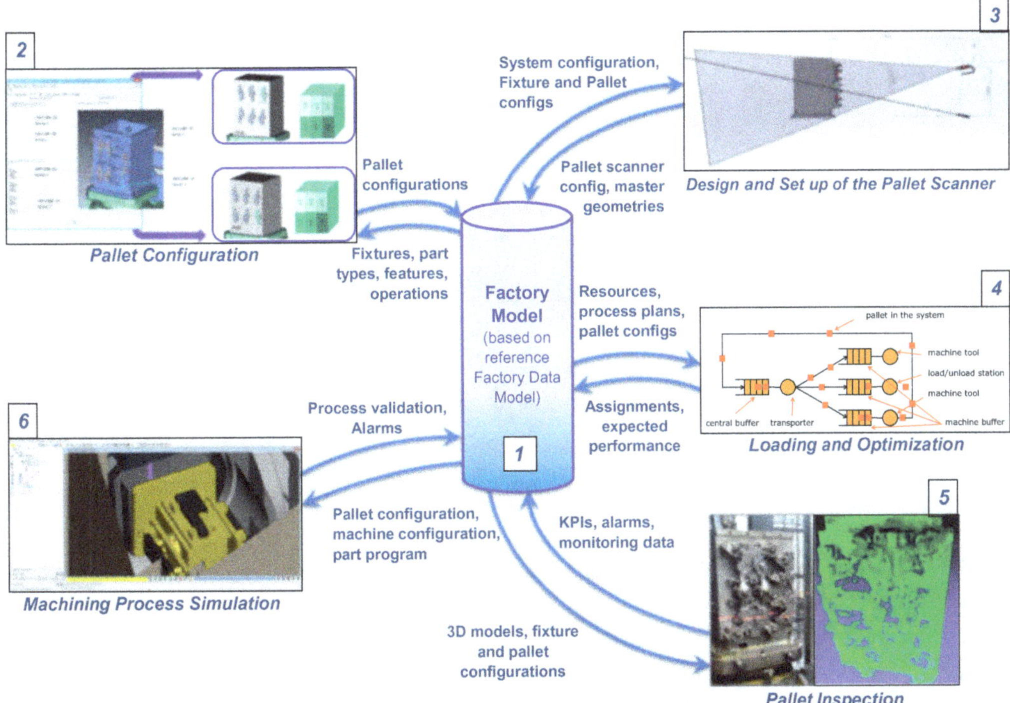

Fig. 3.4 Model-based integration of methods and tools to support the proposed Pallet Design and Management approach

The following sub-sections will delve into methods and tools composing the proposed approach that is shown in Fig. 3.4. The approach resembles a *star network* because the need of interaction and data exchange between the tool is operated through a shared factory model based on a common Factory Data Model (*component 1* in Fig. 3.4) presented in Sect. 3.4.1. The methods and tools (*components 2–6* in Fig. 3.4) are described in Sects. 3.4.2–3.4.6.

3.4.1 Factory Data Model

The data model is a key element to support interoperability between different digital tools, providing the capability to retrieve, store and share information. Therefore, a suitable data model must be able to cover and integrate heterogeneous knowledge domains, while guaranteeing extensibility.

Semantic Web technologies and in particular ontologies can be employed to meet these requirements [29, 30]. Herein, a modular ontology-based Factory Data Model is proposed to formalize the information that is in particular relevant to the design and management of modular reconfigurable pallets. The OWL 2 ontology language [31] is adopted and the work of Pellegrinelli et al. [24] is taken as the basis of the proposed Factory Data Model that aims at representing a detailed pallet structure and

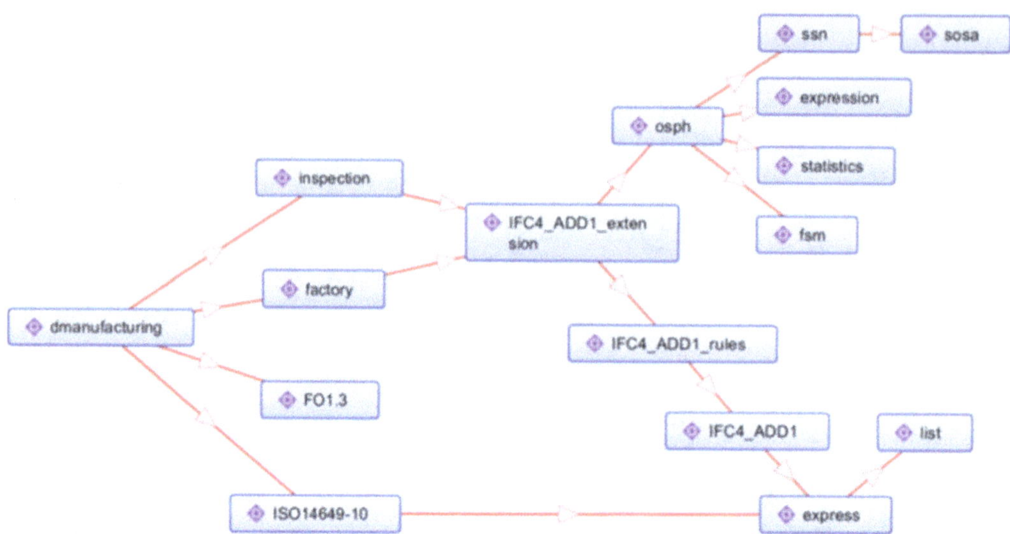

Fig. 3.5 Ontology modules of the Factory Data Model linked by import relations as shown by Protégé software tool [39]

its fixture elements, workpiece setup, pallet inspection systems, and the evolution of the state of the factory objects. The architecture of the data model is shown in Fig. 3.5 and consists of the following ontology modules:

- *list*, an ontology defining the set of entities used to describe the OWL list pattern [32].
- *express*, ontology mapping the concepts of EXPRESS [33] language to OWL [32].
- *IFC_ADD1*, the ifcOWL ontology that is converted from the IFC standard defined in EXPRESS language [26, 32].
- *IFC_ADD1_rules*, an enhancement of the ifcOWL ontology with axioms derived from WHERE rules in the original IFC EXPRESS schema [34, 35].
- *fsm*, an ontology defining the concepts required for modelling finite state machines [36].
- *sosa*, the Sensor, Observation, Sample, and Actuator (SOSA) Core Ontology [37].
- *ssn*, the Semantic Sensor Network Ontology [37].
- *statistics*, an ontology that defines probability distributions and descriptive statistics concepts [38].
- *expression*, an ontology modelling algebraic and logical expressions [38].
- *osph*, an ontology modelling Object States and Performance History, while integrating the ontology modules *fsm*, *statistics*, *ssn*, *sosa*, *expression* [38]
- *IFC_ADD1_extension*, an ontology module integrating *osph* and *IFC_ADD1_rules* modules, while adding general purpose extensions to *IFC_ADD1*.
- *ISO14649-10*, based on the STEP-NC standard [27] converted from EXPRESS schema into OWL ontology according to the pattern defined in [32].
- *factory*, a specialization of *IFC_ADD1* with definitions related to production processes, part types, manufacturing systems and machine tools.

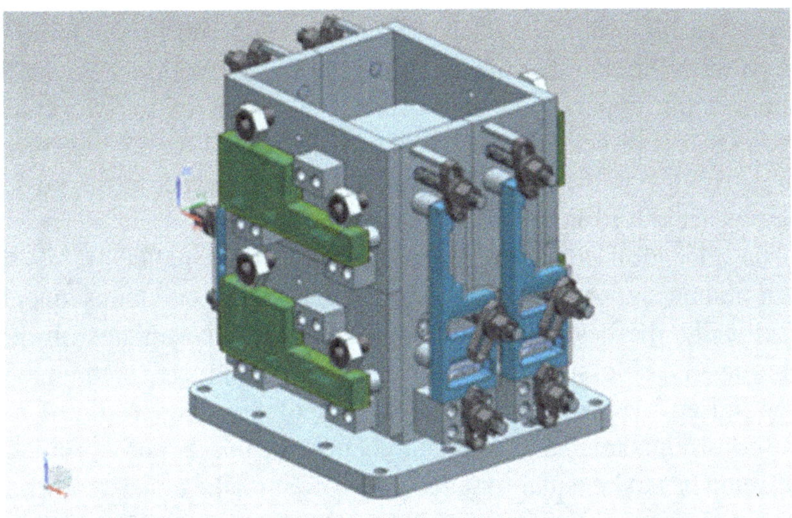

Fig. 3.6 Example of pallet configuration

- *inspection*, an ontology defining the concepts of object inspection and 3D scanner system for data acquisition (e.g. laser scanner and its components).
- *FO1.3*: a fragment of the *FixOnt* ontology presented in [23].
- *dmanufacturing*, ontology module integrating four modules (*ISO14649-10, factory, inspection, FO1.3*), while adding specializations for the discrete manufacturing domain.

The modules *IFC4_ADD1_extension, factory, inspection*, and *dmanufacturing* represent an evolution of the modules *IFC4_ADD1_extension, FactoryDomain, VisualInspectionDomain*, and *DiscreteManufacturingDomain*, respectively, presented in [24].

The Factory Data Model can be instantiated to generate libraries of objects such as part types, operations, fixtures, and pallet components that can be later exploited to define specific pallet configurations, as discussed in the following subsections.

3.4.2 Pallet Configuration

Optimizing the pallet configuration means to identify the best location of the workpiece on the baseplates and their positioning on a given structure, taking into account the setups necessary for machining the workpiece as well as its operations (Fig. 3.6). The proposed method for pallet configuration is able to manage zero-point clamping systems with two sequential steps: accessibility analysis and workpiece allocation, thus supporting activity *A13* in Fig. 3.2.

Accessibility analysis requires as input the identification of the baseplates that can be mounted on the pallet, the dimensions of each baseplate, the setups for each considered workpiece in terms of operation tool access directions and the number of axes of the machine tool. For each workpiece setup and for each baseplate, the admissible patterns (i.e. number of rows and columns of workpieces) are listed. The pattern admissibility depends on the spatial dimensions of the workpiece on the zone, the tool access directions that have to be respected for all the workpieces of the pattern as well as on the position of the face with respect to the position of the zone in the physical face of the pallet.

Workpiece allocation defines the configuration of the pallet by identifying the position and numbers of the workpieces on the baseplates and, consequently, on the pallet. Specifically, the best combination of admissible baseplates on the pallet is selected among all the possible combinations. A mixed integer mathematical model (MIP) is optimised by maximising the number of finished produced parts while meeting the constraints related to pallet balancing (the number of workpieces in each work piece setup has to be equal for each part type) and the consistency between the baseplates, the physical pallet face and the workpiece setups. The model can be re-run multiple times to generate different solutions in terms of pallet balancing and/or placement of the workpieces. More details about the pallet configuration method can be found in [12].

3.4.3 Design and Setup of the Pallet Scanner

The proposed approach performs the inspection of the pallet configuration at the shop floor level by means of a 3D laser scanner that acquires the pallet data as a point cloud. The laser scan technology is promising as a versatile and low cost solution capable to operate under difficult shop floor conditions in term of light sources and dust.

The design of the pallet scanner (i.e. part of activity *A11* in Fig. 3.2) aims at supporting the complete acquisition of the pallet while considering the characteristics and size fixtures that can be used to generate pallet configurations. The following activities are carried out to design the scanner configuration and distance from the inspected object (i.e. pallet) [13, 40]: selection of the camera with optics guaranteeing the appropriate resolution; selection of laser type and optics with fan angle and positioning to cover the required volume; identification of step motors for the selection of the number of laser edges.

The point cloud generated when the scanner inspects a pallet must be elaborated and compared with the desired pallet configuration, i.e. the *master geometry* associated with the correct positions and shapes of all the elements composing the pallet configuration. The generation of the needed master geometries is part of activity *A14* in Fig. 3.2. Two possible ways of generating the master geometries can be foreseen:

1. *Empirical* generation. The possible and relevant pallet configurations are implemented and each face of the pallet is acquired by the laser scanner. The resulting point clouds are elaborated and the master geometries of the various pallet faces (with fixtures and workpieces) are stored in a database for future use. This solu-

tion can be managed if the number of possible pallet configurations (or at least the number of different pallet faces) is not too high; otherwise the workload of the setup phase becomes excessive.

2. *Model-based* generation. The master geometry is calculated by exploiting the 3D CAD models of the pallet together with the topological configuration of the laser scanner. The following comparison with the scanned point cloud can be made easier by computing the area of the master geometry corresponding only to the portion of the pallet that is actually visible and acquired by the laser scanner. Therefore, it is necessary to identify all the mesh elements of the 3D CAD models that are simultaneously visible by the camera and the laser of the scanner [12]. These two sets are obtained by considering the scanner camera and the laser positions to determine the viewing frustum as in 3D computer graphics. Furthermore, critical elements that are almost parallel to the view direction can be removed to reduce the noise during the comparison elaborations.

The proposed approach is able to deal with both ways of generating the master geometries, therefore it is not necessary to impose restrictions.

3.4.4 Loading and Optimization

As anticipated in the introduction, the use of zero-point fixture systems can actually increase the flexibility of an FMS. In particular, the possibility to change the fixtures mounted on a pallet in a short time can lead to:

- Lower number of pallets and fixtures needed to satisfy the same demand mix.
- Better workload optimization for all the involved resources (machine tools, pallets and fixtures).
- Shorter makespan to accomplish the overall required production.

These options open new ways for production planning policies never considered in literature because not applicable with traditional clamping technologies. Therefore, herein the aim is to support activity *A31* (see Fig. 3.3) by optimizing the reconfiguration of physical pallets (i.e. changing the assignment of the available baseplates to the tombstones) and the assignments of machining operations to machine tools. The set of machining operations associated with a pallet configuration can be assigned to more than one machine tool thanks to the non-linear formulation of the part program [41, 42], thus increasing the complexity of the loading problem.

The proposed *Loading and Optimization* method jointly tackles the fixture assignment and loading problem by sequentially solving two sub-models (see Fig. 3.7) to reduce the computational complexity while introducing a certain degree of approximation:

1. *Model 1* assigns the fixtures (i.e. baseplates) to the pallets over the planning horizon, thus determining the needed pallet reconfigurations. The goal is to minimize both the required number of baseplates and the makespan.

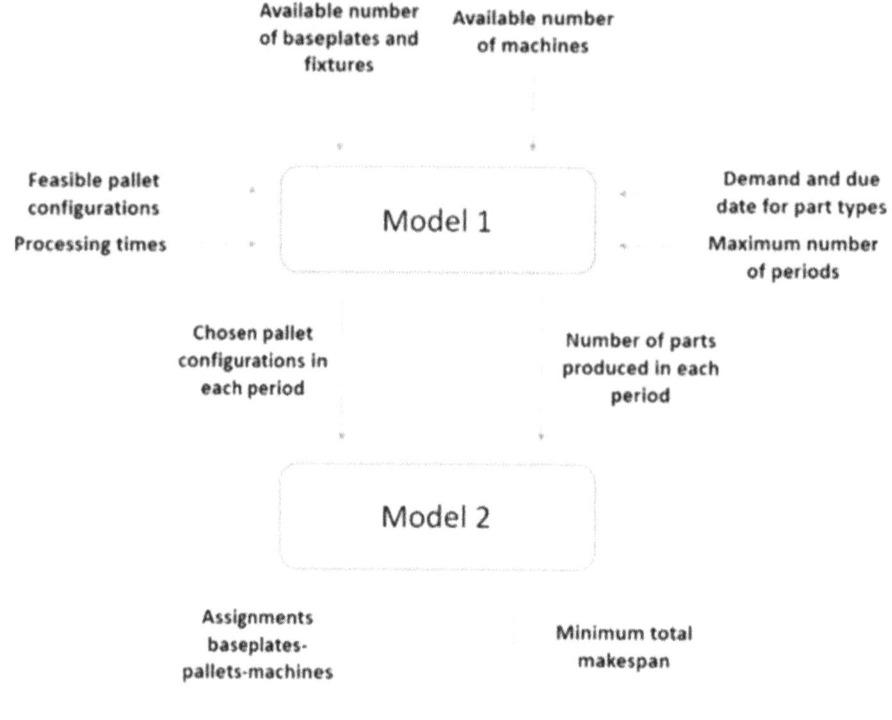

Fig. 3.7 Structure of the Loading and Optimization method

2. *Model 2*, given the outcome of the first model, assigns the machining operations of the baseplates to the available machines (or groups of machines), thus determining the routing of the pallets in the system. The goal is to minimize the makespan while taking into consideration a higher level of details with respect to *Model 1*.

The proposed method has been tested on 10 production problem instances described in Table 3.1. The results of the proposed method have been compared with the results that can be obtained without zero-point clamping technology, i.e. using traditional pallets. The same number of physical pallets has been considered in both cases, therefore the cost for the pallets and fixturing equipment is equivalent, except the additional cost for the baseplates endowed with zero-point clamping technology.

The experiments showed a saving of about 10% on average in terms of makespan thanks to the use of zero-point clamping technology compared with the use of traditional pallets. This means that, if the benefit linked to this time saving is higher than the cost of adopting the zero-point clamping technology, then it is advised to opt for it.

Table 3.1 Production problem instances

Parameter	Value
Max. number of periods/reconfigurations	6
Number of part types	2
Demand volume for each part type	180–600 parts
Types of baseplates	5
Number of possible pallet configurations	7
Number of available pallets/tombstones	5
Number of machines	3
Time horizon	12 shifts

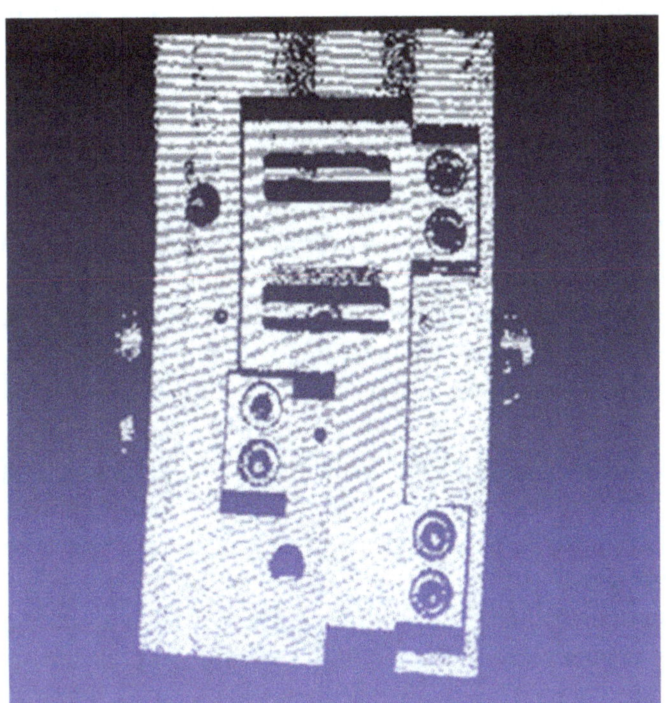

Fig. 3.8 Example of master geometry

3.4.5 Pallet Inspection

The activity *Make Pallet Inspection* (*A32* in Fig. 3.3) is carried out by scanning the actual pallet configuration that must be processed by the machining centres. The pallet scanning returns a point cloud that must be further elaborated to check if the pallet configuration is correct. This check is performed through the evaluation of the difference between the scanned point cloud (i.e. *slave geometry*) and the *master geometry* (Fig. 3.8) generated during the setup of the scanner (see Sect. 3.4.3).

Each point of the slave geometry is analysed and the three closest points in the master geometry are selected to calculate a plane equation. Then, the distance between the point of the slave and the plane is computed. Finally, the minimum square error based on all the computed distances is calculated. After the elaboration, the method returns statistics in terms of maximum, mean, and minimum errors with respect to the reference pallet configuration [12]. These results can be used to support a decision system identifying significant deviations as a symptom of possible errors in the mounting of the parts or be presented to the operator who decides if any intervention is needed.

3.4.6 Machining Process Simulation

The activity *Generate & Validate machining process* (*A33* in Fig. 3.3) consumes the results of the pallet inspection together with the information related to the machining operations and the fixture and pallet configuration. The part program associated with the whole pallet can be automatically generated while adopting the Network Part Program method mentioned in Sect. 3.2. The part program is defined in terms of a partially ordered set of operations exploiting the STEP-NC [27] structure through machining working steps (MWs). A set of MWs is associated with each baseplate, whose toolpaths are referred to the coordinates of the baseplate itself. Once the baseplates are located onto the tombstone and recognized through the pallet inspection, the coordinates of the baseplates are updated accordingly and the actual paths for the tool are derived automatically. Specific algorithms are used to generate the rapid movements between the baseplates taking into consideration their placement as well as the geometry of the fixture, in order to avoid collisions. Advanced algorithms can be employed to optimize the process for the whole system, i.e., clustering operations sharing the same tool to minimize the tool changes.

The commercial software tool Vericut[1] has been adopted to simulate the part program and validate the CNC machining while considering the static and dynamic properties of the machine tools. This process simulation is able to check if the machine tool can perform the required machining operations without collisions.

3.5 Prototype and Testing

The proposed approach has been tested thanks to a hardware and software prototype, named Pro2ReFix (see Sect. 3.5.1), using the information related to a part type derived from an industrial case (see Sect. 3.5.2).

---

[1] http://www.cgtech.com/.

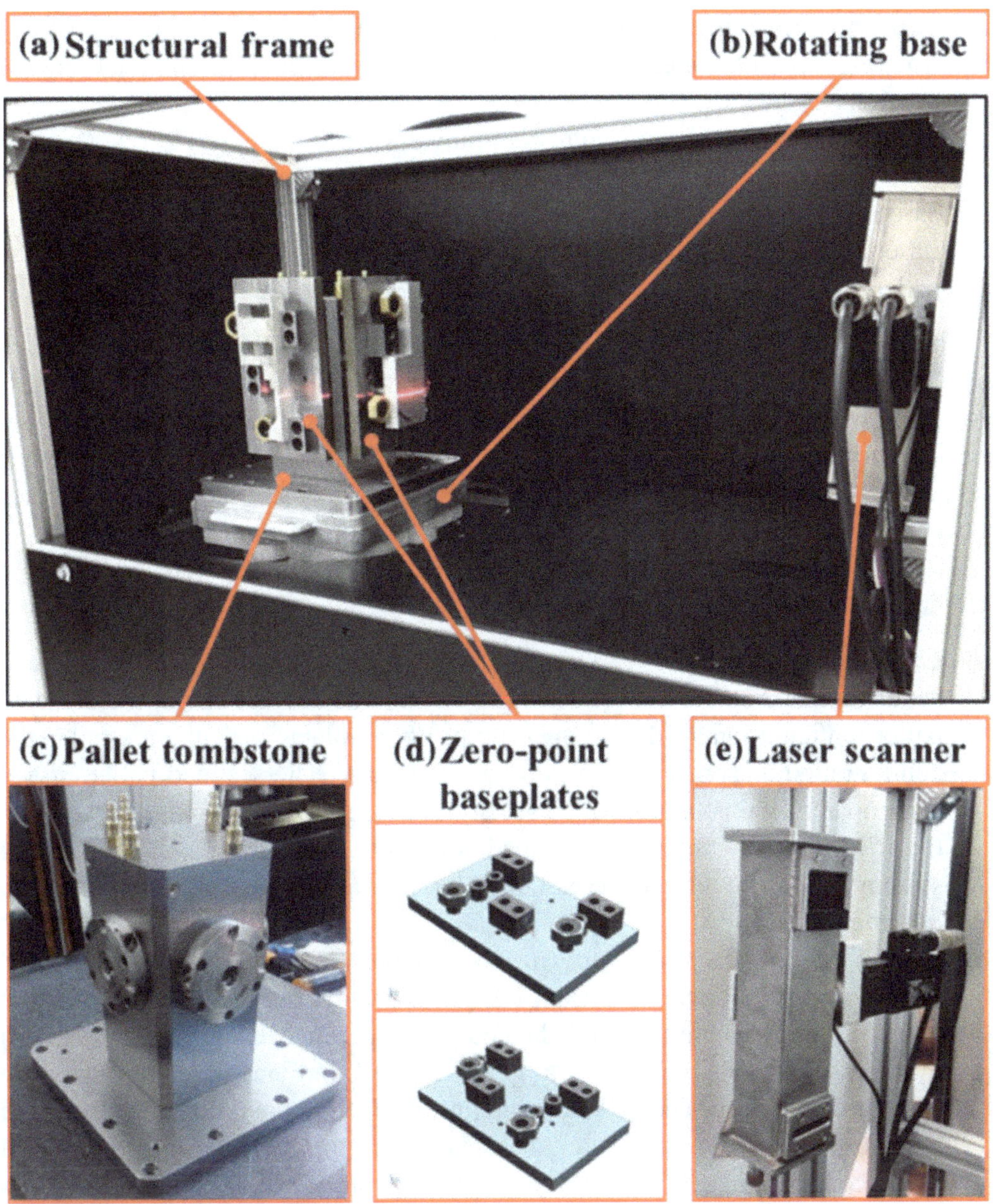

Fig. 3.9 Hardware components of the Pro2ReFix prototype

3.5.1 Pro2ReFix Prototype

The Pro2ReFix prototype is a portable frame demonstrating fast pallet reconfiguration through zero-point clamping systems and the configuration check by using a laser scanner. The prototype consists of various components as shown in Fig. 3.9:

(a) Structural frame hosting the functional components.
(b) Rotating base integrated in the frame to hold the tombstone and enabling the rotation.
(c) Pallet tombstone with zero-point clamping system.
(d) Zero-point baseplates with predefined fixtures configuration hosting different parts.
(e) Laser scanner to acquire the data for the identification and verification of the pallet.

The laser scanner consists of a camera FPGA 4Mpx (2048 dot acquired in each edge) 1 kHz and a laser (Class 3R) emitters with nominal power 100 mW characterized by low speckle. In addition, the design of the laser scanner has taken into consideration requirements like protection level (IP68), safety issues, low device footprint, electric shock. The laser scanner can capture object placed at a distance between 500 and 1500 mm with an average accuracy of 0.1 mm.

The prototype also entails a set of software components that operate the hardware elements and make elaborations, including:

(a) *Laser scanner controller*. The controller takes care of the setup (Sect. 3.4.3), data acquisition and elaboration (Sect. 3.4.5) to provide as final output the detected pallet configuration by identifying the type of baseplates currently present in the configuration and their position on the tombstone. The scanned pallet (*slave*) is compared against a predefined set of 3D models of the available baseplates (*master*). The program was developed in Python language on a Unix operating system and can be directly accessed via monitor and keyboard, or remotely via Ethernet connection.
(b) *Generator of digital pallet configuration*. The results of the laser scanning and the following elaboration are taken as input to obtain a digital model of the pallet as it is currently configured. A dedicated C++ program executes this elaboration by accessing ontology modules to generate OWL individuals representing the pallet configuration. The pallet configuration is defined in terms of a structured assembly of part types, fixtures, baseplates and tombstone. Once the digital pallet configuration is available, this information can be elaborated by other digital tools, e.g. for visualization purposes.
(c) *Generator of pallet part program*. Another C++ program automatically generates the complete machining process by taking as input the current pallet configuration and the part program blocks associated with the machining of the single baseplates. The part program blocks are assembled and integrated with the proper rapid movements. The machining process is generated adopting the G-Code language.
(d) *Validation of the machining process*. Taking as input the digital pallet configuration and the pallet part program, another C++ program was developed to automatically generate a simulation model of the pallet. This model is serialized as an XML file that can be loaded by the commercial software tool Vericut. This tool enables to run a detailed process simulation by executing the part program,

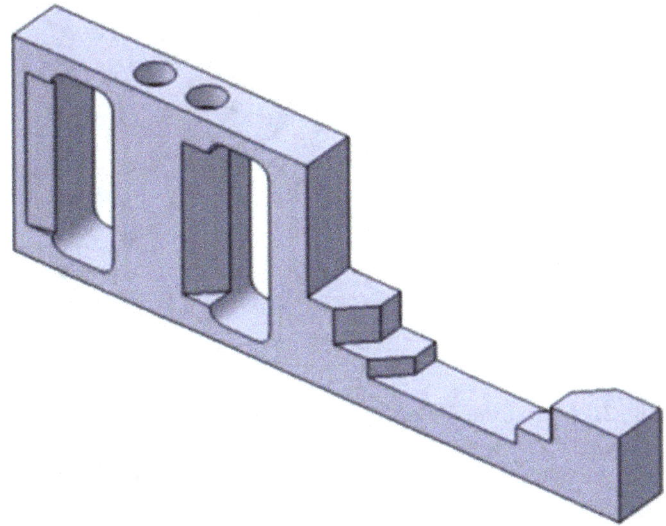

Fig. 3.10 Reference part type wp

Table 3.2 Setup, input/output part type and baseplates

Setup	Input part type	Output part type	Baseplate
Setup01	wp_raw	wp_wip1	bpSetup01
Setup02	wp_wip1	wp_wip2	bpSetup02
Setup03	wp_wip2	wp_finished	bpSetup03

so that the process can be validated and possible errors (e.g. collisions) can be detected before loading the part program on a real machine tool (Sect. 3.4.6).

The formalization of the information related to the prototype is based on the ontology described in Sect. 3.4.1 that facilitates the data exchange between the various software components. The elements characterizing the Pro2ReFix prototype are modelled as OWL individuals, i.e. instances of the ontology.

3.5.2 Experiments

The capabilities of the Pro2ReFix prototype and the zero-point clamping technologies have been tested with a reference part type *wp* (Fig. 3.10) whose process plan consists of three setups (*Setup01, Setup02, Setup03*). Therefore, based on the production stage, the part type can be in four different phases (*wp_raw, wp_wip1, wp_wip2, wp_finished*). Table 3.2 reports which is the input and output part type for each setup.

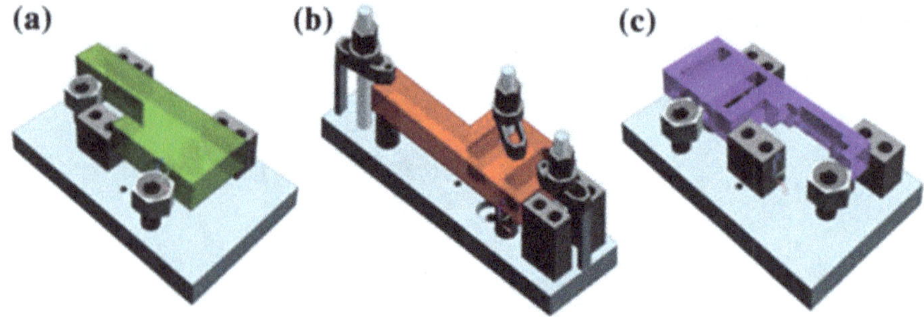

Fig. 3.11 Baseplates, fixtures and input part type in **a** Setup01, **b** Setup02, **c** Setup03

Fig. 3.12 An example of the designed reconfigurable pallet mounting three baseplates

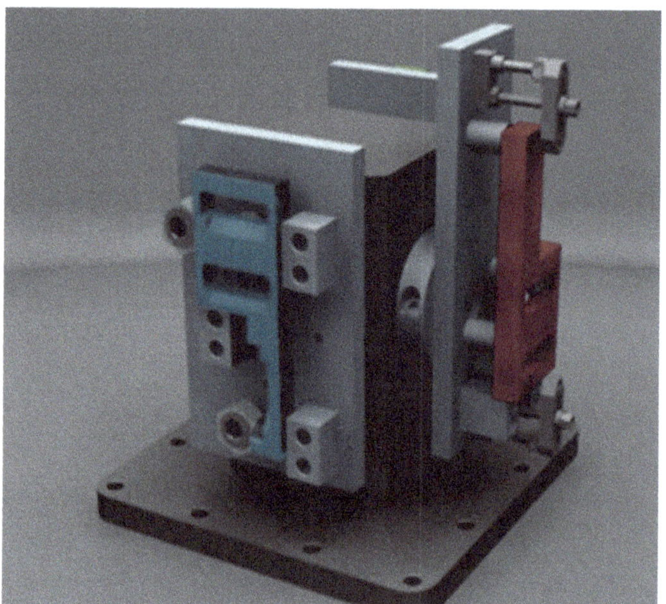

A different baseplate with zero-point clamping devices and appropriate fixtures have been designed and realized for each setup, as shown in Fig. 3.11 and indicated in Table 3.2. An example of pallet configured with different baseplates is shown in Fig. 3.12.

A set of experiments has been carried out to test the laser scanner on various physical pallet configurations including baseplates and part types. Four different configurations of tombstone faces have been considered and the corresponding *master geometries* have been acquired and stored (see Table 3.3). The master geometry *Master0* represents an empty face of the tombstone where no baseplate has been mounted. The other master geometries correspond to the baseplates associated with the three setups. Figure 3.13 shows the point clouds of *Master1* and *Master3*.

After the definition of the master geometries, various physical pallet configurations were realized and the laser scanner acquired the point cloud for each of their faces. Table 3.4 reports the configuration of three tombstone faces together with results provided by the *Laser scanner controller*, i.e. the number of points identified in the slave and the comparison with all available master geometries. This compar-

Table 3.3 Master geometries

Master geometry	Visible elements	Description
Master0	tombstone	Tombstone where no baseplate has been mounted
Master1	tombstone, bpSetup01, wp_raw	Baseplate of Setup01 with part type wp_raw to be processed
Master2	tombstone, bpSetup02, wp_wip1	Baseplate of Setup02 with part type wp_wip1 to be processed
Master3	tombstone, bpSetup03, wp_wip2	Baseplate of Setup03 with part type wp_wip2 to be processed

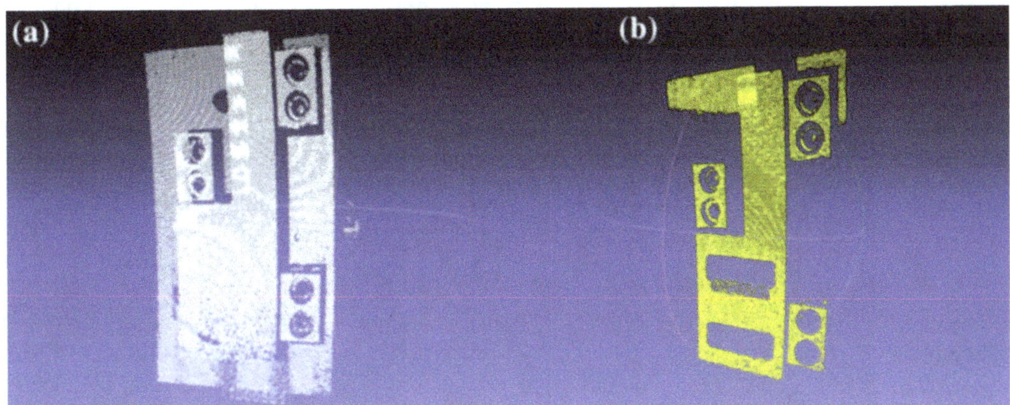

Fig. 3.13 Point cloud of **a** Master1 and **b** Master3

ison returns the number of points of the slave that are matched/not matched on the master, together with information about the error (maximum, mean, minimum, and standard deviation). These statistics are further elaborated by a matching algorithm to identify the best fit among the master geometry, i.e. *Master1, Master2, Master3* for the three experiments in Table 3.4.

The digital model of the physical pallet can be obtained as soon as all its faces are identified. Then, the part program of the whole pallet can be automatically generated and tested using the simulation software Vericut (Fig. 3.14) to identify possible interferences between the tools (and the machine) and the fixture, thus providing a verification for the pallet configuration under study. The simulation was successful for experiments shown in Table 3.4. An interference between the tool and the baseplates was correctly identified for another set of experiments where the baseplates were rotated by 90°, thus labelling the configuration as infeasible.

Table 3.4 Results of the experiments

Slave		Comparisons with masters						
Tombstone face config.	N. points	Master	N. points	Error				
				On master	Not on master	Max	Mean	Min	Std. dev.
tombstone, bpSetup01, wp_raw	64,801	Master0	22,779	42,022	30.358	3.9866	0.0	9.06
		Master1	**54,222**	10,579	32.031	0.7628	0.0	2.60
		Master2	48,931	15,870	30.122	0.6680	0.0	3.08
		Master3	39,696	25,105	30.408	1.3775	0.0	5.1
tombstone, bpSetup02, wp_wip1	58,501	Master0	23,004	35,497	30.503	4.2717	0.0	9.4
		Master1	48,894	9607	31.992	0.7166	0.0	2.05
		Master2	**52,156**	6345	30.034	0.4678	0.0	2.31
		Master3	40,219	18,282	30.373	1.1219	0.0	4.69
tombstone, bpSetup03, wp_wip2	46,098	Master0	22,857	23,241	30.344	2.6317	0.0	7.6
		Master1	39,095	7003	31.616	1.2956	0.0	4.2
		Master2	39,525	6573	30.383	0.9101	0.0	3.9
		Master3	**44,119**	1979	8.4721	0.0401	0.0	0.12

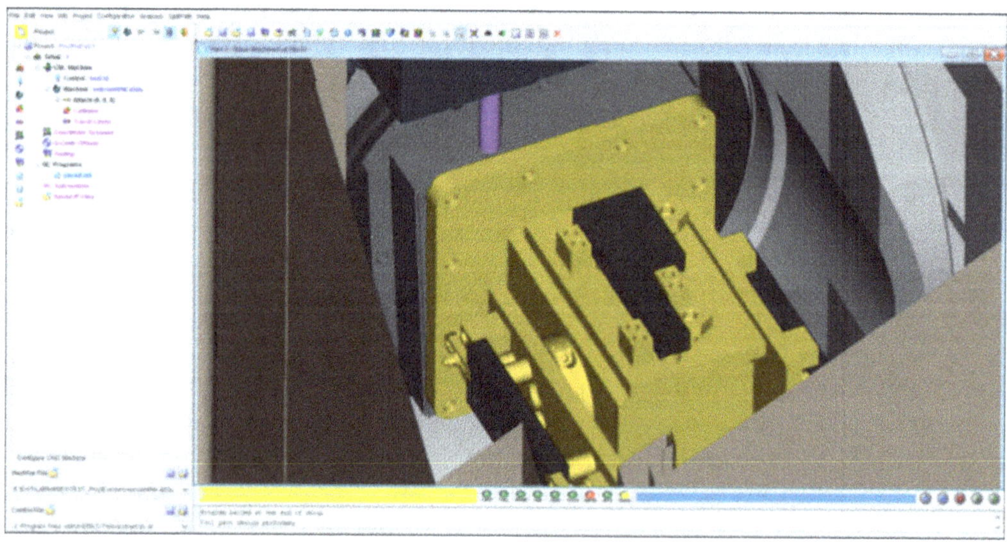

Fig. 3.14 CNC simulation of the new pallet configuration in Vericut©

3.6 Conclusions and Future Research

This chapter presented an integrated approach with methods and tools to support the configuration, planning and execution activities in flexible manufacturing systems, while exploiting zero-point fixtures to support fast pallet (re)configuration.

The interoperability and knowledge integration issues have been addressed with a Factory Data Model based on semantic web technologies. The proposed inspection system consists of a laser scanner and verifies the correctness of the mounted pallet in the shop floor by comparing the scanned elements with the information included in the semantic model. Finally, non-linear process plan technologies have been exploited to support the generation of modular process plans and management policies have been used to take advantage of the degrees of freedom given by zero-point fixtures.

More details about the developed methods and tools have already been presented in conference papers [24, 43, 44] and contributions in journals and books [12, 26, 28, 32, 34, 35]. The methodologies and the developed prototypes have been tested on realistic industrial cases, thus demonstrating the feasibility and potential of the approach. The Pro2ReFix prototype (Sect. 3.5.1) was presented at BI-MU 2016 exhibition in Milan and attracted interest from various machine tool and system producers. The acquired knowledge and developed tools form the basis to expand the partner capabilities to start new industrial and research collaborations.

Future developments aim at reducing the limitations of the proposed methods and tools. Regarding pallet inspection, possible variations of the ideal geometry (e.g. due to allowable differences in part shape and positioning) should be considered. In addition, a broader analysis of the geometric and functional characteristics of the various elements in relations to the potential mounting errors should be performed to provide more precise information to the operator (e.g. wrong part mounted, no part mounted, fixture not in correct position). Based on the new functionalities provided by the software tools, also extensions to the current version of the Factory Data Model will be needed to properly support interoperability.

Acknowledgements This work has been funded by the Italian Ministry of Education, Universities and Research (MIUR) under the Flagship Project "Factories of the Future—Italy" (Progetto Bandiera "La Fabbrica del Futuro") [45], Sottoprogetto 2, research projects Pro2Evo (Product and Process Co-Evolution Management via Modular Pallet configuration) and Pro2ReFix (Product and Process Co-Evolution Management via Reconfigurable Fixtures).

The authors thank the companies Cembre S.p.A. Italia, Schunk Intec S.r.l., and MCM Machining Centers Manufacturing S.p.A. for the useful suggestions and discussions. In addition, Cembre S.p.A. made available their product and production data and Schunk provided the modular pallet for the prototype realization. Particular thanks go to all the colleagues at CNR and POLIMI participating in the definition and development of the work presented in this chapter.

References

1. Tolio T, Ceglarek D, El Maraghy HA et al (2010) SPECIES—co-evolution of products, processes and production systems. CIRP Ann: Manuf Technol 59(2):672–693
2. Wiendahl HP, ElMaraghy H, Nyhuis P et al (2007) Changeable manufacturing – classification, design and operation. CIRP Ann: Manuf Technol 56(2):783–809
3. Terkaj W, Tolio T, Valente A (2009) Focused flexibility in production systems. In ElMaraghy HA (ed) Changeable and reconfigurable manufacturing systems. Springer, pp 47–66
4. Luggen WW (1991) Flexible manufacturing cells and systems. Prentice Hall International
5. Nguyen MV (2016) Zero-point fixture. United States Patent Application 20160375533
6. Gandhi MV, Thompson BS (1986) Automated design of modular fixtures for flexible manufacturing systems. J Manuf Syst 5(4):243–252
7. Perremans P (1996) Feature-based description of modular fixturing elements: the key to an expert system for the automated design of the physical fixture. Adv Eng Softw 25:19–27
8. Hunter R, Rios J, Perez JM, Vizan A (2006) A functional approach for the formalization of the fixture design process. Int J Mach Tools Manuf 46(6):683–697
9. Wu Y, Rong Y, Ma W, LeClair SR (1998) Automated modular fixture planning: accuracy, clamping, and accessibility analyses. Robot Comput Integr Manuf 14(1):17–26
10. Pellegrinelli S, Valente A, Molinari Tosatti L (2014) Energy-efficient distributed part programme for highly automated production systems. Int J Comput Integr Manuf 28(4):395–407
11. Borgia S, Pellegrinelli S, Petrò S, Tolio T (2013) Network part program approach based on the STEP-NC data structure for the machining of multiple fixture pallets. Int J Comput Integr Manuf 27(3):281–300
12. Pellegrinelli S, Cenati C, Cevasco L, Giannini F, Lupinetti K, Monti M, Parazzoli D, Molinari Tosatti L (2018) Configuration and inspection of multi-fixturing pallets in flexible manufacturing systems: evolution of the network part program approach. Robot Comput Integr Manuf 52:65–75
13. Minetola P, Iuliano L, Calignano F (2015) A customer oriented methodology for reverse engineering software selection in the computer aided inspection scenario. Comput Ind 67:54–71
14. Bi ZM, Wang L (2010) Advances in 3D data acquisition and processing for industrial applications. Robot Comput Integr Manuf 26:403–413
15. Shamir A (2008) A survey on mesh segmentation techniques. Comput Graph Forum 27(6):1539–1556
16. Attene M, Falcidieno B, Spagnuolo M (2006) Hierarchical mesh segmentation based on fitting primitives. Vis Comput 22(3):181–193
17. Liu ZB, Bu SH, Zhou K, Gao SM, Han JW, Wu (2013) A survey on partial retrieval of 3D shapes. J Comput Sci Technol 28(5):836–851
18. Stecke KE, Kim Y (1988) A study of FMS part type selection approaches for short-term production planning. Int J Flex Manuf Syst 1:7–29
19. Grieco A, Semeraro Q, Tolio T (2001) A review of different approaches to the FMS loading problem. Int J Flex Manuf Syst 13:361–384
20. Terkaj W, Tolio T (2006) A stochastic approach to the FMS loading problem. CIRP J Manuf Syst 35(5)
21. Bugtai N, Young RIM (1998) Information models in an integrated fixture decision support tool. J Mater Process Technol 76(1):29–35
22. Ameri F, Summers JD (2008) An ontology for representation of fixture design knowledge. Comput Aided Des Appl 5(5):601–611
23. Gmeiner T, Shea K (2013) An ontology for the autonomous reconfiguration of a flexible fixture device. J Comput Inf Sci Eng 13(2):021003
24. Pellegrinelli S, Terkaj W, Urgo M (2016) A concept for a pallet configuration approach using zero-point clamping systems. Procedia CIRP 41:123–128
25. Terkaj W, Pedrielli G, Sacco M (2012) Virtual factory data model. In: Workshop on ontology and semantic web for manufacturing OSEMA 2012, CEUR workshop proceedings, vol 886, pp 29–43

26. Pauwels P, Terkaj W (2016) EXPRESS to OWL for construction industry: towards a recommendable and usable ifcOWL ontology. Autom Constr 63:100–133
27. ISO 14649-10 (2003) Industrial automation systems and integration—physical device control—data model for computerized numerical controllers. Part 10: General process data, ISO 14649-10:2004
28. Urgo M, Terkaj W, Cenati C, Giannini F, Monti M, Pellegrinelli S (2016) Zero-point fixture systems as a reconfiguration enabler in flexible manufacturing systems. Comput Aided Des Appl 13(5):684–692
29. Blackburn MR, Denno PO (2015) Using semantic web technologies for integrating domain specific modeling and analytical tools. Procedia Comput Sci 61:141–146
30. Ekaputra FJ, Sabou M, Serral E, Kiesling E, Biffl S (2017) Ontology-based data integration in multi-disciplinary engineering environments: a review. Open J Inf Syst (OJIS) 4(1):1–26
31. W3C (2012) OWL 2 web ontology language document overview, 2nd edn. https://www.w3.org/TR/owl2-overview/. Accessed 20 Apr 2018
32. Pauwels P, Krijnen T, Terkaj W, Beetz J (2017) Enhancing the ifcOWL ontology with an alternative representation for geometric data. Autom Constr 80:77–94
33. International Organization for Standardization (2004) ISO 10303-11 Industrial automation systems and integration—product data representation and exchange. Part 11: Description methods: the EXPRESS language reference manual. http://www.iso.org/iso/iso_catalogue/catalogue_tc/catalogue_detail.htm?csnumber=38047. Accessed 20 Apr 2018
34. Borgo S, Sanfilippo EM, Sojic A, Terkaj W (2015) Ontological analysis and engineering standards: an initial study of IFC. In: Ebrahimipour V, Yacout S (eds) Ontology modeling in physical asset integrity management. Springer, pp 17–43
35. Terkaj W, Sojic A (2015) Ontology-based Representation of IFC EXPRESS rules: an enhancement of the ifcOWL ontology. Autom Constr 57:188–201
36. Dolog P (2004) Model-driven navigation design for semantic web applications with the UML-guide. In: Proceedings of ICWE, pp 75–86
37. Haller A, Janowicz K, Cox S, Phuoc DL, Taylor K, Lefrançois M (2017) Semantic sensor network ontology. https://www.w3.org/TR/vocab-ssn/. Accessed 20 Apr 2018
38. Terkaj W, Schneider GF, Pauwels P (2017) Reusing domain ontologies in linked building data: the case of building automation and control. In: Proceedings of the 8th workshop formal ontologies meet industry, joint ontology workshops 2017, CEUR workshop proceedings, vol 2050
39. Musen MA (2015) The protégé project: a look back and a look forward. AI Matters 1(4):4–12
40. Brosed FJ, Santolaria J, Aguilar JJ, Guillomía D (2012) Laser triangulation sensor and six axes anthropomorphic robot manipulator modelling for the measurement of complex geometry products. Robot Comput Integr Manuf 28(6):660–671
41. Contini C, Tolio T (2004) Computer-aided set-up planning for machining centres configuration. Int J Prod Res 42:3473–3491
42. Yao S, Han X, Yang Y, Rong Y, Huang SH, Yen DW, Zhang G (2007) Computer aided manufacturing planning for mass customization: part 2, automated setup planning. Int J Adv Manuf Technol 32(1–2):205–217
43. Terkaj W, Urgo M (2014) Ontology-based modeling of production systems for design and performance evaluation. In: Proceedings of 12th IEEE international conference on industrial informatics
44. Borgo S, Sanfilippo EM, Sojic A, Terkaj W (2014) Towards an ontological grounding of IFC. In: Proceedings of the 6th workshop on formal ontologies meet industry (FOMI 2014), CEUR workshop proceedings, vol 1333
45. Terkaj W, Tolio T (2019) The Italian Flagship project: factories of the future. In: Tolio T, Copani G, Terkaj W (eds) Factories of the future. Springer

Chapter 4
Knowledge Based Modules for Adaptive Distributed Control Systems

Andrea Ballarino, Alessandro Brusaferri, Amedeo Cesta, Guido Chizzoli,
Ivan Cibrario Bertolotti, Luca Durante, Andrea Orlandini,
Riccardo Rasconi, Stefano Spinelli and Adriano Valenzano

Abstract Modern automation systems are asked to provide a step change toward flexibility and reconfigurability to cope with increasing demand for fast changing and highly fragmented production—which is more and more characterising the manufacturing sector. This reflects in the transition from traditional hierarchical and centralised control architecture to adaptive distributed control systems, being the latter capable of exploiting also knowledge-based strategies toward collaborating behaviours. The chapter intends to investigate such topics, by outlining major challenges and proposing a possible approach toward their solution, founded on autonomous, self-declaring, knowledge-based and heterarchically collaborating control modules. The benefits of the proposed approach are discussed and demonstrated in the field of re-manufacturing of electronic components, with specific reference to a pilot plant for the integrated End-Of-Life management of mechatronic products.

4.1 Scientific and Industrial Motivations

Since several years, the manufacturing industry has been facing a number of technological and production challenges related to the increasing variability of mix and demand of products driven by a short product life cycle. Nevertheless, the growing industrial demand for increased levels of reconfigurability—having impact both in production process automation, supervision and data collection and aggregation systems—is not properly fulfilled, mainly due to the lack of widely accepted solutions

A. Ballarino (✉) · A. Brusaferri · G. Chizzoli · S. Spinelli
CNR-STIIMA, Istituto di Sistemi e Tecnologie Industriali Intelligenti per il Manifatturiero
Avanzato, Milan, Italy
e-mail: andrea.ballarino@stiima.cnr.it

A. Cesta · A. Orlandini · R. Rasconi
CNR-ISTC, Istituto di Scienze e Tecnologie della Cognizione, Rome, Italy

I. C. Bertolotti · L. Durante · A. Valenzano
CNR-IEIIT, Istituto di Elettronica e di Ingegneria dell'Informazione e delle Telecomunicazioni,
Turin, Italy

© The Author(s) 2019

T. Tolio et al. (eds.), *Factories of the Future*,
https://doi.org/10.1007/978-3-319-94358-9_4

and integrated reference models. In fact, implemented solutions are today typically characterized by rigid hierarchical structures, organized into strictly coupled layers.

Centralized PLCs (Programmable Logic Controller) architectures are often deployed, integrating the overall control logic of the production line for coordinating the execution of single mechatronic devices and machines. Such monolithic approach represents one of the major obstacles to achieve short-time reconfigurations. In fact, without a proper modularization and standardization of the control application entities, adaptations of automation system behaviour—to properly support required changes of factory production assets—result to be strongly time consuming for control system engineers and system integrators. Furthermore, the lack of modularization and standardization critically impacts on readability, portability and integration of single control modules across different applications, thus preventing the capitalization and re-use of company specific know-how on production process.

Further to this, modifications or extensions within the programs of the production line PLC possibly leads to the introduction of code errors, thus requiring extensive validation of complete automation system before deployment. Nevertheless, such activity is fundamental to avoid unpredicted deviations of the process execution, which could cause damages to devices/machines or, more critically, to human operators. As a result, existing industrial control systems are often very inefficient in facing the ever increasing demand for flexibility, expandability, agility and reconfigurability, which are typical requirements of advanced manufacturing system solutions [1, 2].

The previous requirements are fundamental to ensure a modern approach to efficient manufacturing, whose major competitiveness pillars rely upon: high value added goods, knowledge intensive production processes, and efficient and safe operating environments. This extremely articulated context is a very promising and appealing opportunity for countries heavily involved in the production of capital goods, machine tools and sophisticated solutions for manufacturing systems. A key capability of the control system is to evolve over time in order to anticipate and persistently adapt the control logics, functions and architectures to the evolving production scenarios. Therefore, flexibility in control infrastructure shall be prosecuted at the level of single control unit to leverage the hardware reconfigurability by exposing a set of (automation) functions/tasks that the single machine or device can execute to support production objectives. This latter aspect could effectively help in transforming monolithic and hardcoded solutions toward service oriented approaches. Yet this transformation cannot be fully tackled without considering that the execution of single task or function on a machine could either address the interaction with physical actuators, or imply a synchronization problem between different devices. Therefore, a real-time software infrastructure and communication framework shall be developed to properly support reliability within industrial automation field.

This chapter addresses the challenges related to the conception and development of next generation control systems, with reference to their application in Reconfigurable Manufacturing Systems (RMSs), as a viable solution while addressing varying production conditions within the manufacturing industry. The proposed approach, named *Generic Evolutionary Control Knowledge-based mOdule* (GECKO), aims at developing a flexible distributed control infrastructure, based on the interactive coop-eration of control modules. GECKO enables each device (e.g. end-effector, complex machine equipment, integrated cell or even system) to evolve from a stand-alone, rigidly and hierarchically managed condition to an autonomous, self-declaring, hier-archically interacting and collaborating scenario. GECKO modules are conceived to detect and interpret the production environment characteristics and to adapt its capabilities on the basis of specific requirements by automatically accomplishing local and global objectives.

In consideration of aforementioned topics, the next sections will present the rele-vant state of the art (Sect. 4.2) and then the context and reference problem (Sect. 4.3). The overall approach toward an adaptive control infrastructure is first described (Sect. 4.4), and then discussed with particular reference to its realization via the GECKO solution (Sect. 4.5). A specific focus on Reconfigurable Transportation Systems (RTSs) will be kept as target application for proposed GECKO solution (Sect. 4.6), since RTSs can play a relevant role in embedding manufacturing systems with the capability of adapting their structure and functionalities to the evolving needs of production processes. Finally, the conclusions are presented in Sect. 4.7.

4.2 State of the Art

Traditional control systems based on hierarchical and centralized control structures hold good performance in terms of productivity over a limited and specific range of conditions. Nonetheless, such large monolithic software packages typically need relevant modifications of the control code to cope with any sort of (even minimal) system adaptation and reconfiguration. As a result, they are very inefficient to face the current requirements of flexibility, expansibility, agility and re-configurability required by advanced manufacturing system solutions.

When considering the existing scientific and industrial practices to evolve from current centralised approaches, three major topics shall be addressed, namely inter-operable and pluggable (mostly defined as *plug and play*, P&P) control solutions, design of automatically reconfigurable control solutions as well as implementation of optimization strategies and learning mechanisms in the control.

In the field of P&P solutions, a number of EU projects [3–5] and scientific papers [6, 7] propose advanced solutions mostly related to specific devices (such as robots) or to specific enablers (such as the communication layers). The industrial manufacturing practice limits the usage of P&P concepts to very basic applications mostly related to the connection of devices preliminarily connected to a communication network, mainly disregarding the phases of automatic discovery and capability assessment, as well as of dynamic cooperation. Yet control entities—conceived to control mecha-tronic devices so that the former can be automatically run as soon as the device is physically plugged in the system dorsal—still represent an objective to be achieved. This basically implies addressing a number of challenging automation solutions under specific perspectives such as:

- standard architecture designed as a component-based software solution structured with standardized interconnected components, each one dealing with different tasks and goals;
- cooperation and interaction mechanisms between different control entities rely upon both standardized physical and logic interfaces—ensuring the possibility for modules to operate in multiple configurations and to be nested on different resources—and a common vocabulary, an open ontological classification of devices and their functionalities as well as a shared knowledge representation system [8, 9]—guaranteeing that the different components of the system use a consistent and shared representation of products, processes and resource data, as well as semantics and rules characterizing the information interrelations [10, 11];
- an open communication network enabling both the physical and logical connections of the control module to the plant, providing Timeliness, Security and Energy Efficiency properties.

Regarding the first topic, the conception and deployment of reconfigurable solutions have mostly concentrated on the mechanical aspects of the devices, leading to the realization of reconfigurability enablers targeting several applications, from machine tools and robots, to transportation systems and fixturing systems based on standard interfaces and flexible CNC [12]. Challenging industrial applications of these reconfigurability concepts [13] have been realized by machine tool builders (e.g. Panasonic, Zevac, Mori Seki), by providers of robotic solutions (e.g. Kuka, Mitsubishi, Robotnik) and by providers of modular equipment (e.g. Festo and Flexlink). These automation reconfigurability options basically consist in a set of predefined optional add-ons or a set of deterministic capabilities that are acquired since the beginning, thus endowing the resources with very high flexibility instead of reconfiguration enablers [14].

As already mentioned, the majority of automation applications is characterised by traditional centralized control architecture, with consequent sensible limitations in the achievement of agile adaptation to run-time production changes. Moreover, such limitations become even more critical when the complexity of the process to be controlled and the functionalities to be automated increase.

The challenging aspect in the reconfigurability concept consists in its concurrent applicability to the control with regard both to logic and physical aspects. Logic reconfigurations of the control should stem from the need to modify the control functions and strategies in response to a production change (e.g. new product), events altering pieces of equipment (e.g. resource failures or abnormal behaviours) or revised production goals (e.g. minimization of energy consumption vs. idle times). On the other side, physical reconfigurations imply that the control physical device and the related control software interact with a modified control system, where new entities have been integrated or dismissed. This firstly requires a more complex reconfiguration process that goes beyond the control logic features and deals with physical interconnections, communications and bindings. Beside this, a new functional-oriented architecture enabling modular automation is needed to realize the encapsulation and (re-) use of self-contained functions and/or automation tasks as services.

SOA approaches can help to fulfil these requirements [15]. SOA is meant to provide services through a set of black box components exposing interfaces based on communication services [16]. In the field of modular control systems, single modules can be represented by such black box components, aimed at accomplishing functions by encapsulating services. The OASIS reference model was published for SOA, containing the definition of a service as a "…mechanism to enable access to one or more capabilities…" [17]. Services exposed by an entity are meant to be used by other entities. On top of this, the European research and development project named SOCRADES[1] addressed a novel manufacturing paradigm deeply based on SOA, and particularly on web services [18]. According to SOCRADES framework the intelligence of the manufacturing system is implemented by an agent-based approach embedded in smart devices. These devices collaboratively act at the same hierarchical level by using web services as an interface for mutual communication [19]. The autonomous units therefore operate cooperatively, each being intelligent and proactive. In addition to the more general SOA approaches described, Mendes et al. [20] proposed an approach for the implementation of service-oriented control in process industry based on adoption of high level Petri nets.

The separation of control functionalities into services supporting process control and intelligence can be found in [21]. The proposed hierarchical structure is similar to the one described in ISA 106 [22]. In [23] a SOA implementation approach based on the adoption of function blocks [24] is introduced. Basic or atomic services are developed as encapsulated elements, to be invoked either directly by request-response messaging or indirectly, when contained in other complex high level services. Furthermore, both actuators and sensors signals require the access to field data. Such access is also implemented by using services. While mutual connectivity between all devices is required, services do not need to be allocated to particular hardware. Function block types deployed according to [24] are used as service types. Connections between function blocks realize the messaging mechanism, and therefore, the communication between corresponding services. Communication contains message types and parameters, where the former are achieved by event connections and the latter by data connections. Therefore, the introduced approach decomposes complex manufacturing tasks to the very basic level. Each basic task is implemented as an atomic service. Atomic services can be combined into aggregated ones to achieve more complex functionalities: the resulting set of basic and aggregated services can be orchestrated by a central coordinator or by autonomous choreography [23]. A high control level can invoke a service by messaging and the invoked service can invoke downstream subservices in a specific order.

When referring to the second of the mentioned topics (i.e. optimization strategies and learning mechanisms), the goal of the control optimization process traditionally pertains to a number of aspects related to the products quality to be achieved, the technologies to be exploited, along with the physical equipment usage over time.

---

[1] http://www.socrades.net.

Coherently with the traditional way of modelling the factory as a number of linked layers, these optimization paths are handled at the Production Management and Manufacturing Execution System (MES) levels by the scientific and industrial communities [25–31]. In this view, the system logic control is conceived with a very little awareness of the shop-floor and limited options to perform dynamic control adaptations. This is evident also when focussing on part routing problems in manufacturing transportation systems, where context recognition and optimization issues become crucial. Existing control solutions are based on centralized/hierarchical control structures that offer good performance but they require a big effort to implement, maintain or reconfigure control applications when a high-level of flexibility is needed. Thus, classical approaches usually do not meet requirements of manufacturing systems in terms of flexibility, expansibility or reconfigurability. Current R&D efforts focus on the development of novel control systems characterized by distributed intelligence, robustness and adaptation to the changes in the environment and exogenous factors [32]. The multi-agent paradigm aims at addressing control objectives by introducing modularity, decentralization, autonomy, scalability and re-usability as main features. Even if the definition of agent concept is neither unique nor shared [33], its most important properties are the autonomy, intelligence, adaptation and cooperation [34].

Although Artificial Intelligence (AI) based approaches have been considered as local enablers in some RMSs [35], the design of control models considering all the possible failures and changing situations remains a crucial problem. Moreover, relevant structural modifications in an agent configuration entail a re-design of the control strategies, which is hard to manage on the fly. In manufacturing, knowledge-based approaches exploiting ontologies have been applied to increase flexibility in modelling and planning of mechatronic devices [36], resources in collaborative environments [37], automation and control systems [38], and to manage information of distinct types [39]. In these cases, planning specifications, if considered, are fixed and neither automatically generated nor subsequently adapted. In robotics, ontologies have been more widely exploited in the knowledge framework OMRKF [40], KnowRob [41], ORO [42, 43]. Nevertheless, none of the previous approaches addresses the issue of dynamically adapting the planning models. Other AI approaches based on Answer Set Programming have been proposed [44, 45].

From the software infrastructure point of view, the use of a real-time operating system (RTOS) as an execution framework in control applications is now becoming more and more popular with respect to full-custom designs, due to its clear advantages in terms of software development time and cost. RTOS can be divided into two main categories according to their application programming interface (API) that directly affects application development and portability:

- operating systems providing a full-fledged POSIX API, sometimes tailored to the specific requirements of embedded systems, as specified by the IEEE Standard 1003.13 [46]. For instance, this is the case of Linux;
- operating systems providing their own proprietary, and usually simpler and more efficient, API. A typical example belonging to this category is the FreeRTOS open-source, real-time operating system [47].

Among RTOS solutions, the POSIX interface is to be preferred when aiming at portable software and firmware suitable for long-term reuse and maintainability, because it is backed up by an international standard. As an additional benefit, it is widespread and well-known to programmers. Focusing on Linux, this option is made even more appealing by the increasing maturity of real-time extensions for the mainstream kernel, like the preemption Real-Time patch (Preempt_RT),[2] Real-Time Application Interface patch (RTAI),[3] and Xen-based virtualization support.[4]

POSIX interface is also in line with International Standards of the International Electrotechnical Commission (IEC) IEC 61158 and IEC 61784 which enable interoperability between devices of different manufacturers. Three types of industrial networks are currently available: Fieldbuses, Real-Time Ethernet Networks (RTE) and wireless networks. Moreover, these networks may be used in a combined, resulting in hybrid systems [48–50]. In addressing the specific choice for each application, a key requirement to be considered is the capability to guarantee traffic separation, a critical feature needed to avoid interference between traffic flows with different timing requirements.

4.3 Problem Statement

A new approach to the design of automation system is required to address the aforementioned requirements and challenges, while exploiting advanced engineering practices for control modularization and distribution. Novel integrated platforms for automation system development, supporting distributed control system engineering, represent a fundamental enabling technology, where openness, interoperability and compliancy with international industrial standards shall be considered as fundamental features to be guaranteed.

The adoption of a distributed automation approach can help to increase the reconfigurability and robustness of the automation system. Traditional centralized and hierarchical architectures should be replaced by new modular control solutions that can better support a fast integration of new functional components and run-time adaptation of the system to the dynamic change of production demand and requirements. The modularization concept enables the organization of the control solution into software entities, encapsulated within standard interfaces, that can be directly connected to functional components of the system to be controlled, so that the re-usability of the applications can be enhanced.

The following core requirements must be tackled by a novel approach based on distributed automation:

- *Interoperable and pluggable control solution*. Compared to the existing solutions, control entities shall be conceived to control mechatronic devices so that the for-

---

[2] https://rt.wiki.kernel.org/index.php/Main_Page.

[3] https://www.rtai.org/.

[4] https://www.xenproject.org/.

mer can be automatically run as soon as the device is physically plugged in the system dorsal. This P&P feature enables the automatic recognition and configuration process of the control modules; moreover, the interoperability characteristics should allow applying this procedure to every resource type and/or tasks, thus ensuring the possibility to interact with all the resources of the shop-floor.

- *Automatically reconfigurable control solutions.* As mentioned, the reconfigurability concept should handle both changes in the control logic and in physical aspects. The first aspect is aimed at matching the need to adjust the control functionalities as a consequence of change in production goals. The second aspect should tackle situation where the (control) system modified, since new entities have been integrated or dismissed. This requires a more complex process that manages both known a priori reconfigurations—by implementing pre-determined portions of the control which are not utilized until the occurrence of the activation event—and brand new reconfigurations—where the physical integration of a new entity recalls for system recognition phase to determine the automatic nesting of the new portion of the control within the overall architecture, as well as the updating of the knowledge of existing control module about system manufacturing capabilities and status.

- *Optimization processes and learning mechanisms.* The realization of distributed control solutions based on cooperating modules enables the implementation of enriched and comprehensive optimization strategies at control level, aimed at adapting the control functions and strategies on the basis of the actual status of the controlled device(s) and the knowledge about the nominal behaviour. In order to ensure the accomplishment of productivity, quality and energy efficiency targets as well as matching the frequent changes of the production environment, the optimization component should incorporate local and global optimization goals to be balanced coherently with the production scenario. A further feature of the optimization process concerns the integration of learning mechanisms. The learning mechanism incorporated in the optimization component should be aimed at structuring a basic knowledge layer and providing learning functions to support the analysis and interpretation of information and adjust the control behaviour over time.

Nevertheless, this design methodology requires an advanced development environment, capable of supporting the design of the control code according to international standards—for compliancy and interoperability—and of implementing run-time information exchange among different modules of the distributed control systems—to achieve global goal by means of real time collaborative strategies.

4.4 Proposed Approach

The proposed approach, named *Generic Evolutionary Control Knowledge-based mOdule* (GECKO), consists in a control infrastructure based on an interactive cooperation of control modules to cope with the requirements defined in the previous section. The GECKO approach enables the single devices—from end-effector, complex machine equipment up to the integrated cell and system—to evolve from stand alone, rigidly and hierarchically managed components into autonomous, self-declaring, heterarchically interacting and collaborating components. GECKO is conceived to detect and interpret the production environment features and adapt its capabilities on the basis of the specific requirements by automatically accomplishing local and global objectives. Under this perspective, opposite to pre-determined production system architecture, GECKO presents an adapting behaviour resulting from an advanced reconfiguration mechanism together with a form of intelligence and knowledge enabling the recognition of the products, events and other entities operating in the shop-floor.

GECKO is designed as a complex control architecture consisting of a number of cooperating software components:

- The *Communication* component (Sect. 4.5.1) is responsible for exchanging information and signals with the shop-floor. This module is in charge to: (i) enforce interoperability with other entities (included GECKO modules), (ii) collect data from the sensor infrastructure, (iii) manipulate data in order to assess the production environment over time and, finally, (iv) publish both the internal status and the capabilities of the entity itself.
- The *Control* component (Sect. 4.5.2) activates the control functions based on the capabilities to be accomplished by also dynamically managing events that can be both exogenous (e.g. new product features or demand volumes) and endogenous (e.g. failures of machine tools).
- The *Capability Assessment* component (Sect. 4.5.3) aims at matching the context reproduction with the GECKO nominal capabilities and determining the more suitable GECKO features and behaviours to be exploited in a specific time.
- The *Context Recognition* component (Sect. 4.5.4) reproduces an abstraction of the production context relying on an internal knowledge base elicited from the information collected by the communication component and exploiting intelligent reasoning and learning techniques in order to dynamically recognize the actual production context as well as to increase its knowledge.
- The *Optimization* component (Sect. 4.5.5) will determine the local and global optimization strategies driving the control module together with an evolutionary mechanism enabling the GECKO entity to learn and improve its functionalities over time.

These components provide functionalities that are integrated through a software infrastructure enabling the persistent internal communication. This infrastructure constitutes the physical and logic foundations to enable the joint functioning of all the GECKO modules. The production environment would consequently result in a community of GECKO entities encapsulated in the pieces of equipment that communicate, cooperate and negotiate to achieve the production goals.

4.5 Developed Methodologies and Tools

While further describing the software solutions developed for the GECKO approach, the attention will be focused on Reconfigurable Transportation Systems (RTSs) as target application, in consideration of their modular nature in implementing alternative inbound logistic systems' configurations. The transportation modules of a RTS are designed as mechatronic devices equipped with standard interfaces that allow modules to connect each other and embed logic controllers for actuators/motors and sensors. A central problem of RTSs is the Online Part Routing Problem (OPRP) addressing the formalization and synchronization of RTS mechatronic modules to transport all the parts in the manufacturing system according to their destinations [51]. A solution for the OPRP must ensure that all the parts are properly worked, the routings must be collision-free and routings must be efficient. RTS therefore represent a suitable bench-mark for proposed GECKO solution.

Since openness, interoperability and international industrial standards compliancy are fundamental requirements to be guaranteed, the hardware architecture for the deployment of GECKO control module is based on embedded industrial PC systems.

With reference to the operating system, as a foundation for the design choices, three different approaches to real-time execution support have been evaluated. Stand-alone RTOS, Linux plus RT Patch, and Xen-based virtual machines have been compared with respect to the following metrics: performance and determinism of execution at the operating system level, performance and determinism of real-time network communication, coexistence with non real-time applications, language and execution environment support, and communication among software modules. Linux and RT Patch have been chosen as base execution framework for the GECKO control module.

The GECKO components, as designed for the RTS case, are represented in Fig. 4.1 and will be described in the following subsections.

4.5.1 Communication Component

The communication architecture for GECKO modules has to accomplish two different tasks: data exchange with sensors/actuators distributed on the shop floor and communication between GECKO modules. The first task represents a typical feature of the networks deployed at the lowest levels of factory automation systems and is characterized by the fast and timely transmission of limited amounts of data, requiring both real-time and deterministic behaviours of the underlying networks. The second task serves to the purpose of coordinating modules activity and is typically characterized by higher traffic volume and more relaxed timing constraints than in the previous case. Considering the aforementioned traffic types, the best choice for the GECKO modules network is represented by Real-time Ethernet networks. Traffic separation, a critical feature needed to avoid interference between traffic flows

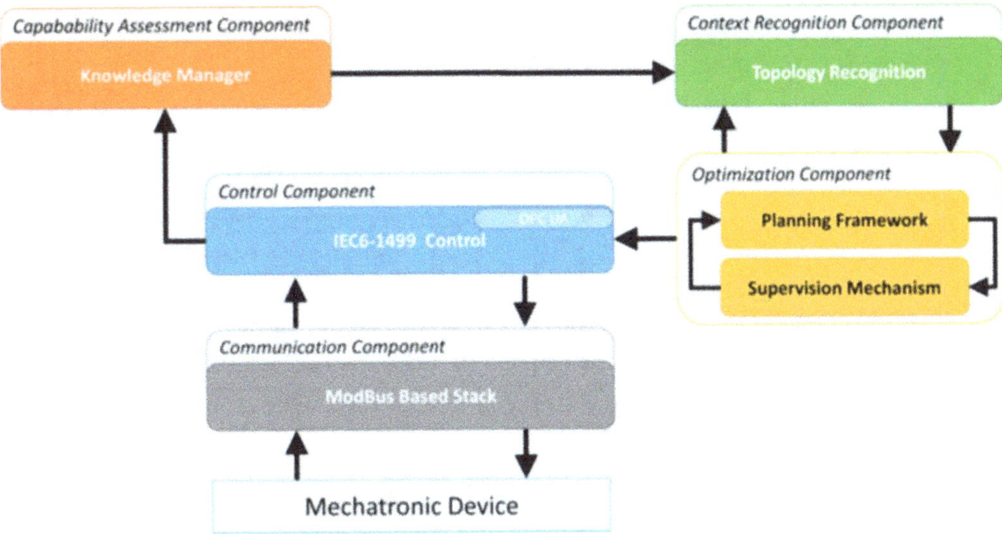

Fig. 4.1 GECKO module components

with different timing requirements, can be achieved in two ways: a single physical network, using a time-division multiple access (TDMA) method, and a physical
traffic segregation that requires GECKO modules to be equipped with two network
interfaces.

With reference to the communication protocol to access field I/O, Modbus was
chosen, since the end-to-end round-trip delays proved to be compatible with execution times.

The implementation of the prototype (see Fig. 4.2) targeted the microcontrollers
typical of low-cost embedded and industrial applications, namely the NXP LPC1768
and LPC2468, and required the configuration and integration of existing code[5] (light
grey blocks in Fig. 4.2), specifically (i) a custom-built Modbus TCP node, leveraging
the FreeRTOS RTOS; (ii) the lwIP TCP/IP protocol stack; (iii) an open-source Modbus TCP protocol stack (for slave nodes) and a commercial Modbus TCP protocol
stack (for masters). Moreover, it was necessary to develop dedicated modules of code
from scratch (white blocks in Fig. 4.2), better described in the following:

- Two different lwIP-specific adaptation layers between the network protocol stack
  interfaces provided by the open-source and commercial Modbus TCP libraries and
  the sockets-based lwIP interface.
- A suite of test programs to collect experimental data using the Modbus TCP
  protocol stacks on the master and slave sides.

---

[5] https://www.freemodbus.org/.

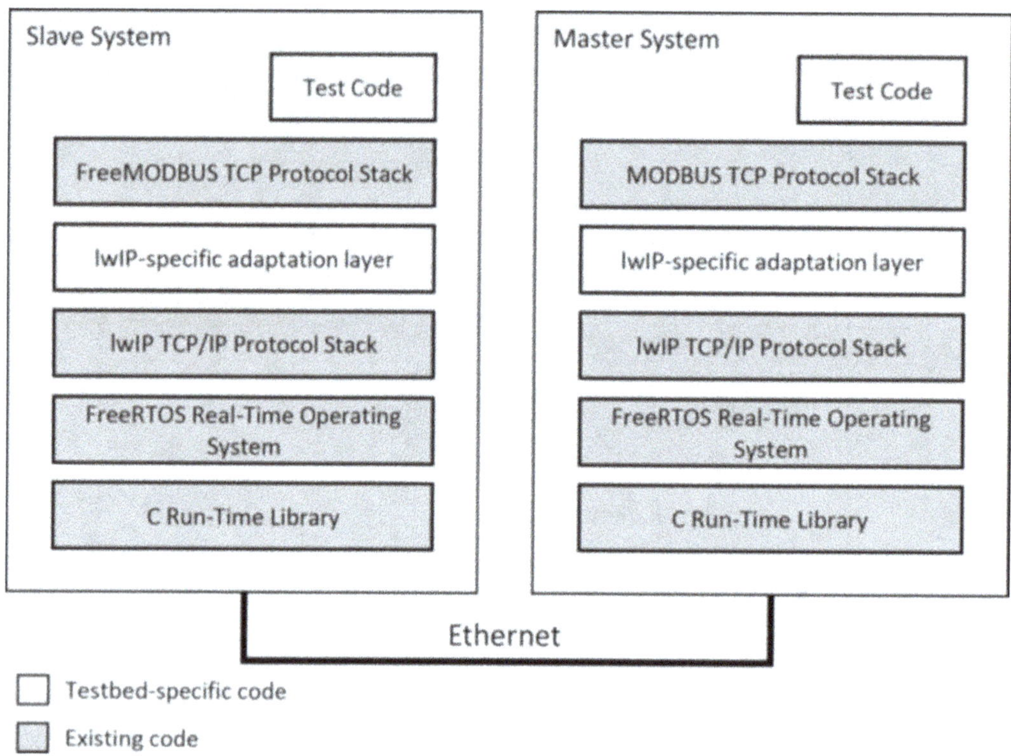

Fig. 4.2 GECKO communication and software infrastructure prototype

4.5.2 Control Component

IEC 61499 modelling standard has been adopted to develop a modular control system. According to the standard, the function block (FB) represents a unit of control software that is associated with a hardware component in the controlled system. It is therefore aimed at defining the control logic for each basic functional component of the system, meant as a self-contained module. Each function block exposes: (i) an event type input request corresponding to each task the module is capable of performing (i.e. automation function); (ii) a data type input for each configuration parameter related to the execution of module tasks; (iii) an event type output to acknowledge the execution of each module task (including un-nominal and failure conditions) and to request task execution towards downstream modules; (iv) a data type output for each variable representing the module internal state and tasks execution [52]. The execution of the control logic is regulated by the Execution Control Chart (ECC): ECC is basically a Moore automaton, consisting of states, event-conditioned transitions and actions, as well as a set of algorithms, associated with the ECC states, to be executed during specific operating conditions.

The development of IEC 61499 compliant control application takes place into a softPLC environment, named IsaGRAF,[6] natively supporting the IEC 61499 stan-

---

[6]http://www.isagraf.com/index.htm.

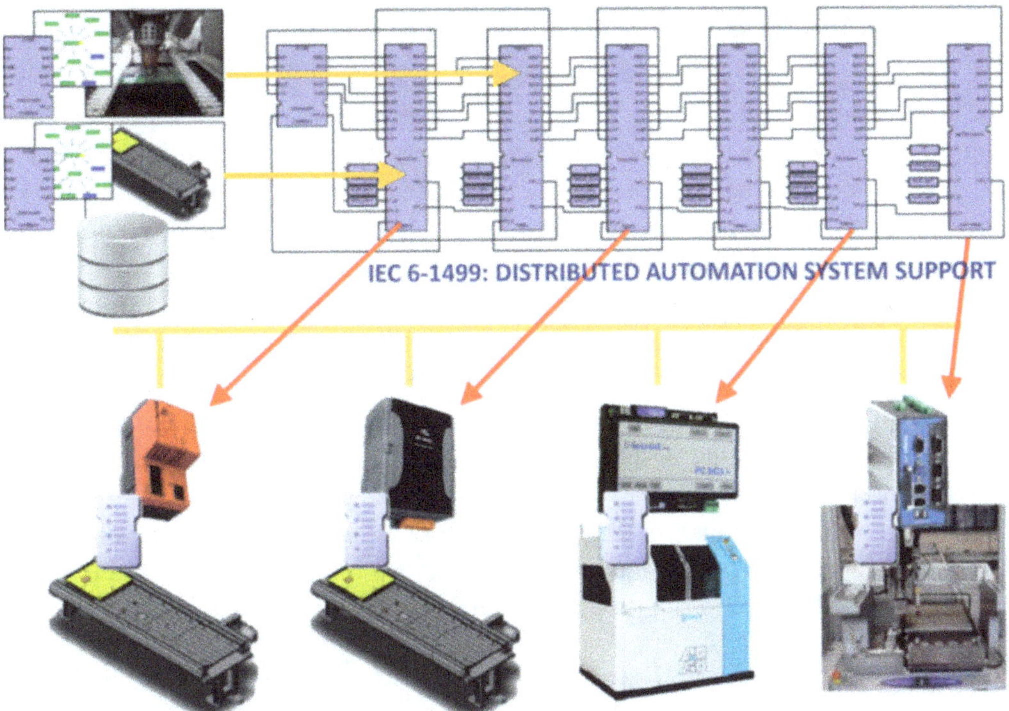

Fig. 4.3 IEC61499 based distributed control of an electronic component remanufacturing line consisting of conveyors and specialised machines

dard. As a result, a structured modular design of the overall system can be addressed through composition, i.e. by aggregating simpler mechatronic devices in a recursive way, from the level of *atomic* devices, such as sensors/actuators, till more complex equipment. In fact, by IEC 61499 reference model, a complex control application is defined by connecting FBs, as well as encapsulating basic function blocks into aggregated (composite) function blocks.

Figure 4.3 shows the distributed control engineering framework schema based on IEC 61499.

In terms of internal connections with other GECKO components, the control component includes an OPC-UA server, linked to the IEC 61499 run-time system, for exposing (as software services) the capabilities provided by the controller, thus supporting dynamic discovery, assessment, configuration and interoperability. The address space nodes are structured according to an object oriented approach based on OPC-UA. A remarkable element to be mentioned is that the IEC 61499 modelling of the control code results in a consequent identification and *clustering* of data/parameters/variable as inputs/outputs of the FB, being therefore propaedeutic to their handling according to object oriented data exchange standards. Furthermore, function driven control development is intuitively oriented and absolutely prone to the integration with Service Oriented Computing, therefore resulting in a promising hybrid paradigm.

4.5.3 Capability Assessment Component

The large amount of information linked to the capabilities exposed by various control components as well as to the technological process has led to the creation of the so called Knowledge Manager, a dedicated component capable of managing and assessing such information. The Knowledge Manager is based on a suited ontology that models the general knowledge of manufacturing environments, classifying relevant information in three distinct contexts (i.e. Global, Local and Internal) and considering a Taxonomy of Functions which classifies the set of functions the GECKO modules can perform according to their effects in the environment [53]. The Knowledge Manager exploits the ontology to build and manage a Knowledge Base (KB) that represents an abstract description of the structure and the capabilities of the GECKO modules.

In terms of implementation, the Web Ontology Language (OWL), a well-known technology for semantic data modelling, has been exploited to represent the KB. The ontology editor Protégé [54] has been used for KB design and testing.

4.5.4 Context Recognition Component

Based on contents stored in the capability assessment component, a Rule-based Inference Engine analyses the KB information and infers the advanced functional capabilities the overall control module is actually able to perform by composing the atomic capabilities exposed by the control component. The result of such an inference is a *planning* model generated from the KB to be used by the optimization component for actually control the behaviour of the mechatronic device [55]. The planning model contains an abstraction of the device, the environment's parameters in which it operates, and all the relevant constraints necessary to guarantee physical consistency. In the specific case of RTS, the context to be recognised is the topology of the plant, which shall be maintained and/or updated dependently on reconfigurations and/or failure of single modules. Thanks to the information exchanged with other neighbouring modules, every module dynamically re-builds (and constantly keeps updated) a local map of the shop-floor topology. Each module acquires a complete layout of the RTS including the connections of all the modules [56].

In terms of implementation, the Knowledge Processing Mechanism relies on Ontology and RDF APIs and Inference API provided by the Apache Jena Software Library.[7]

---

[7] http://jena.apache.org.

4.5.5 Optimization Component

The optimization component determines the global optimization strategies as response to the plant specific optimization problem, therefore driving the control module in its evolution.

In the context of RTS, the routing problem—as the main problem to be solved—is addressed within the optimization component as a multi-agent problem where a set of agents interact and share information in order to transport pallets (or parts) to their final destination within a RTS. Specifically, a distributed auction-based algorithm for part routing has been developed that coordinates the transport modules [51]. Such algorithm, denominated Planning Framework, establishes the assignment (i.e. routing) of the parts to the most suitable agent (i.e. module) as a result of a negotiation process by executing an auction-based mechanism with an associated multi-objective function [56].

Further to this, a Supervision mechanism continuously checks the achievement of each task via a flexible and dynamic execution monitoring system. This system analyses the pallet routings to detect possible redundancies and deadlock situations. If necessary the Planning Framework is re-executed and the path updated. The same happens in case of unforeseen events, e.g. a RTS module failure [56].

The part routing problem has been implemented in Java SE 7 by developing a multi-agent framework within the JADE platform. Both Planning Framework and Supervision mechanisms use timeline-based technology and were framed by means of the EPSL architecture [57].

4.6 Testing and Validation of Results

4.6.1 Industrial Case

The De-Manufacturing Pilot Plant sited in the lab of CNR-STIIMA (ex CNR-ITIA) [58] is dedicated to the integrated End-Of-Life processing of mechatronic products (specifically, printed circuit board, or PCB) using modular technological solutions to process heterogeneous work pieces (i.e. the PCBs) while requiring limited hardware and software reconfigurations. The first goal of the plant is to repair the PCB via remanufacturing and, if this is not suitable or technically feasible, by recovering valuable components through disassembly. Eventually, the last option is to recover the raw material by shredding and separation.

The PCB enters the De-Manufacturing Pilot Plant by means of a robotized station and then it is mounted on a pallet and loaded on an automatic conveyor. Such conveyor transports PCBs to the workstations (see Fig. 4.4) that can execute the main operations: PCB circuit analysis (*Test* station) to identify possible failed components on the board; PCB rework (*Rework* station) to replace failed components; PCB or components that cannot be repaired for technical reason (e.g. overall circuit damage),

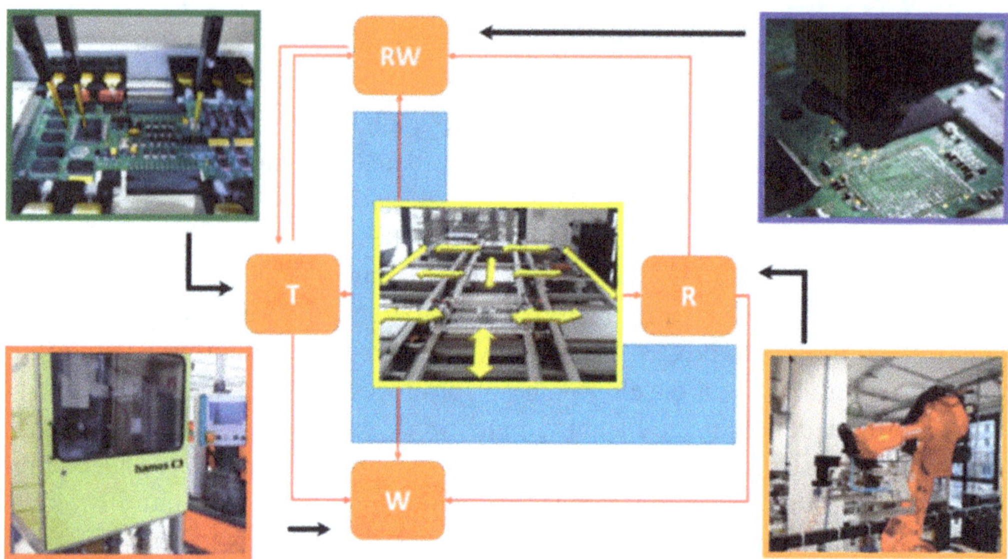

Fig. 4.4 De-manufacturing process architecture: *RW* Rework, *T* Circuit Test, *R* Robotic disassembly, *W* material recovery cell

or which economic value does not justify remanufacturing are sent to a further cell (*Disassembly* station), and possibly to a shredding cell (*Material Recovery* station) aimed at disassembling the PCB and recovering precious raw materials, respectively.

If the PCB entering the system has an economic value justifying remanufacturing, then it is first sent to the PCB analysis process for identifying eventual corrupted components or board damages. If no unrecoverable damage is identified, the PCB is sent back to the first robot station and exits the system as reusable product. If the analysis identifies a not repairable failure on the board, then the PCB is sent to the material recovery cell. If the analysis identifies a failure on one (or more) components, then the PCB is sent to the rework process for substituting the components before undergoing again the PCB analysis for checking conditions. If the test is passed, then the PCB is moved to the first station, otherwise the loop is repeated (with a configurable number of iterations). As soon as the maximum number of iterations is reached, the PCB is sent to the material recovery process. Before such stage, if the PCB mounts valuable components, it is sent to the rework station for chips de-soldering. If the PCB has not an economic value justifying remanufacturing, it is moved directly to the material recovery cell. In case some components of the PCB are valuable for reuse, or have to be removed before shredding for safety reason (i.e. hazardous material content), they are sent to the rework station to disassemble such components before shredding.

The overall conveyor system is composed of a set of mechatronic modules to provide the maximum flexibility and scalability. All modules have the same dimensions, are composed of three intermediate positions for pallet hosting, each of them supporting two main (straight) transfer services and in some cases cross module transfer services. From a mechatronic point of view, the main sensing and actuating elements

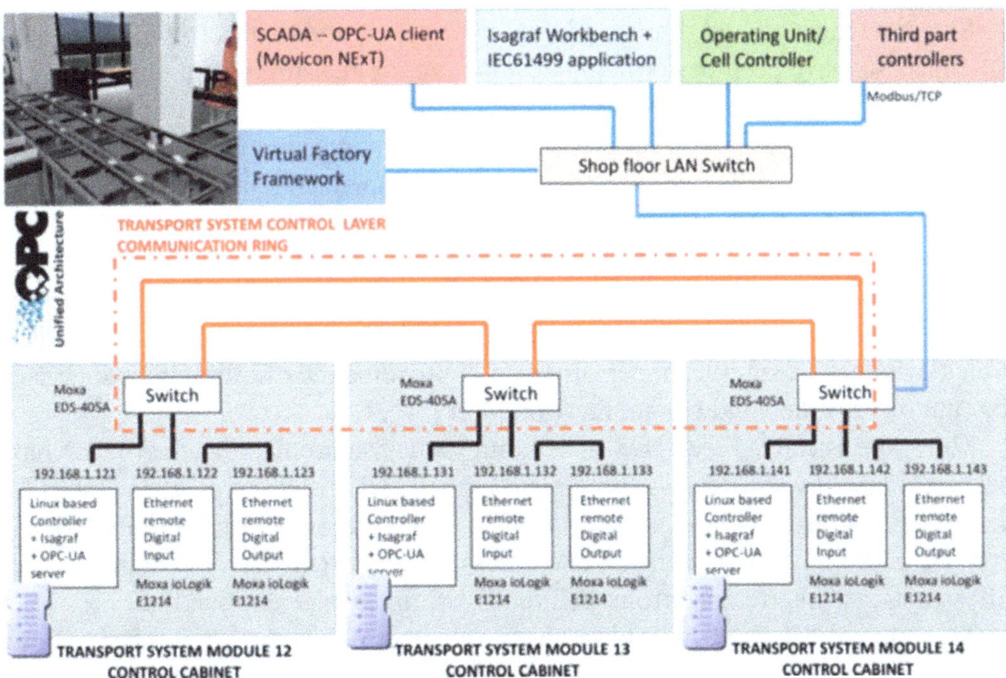

Fig. 4.5 RTS transportation module—Implemented control architecture

for each of the intermediate positions (and consequently for the whole module) are two-way motors—powering the belts for pallet transferring—indexing pins for pallet sequencing and proximity sensor for pallet detection. Each mechatronic module is mechanically, electronically and pneumatically self-contained, and designed so that connectors from one module to the neighbouring carry both power/air supply and Ethernet. Therefore, it is possible to dynamically design and deploy several configurations, depending on the particular needs of the transportation system and/or on the possible online events to occur during the production (e.g. integration of new machines in the line). As an example, a configuration can support forward and backward movements and one or more specific positions with right and left cross transfer capabilities.

4.6.2 Experiments, Results and Analysis

The GECKO control architecture has been validated within the De-Manufacturing plant described in Sect. 4.6.1 (hereinafter also referred to as *GECKO pilot plant*), by realizing a novel automation system for the RTS transportation modules. Following the GECKO approach, the control solution governing each module of the conveyor system has been re-designed to expose the dedicated services. Figure 4.5 summarizes the implemented control architecture for the RTS transportation modules of the de-manufacturing plant.

Since the most critical timings of a control system often concern I/O operations, the first tests have been performed by evaluating the real-time communication performance of the runtime by means of a custom Modbus TCP slave, implemented as described in Sect. 4.5. The tests have been carried out considering different increasing computational loads designed to reproduce the characteristics of the other software components expected to run on the system in the real plant. Moreover, existing knowledge about the internals of the RT Patch extension of the Linux kernel has been leveraged to bring the system towards its worst case behaviour. During the experiments, a custom Modbus TCP slave collected traffic statistics for later analysis, which evidenced a significant performance improvement by limiting the worst-case amount of activation and communication jitter [59].

More specifically, FreeRTOS scheduling and synchronization overheads have been evaluated by considering three main aspects, all of them typical of embedded distributed applications: (i) task scheduling overhead upon semaphore synchronization; (ii) task scheduling overhead upon message queue (mailbox) synchronization; (iii) worst-case interference from the FreeRTOS timer interrupt handler.

The internal delays of lwIP have been evaluated by inserting a number of high-resolution timestamping points into the protocol stack code, whose location has been determined by code analysis, and developing test code to generate the appropriate traffic patterns to be analysed. The experiments have been carried out using a frame size of 242 bytes, that is, close to the maximum frame size of Modbus TCP. The final part of the evaluation was by far the most complex one, because it involved, first of all, the implementation of a 2-node experimental testbed consisting of a Modbus TCP master/slave pair. Overall, the main outcome was the complete breakdown of Modbus TCP communication delays into simpler components, contributed by different software modules. For what concerns jitters, the most important discovery was to identify the role played by TCP acknowledgements, also depending on whether these acknowledgments are piggybacked onto other TCP segments or transmitted on their own.

An additional outcome has been a comprehensive comparison between the code execution performance of the two microcontrollers considered in the experiments. More specifically, besides the CPU clock speed, the two main contributing factors to the different performance of the two architectures are:

- Memory bus width. On the LPC1768 the static RAM is internal to the microcontroller and is connected to the CPU through a 32-bit bus. On the LPC2468 the RAM is dynamic for the most part, is external to the microcontroller, and is connected to it by means of a 16-bit bus originating from the on-chip External Memory Controller (EMC). For word-sized transfers, the most common kind of transfer, this implies a time scaling factor of 2 when clock frequencies are assumed to be the same in both cases, because two bus cycles are needed to access the RAM on the LPC2468, versus one on the LPC1768.
- DRAM/SRAM access. The access methods of static versus dynamic RAM are very different. SRAM access is simpler, because that kind of memory can consistently provide one 32-bit word of data every clock cycle. On the contrary, DRAM access

time is variable and depends on a multitude of factors, for instance, the length and likelihood of intervening refresh cycles, the current state of the memory controller state machine, and write buffer status. Due to the difficulty of calculating this factor analytically, its value is best derived from experimental data.

The next step aimed at verifying the real-time behaviour of the ISAGRAF IEC 61499. Functional tests have been conducted on a set of modules representative of the GECKO pilot plant, with runtime component hosted on the plant PLCs, so as to establish that this latter is not adversely affected by the concurrent execution of additional, interfering tasks on the same embedded control system. The output of such tests proved the operating system to be robust enough to be able to effectively isolate the IEC 61499 runtime component from the execution timing point of view.

The GECKO distributed control approach has been then tested both on the real plant and on a set of different simulated layouts to verify the effectiveness of the approach independently from the specific plant configuration. The design of experiment aims at (i) evaluating RTS throughput and parts lead time considering different layouts, (ii) comparing the features of the proposed control approach with respect to a well-defined benchmark case and (iii) assessing the benefits of the distributed approach when temporary and dynamic changes occur in the RTS. The observing horizon was set equal to a work shift (i.e. 8 h), while the rate of parts entering the system from the load station is kept constant. A complete and exhaustive description of such experiments can be found in [51].

Figure 4.6 shows the different RTS layouts used in simulated experiments. All layouts are composed of 20 transportation modules, where each module has a main movement direction and is equipped with cross-transfer devices (represented in the picture by the bold black segments) enabling the cross-transportation between adjacent transportation modules. Machines and I/O modules are represented, respectively, by red and green blocks.

The experiments have been carried out in two phases to (i) evaluate RTS throughput and parts lead time while considering different layouts (first phase), and (ii) assess the benefits of the distributed approach when changes occur in the RTS (second phase).

The experimental runs of the first phase proved the benefits associated with the distributed auction-based approach to support the design of RTS configurations. After the evaluation of the system throughput and the part lead time for the different RTS layouts, the system designer can select the most efficient configuration according to the production and technology requirements. Considering average throughput values and part lead time in each layout, the results of the experiments are reported in Table 4.1.

Figure 4.7 shows the trends of finished parts, i.e. the number of processed parts (y-axis) sampled every 30 min (x-axis). As shown in the figure, the RTS *Cross* configuration is the best topology among the tested ones, if the average throughput and part lead time are considered.

The experimental runs of the second phase targeted the capability of the distributed control approach to match temporary or permanent changes (e.g. module failures)

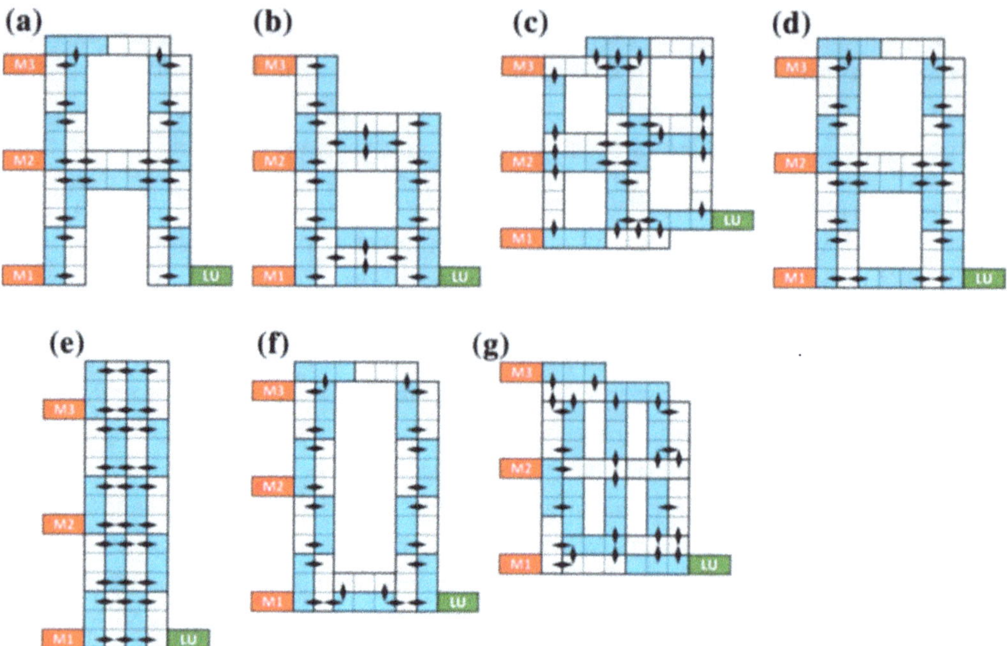

Fig. 4.6 Examples of RTS layouts considered for the experiments: **a** RTS *A*, **b** RTS *B*, **c** RTS *Cross*, **d** RTS *Eight*, **e** RTS *Full*, **f** RTS *Ring*, **g** RTS *Star*

Table 4.1 Average throughput values and part lead time of tested RTS layouts

Layout	Average throughput (part/h)	Part average lead time (s)
A	19	933
B	20	889
Cross	28	627
Eight	20	848
Full	21	840
Ring	20	840
Star	23	746

of the RTS layout. A failure leads to a temporary unavailability of the corresponding module, thus triggering the generation of a new RTS layout corresponding to the new reality of the system, and allowing to resume the production of new routing policies until the original layout is (possibly) restored. The interested reader can find further details in [51].

Additional tests have been carried out to evaluate the knowledge processing mechanism and the capability of a single GECKO agent to build a consistent abstraction of the production context and synthesize timeline-based models accordingly. Experiments have been designed to evaluate—across different configurations—the inference time needed to build the knowledge base of an agent as well as the capability of updating such a knowledge every time that a physical change (i.e. a reconfiguration of the transportation module) occurs in the module. All the different physical

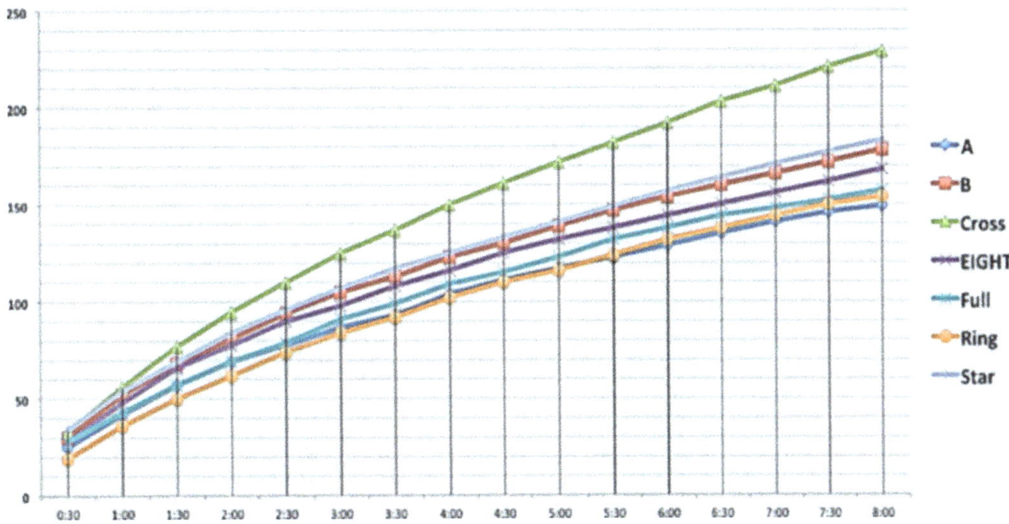

Fig. 4.7 Production volumes associated to examples of RTS layouts

configurations of a transportation module have been considered, from unidirectional module (with no cross-transfer device) to module with three cross-transfer devices. These configurations are referred to as *simple*, *single*, *double* and *full*, respectively. A further possible configuration entails a different number of connected neighbouring modules. Clearly, the more complex scenario is the superimposed one with the highest number of cross-transfers (full configuration) and neighbours. In order to pursue such complex scenario, reconfiguration scenarios have been addressed, based on occurrence on different external events, particularly an increasing number (from 1 to 3) of transportation module neighbours momentarily unable to exchange pallets, as well as on occurrence of internal failures to a cross-transfer device engine to a specific port.

Figure 4.8 shows the results concerning the time needed to infer the knowledge base as a response to the new reconfigured scenario and the time needed to generate a corresponding timeline-based model. The experimental results prove the practical feasibility of the knowledge processing mechanism in increasingly complex configurations of a transportation module. The cost of such a reasoning mechanism has low impact on the performance of a module during its operations. Further details about the inference process and the experimental results can be found in [55].

4.7 Conclusions and Future Research

The proposed GECKO multi-agent distributed control approach [59] enables autonomous agents to dynamically recognize the working settings, identify their capabilities, and collaborate to control the production and support its reconfiguration [53, 55]. The proposed solution is capable of automatically dealing with structural

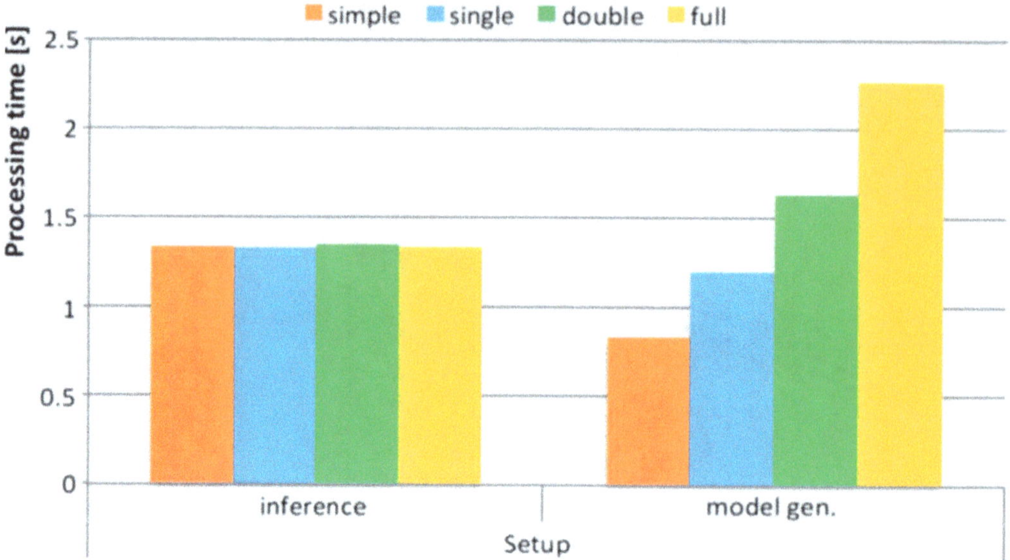

Fig. 4.8 Knowledge base inference time and timeline-based model generation time

changes in the topology of the plant as well as the capabilities of the transportation modules without affecting the production, so that it is possible to cope with changes in the demand and/or unforeseen events/failures at shop floor level. Indeed, unlike classical centralized control approaches, the GECKO distributed control approach can dynamically optimize the paths that pallets follow, by reasoning online on the actual state of the plant. It is worth highlighting that this reasoning is performed without suspending the production.

The adoption of the GECKO approach for design and development of novel solutions to automation systems has been discussed with reference to the specific case study of a de-manufacturing plant, showing how such a decentralized control approach and the integration of planning and knowledge reasoning technologies can be beneficial to improve the efficiency and the flexibility of complex systems like RTSs [56].

The integration and (re)use of the control solutions is enhanced by the encapsulation of the control functionalities into a network of interconnected function blocks. Moreover, considering the development of the overall control system, the activity of the control engineer benefits from the availability of control modules of mechatronic devices, as single components, already tested and validated.

Distributed control system methodologies and prototype tools presented in this work are further extended towards a higher TRL (Technology Readiness Level) within the framework of two Horizon 2020 projects. In particular, the Daedalus project[8] deals with the extension of the modular and distributed control system platform, orienting towards the Industry 4.0 CPS (Cyber Physical System) paradigm. Moreover, the MAYA project[9] addresses the integration of the distributed automa-

---

[8] http://daedalus.iec61499.eu/index.php/en/.

[9] http://www.maya-euproject.com/.

tion system interoperability framework to support dynamic fitting of factory digital models with data gathered from the real plants, so to improve the reliability of the forecasting and optimizations.

Acknowledgements This work has been funded by the Italian Ministry of Education, Universities and Research (MIUR) under the Flagship Project "Factories of the Future—Italy" (Progetto Bandiera "La Fabbrica del Futuro") [60], Sottoprogetto 1, research project "Knowledge based modules for adaptive distributed control systems" (GECKO). The authors greatly acknowledge the industrial partners of the GECKO project, namely FF Falconi Snc, Magneti Marelli Spa and Tecnint HTE for the technical support to the conception of requirements, development ideas and validation scenarios. Authors from ISTC-CNR would like to thank their colleagues Stefano Borgo and Alessandro Umbrico for their valuable work during the project.

References

1. Barenji RV, Barenji AV, Hashemipour M (2014) A multi-agent RFID-enabled distributed control system for a flexible manufacturing shop. Int J Adv Manuf Technol. https://doi.org/10.1007/s00170-013-5597-2
2. Heilala J, Voho P (2001) Modular reconfigurable flexible final assembly systems. Assem Autom 21(1). https://doi.org/10.1108/01445150110381646
3. Project SMErobot (2006) The European robot initiative for strengthening the competitiveness of SMEs in manufacturing. EU FP6 2006
4. Project COMET (2010) Plug-and-produce COmponents and METhods for adaptive control of industrial robots enabling cost effective, high precision manufacturing in factories of the future. EU FP7 2010
5. Project DYNEXPERTS (2010) Plug and produce components for optimum dynamic performance manufacturing systems. EU FP7 2010
6. Zimmermann U, Bischoff R, Grunwald G, Plank G, Reintsema D (2008) Communication, configuration, application—the three layer concept for plug-and-produce. In: Proceedings of the fifth international conference on informatics in control, automation and robotics, Funchal, Madeira, 11–15 May 2008. INSTICC Press, Setúbal, pp 255–262
7. Reinhart G, Krug S, Hüttner S, Mari Z, Riedelbauch F, Schlögel M (2010) Automatic configuration (plug & produce) of industrial ethernet networks. In: 9th IEEE/IAS international conference on industry applications, 8–10 Nov, Sao Paulo, Brazil
8. Guarino N (1998) Formal ontology and information systems. FAIA 46, IOS Press, pp 3–15
9. Borgo S, Leitao P (2007) Foundations for a core ontology of manufacturing. In: Sharman R, Kishore R, Ramesh R (eds) Ontologies: a handbook of principles, concepts and applications in information systems, volume 14 of Integrated series in information systems. Springer, pp 751–776
10. Kuijpers EA, Carotenuto L, Cornier C, Damen D, Grimbach A (2010) The Ulisse environment for collaboration on ISS experiment data and knowledge representation. In: Proceedings of 61st international astronautical congress (IAC'10), Prague, CZ
11. Kuijpers EA, Carotenuto L, Malapert JC, Markov-Vetter D, Melatti I, Orlandini A, Pinchuk R (2012) Collaboration on ISS experiment data and knowledge representation. In: Proceedings of 63rd international astronautical congress (IAC'12), Naples, Italy
12. Koren Y (1999) Reconfigurable machine tools. Patent number: 5943750
13. Koren Y, Heisel U, Jovane F, Moriwaki T, Pritschow G, Ulsoy G, Van Brussel H (1999) Reconfigurable manufacturing systems. CIRP Ann Manuf Technol 48(2):527–540
14. Terkaj W, Tolio T, Valente A (2009) Focused flexibility in production systems. In: ElMaraghy HA (ed) Changeable and reconfigurable manufacturing systems. Springer, pp 47–66

15. Bloch H, Fay A, Hoernicke M (2016) Analysis of service-oriented architecture approaches suitable for modular process automation. In: 2016 IEEE 21st international conference on emerging technologies and factory automation (ETFA), Berlin, pp 1–8
16. Henneke D, Elattar M, Jasperneite J (2015) Communication patterns for cyber-physical systems. In: 20 th IEEE international conference on emerging technologies and factory automation
17. MacKenzie CM, Laskey K, McCabe F, Brown PF, Metz R (2015) Reference model for service oriented architecture 1.0. http://docs.oasis-open.org/soa-rm/v1.0/soa-rm.html. Accessed 15 Oct 2015
18. European IST FP6 project, SOCRADES—service-oriented cross-layer infrastructure for distributed smart embedded devices. www.socrades.net
19. Cannata A, Gerosa M, Taisch M (2008) SOCRADES: a framework for developing intelligent systems in manufacturing. In: 2008 IEEE international conference on industrial engineering and engineering management (IEEM), pp 1904–1908
20. Mendes JM, Leitao P, Colombo AW, Restivo F (2008) Service oriented process control using High-Level Petri Nets. In: 2008 6th IEEE international conference on industrial informatics (INDIN), pp 750–755
21. Karnouskos S, Colombo AW, Bangemann T, Manninen K, Camp R, Tilly M, Sikora M, Jammes F, Delsing J, Eliasson J, Nappey P, Hu J, Graf M (2014) The IMC-AESOP architecture for cloud-based industrial cyber-physical systems. In: Colombo AW, Bangemann T, Karnouskos S, Delsing J, Stluka P, Harrison R, Jammes F, Lastra JLM (eds) Industrial cloud-based cyber-physical systems: the IMC-AESOP approach. Springer International Publishing, pp 49–88
22. Procedure Automation for Continuous Process Operations, ISA-TR 106.00.01
23. Dai WW, Christensen JH, Vyatkin V, Dubinin V (2014) Function block implementation of service oriented architecture: case study. In: 12th IEEE international conference on industrial informatics (INDIN), pp 112–117
24. Function blocks—Part 1: Architecture, IEC 61499-1 (2012)
25. Lotter B, Wiendahl H-P (2009) Changeable and reconfigurable assembly systems. In: Elmaraghy H (ed) Changeable and reconfigurable manufacturing systems. Springer, London, pp 127–142
26. Nonaka Y, Erdıs G, Kis T, Nakano T, Váncza J (2012) Scheduling with alternative routings in CNC workshops. CIRP Ann Manuf Technol 61:449–454
27. Azab A, ElMaraghy HA (2007) Mathematical modeling for reconfigurable process planning. CIRP Ann Manuf Technol 56:467–472
28. Nyhuis P, von Cieminski G, Fischer A (2005) Applying simulation and analytical models for logistic performance prediction. Ann CIRP 54:417–420
29. Tolio T, Ceglarek D, ElMaraghy HA, Fischer A, Hu SJ, Laperriere L, Newman ST, Vancza J (2010) SPECIES—co-evolution of products, processes and production systems. CIRP Ann Manuf Technol 59(2):672–693
30. Hu J, Camelio J (2006) Modeling and control of compliant assembly systems. CIRP Ann Manuf Technol 55:19–22
31. Terkaj W, Tolio T, Valente A (2009) Design of focused flexibility manufacturing systems (FFMSs). In: Tolio T (ed) Design of flexible production systems—methodologies and tools. Berlin, Heidelberg, pp 1–18
32. Quiao, B, Zhu J (2000) Agent-based intelligent manufacturing system for the 21st century. In: Proceedings of the international forum for graduates and young re-searchers of EYPO, The World Exposition in Germany, Hannover
33. Ferber J (1999) Multi-agent systems: an introduction to distributed artificial intelligence, vol 1. Addison-Wesley, Reading
34. Heragu S, Graves R, Kim B et al (2002) Intelligent agent-based framework for manufacturing systems control. IEEE Trans Syst Man Cybern Part A Syst Hum 32(5):560–573. https://doi.org/10.1109/TSMCA.2002.804788
35. Crawford LS, Do MB, Ruml W, Hindi H, Eldershaw C, Zhou R, Kuhn L, Fromherz MP, Biegelsen D, de Kleer J et al (2013) Online reconfigurable machines. AI Mag 34(3):73–88

36. Balakirsky S (2015) Ontology based action planning and verification for agile manufacturing. Robot Comput Integr Manuf 33(0):21–28. Special Issue on Knowledge Driven Robotics and Manufacturing
37. Solano L, Rosado P, Romero F (2013) Knowledge representation for product and processes development planning in collaborative environments. Int J Comput Integr Manuf 27(8):787–801
38. Terkaj W, Schneider GF, Pauwels P (2017) Reusing domain ontologies in linked building data: the case of building automation and control. In: Proceedings of the 8th workshop formal ontologies meet industry, joint ontology workshops 2017, CEUR workshop proceedings, vol 2050
39. Hristoskova A, Aguero E C, Veloso M, De Turck F (2013) Heterogeneous context-aware robots providing a personalized building tour. Int J Adv Robot Syst
40. Suh I H, Lim G H, Hwang W, Suh H, Choi J H, Park Y T (2007) Ontology-based multi-layered robot knowledge framework (OMRKF) for robot intelligence. In: IEEE/RSJ international conference on intelligent robots and systems. IROS 2007, pp 429–436
41. Tenorth M, Beetz M (2015) Representations for robot knowledge in the KnowRob framework. Artif Intell. http://dx.doi.org/10.1016/j.artint.2015.05.010
42. Lemaignan S, Ros R, Mosenlechner L, Alami R, Beetz M (2010) ORO, a knowledge management platform for cognitive architectures in robotics. In: 2010 IEEE/RSJ international conference on intelligent robots and systems (IROS), pp 3548–3553
43. Hartanto R, Hertzberg J (2008) Fusing DL reasoning with HTN planning. In: Dengel A, Berns K, Breuel T, Bomarius F, Roth-Berghofer T (eds) KI 2008 advances in artificial intelligence, volume 5243 of Lecture notes in computer science. Springer, pp 62–69
44. Yang F, Khandelwal P, Leonetti M, Stone P (2014) Planning in answer set programming while learning action costs for mobile robots. In: AAAI Spring 2014 symposium on knowledge representation and reasoning in robotics (AAAI-SSS)
45. Terkaj W, Urgo M, Andolfatto D (2017) Answer set programming for modeling and reasoning on modular and reconfigurable transportation systems. In: Proceedings of the 2017 federated conference on computer science and information systems
46. IEEE, IEEE Std 1003.13 (2003) Edition, standard for information technology—standardized application environment profile (AEP)—POSIX realtime and embedded application support
47. Barry R (2010) Using the FreeRTOS real time kernel. Raleigh, North Carolina
48. Decotignie JD (2005) Ethernet-based real-time and industrial communications. Proc IEEE 93(6). https://doi.org/10.1109/jproc.2005.849721
49. Dzung D, Naedele M, Von Hoff TP, Crevatin M (2005) Security for industrial communication systems. Proc IEEE 93(6). https://doi.org/10.1109/jproc.2005.849714
50. Moyne JR, Tilbury DM (2007) The emergence of industrial control networks for manufacturing control, diagnostics, and safety data. Proc IEEE 95(1). https://doi.org/10.1109/jproc.2006.887325
51. Carpanzano E, Cesta A, Orlandini A, Rasconi R, Suriano M, Umbrico A, Valente A (2016) Design and Implementation of a distributed part-routing algorithm for re-configurable transportation systems. Int J Comput Integr Manuf 29(12):1317–1334. https://doi.org/10.1080/0951192X.2015.1067911
52. Brusaferri A, Ballarino A, Cavadini FA, Manzocchi D, Mazzolini M (2014) CPS-based hierarchical and self-similar automation architecture for the control and verification of reconfigurable manufacturing systems. In: Proceedings of the 2014 IEEE emerging technology and factory automation (ETFA), Barcelona, pp 1–8
53. Borgo S, Cesta A, Orlandini A, Umbrico A (2015) An ontology-based domain representation for plan-based controllers in a reconfigurable manufacturing system. In: Proceedings of 28th Florida artificial intelligence research society conference (FLAIRS-15)
54. Musen MA (2011) The protégé project: a look back and a look forward. AI Matters 1(4):4–12
55. Borgo S, Cesta A, Orlandini A, Umbrico A (2016) A planning-based architecture for a reconfigurable manufacturing system. In: Proceedings of 26th international conference on automated planning and scheduling

56. Carpanzano E, Cesta A, Orlandini A, Rasconi R, Valente A (2014) Intelligent dynamic part routing policies in Plug&Produce reconfigurable transportation systems. CIRP Ann Manuf Technol 63(1):425–428
57. Umbrico A, Orlandini A, Cialdea Mayer M (2015) Enriching a temporal planner with resources and a hierarchy-based heuristic. In: Gavanelli M, Lamma E, Riguzzi F (eds) AI*IA 2015 advances in artificial intelligence. Lecture notes in computer science, vol 9336. Springer, Cham
58. Tolio T, Copani G, Terkaj W (2019) Key research priorities for factories of the future—Part II: Pilot plants and funding mechanisms. In: Tolio T, Copani G, Terkaj W (eds) Factories of the future. Springer
59. Borgo S, Cesta A, Orlandini A, Rasconi R, Suriano M, Umbrico A (2014) Towards a cooperative knowledge-based control architecture for a reconfigurable manufacturing plant. In: Proceedings of 19th IEEE international conference on emerging technologies and factory automation (ETFA). IEEE
60. Terkaj W, Tolio T (2019) The Italian flagship project: factories of the future. In: Tolio T, Copani G, Terkaj W (eds) Factories of the future. Springer

Chapter 5
Highly Evolvable E-waste Recycling Technologies and Systems

Giacomo Copani, Nicoletta Picone, Marcello Colledani, Monica Pepe
and Alessandro Tasora

Abstract Materials recycling is a key process to close the loop of materials in the direction of circular economy. However, the variability of waste and the high volatility of the price of recovered materials are posing serious challenges to the current rigid design of mechanical recycling systems. This is particularly true for Waste Electric and Electronic Equipment (WEEE), whose volume is growing more than other waste streams in Europe due to the diffusion of electronic products and to their short technology cycles. This study is aimed at the development of new flexible recycling systems through the implementation of a Hyper Spectral Imaging system and a simulation model enabling the real-time characterisation of shredded particles and the dynamic optimisation of process parameters for efficient sorting. A hardware and software prototype was realised and tested at the De- and Re-manufacturing pilot plant of CNR-STIIMA. The positive economic impact of flexible recycling systems enabled by new technologies was assessed through scenario analysis.

5.1 Scientific and Industrial Motivations

De-manufacturing acquired high relevance in last few years due to increasing raw material costs and to laws in many countries aimed at improving material recycling rates and at limiting supply shortage risks for rare metals used in high-tech applications. While manufacturing transforms raw materials into products meeting the

G. Copani (✉) · N. Picone
CNR-STIIMA, Istituto di Sistemi e Tecnologie Industriali Intelligenti per il Manifatturiero Avanzato, Milan, Italy
e-mail: giacomo.copani@stiima.cnr.it

M. Colledani
Dipartimento di Meccanica, Politecnico di Milano, Milan, Italy

M. Pepe
CNR-IREA, Istituto per il rilevamento Elettromagnetico dell'Ambiente, Milan, Italy

A. Tasora
Dipartimento di Ingegneria e Architettura, Università di Parma, Parma, Italy

© The Author(s) 2019
T. Tolio et al. (eds.), *Factories of the Future,*
https://doi.org/10.1007/978-3-319-94358-9_5

customer requirements, de-manufacturing transforms post-consumer products into valuable materials meeting the customer requirements for secondary use. However, manufacturing and de-manufacturing objectives are strictly dependent. In fact, products that are currently manufactured will constitute the waste that will be treated by de-manufacturing system in the next years.

De-manufacturing can be defined as the set of processes, technologies, tools and knowledge-based methods to re-use, re-manufacture and recycle materials and components from End-of-Life products [1]. The de-manufacturing process includes the following steps:

1. *Disassembly* that aims at separating (i) hazardous components that have to be treated separately, (ii) re-usable components with high residual value, (iii) components needing to go through a special recycling process due to their material composition. In general, in the current state of the art, most of disassembly operations are performed manually.
2. *Mechanical pre-treatment* that consists in shredding, reduction of particles' size and separation of different materials flow. The goal of separation stages is to split a mixed input material flow into two or more output flows that are characterised by a concentration of target materials higher than in the input flow. The size-reduction of the particles can help to isolate target material from other type of materials, thus increasing the quality of the downstream separation process.
3. *Material recovery or end-refining* by means of chemical and thermal processes separating target materials at very high grade levels to obtain compounds that can be re-used for the production of other materials.

According to recent studies, the major cause for losses of key metals (from 40 to 100% of material) in de-manufacturing process is the low efficiency of the mechanical pre-treatments step, while only marginal losses are due to the collection and to the downstream chemical end-processes. Therefore, the improvement of mechanical recycling technologies and systems is important for the recycling and high-tech product manufacturing sectors.

Currently, recyclers perform the mechanical pre-treatment of wastes, including metal and non-metal fraction, by:

- manual dismantling processes, to isolate hazardous components and easily disassemble parts;
- mechanical shredding and separation processes, aimed at reducing the size and refining the input material into purified material flows that can be sold either in the market (aluminium, copper, steel) or to end-processing plants for further purification (such in the case of Printed Circuit Boards).

In Europe, 85% of recycling companies are Small Medium Enterprises (SMEs). Due to the space and budget limitations, it is extensively challenging for these companies to operate dedicated treatment lines for mechanical shredding and separation processes for each specific product flow. Batch production and high utilization of a single or few recycling lines is a common solution. The process parameters are selected as a compromise between the different product types, thus leading to un-optimised performance. Moreover, in spite of the high variability of post-use products and material to be treated, currently available mechanical recycling systems are extremely rigid and they cannot be adapted to the different type of waste which is generated over time according to the technology cycle of products. Their design and technology does not allow to dynamically modify process parameters and material routing based on the specific type of mixture to be treated. In addition, the integration of new process technologies in the system is extremely time-consuming and expensive, since the rigid conveyor systems need to be re-arranged for the specific purpose with long ramp-up times. For these reasons, aligning recycling with the evolution of products and waste is a significant challenge. The system rigidity, coupled with the high variability in the input material composition and in the targeted output performance, ultimately causes many shortcomings of the process including:

- low recycling rates, especially for key-metals;
- excessive recourse to landfilling also for materials that could be potentially recycled, such as high value-added plastics;
- lack of competitiveness of the Small Medium Enterprises that typically apply mechanical recycling processes, and
- untapped market potentials for the recycling industry.

Thus, the development of flexible and adaptable systems for the recycling industry is a promising avenue for the improvement of material recovery and for the adoption of recycled material in high-quality and high-value applications [2, 3]. The concepts of flexibility, re-configurability and adaptability have already been widely investigated in the scope of manufacturing systems [4–7], but they have been poorly addressed until now in the de-manufacturing area. Grounding on a solid knowledge base derived from manufacturing industry and getting inspiration from changeability enablers adopted in manufacturing processes, research activities should aim at the development of innovative technologies for supporting the transition towards smarter de- and re-manufacturing systems characterized by: (i) high adaptability to different waste to be treated and to changing market conditions, (ii) high automation level, (iii) availability and traceability of information, and (iv) intelligent decision algorithms based on data analytics and cyber-physical systems.

To this aim, a set of *hardware and software systems as well as business model applications* are required to be integrated, enabling a feedforward control to manage the system evolution leading to a more coherent matching between the fast dynamics and large variability of the End-of-Life product life-cycle and the system life-cycle, also called *co-evolution* [8]. Co-evolution is defined as the ability of a company to address engineering change according to context modifications, strategically and operationally, in order to gain and maintain competitive advantage [8]. The change propagation acts as a cause-effect wave across the various company domains, spanning from corporate strategy to plants and technologies. The co-evolution problem has been traditionally studied within the manufacturing industry (e.g. within the

EU funded projects RobustPlanet,[1] RLW Navigator,[2] DEMAT[3]) but has never been addressed within the de-manufacturing sector.

This work investigates the implementation of the co-evolution principle for de-manufacturing systems. New flexible and adaptive recycling systems are proposed while addressing the process technology and ICT development. Furthermore, new business and financial models are developed to enable the industrial uptake and large-scale diffusion of the new processes by dynamically suggesting the best strategies to adopt and guaranteeing process and business dynamic adaptation to waste flows evolution over time.

The chapter is organised as follows. Section 5.2 presents the state of the art on flexible recycling systems. Section 5.3 outlines the research scope and approach, whereas Sect. 5.4 presents the different results of the research (i.e. hardware and software solutions, business model validation and prototypes). The results of industrial testing are discussed in Sect. 5.5 and, finally, the conclusions together with indications for future research are presented in Sect. 5.6.

5.2 State of the Art

The literature about recycling systems mainly concentrates on process planning [9] and economic assessment of the processes [10]. However, recent literature showed that in order to efficiently treat different material mixtures in the same mechanical material separation plant, the system itself should be designed to enable on-line modifications of the material flows [11]. Figure 5.1 shows how different routing configurations in the same system can bring to significantly different recovery and grade performance. In particular, treating multiple products would support coping with low revenues from material treatment, high variability of products and limited availability of specific post-use products, thus the capacity constraints [12, 13]. Therefore, the system architecture should be modified depending on what products are treated and which is the target recycling rate of the materials. If the system configuration and the process sequence is not modified when the mixture under processing changes, then the performance of the entire system drastically drops making the recovery process unsustainable for some products.

The volatility of material price and the high variability in composition of the input material represent considerable challenges for the sustainability of recycling systems due to their rigid designs. The major limitations in terms of flexibility of state of the art recycling technologies regard, among all, the two following aspects:

---

[1]ROBUSTPLANET—"Shock—robust Design of Plants and their Supply Chain Networks", FoF. NMP. 2013–9 Grant Agreement no 609087.

[2]FP7-2011-NMP-ICT-FoF EU Project, RLW Navigator "Remote Laser Welding System Navigator for Eco & Resilient Automotive Factories", Grant Agreement no 285051.

[3]DEMAT—"Dematerialised Manufacturing Systems: A new way to design, build, use and sell European Machine Tools", FP7. NMP. 2007–2013 Project ID 246020.

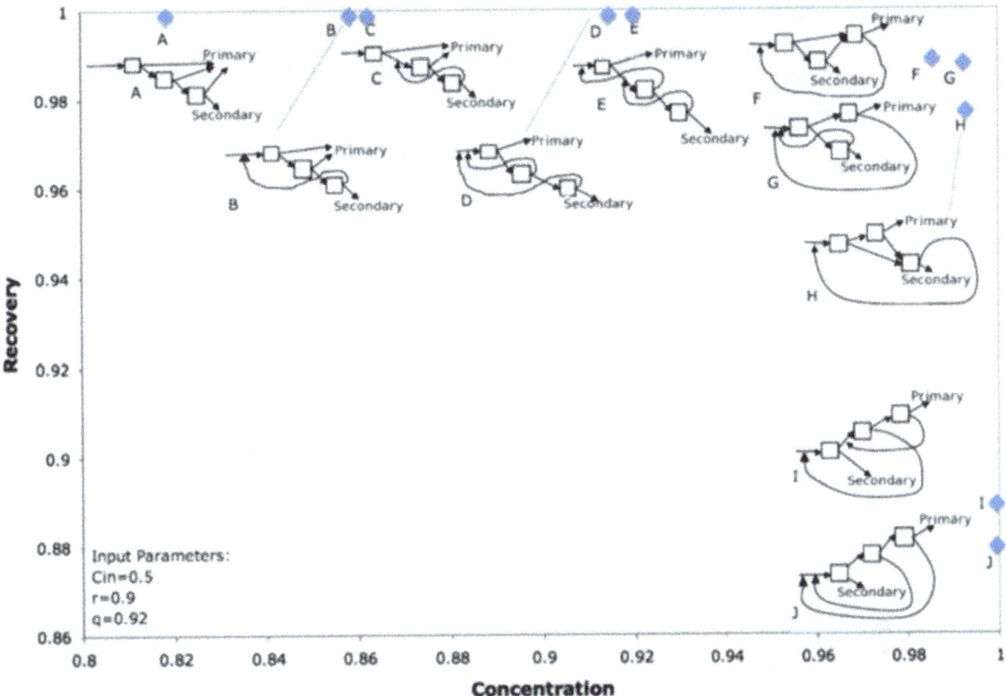

Fig. 5.1 Optimal system configurations as a function of recovery (y-axis) and grade (x-axis) targets [11]

- Modelling, control and optimisation of the recycling process.
- Low performance of hyperspectral imaging to identify metal and non-metallic materials.

Within the research on recycling process modelling and control, separation process modelling has attracted interest from researchers, especially in the mineral technology area [14]. Physical models for particle trajectories in Corona Electrostatic Separation (CES) have been developed [15]. However, these models are always focused on single particle trajectories and do not support the modelling of two relevant causes determining the low performance of separation process, i.e. (i) particle impacts and interactions, and (ii) the presence of unliberated particles in the material flow. Discrete granular material flow models could handle these issues. Very recently, multi-particle and multi-body simulation models have been developed to analyse the dynamics of the impacts within recycling processes (Fig. 5.2) [16]. Although these models outperform existing ones to explain the process physics, they are relatively time consuming tools and hard to be used to support process optimization and control. Moreover, their validation in real settings is still ongoing.

Regarding hyperspectral imaging for metal and non-metallic materials, current methods to distinguish and characterise the composition of parts and shredded particles are based on off-line inspection, colour inspection systems, X-Ray, X-Ray-Backscatter or even spectroscopy over thermally stressed materials [17]. However, these methods are inaccurate with chromatic materials (mainly metals) or their adop-

Fig. 5.2 Multi-particle CES simulation model

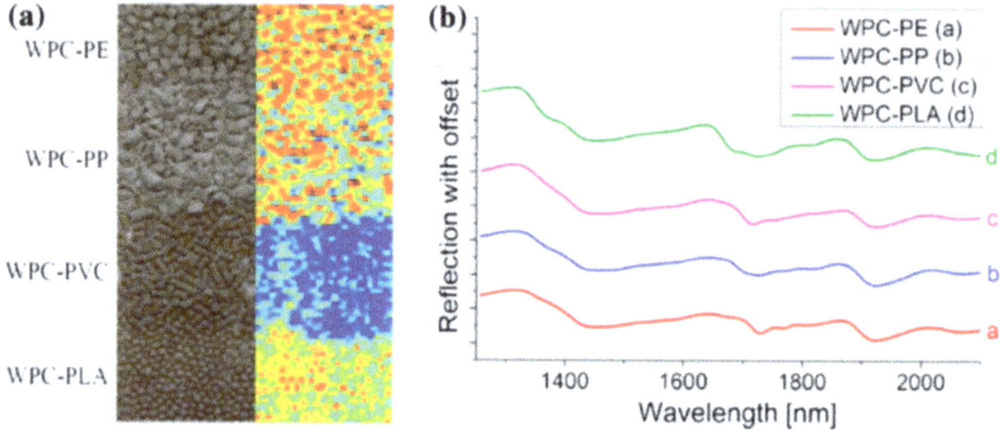

Fig. 5.3 Application of hyperspectral imaging for identification of plastics products: **a** classified images, **b** related reference spectra [22]

tion is not feasible in industry where on-line inspection procedure is becoming extremely important [18–20]. As a matter of fact, although attempts have been made to extend this technology to metallic particles (e.g. EU project Sormen[4]), research contributions are mainly available for plastics particles (Fig. 5.3) [21, 22].

Commercial equipment for automatic optical sorting based on hyperspectral cameras can be found in the market. Although hyperspectral technologies can help characterizing the composition of metallic materials, current hyperspectral processing approaches involve poor characterization of the inherent variability of these materials and their visual appearance (optical texture, geometry, and pattern) of the particles

---

[4]SORMEN—Innovative Separation Method for Non Ferrous Metal Waste from Electric and Electronic Equipment (WEEE) based on Multi- and Hyperspectral Identification project, 6FP Horizontal Research Activities Involving SMES Co-Operative Research, 2006.

[23]. Besides this, the high amount of data that is inherent to hyperspectral imaging implies slow data processing rate, which is not acceptable for real-time recycling applications [24–26]. Thus, new real-time hyperspectral models enclosing texture, optical pattern, component variability and adaptability to improve the overall recycling process are necessary.

5.3 Problem Statement and Research Approach

An approach to deal with the product, process and system co-evolution problem for recycling systems needs to satisfy the following requirements:

- in-line characterization of metal and non-metal materials composing the relevant waste flow;
- knowledge-based optimization and control of recycling processes, in view of the adjustment of the system parameters for the specific mixture under treatment;
- flexible and modular automation to increase shock-robustness via reconfiguration and modularity of machinery and systems;
- integrated software tools to support co-evolution related decision-making in industry;
- assessment of the sustainability of new technologies and proposal of suitable recycling business models for risk mitigation.

The proposed approach consists of the following technological and ICT-based solutions (hardware and software) to meet the previous requirements:

- in-line hyperspectral imaging technology to provide field data about the particles size, shape and material composition (Sect. 5.4.1);
- new process optimization and control models, software and code for the mechanical separation technologies based on CES (Sect. 5.4.2);
- scenario-based analysis techniques to decide under uncertainty what products to treat in the recycling system in a decision period and what are the target grade and recovery levels to be achieved for each material (Sect. 5.4.3);
- a new flexible and automated station to separate upfront the highest fraction of mixed metals and non-metals, thus enabling a more efficient process-chain (Sect. 5.4.4);

The developed solutions were demonstrated in prototypal applications (Sect. 5.4.4) and the applicability of the approach was tested in a real industrial case (Sect. 5.5), i.e. the De- and Re-manufacturing Pilot Plant at CNR-STIIMA (ex CNR-ITIA). The pilot plant is a fully automated, modular and reconfigurable solution involving integrated technologies for the disassembly, testing, reworking, shredding and mechanical separation of mechatronic products. The facility supports industrial research and innovation with the main purpose of recovering both functionality (when possible) and target materials from End-of-Life products.

5.4 Hardware, Software, Business Models and Prototypes

5.4.1 Hardware System

The developed Hyper Spectral Imaging (HSI) system enables the in-line recognition and classification of shredded products in terms of mixture composition (percentage of metal and non-metal fractions, shape and dimensional distribution) [27], with the aim to adapt the configuration and the working parameters of the downstream separation processes. In particular, the system provides feedback to trigger the control model of CES (i.e. drum speed, electrode voltage, feed rate and output stream splitter position), according to the specific mixture under treatment.

From a technological point of view, the main challenge consisted in the development of a robust and flexible imaging system for the in-line characterization of the metal and non-metal fraction of waste. The underpinning knowledge regarding spectral signatures of pure materials is derived from end-member collections, with cross-check performed by Scanning Electron Microscope (SEM) measurements.

The information about composition and geometry of the particles in the mixture made available by the developed vision system is used to feed a multi-body simulation model and an optimization and control module to get optimal separation parameters and improve the configurability of the recycling system.

The implemented vision system includes three main elements: (1) the hyperspectral camera, (2) the illumination system and (3) the transportation system. The conveyor speed, the image frame rate and binning are mutually aligned for the effective performance of the system.

5.4.2 Software System

The software tool controlling the vision system was implemented ad hoc in Matlab. The group of scripts is able to automatically control the camera acquisition and characterize the mixture in terms of composition, geometry and dimensional distribution. The software tool sends then this information to the separator to optimize the sorting control parameters. In particular, the following steps have been coded in the script:

1. radiometric calibration to relative reflectance;
2. removal of bad bands at the spectral edges;
3. bands compression using triangular fuzzy set;
4. background removal (conveyor belt);
5. compensation for the illumination;
6. spectral subset in order to focus only to a specific spectral region (optional);
7. waste mixture classification;
8. geometrical attributes extraction;
9. calculation of mixture composition statistics (percentage).

For the purpose of the simulation developments, granular flows involved in waste processing exhibit complex dynamical phenomena. Therefore, a custom simulation software based on multi-body dynamics was developed. Such a tool simulates the trajectory of thousands of processed particles, including mutual collision effects and their interaction with electric and magnetic fields in the CES and Eddy Current Separation (ECS) machines.

The large-scale nature of this type of simulation, along with the complexity of collision detection between heterogeneous shapes in three-dimensional space, advocated the adoption of innovative and efficient simulation algorithms backed by the theory of non-smooth dynamics, Measure Differential Inclusions (MDI) and Differential Variational Inequalities (DVI) [28]. The simulation software, available as open source C++ software, is written in C+ language and is dynamically linked with the multi-physics ProjectChrono library, whose features have been customized to meet the needs of this work [29]. A simplified 2D version of the simulation tools has been developed in Matlab as well [30].

Two approaches were implemented to provide a source of virtual waste flow in the simulation tool. The first approach consisted in pre-processing a sample of waste flow using the hyper-spectral video camera, hence obtaining a set of faceted particle shapes and corresponding materials that are sorted and replicated continuously within the virtual inlet during the simulation. The second one consisted in the generation of particles as random samples from user-defined multivariate distributions; to this end, an innovative system based on a hierarchical description of the probability space was designed [31].

An add-in for the SolidWorks CAD, using C+ language, was also designed to export coordinates, geometries, references and collision boundaries of the separating machines into a neutral file format to be processed by the simulator in a following phase. A further advancement came from the adoption of parallel computing hardware for accelerating the collision detection and the time integration [15].

5.4.3 Business Model Validation

Thanks to the new hardware and software technologies, the resulting flexible recycling systems are able to dynamically adapt the implemented recycling strategy to business conditions. This means that it will be possible not only to adapt process parameters to variations of volumes and characteristics of waste that will be required to treat over time, but also to select strategically—and source accordingly—the type of waste that should be treated to maximise the profit based on material pricing and availability of waste. However, a wider range of options and the uncertainty that is typical of the recycling context makes difficult to companies to fast decide what is the best strategy to be pursued. Thus, decision support tools able to indicate suitable business models and to predict economic performance under uncertain conditions are needed to dynamically address business operations and maximise profit.

A scenario analysis model was implemented for assessing the economic feasibility of the reconfigurable recycling system treating multiple waste product streams. The scenario analysis is capable of analysing possible future events by considering alternative probabilistic outcomes [32]. A base-case, optimistic and pessimistic scenario were defined and economic performances were assessed under each scenario calculating the Net Present Value (NPV) and the Pay Back Time (PBT).

Based on interviews to recycling companies and experts, two variables were selected as the main determinants of recycling economic performance: the relative volume of waste products to be treated in a time period (waste mix) and the market price of materials. Waste volume and mix may vary unpredictably due to fast products technology cycles, success (or failure) in regional tenders for treating specific waste over fixed time periods, new regulation, etc. Different types of waste contain different percentages of valuable materials and require different type of treatment processes which impact on costs. On the other hand, the market price of materials depends on their periodic demand and availability, thus determining the revenue that is made by selling recycled materials on the market. Scenarios were designed by making the hypothesis of average, favourable and unfavourable waste mixes according to their content of precious materials and of recycling cost. Such waste mixes were combined with three different levels of materials costs, considering positive and negative variations of prices from current levels (see Sect. 5.5.2).

5.4.4 Prototypes

A CPS platform was realised to integrate two workstations including *Vision System, Classification/Control Module* and *CES, Multi-Body Simulation Model and Optimization/Control Module* (Fig. 5.4).

The vision system workstation integrates an ImSpector V10E (Specim) working from 400 to 1000 nm, with a spectral sampling range from 0.78 to 6.27 nm/pixel, coupled with a CMOS sensor (1312 spatial $\times$ 768 spectral pixels). The camera is equipped with an OLE 23 fore objective lens with a focal length of 23 mm and a FOV of 25.7°. The HSI sensing device is a push-broom system working as a line scan camera providing full contiguous spectral information for each pixel in the line illuminated by two opposite slots housing 3 halogen lamps each, which emit in the spectral range 380–2200 nm. The implementation of the HSI system deals with the need to recognize the materials on-line, enabling the recycler to perform a continuous quantitative characterisation of mixture composition, thus avoiding time-consuming off-line sampling material characterization procedures.

The separation workstation integrates a Corona Electrostatic Separator Hamos KWS. The quality of the separation process is influenced by five controllable parameters: (i) electrode voltage and position; (ii) drum speed; (iii) output stream splitters position; (iv) material feed rate; and (v) feeder vibration. However, due to the complexity of the process, non-controllable effects, such as particle-particle interactions and impacts among particles and machine, influence the efficiency of the pro-

Fig. 5.4 Prototype for feed-forward control strategies: HSI workstation (right) and electrostatic separator (left). Implemented at the de- and re-manufacturing pilot plant of CNR-STIIMA (ex CNR-ITIA)

cess, providing contaminated output flows. For this reason, a multi-body simulation model of electrostatic separator process, considering particles attributes, trajectories, interaction and collision, has been developed. The whole system has been validated with experimental analyses, under different operational conditions (i.e. mixture type, machine working parameters).

Thanks to the integration between CES simulation model (that gives a prediction of particles trajectories) and the optimization and control modules, the optimal separation parameters are suggested to the operator for the mixture under treatment. A two-layer architecture integrated within the recycling process-chain supports the control, operation and reconfiguration of advanced multi-material de-manufacturing plants. This two-level approach is focused both on the process level and on the system level and considers the interconnections between physical and virtual entities (Fig. 5.5).

Data about the mixture (composition, geometrical attributes and dimensional distribution) gathered in real-time by HSI vision system are sent to the two-level modelling platform to enable the control of technologies and the reconfiguration of the recycling process chain according to the specific system and material state. A new software platform, controlled by the customized user interface shown in Fig. 5.6 has been designed and implemented to manage the information flow between software and hardware at system level. In this way, the whole process is automated and supervised. However, the operator can easily intervene in manual mode, if necessary.

The implementation of a CPS platform significantly improves the adaptability of the recycling system and the recycling rates. It avoids losses of high-value materials, improves the information about the quality of recycled material and, as a consequence, its market value.

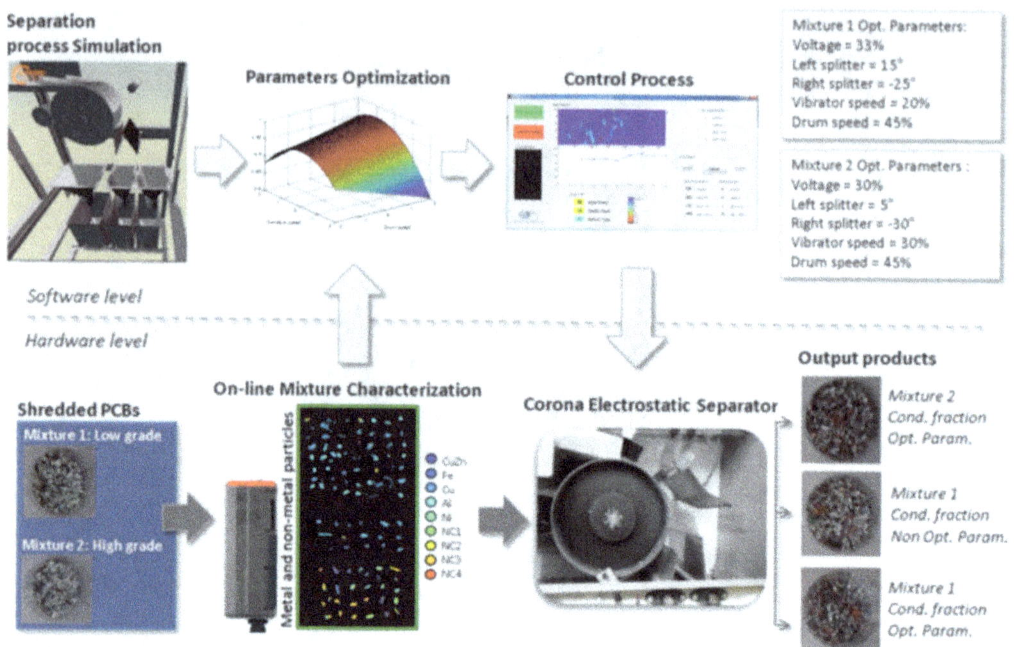

Fig. 5.5 Architecture of the CPS for online adaption of recycling processes

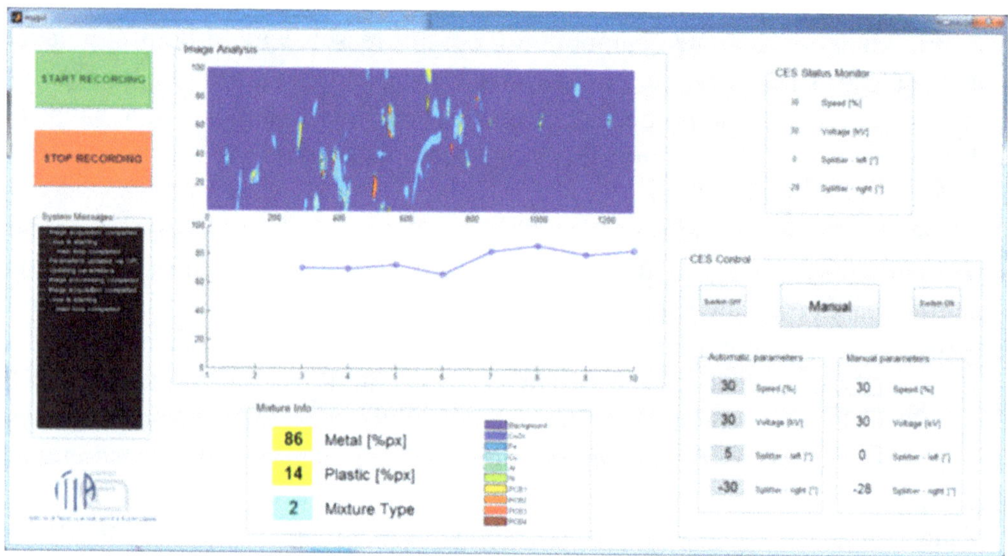

Fig. 5.6 Graphical user interface for process control

The realised prototype demonstrates the benefits of the pursued integration approach in recycling systems, where high-cost and sophisticated control solutions would result inadequate due to technical (particles of different shape, size and composition) and economic reasons. The prototype represents a step towards the implementation of the HSI technology for a highly accurate classification of independent sets of wastes including metal and non-metal fraction undergoing mechanical treatment.

5.5 Industrial Testing and Results

5.5.1 Industrial Case

One of the most challenging modern recycling applications is the treatment of Waste from Electric and Electronic Equipment (WEEE) that represents the European waste with highest yearly average increase rate (5%). Such an increase was even higher in some European nations, e.g. in Italy the increase amounted to 30% in 2010 compared to the previous year (in terms of collected WEEE). Due to its material composition, WEEE represents an important source of recycled key-metals to be used for the production of advanced technological products. E-waste is a highly variable waste flow because of the wide range of products and material mixtures, which are in continue evolution due to fast technology cycles. For example, in small WEEEs (which in Italy are identified as "RAEE class 4"), Cathode Ray Tube (CRT) TVs, cellular phones and traditional HDD are rapidly being replaced by LCD/LED TVs, smart phones and SSD (Solid State Disks), respectively. Other new waste flows are rapidly entering the recycling streams, such as for example solar panels and tablet PCs. For the successful treatment of WEEE, besides being flexible, recycling systems must guarantee a high efficiency, since critical metals are present in small quantities (e.g. indium, palladium, ruthenium, gallium, tantalum, and platinum). For instance, indium is found in WEEE with an average weight of 39 mg in Notebooks, 79 mg in computer monitors and 254 mg in LCD TVs.

Considering these challenges, the scenario of WEEE was chosen as industrial case for the demonstration of the results of this research. In particular, Printed Circuit Boards (PCB) were chosen as reference WEEE products. PCBs are one of the key and most common components present in the majority of WEEEs. Critical and high-value metals (such as copper, tin, nickel, gold and silver) account for 25–40% of their total weight composition. The rest of materials (60–75% in weight) are ceramics, plastics and glass fibres. PCBs, which are the major constituent of the obsolete and discarded electronic scrap, account approximately for 3–6% of the total WEEE mass [33, 34].

PCBs are evolving fast from the technological point of view (from large high-grade matrixes into small, compact and medium grade products). For this study, PCBs disassembled from End-of-Life household appliances (e.g. washing machines) have been used and a representative sample of different models has been selected to be processed (Fig. 5.7).

5.5.2 Vision System and Material Separation

The experiments have been carried out in a real life demonstrator including two workstations: the vision system and the separation machine (see Sect. 5.4.4).

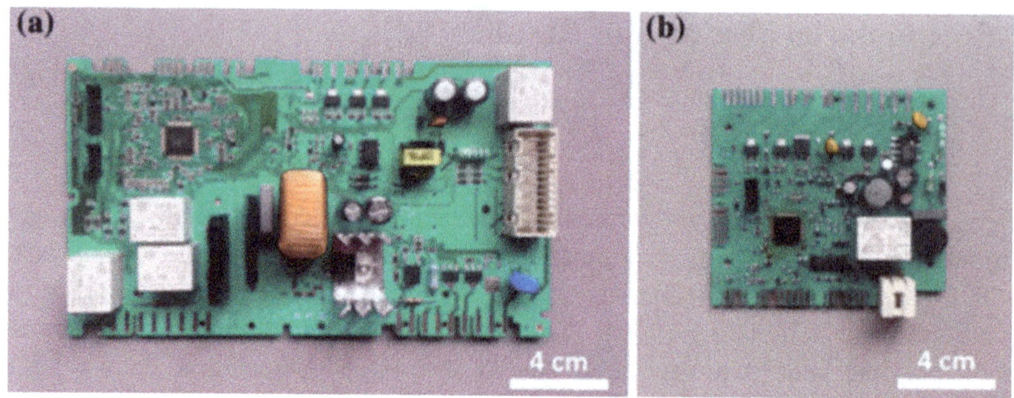

Fig. 5.7 Selected PCBs from EoL washing machines of different brands (**a**) and (**b**)

Two different sample sets have been prepared to: (i) train and test the HSI system and (ii) test the HSI system for the feed-forward control of separation processes. For both sets, the first part of the preparation procedure consisted of a two-stage shredding by using hammer and cutter mills, followed by sieving treatment. The liberated fraction (average diameter below 1 mm) was thus separated by the CES to obtain two fractions: conductive and non-conductive. For the HSI system training and validation, several metal and non-metal particles have been manually sorted from those output fractions according to their colours (conductive: yellowish, greyish or reddish; non-conductive: greenish), to be further analysed by using Scanner Electron Microscope (SEM).

An experiment was designed to assess the performance of the vision system for the mixture characterization. Fine metal particles with average size smaller than 2 mm were manually selected from shredded PCBs, sorted based on their colour and size, and placed on two different 5 cm × 5 cm tiles in a 9 by 7 grid configuration (Fig. 5.8) [27].

The particles were also analysed with a scanning electron microscope (SEM) which provided information on their chemical composition. This information was used for the choice of training samples and for the accuracy assessment of classifications. The particles were classified according to the procedure, testing several different combinations of both classification algorithms and methods for the illumination compensation. The best results were achieved using algorithms which take into accounts spectral correlations through a covariance matrix, such as Mahalanobis Distance and Maximum Likelihood [35], and with the method for illumination compensation proposed by Stokman and Gevers [36]. The best classifications showed values of overall accuracy (OA) and kappa coefficient (KC) greater than 95% and 0.95, respectively.

Finally, the developed multi-body particle simulation tools were used to perform off-line optimization of the on-line control algorithms of the separation processes [37]. A virtual inlet was defined to generate a continuous flow of particles in the 3D simulated environment (given input statistical distribution about particle size, density, material and shape, according to the approach that presented in [31]). Gen-

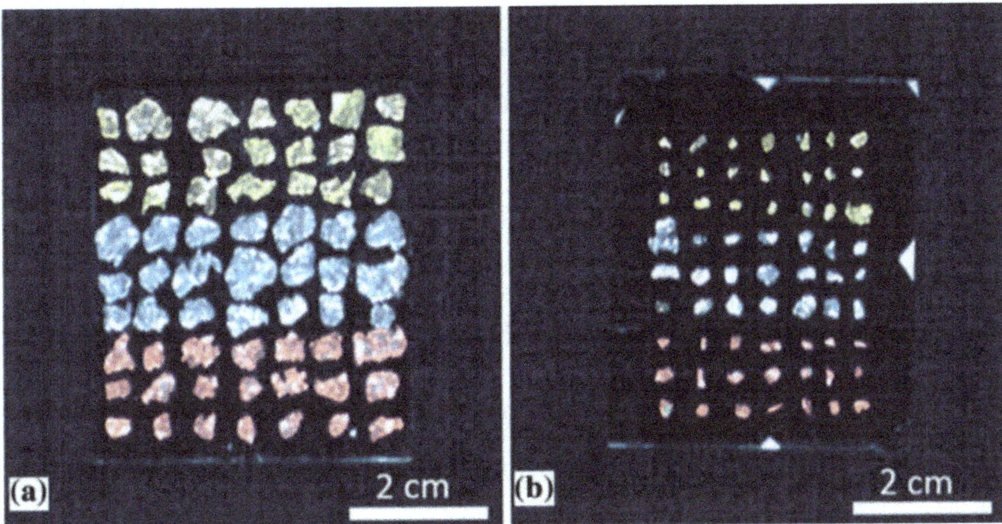

Fig. 5.8 True colours images of shredded PCBs of two selected metal fractions (**a**) and (**b**)

erated particles, in the order of tenths of thousands, underwent the simulated CES or ECS separation and finally the performance of the separation was evaluated by computing the statistical distribution of the ballistic separation in a post-processing stage, similarly to a Montecarlo method. Results of the processed multibody simulations, namely the probability density functions of the lengths of the trajectories for different materials, were discussed in detail in [28, 29] where the comparison against experimental data (sampled by hand using sieves and bins) shows that the multibody model can be used as a digital twin of the real separation process. The simulation of even five seconds of separation might involve large amounts of particles, thus leading to a computational bottleneck that, in the presented scenario, required up to one hour of computation. Therefore new classes of DVI solvers were used to improve the computational efficiency [38–40]. In addition, GPU parallel computing was adopted as discussed in [15].

5.5.3 Business Sustainability

The business sustainability of new flexible recycling technology (with the capacity of 400 kt/year) was simulated considering multiple waste flow that can be optimally treated in different time periods.

Besides PCBs, other selected waste flows were cellular phone, non-hazardous LED lamps and incandescent lamps. Considering material value, within the three types of product, PCB and cellular phone are characterised by higher material value compared to lamps.

The following scenarios were set for waste mix:

- *Optimistic scenario*, waste composed of 45% PCBs, 45% cellular phones and 10% lamps.
- *Base scenario*, waste composed of 33.3% PCBs, 33.3% cellular phones and 33.3% lamps.
- *Pessimistic scenario*, waste composed of 70% lamps, 15% cellular phones and 15% PCBs.

The following scenarios were set for materials price:

- *Optimistic scenario*, material price is 20% higher than material price at the time of the research.
- *Base scenario*, material price remains equal to the material price at the time of the research.
- *Pessimistic scenario*, material price is 20% lower than material price at the time of the research.

The combination of the possible outcomes leads to nine global scenarios. A hypothetic recycling company already operating in the market was taken in consideration to assess the acquisition of the new flexible technologies for business expansion. Investment costs, operation costs and performance parameters (such as system capability, lead time, grade, etc.) were estimated through experimental tests carried out at the CNR-STIIMA (ex CNR-ITIA) pilot plant [41] equipped with the new technologies and relying on data from recycling literature.

Table 5.1 reports the results of the Scenario Analysis for the described industrial case.

As it can be noticed, under the hypotheses of the study, the investment in new flexible technologies able to efficiently treat multiple waste streams appears sustainable in all scenarios (NPV at 10 years ranges from 7,457,343 euro in the pessimistic scenario to 13,502,507 euro in the optimistic one) and is repaid after the first year of operation. These results confirm the high business potential of flexible recycling plants.

Table 5.1 Scenario analysis of the business model validation

			Volume of the mix [euro]		
			Pessimistic scenario	Base scenario	Optimistic scenario
Material price	Pessimistic scenario	NPV(10)	7,457,343	9,155,943	10,362,141
		PBT	1	1	1
	Base scenario	NPV(10)	8,424,074	10,493,888	11,958,429
		PBT	1	1	1
	Optimistic scenario	NPV(10)	9,232,109	11,739,148	13,502,507
		PBT	1	1	1

5.6 Conclusions and Future Research

Circular Economy represents an innovative paradigm for sustainable manufacturing [42]. Relying on technology and business model innovation, the Circular Economy concept aims at maximising the use of products and materials. Recycling is one of the strategies that can be pursued in circular economy, together with re-use and re-manufacturing. Through recycling, critical materials and key-metals can be recovered from post-use products when they can not be re-used or remanufactured.

Recycling systems are nowadays very rigid in their design and they do not allow dynamic adaptation of process parameters to different types of waste. For this reason, the performance of recycling plants is often sub-optimal. With the goal of developing a new generation of flexible recycling systems, new hardware and software solutions were developed in this study for the dynamic adaptation of process parameters to different type of waste. The Hyper Spectral Imaging (HSI) system and the software tool for the control of the sorting parameters based on a multi-body simulation model were demonstrated in a prototype at the De- and Re-manufacturing pilot plant of CNR-STIIMA (ex CNR-ITIA).

Business sustainability of flexible recycling plants based on innovative technical solutions was assessed as well. The selected industrial scenario is the recycling of WEEE, which poses significant challenges because of the high content of critical and key-metals and of the variability generated by rapid technology cycles. A scenario analysis model was developed to preliminary assess the economic sustainability of flexible technologies under the uncertainty deriving from the variability of future waste mixes and the turbulence of material prices. Under the hypotheses of the analysis, the investment in flexible technologies appears sustainable and very promising in all the simulated industrial scenarios, confirming the high business potential of flexible recycling technologies.

The conclusions of this research are aligned with some studies on other waste streams (e.g. in the recycling of construction and demolition waste [43] and of vehicle waste [44]). Results of the scenario analysis support the strategic importance of establishing reconfigurable recycling system due to their higher profitability, which should be achieved through the increase of quality of recovered materials [44].

This study has some limitations. Firstly, developed technologies were tested with a limited number of waste types and the wider capability of real-time adaptation to rapidly-changing waste flows was not extensively tested. Secondly, the economic assessment relied on data coming from literature and on the estimation of some costs derived from experimental tests and pilot infrastructure investment, which could differ from real industrial environment conditions. Further details on this research are reported in [31, 45–47].

This study provided first prototype results towards the introduction of flexible recycling systems in industry through HSI technology for the classification of independent sets of WEEE with high level of accuracy. Future technology developments will be aimed at:

- improving the spectra database, increasing the number of reference spectra both for metals and non-metals;
- addressing the problem of dark particles, which were not considered in this work;
- integrating a wider set of algorithms for materials identification to improve the workflow flexibility and the process performance;
- improving the data elaboration step to speed-up the procedure and enable the real-time/on-line applications.

In addition, the business assessment of new technologies will be improved through the introduction of more accurate cost and revenue structures based on a wider set of experiments and the introduction of a higher number of industrial scenarios.

Acknowledgements This work has been funded by the Italian Ministry of Education, University and Research (MIUR) under the Flagship Project "Factories of the Future—Italy" (Progetto Bandiera "La Fabbrica del Futuro" [48], research projects "WEEEReflex – Highly Evolvable E-waste Recycling Technologies and Systems" and "WEEE ReFlex CPS—Cyber-Physical System (CPS) for reconfigurable e-waste recycling processes". Authors are grateful to Dr. Sarah Behnam for her support to the finalisation of the research and of this publication.

References

1. Copani G, Brusaferri A, Colledani M, Pedrocchi N, Sacco M, Tolio T (2012) Integrated de-manufacturing systems as new approach to End-of-Life management of mechatronic devices. In: Proceedings of the 10th global conference on sustainable manufacturing, Istanbul, 31 Oct–2 Nov 2012
2. Colledani M, Copani G, Tolio T (2014) De-manufacturing systems. In: ElMaraghy H (ed) Procedia CIRP, vol 17, Elsevier, p 14–19
3. Barwood M, Li J, Pringle T, Rahimifard S (2015) Utilisation of reconfigurable recycling systems for improved material recovery from e-waste. Procedia CIRP 29:746–751
4. Wiendahl H-P, ElMaraghy H, Nyhuis P, Zaeh M, Wiendahl H-H, Duffie N, Brieke M (2007) Changeable manufacturing—classification, design and operation. CIRP Ann Manuf Technol 56(2):783–809
5. Koren Y, Heisel U, Jovane F, Moriwaki T, Pritschow G, Ulsoy G, Brussel HV (1999) Reconfigurable Manufacturing Systems. CIRP Ann Manuf Technol 48(2):527–540
6. Terkaj W, Tolio T, Valente A (2009) A review on manufacturing flexibility. In Tolio T (ed) Design of flexible production systems. Springer, p 41–61
7. Terkaj W, Tolio T, Valente A (2009) Focused flexibility in production systems. In ElMaraghy HA (ed) Changeable and reconfigurable manufacturing systems. Springer, p 47–66
8. Tolio T, Ceglarek D, Elmaraghy HA, Fischer A, Hu SJ, Laperriere L, Newman ST, Vancza J (2010) SPECIES-Co-Evolution of products, processes and production systems. CIRP Ann Manuf Technol 59(2):672–693
9. Rahimifard S, Abu Bakar MS, Williams DJ (2009) Recycling process planning for the End-of-Life management of waste from electrical and electronic equipment. CIRP Ann Manuf Technol 58(1):5–8
10. Van Schaik A, Reuter MA (2010) Dynamic modeling of e-waste recycling system performance based on product design. Miner Eng 23(3):192–210
11. Colledani M, Wolf M-I, Gutowski T, Gershwin S-B (2013) Design of material separation systems for recycling. IEEE Trans Autom Sci Eng 10(1):65:75
12. Meacham A, Uzsoy R, Venkatadri U (1999) Optimal disassembly configurations for single and multiple products. J Manuf Syst 18(5):311–322

13. Lu Q, Williams JAS, Posner M, Bonawi-Tan W, Qu X (2006) Model-based analysis of capacity and service fees for electronics recyclers. J Manuf Syst 25(1):45–57
14. Lynch AJ, Elber L (1980) Modelling and control of mineral processing plants. IFAC Proc 13(7):25–32
15. Negrut D, Tasora A, Mazhar H, Heyn T, Hahn P (2012) Leveraging parallel computing in multibody dynamics. Multibody Syst Dyn 27:95–117
16. Critelli I, Degiorgi A, Tasora A, Colledani M (2014) A simulation model of Corona Electrostatic Separation (CES) for the recycling of Printed Circuit Boards (PCBs). Symposium of Urban Mining, Bergamo
17. Tirez K, Vanhoof C, Brusten W, Beutels F, Gilliot C, Spooren J, Debaene L, Umans L (2014) A multi-element determination of critical elements (REE, PGM, …) in solid (waste) materials by SFICP-MS and EDXRF. In: Proceeding of second symposium on urban mining, 19–21 May 2014, Bergamo, Italy
18. Wu D, Shi H, Wang S, He Y, Bao Y, Liu K (2012) Rapid prediction of moisture content of dehydrated prawns using online hyperspectral imaging system. Anal Chim Acta 726:57–66
19. Gowen AA, O'Donnell CP, Cullen PJ, Downey G, Frias JM (2007) Hyperspectral imaging—an emerging process analytical tool for food quality and safety control. Trends Food Sci Technol 18(12):590–598
20. Vermeulen PH, Fernández Pierna JA, van Egmond HP, Dardenne P, Baeten V (2012) Online detection and quantification of ergot bodies in cereals using near infrared hyperspectral imaging. Food Addit Contam Part A 29(2):232–240
21. Gosselin R, Rodrigue D, Duchesne C (2011) A hyperspectral imaging sensor for on-line quality control of extruded polymer composite products. Comput Chem Eng 35:296–306
22. Mauruschat D, Plinke B, Aderhold J, Gunschera J, Meinlschmidt P, Salthammer T (2016) Application of near-infrared spectroscopy for the fast detection and sorting of wood–plastic composites and waste wood treated with wood preservatives. Wood Sci Technol 50(2):313–331
23. Mirzapour F, Ghassemia H (2015) Improving hyperspectral image classification by combining spectral, texture, and shape features. Int J Remote Sens 36 (4):1070–1096
24. Picón A, Ghita O, Bereciartua A, Echazarra J, Whelan PF, Iriondo PM (2012) Real-time hyperspectral processing for automatic nonferrous material sorting. J Electron Imaging 21(1):013018 (SPIE and IS&T)
25. Koyanaka S, Kobayashi K (2011) Incorporation of neural network analysis into a technique for automatically sorting lightweight metal scrap generated by ELV shredder facilities. Resour Conserv Recycl 55:515–523
26. Ferrari C, Foca G, Calvini R, Ulrici A (2015) Fast exploration and classification of large hyperspectral image datasets for early bruise detection on apples. Chemometr Intell Lab Syst 146:108–119
27. Picone N, Candiani G, Colledani M, Pepe M (2016) Fine mixture characterization by hyperspectral imaging (HSI) in WEEE demanufacturing plants. In: Proceeding of 3rd symposium on urban mining (SUM), Bergamo, 23–25 May 2016
28. Critelli I, Tasora, A, Degiorgi A, Colledani M (2014) Particle simulation of granular flows in electrostatic separation processes. In: SIMUL 2014 sixth international conference on advances in system simulation, Nice, 12–16 October 2014
29. Critelli I, Degiorgi A, Colledani M, Tasora A, (2014) A multi-body simulation model for a corona electrostatic separator machine. In: ECT2014 the ninth international conference on engineering computational technology, Naples, 2–5 September 2014
30. Critelli I, Degiorgi A, Colledani M, Tasora A (2014) A simulation model of Corona Electrostatic Separation (CES) for the recycling of Printed Circuit Boards (PCBs). In: SUM 2014 second symposium on urban mining, Bergamo, 19–21 May 2014
31. Tasora A, Critelli I, Mazhar H (2015) A parametric approach to the generation of multidisperse granular flows for particle simulations. In: ECCOMAS multibody dynamics 2015, Barcelona, 29 June–2 July 2015
32. Copani G, Rosa P (2014) DEMAT: sustainability assessment of new flexibility-oriented business models in the machine tools industry. Int J Comput Integr Manuf 28(4):1–10

33. Ongondo FO, Williams ID, Cherrett TJ (2011) How are WEEE doing? A global review of the management of electrical and electronic wastes. Waste Manag 31:714–730
34. Das A, Vidyadhar A, Mehrotra SP (2009) A novel flowsheet for the recovery of metal values from waste printed circuit boards. Resour Conserv Recycl 53:464–469
35. Richards JA (1999) Remote sensing digital image analysis, vol 3. Springer, Heidelberg
36. Stokman HM, Gevers T (1999) Detection and classification of hyper-spectral edges. In: Proceedings of the tenth British machine vision conference (BMVC), Nottingham, 13–16 September 1999
37. Colledani M, Critelli I, Diani M, Tasora A (2015) A computer-aided methodology for the design of de-manufacturing process for waste recycling. In: Paper presented at international CAE conference, Pacengo del Garda, Verona, 19–20 October 2015
38. Negrut D, Serban R, Tasora A (2017) Posing multibody dynamics with friction and contact as a differential complementarity problem. J Comput Nonlinear Dyn
39. Heyn T, Anitescu M, Tasora A, Negrut D (2013) Using Krylov subspace and spectral methods for solving complementarity problems in many-body contact dynamics simulation. Int J Numer Methods Eng 95(7):541–561
40. Mangoni D, Tasora A, Garziera R (2018) A primal-dual predictor-corrector interior point method for non-smooth contact dynamics. Comput Methods Appl Mech Eng 330:351–367
41. Tolio T, Copani G, Terkaj W (2019) Key research priorities for factories of the future—part II: pilot plants and funding mechanisms. In: Tolio T, Copani G, Terkaj W (eds) Factories of the future. Springer
42. Tolio T, Copani G, Terkaj W (2019) Key research priorities for factories of the future—part I: missions. In: Tolio T, Copani G, Terkaj W (eds) Factories of the future. Springer
43. Duran X, Lenihan H, O'Regan B (2006) A model for assessing the economic viability of construction and demolition waste recycling—the case of Ireland. Resour Conserv Recycl 46(3):302–320
44. Farel R, Yannou B, Ghaffari A, Leroy Y (2013) A cost and benefit analysis of future end-of-life vehicle glazing recycling in France: a systematic approach. Resour Conserv Recycl 74:54–65
45. Candiani G, Picone N, Pompilio L, Pepe M, Colledani M (2017) Characterization of fine metal particles derived from shredded WEEE using a hyperspectral image system: preliminary results. Sensors 17(5):1117
46. Candiani G, Picone N, Pompilio L, Pepe M, Colledani M (2015) Characterization of fine metal particles using hyperspectral imaging in automatic WEEE recycling systems. In: Paper presented at IEEE workshop on hyperspectral image and signal processing: evolution in remote sensing, Tokio, 2–5 June 2015
47. Tasora A, Critelli I (2015) A method for advanced stochastic generation of particles in granular flow simulations and its practical applications in industry. In: Paper presented at international CAE conference, Pacengo del Garda, Verona, 19–20 October 2015
48. Terkaj W, Tolio T (2019) The Italian flagship project: factories of the future. In: Tolio T, Copani G, Terkaj W (eds) Factories of the future. Springer

Part III
Sustainable Factory

Chapter 6
Innovative and Sustainable Production of Biopolymers

Simona Ortelli, Anna Luisa Costa, Cristian Torri, Chiara Samorì,
Paola Galletti, Claudia Vineis, Alessio Varesano, Luca Bonura
and Giacomo Bianchi

Abstract This work aims to develop methodologies and tools to support the design
and management of sustainable processes for the production of biodegradable poly-
hydroxyalcanoates (PHAs) biopolymers. PHAs are linear polyesters produced in
nature by bacteria through aerobic fermentation of many carbon sources, completely
biodegradable and biocompatible. We carried out a study inherent to the advancement of an innovative, cost-effective and environmentally sustainable technology
for isolating PHAs from bacteria mixed cultures by combining: (a) innovative cells'
pre-treatments and polymer purification's strategy by means of TiO_2/UV or Ag^0
nanostructured materials; (b) polymer extraction through a green and safe system
directly applicable to bacterial cultures, which combines the advantages of solvent
extraction and these of dissolution of the non-PHAs cellular matrix through surfactants; (c) monitoring and control tools for process energy and efficiency management.
The outcomes put the basis for the design and subsequent building of a working pilot
system for the production of completely biodegradable and biocompatible PHAs. The
efficiency can be improved and the investments and operating costs can be decreased
thanks to the optimization of the production process with the introduction of safe
and cheap PHAs extraction route without use of toxic and harmful chemicals and
the integration of monitoring and automation tools. The engineering and integration
of nano-TiO_2 phase within textile fibres and their use as photocatalytic active media
for bacteria pre- and post-treatment of waste water added a new opportunity for
improving process efficiency and sustainability.

S. Ortelli (✉) · A. L. Costa
CNR-ISTEC, Istituto di scienza e tecnologia dei materiali ceramici, Faenza, RA, Italy
e-mail: simona.ortelli@istec.cnr.it

C. Torri · C. Samorì · P. Galletti
Dipartimento di Chimica, Università di Bologna, Campus di Ravenna, Ravenna, Italy

C. Vineis · A. Varesano
CNR-ISMAC, Istituto per lo studio delle macromolecole, Biella, Italy

L. Bonura · G. Bianchi
CNR-STIIMA, Istituto di Sistemi e Tecnologie Industriali Intelligenti per il Manifatturiero
Avanzato, Milan, Italy

© The Author(s) 2019
T. Tolio et al. (eds.), *Factories of the Future*,
https://doi.org/10.1007/978-3-319-94358-9_6

6.1 Scientific and Industrial Motivations

Plastic materials from non-renewable feedstock (e.g. petroleum) have a crucial and undeniable role in our everyday life. However, the chemically engineered durability and the slow rate of biodegradation of most of the fully fossil-based plastics [e.g. polyethylene (PE), polyethylene terephthalate (PET), polypropylene (PP), polyvinyl chloride (PVC)], enable these synthetic polymers to withstand the ocean environment and terrestrial ecosystems for several years, thus affecting organisms at multiple trophic levels [1]. Polyhydroxyalkanoates (PHAs) are bio-based biodegradable polyesters produced by aerobic bacteria from various carbon sources. According to the carbon source, different monomers are produced, whose combination provides different chemo-physical properties. Thus, PHA can have properties ranging from polypropylene to polyethylene, being at the same time biodegradable and bio-based. This brings the interest towards the application of PHAs as packaging films and containers, biodegradable carriers for controlled chemical and/or drug release, disposable items, surgical pins and sutures, wound dressings, and bone replacements [2]. With respect to other bioplastics which are already commercially available and currently produced on industrial scale, e.g. polylactic acid (PLA) or starch-based polymers (Mater-Bi®), PHAs are still not widely used because of the high production costs, more expensive than usual petrochemical plastics (~5–6 vs. 0.5–2 \$/kg) [3]. The high costs' production is mainly due to a combination of three factors: (i) absolutely sterile operation conditions are necessary for pure microbial fermentation; (ii) the use of expensive substrates as carbon source which represents approximately 40% of total production costs; (iii) the extraction and purification process for obtaining PHAs. Thus, the commercialization of PHAs is now limited to applications with high added value. Given the need to reduce costs, a multilevel approach to tackle all bottlenecks is mandatory to develop more cost-effective processes for PHAs production. In principle, the use of microbial mixed cultures (which do not require sterility) combined with the utilization of low value substrates, as agro-industrial waste and by-products, and with safe and cheap extraction and purification methodologies would allow lower investments and operating costs for the whole production process.

This work aimed to develop a new, cost-effective, and environmentally sustainable technology for isolating PHAs from bacteria mixed cultures, by combining:

1. the use of green carbon sources and microbial mixed cultures;
2. a strategy for innovative bacteria cells' pre-treatments and polymer purification by means of TiO_2/UV or Ag^0 nano-phases supported on process compatible substrates;
3. a polymer extraction methodology through a green and safe system based on the use of non-volatile organic solvents;
4. automation tools to minimize human operator cost and improve production efficiency through management strategies adapted to the natural variability of substrate and bacteria conditions.

The current technology limitations are tackled by optimizing the main process steps: microbial fermentation for PHAs purification, biomass pre-treatment, PHAs extraction and purification. An effort to implement the whole process was done by improving the monitoring and automation tools. Finally, on the basis of the results obtained by the process optimization, the study of a scaled-up prototype was performed leading to the realization of PHAs production semi-pilot plant as well as the construction of photocatalytic reactors for bioreactor wastewater purification with up to 100 L capacity.

6.2 State of the Art

Polyhydroxyalkanoates (PHAs) are bio-based and biodegradable polyesters produced by aerobic bacteria from various carbon sources. These polymers have tuneable elastomeric/thermoplastic properties according to the monomer composition. Despite the high potential and the efforts to put forward the development of cost effective fermentative systems, the PHA production costs remain considerably high (~5–6 $/kg), thus hampering the exploitation of these biopolymers as commodity materials [4]. The main limitations could be removed by:

- changing the cultivation approach, thus using mixed microbial cultures (MMCs) instead of pure cultures;
- using cheap carbon sources instead of selected and pure substrates;
- improving downstream steps (extraction/purification) using green and sustainable extraction tool aids instead of toxic and not-environmental friendly solvents and additives.

Mixed microbial population (MMCs) can be exploited for producing PHAs in a cheaper and more sustainable way [5]. In fact, compared to single strains (e.g. *Cupriavidus necator* or genetically modified *Escherichia coli*), MMCs do not require sterile conditions: this drastically reduces equipment costs and allows the exploitation of many cheaper substrates (e.g. wastes) thanks to the wider metabolic potential of MMCs in comparison to single strains such as waste biomass (e.g. fermented molasses), wastewater and pyrolysis products. Currently, in the literature there are few publications on PHAs recovery from MMCs, since the research effort is mainly focused on the set-up of efficient and sustainable cultivation systems. Little information is available on the characterization of PHAs extracted from MMCs and, above all, on the efficiency of the polymer recovery in the extraction step [6]. As it is well described in the literature for single strain cultures, obtaining the polymer from bacteria through a series of downstream steps (e.g. microbial biomass pre-treatment, polymer extraction, and post-treatment purification), could be challenging [7]. Indeed, it is reported that the downstream costs can represent up to 50% of the total production cost because non-recyclable chemicals/materials and high energy demands are required. However, since MMCs are more resistant than single strains, the extraction/purification of PHA can become challenging. The protocols reported in the literature [8, 9] for the extraction of PHA from MMCs include (i) an optional cell pre-treatment under acidic conditions followed by a solvent extraction (e.g. with acetone or chlorinated compounds) to dissolve the polymer or (ii) a treatment with additives that break microbial membranes enabling the release of the polymer inside the cells. The use of organic solvents helps in extracting high molecular weight PHAs but with a low yield (18–30%), whereas strong oxidants decrease the molecular weight but provide high extraction yields (close to 100%). Feed-back control strategies have already been studied for fed-batch systems, mainly to determine on-line when to feed the next pulse of substrate in the production reactor. Continuous feeding is also possible but it is used most successfully with single substrates and high cell densities [10]. Whiffin et al. [10] have proposed an algorithm that alternates continuous feeding and famine periods in order to prevent acetate accumulation and allow shorter feedback control intervals. The dissolved oxygen (DO) is closely related to the substrate concentration so that when the substrate is exhausted and oxygen consumption decreases, the DO level increases rapidly, suggesting to exploit the DO signal to control pulse feeding in the fed-batch [11, 12]. Westerberg [12] showed that inhibition of the biomass by the stored polymer is an important factor reducing the substrate and oxygen uptake rates and modelled the DO response to repeated substrate pulses. Another recent approach determines when to feed a pulse of substrate by using it as a control variable to regulate the pH [13]. Popova [14] designed a Kalman filter to estimate unmeasured variables in a mixed culture system, based on measurements of the substrate concentration. Vargas et al. [15] have explored the use of a model-based control scheme to enhance the productivity of polyhydroxyalkanoate (PHA) production in a mixed culture two-stage system. Oehmen et al. [16] operated a three-stage bioreactor with anaerobic fermentation, culture selection and PHA accumulation. Some adopted sensory systems, e.g. on-line gas chromatography in [17], strongly affect plant economic sustainability. In this work, a plant supervisor and controller has been developed exploiting the actual trend toward intelligent and interconnected production systems that integrate low cost distribute processing.

6.3 Problem Statement and Proposed Approach

A systematic analysis and consequent improvement strategies of the PHAs production process is needed to reduce costs and limit the use of hazardous solvents for extraction and purification technologies. The main actions to enhance sustainability and cost-effectiveness are schematised in Fig. 6.1 and better described in the remaining part of this section.

Wastes such as starch, cellulose and hemicellulose deriving materials, plant oils, molasses, whey, and various industrial by-products have been proposed as potentially useful and cheap carbon sources. As a matter of fact, a wide range of agriculture and forestry wastes can be converted into carbon sources for microbial fermentation if coupled with suitable pre-treatments like thermochemical conversions (e.g. pyrolysis or hydrothermal liquefaction) or acidogenic fermentation steps. These specific pre-treatments allow to convert substrates which could not be directly utilized by

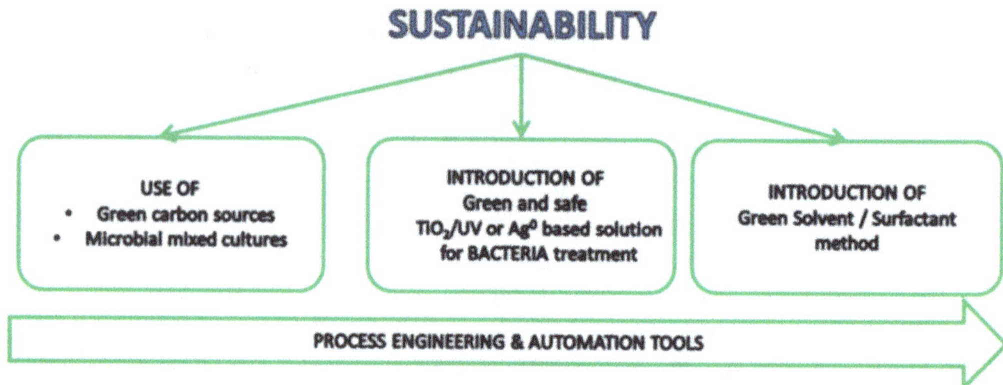

Fig. 6.1 Sustainability tools proposed and developed

microorganisms to produce PHAs, into PHAs chemical building blocks (volatile fatty acids, VFA), with a consequent abatement of the costs related to the carbon source and the accessibility to a widespread pool of substrates without any competition with food resources.

As regards microbial fermentation for PHA production, a feast (carbon excess) and famine (substrate exhaustion) approach was followed for growing MMCs. During the feast phase, bacteria able to accumulate PHA are selected since during the famine phase PHA can be exploited as energy and carbon source.

Highly skilled workforce for monitoring the system state and solving possible problems are required to ensure the achievement of satisfactory yield within an ecologically stable process. For this reason, a key aspect is to develop a complete monitoring of the process variables (physical and biological) with programmable logic controller that can readily react to the change observed (e.g. natural oscillation of the bacteria population, bulking problems, clogging) providing feedback in completely automated way and without the intervention of skilled labour. Therefore, the design and manufacturing of a compact all in one system able to convert a specific waste into ready-to-use bioplastic has been conceived. Another aspect investigated is the substitution of standard equipment series (usual approach of chemical plant) with one that can be tuned to provide multi-purpose function (e.g. bacteria growth, harvesting and extraction in the same apparatus), in order to reduce the cost (per unit of output).

As for the biomass pre-treatment, the challenge is the achievement of satisfactory yield in the PHAs extraction process despite the higher resistance of MMC than pure culture. For this reason, after the fermentation step, the bacterial broth is usually pre-treated by means of a freezing/heating approach or salt/alkaline treatment, aiming at denaturing genetic material and proteins, and thus enhancing the cell disruption and polymer recovery.

The novelty, at this point, is the application of a nanoparticles (NPs)-based pre-treatment step to weaken bacterial membranes without degrading the polymer stored inside the cells, exploiting photocatalytic TiO_2 NPs and antibacterial Ag^0 NPs. This pre-treatment requires lower energy consumption than the traditional approaches (freezing and heating) and a lower consumption of chemicals by using supported recyclable nanomaterials.

The PHAs extraction from cultures can be done by adopting a solvent extraction approach (mainly with chlorinated solvents, highly effective in solubilizing PHA granules) or a cellular-digestion approach thus breaking cellular constituents apart with additives. The final polymer purity is higher when solvents are used, however both approaches present some environmental/economic issues such as toxicity and non-recyclability. A new and green extraction process with the aim of increasing the sustainability of the entire process was designed through: (i) an abatement of the use of toxic and hazardous chemical compounds by using green solvents; (ii) an increase in the recyclability of the chemicals used in the process and a reduction in downstream flows which must be treated; (iii) an improvement in the polymer quality ensuring the retention of mechanical properties.

Moreover, PHAs purification was required since MMCs are composed of gram-negative bacteria that can release endotoxins in the form of lipopolysaccharides, from the outer cell wall. Endotoxin can cause several diseases if introduced into the bloodstream of humans or other animals, and for this reason, in view of biomedical applications, the extracted polymer must be purified. Common methods of purification involve sodium hypochlorite or hydrogen peroxide treatment combined with the action of enzymes or chelating agents. An alternative procedure was taken into consideration for the treatment of the biopolymer contaminated by dangerous cellular residues, based on heterogeneous photocatalysis, involving photoinduced redox reactions at the surface of semiconductor ceramics. Furthermore, this approach can reduce the use of compounds that could affect the quality of the resulting biopolymer and the costs of the purification process thanks the application of a mature technology in the field of organic compounds detoxification processes.

6.4 Developed Technologies, Methodologies and Tools

On the basis of the sustainability principles, we ad hoc designed two differently sized plants. As starting point, we developed a working lab-scale system (1 g PHA/day) to produce completely biodegradable and biocompatible PHAs biopolymers from volatile fatty acids (VFA) obtainable from different waste sources.

Subsequently, the lab scale prototype was scaled up to 100 g PHA/day demonstration micro-plant. The system was operated for 150 days showing reliable operations with minimal maintenance. The extraction system showed to be effective in obtaining high purity PHA from bacterial slurries with acceptable yield. PHA can be recovered from MMCs by using dimethyl carbonate (DMC) or DMC plus a pre-treatment with sodium hypochlorite (NaClO). In the first case an overall polymer recovery of 66% can be achieved, with a very high polymer purity and molecular weight (98% and 1.2 MDa); in the second case, NaClO helps in increasing the recovery (82%) but decreases the molecular weight to 0.6–0.2 MDa.

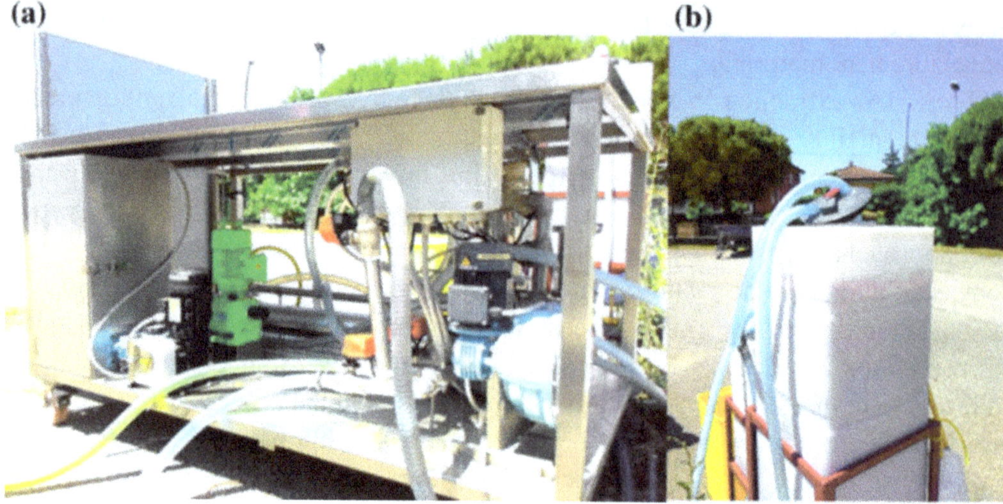

Fig. 6.2 Plant picture during operation, **a** pumps and controller and **b** the bioreactor

The scaled-up prototype consists of a module formed by two pieces of equipment that can be added downstream to a wastewater treatment plant or to a PHA producing bioreactor. The module is composed of (i) a biological reactor which allows the production of a PHA rich sludge (25–40 g_{PHA}/kg_{slurry}) and (ii) a complete dimethyl carbonate (DMC) extraction system able to provide high purity PHA (Fig. 6.2).

The micro-plant controller has been designed in order to assure autonomous operations, with local data logging, and to support remote supervisory interaction via a standard Internet browser. In order to deliver such functionalities with a limited plant cost, a solution based on open source micro-controllers has been developed. Given the required internet connectivity and control logic, the Arduino TIAN™ micro-controller has been chosen. The control architecture is depicted in Fig. 6.3. Different types of sensor are foreseen for each plant tank: Dissolved Oxygen (DO) and pH sensors (connected via an I2C digital bus), temperature sensors (connected via One-wire digital bus). Digital buses allow the addition, for future developments, of additional sensors, identified by a unique address on the same bus line, without the need to upgrade the controller input/outputs. Plant valves are controlled via 18 digital outputs, whereas plant discrete states are read via 20 digital inputs. The Arduino TIAN has been selected because it joins a deterministic SAMD21 Atmel Cortex® M0+ 32-bit microcontroller (MCU) with 32 KB of SRAM program memory and a Qualcomm Atheros AR9342 processor running a Linux operative system (OpenWrt) that supports networking capabilities. A user interface based on a web page, that shows the plant state and allows the user to enter all required reference values and commands, has been implemented by exploiting Arduino libraries for web server creation. Given that web server functionalities are still not available on the Arduino TIAN two alternative solutions have been developed for the user interface. The first one via a web browser, using an additional Arduino YUN, connected to the TIAN, acting as web server, and the second one where the user interacts with the TIAN via a serial connection to a local micro-PC, located in the plant electric cabinet.

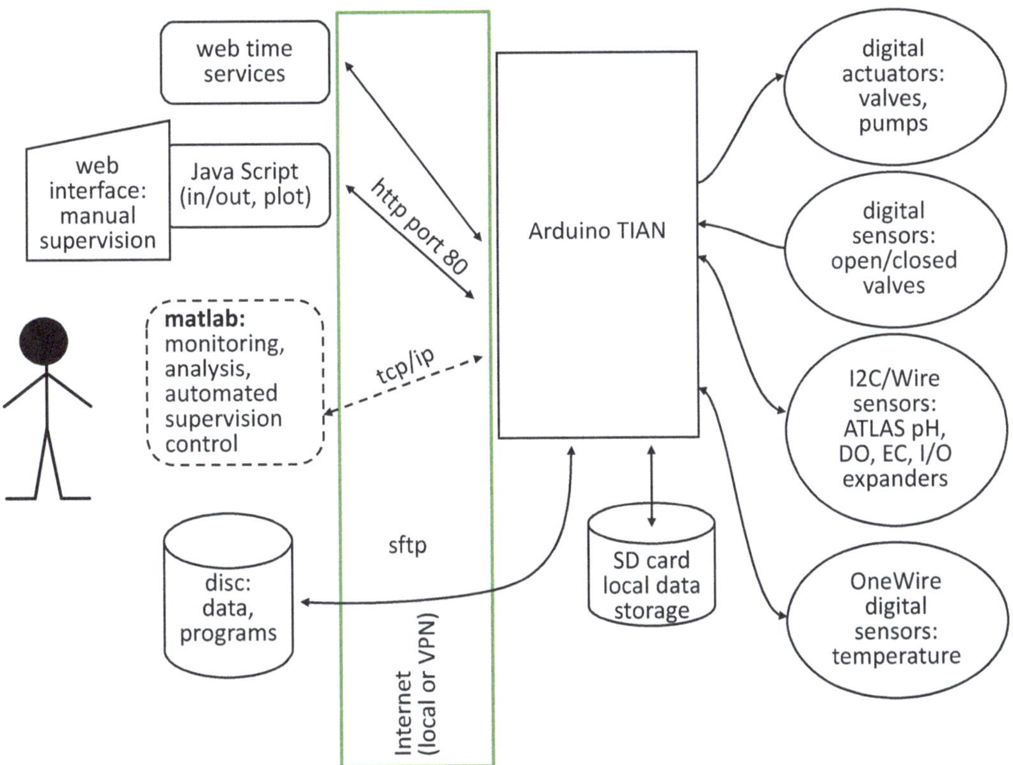

Fig. 6.3 Plant controller architecture: version with internet connection

The controller is implemented as a Discrete-State Controller. For each possible state of the plant, a control state is defined. When the state is activated, all requested control actions are performed (e.g. opening/closing valves, running a pump), then the associated activities (e.g. monitoring the solution DO or pH or counting a predefined time lap) are performed until the exiting conditions are satisfied and a new state become active. The defined states are: VFA dosing to R1 (SBR), Nutrients dosing to R2, Dilution with tap water, Mixing, VFA dosing to R2 (Accumulation Reactor), Mixing, Sedimentation, Collect the biomass, Remove excess water, Backflush for filter cleaning, and Standby. In a concurrent, parallel, control loop, the reactors are temperature controlled by reading their temperatures and switching on/off the corresponding heaters.

6.5 Experiments

The main experimental activities were carried out to:

- optimize the PHA production (Sect. 6.5.1) and the PHA extraction (Sect. 6.5.2);
- design and prepare TiO_2 and Ag^0 based nanostructured materials (Sect. 6.5.3) suitable to be applied during the pre- and post-bacteria treatments (Sect. 6.5.5) and purification wastewater (Sect. 6.5.6);

- investigate the opportunity to develop model-based observers able to estimate, from the available feedback signals, crucial quantities (like bacteria mass and their activity, polymer concentration) that are not directly measurable.

6.5.1 Optimization of PHA Production

Activated sludge from a municipal wastewater treatment plant was used as inoculum for a lab-scale sequencing batch reactor (SBR, 5 L). The culture was fed with synthetic organic acids (acetic and propionic acids) and subjected to 12 h cycles each composed of an initial feed phase, an aerobic reaction phase, a sedimentation phase, and a withdrawal of depleted water. After that the microbial culture was sent to an accumulation reactor (working volume of 1.5 L) and fed with synthetic organic acids (acetic and propionic acids) for increasing the amount of PHA inside the cells [9]. After 3-months, this cultivation system provided a microbial biomass with 40 ± 5 wt% of polymer content, with a 3HB and 3HV molar ratio of 76 and 24, respectively.

6.5.2 Optimization of PHA Extraction from MMC

Various pre-treatments (chemical, thermal, mechanical) were explored for increasing the extraction efficiency of DMC [18]. So far the extraction of PHAs from MMCs with organic solvents has been accomplished mainly with chlorinated compounds or a mixture of chlorinated and not chlorinated solvents. Patel et al. reported that the extraction efficiency of organic solvents reaches a maximum of 30%, against an almost complete polymer recovery from single strains [19].

The results achieved with various extraction systems are summarized in Table 6.1. DMC and CH_2Cl_2 provided the same recovery efficiency (49 ± 2 and $52 \pm 1\%$, respectively, corresponding to PHA extraction yields of 20 ± 1 and 22 ± 1 wt%).

Among the pre-treatments, both the thermal (temperature > 150 °C) and the mechanical (glass beads and sonication) ones did not improve the DMC extraction performance. However, a pre-treatment with NaClO applied before the extraction with DMC increased PHA recovery up to $76 \pm 4\%$ (NaClO for 5 min at 100 °C). Harsher oxidizing/bleaching conditions (NaClO treatment at 100 °C for 1 h) yielded the highest recovery equal to $82 \pm 3\%$.

Very good polymer purities were obtained after the treatment with DMC or CH_2Cl_2 (entries 1 and 3 in Table 6.1, 98 and 94%, respectively). The purity of PHA recovered through DMC after a pre-treatment with NaClO was generally high (88–98%). Promising results in PHA extraction from MMCs are reported in [20].

Table 6.1 Physical characteristics of PHAs extracted from MMCs biomass

Entry	Treatment	Purity (%)[a]	$T_{max\ deg}$ (°C)	$\overline{M_w}$ (MDa)	PDI
1	DMC (1 h@90 °C)	98	254	1.3	1.9
2	DMC (24 h@soxhlet)	87	280	0.5	2.9
3	CH_2Cl_2 (4 h@50 °C)	94	246	1.4	2.0
4	NaClO (1 h@100 °C)	77	269	0.3	2.7
5	NaClO (5 min@rt) then DMC (1 h@90 °C)	92	275	0.6	2.6
6	NaClO (15 min@rt) then DMC (1 h@90 °C)	98	270	0.8	2.4
7	NaClO (1 h@rt) then DMC (1 h@90 °C)	88	238	0.6	2.3
8	NaClO (5 min@100 °C) then DMC (1 h@90 °C)	89	262	0.5	2.8
9	NaClO (15 min@100 °C) then DMC (1 h@90 °C)	92	281	0.5	2.4
10	NaClO (1 h@100 °C) then DMC (1 h@90 °C)	93	281	0.2	2.5

[a]evaluated by TGA mass loss intensity The data are expressed as mean ± standard deviation of four independent replicates of each extraction condition ($T_{max\ deg}$: maximum decomposition temperature, $\overline{M_w}$: mean molecular weight; PDI: polydispersity index)

6.5.3 Design and Prepare TiO$_2$ and Ag^0 Based Nanostructured Materials

To prepare TiO$_2$- and Ag-coated textiles, standard cotton textiles were used without any pre-treatment, while standard polyamide textiles were made more hydrophilic and the affinity for the nanosols was improved, by dipping samples in flasks containing ethanol (80% v/v) and demineralised water at 70 °C for 30 min. The liquor ratio was 80:1 ml/g (volume of solution to weight of fabric). The samples were dried at room temperature. The functionalization of textile was performed by dip-dapping-curing method. Standard cotton and pre-treated polyamide textile were dipped at room conditions (temperature = 21°C and humidity of air = 45%) in the titania (1.5 and 3 wt%) or silver nanosol (0.05%). Fabrics soaked for three minutes are then passed through a two-roller laboratory padder. The resulting coating is fixed on the surface after drying it in an oven at 100 °C and curing at 130 °C for 10 min. The excess of nanoparticles not adsorbed on the surface is removed by water washing coated fabrics in an ultrasound bath for 15 min and then drying at air. This process is called post-washing treatment.

TiO$_2$- and Ag-coated textile samples were used to carry out antibacterial tests and preliminary tests in real experimental condition during pre-treatment step.

Table 6.2 Bacterial reduction % on *Klebsiella pneumoniae* and Mixed microbial culture using TiO_2- and Ag-coated fabrics

Bacterial reduction (%)	Growth	Untreated	TiO_2 1.5 wt%	TiO_2 3 wt%	Ag 0.05 wt%
Klebsiella pneumoniae (Gram negative)	24 h, 37 °C	20	77	77	100
Mixed microbial cultures	48 h, 27 °C	50	50	80	50

6.5.4 Antibacterial Tests

Antibacterial tests against pure microbial cultures of *Klebsiella pneumoniae* (ATCC 4352, Gram negative) and mixed microbial cultures (AATCC 100 Test Method) on different functionalized textile by silver and titanium dioxide were performed. Bacterial reduction is calculated using Eq. (6.1), where A is the number of colonies of the fabrics with titanium dioxide or silver, and B is the number of colonies of the untreated samples (references).

$$\% \text{ bacterial reduction } = \frac{B - A}{B} \cdot 100 \tag{6.1}$$

These antibacterial tests were carried out on both cotton and polyamide fabrics and the results are reported in Table 6.2.

The antibacterial tests against *Klebsiella pneumoniae* showed fairly good data with UV-activated TiO_2-coated samples, while excellent results were obtained by Ag-coated fabrics. In general, the antibacterial activity shown by our coated textile samples seems insufficient to obtain a good bacterial reduction. However, bacteria membrane alteration could be detected as irregular profile as in the spot highlighted. The preliminary observation by SEM of membrane morphology after exposure to photocatalytic component was made, nevertheless further investigation needed to experimentally verify their effects on bacteria membranes degradation. The establishment of a physical correlation between exposure to photocatalytic agents and the direct effects on bacteria cellular membrane would support their integration within PHA bio-reactors, facilitating the performance evaluation of bacteria pre- and post-treatments.

6.5.5 Pre- and Post-treatments with TiO_2 Textiles

The biomass slurry was too much *dense* and it prevented the UV-light to pass through. Unfortunately, the recovery of PHA after the pre-treatment was similar to the results achieved without any treatment (~50%). In post-treatment, PHA (standard or extracted from bacteria, 10 mg) was re-suspended in water and put in contact with a disk of TiO_2-textile (20 mg) in order to remove any residual fraction of bacterial charge. The samples were irradiated for 1 or 2 h under UV.

In this case we experienced an unexpected and detrimental absorption of PHA on the textile that caused a loss of polymer after the treatment.

6.5.6 Purification of Wastewater

Despite the developed photocatalytic tools showed a high reactivity against pure bacteria and fair results against MMC, some difficulties in their integration within PHAs production process were found. Thus, as an alternative to the use of catalytic tools in pre-treatments and polymer purification steps, the developed TiO_2 coated textiles were exploited in purification step of wastewaters, originated from PHAs production process. These wastewaters contain mainly organic pollutants, which could be easily degraded by UV activated nano-TiO_2 based photocatalytic materials. Thus, with the aim to purify wastewater and controlling consuming and environmental sustainability, the integration of TiO_2 based photocatalytic nanostructured surfaces was investigated as final purification step of PHA production process. The photodegradation of organic pollutants catalysed by nano-TiO_2 occurs after the absorption of near-UV light by the surface that induces charge separation and the formation of electron (e^-) and positive hole (h^+) pair. This charge displacement at contact with water and air promotes the formation of reactive oxygen species (ROS) like hydroxyl and singlet oxygen radicals, responsible for the oxidative degradation of organic molecules [21]. As many papers demonstrate [22–25], photocatalytic oxidation is a promising approach in the purification of waters contaminated with organic pollutants.

In this work, the purification treatment of wastewaters with nano-TiO_2 coated textiles was studied. The organic loading abatement of wastewater using photocatalytic activity was tested through TOC (total organic carbon) determination. In general, good photocatalytic results were found, in lab-scale systems (see Fig. 6.4). Specifically, with two hours of UV irradiation, an excellent TOC reduction was measured from the initial amount of organic C (about 0.45 g/L) to about 0.06 g/L (Table 6.3).

6.5.7 Automation Tools for Process Energy and Efficiency Management

The controller has been developed on the Arduino platform, in the different versions described in Sect. 6.4. Given the limitations of the available software libraries for the Arduino TIAN, the Serial User Interface was preferred: the full web interface will be supported by future versions of the Arduino development environment (IDE).[1] The adopted software architecture allows a simple addition of new control states and

---

[1] www.arduino.cc.

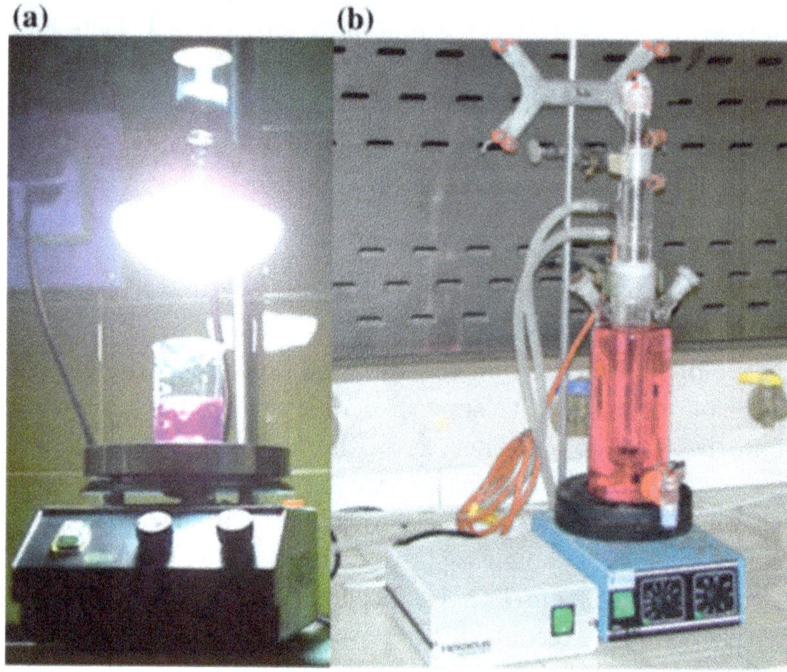

Fig. 6.4 Nano TiO$_2$ coated textile introduced as floating membrane, under UV irradiation, in **a** lab-scale system and **b** lab-scale (1 L capacity) plant

Table 6.3 Photocatalytic test performed with bioreactor wastewater in lab-scale system

Sample	Total C (ppm)	Inorganic C (ppm)	Organic C (ppm)
Test 1—control	579.94	475.05	104.9
Test 2	622.3	466.1	156.2
Test 4—control	809.36	641.15	168.22
Test 5	451.88	267.86	184.02
Test 7—control	551.98	434.09	117.89
Test 8	508.5	444.42	64.08[a]
Waste water	1371.24	912.98	458.26

[a]the results show a photocatalytic efficiency up to 85% with a COD reduction from 450 a 60 mg/L in the condition of Test 8 (2 h of UV irradiation)

strategies: each state can access the whole data structure. The controller reads the inputs and proposes updated control actions by a timing defined by the user, down to 5 s. All plant data are transmitted via the Serial interface and saved by the connected micro-computer.

6.5.8 Scale-up of Photocatalytic Advanced Oxidation Process

The development and scale-up of photocatalytic plants designed for treating downstream PHA waste water (PHAWW) was carried out to decontaminate and recycle the bioreactor wastewater, thus adding value to the whole process. A pilot-scale UV/TiO$_2$ photoreactor with capacity of 1 L/day of PHAWW was assembled (Fig. 6.4b), using lab scale system (Fig. 6.4a) as model and validated it experimentally, in order to obtain data for the evaluation of industrial scale-up. A further effort was made in the scale-up, achieving semi-pilot plant (Fig. 6.5a) and pilot plan (Fig. 6.5b) with 6 and 100 L capacity, respectively. Two main configuration systems: at fluidized (Fig. 6.4) and fixed bed (Fig. 6.5), were proposed in order to improve the versatility and provide the real environment where testing photocatalytic components in the advanced oxidation treatment of waste water.

Encouraging results, in terms of photocatalytic efficiency, were obtained using different TiO$_2$-coated fabric as active membrane in semi-pilot plant (Table 6.4). Further study has been carried out to improve the photocatalytic efficiency in semi-pilot fixed bed system [26].

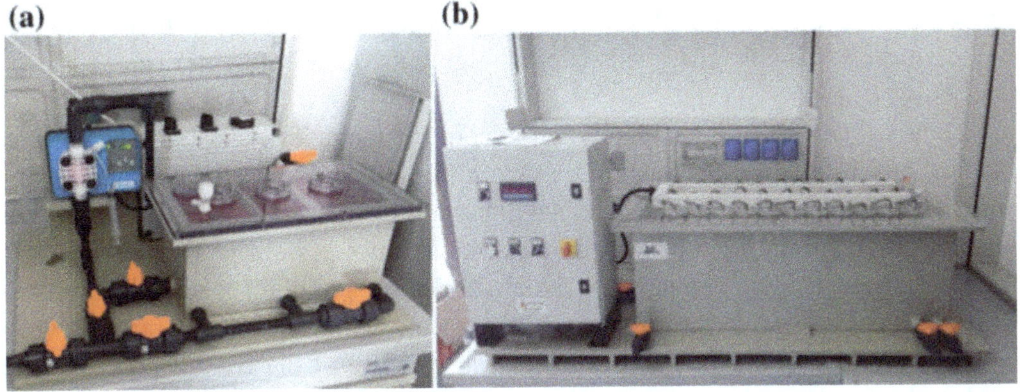

Fig. 6.5 Nano TiO$_2$ coated textile introduced as permeable fixed membranes irradiated by immersed UV lamps, in **a** semi-pilot plant, 6 L capacity and **b** pilot plant, 100 L capacity

Table 6.4 Test performed with synthetic pollutant (Rhodamine B) in semi-pilot plant

Textile	Photocatalytic efficiency (%)	k^a (h^{-1})
Blank	1.25	<0.005
Untreated	1.19	<0.005
Textile 1	85.7	0.6149
Textile 2	86.7	0.6676
Textile 3	93.3[b]	0.8558

[a]kinetic constant
[b]the pollutant degradation was for up to 90%

6.6 Conclusions and Future Research

The main goal addressed in this work about the PHA production, including the scale-up study, was to develop an original, cost-effective, fully-automated and industrial competitive downstream process for obtaining PHA from microbial cultures. This is in fact still an open challenge in the lack of automated PHA extraction processes and integrated advanced oxidation photo-catalytic systems.

The already existing patents related to PHA (more than 150) mainly regard the production of the polymer, ranging from original approaches for microbial cultivation to novel extraction methodologies. Among the latter, various pre-treatments (e.g. physical treatment, alkali/acids, freezing or heating), polymer recovery (e.g. through cyclic carbonates, ionic liquids or surfactants) and polymer purification (e.g. through bleaching solution) have been investigated (the most relevant examples are reported in [27–31]).

No patents or scientific publications are available for the specific applications related to PHA involving the proposed technologies: (i) the applicability of nanoparticles to destabilize bacterial membranes and purify the extracted PHA, and (ii) Pulse Electric Field (PEF)-assisted extraction processes. The available patents regard the use of TiO_2 or Ag^0 nanoparticles for wastewater treatment, purification and decontamination, and the use of PEF for achieving the permeabilization of membranes to extract various valuable chemicals (lipids, pigments, juice, proteins) from plant or algal tissues.

The use of developed photocatalytic tools, in particular TiO_2 coated textiles, was tested at lab scale in the purification treatment of PHAWW. With the aim of improving environmental sustainability and saving water/energy consuming of the whole system, the integration of the waste-to-bioplastic pilot system is strongly encouraged. From a life cycle perspective, the abatement of pollutant load from the very large amount of water disposed during the concentration of biomass (around 98% by volume) represents a key step for the sustainability of the overall process. The produced know-how allowed the scale up of both bio and photocatalytic reactors at pilot scale level, ensuring the achievement of expected TRL 5/6 and providing a big chance for transferring plants and knowledge in a European research and industrial application environments.

The main outcomes achieved in this work beyond the state of the art contributed to make the overall process more attractive for the companies and even transferable on an industrial level. The process optimization provided safe and cheap PHAs extraction without toxic and harmful chemicals, moreover the improvement performed on the monitoring and automation tools represented a good starting point to lower investments and operating costs. Specifically, in the PHA production system, the biomass production showed to be effective in producing PHA enriched biomass with MMC. The biological reactor is easily scalable with acceptable cost.

Nonetheless, the extraction should be improved in order to obtain higher yield of PHA without significant degradation. This could be done by fine tuning the extraction or use of additives. Another option could be to thermally decompose the PHA inside the cell for the obtainment of monomeric chemicals (e.g. crotonic acid or propylene). Moreover, a complete system including the fermentation of raw wastewater to volatile fatty acids (VFA) should be demonstrated. Due to the difficulties found in the integration of TiO_2 and Ag based catalytic tools in PHAs production process, TiO_2 coated textiles were exploited as photocatalyst in purification process of wastewaters, originated from PHAs production process. Photocatalytic tests showed a great reduction of organic C, measured by TOC analysis, after 2 h of UV irradiation in the presence of TiO_2 coated textiles, as expected by a very efficient photocatalytic process. Thus, the photodegradation of organic pollutants present in wastewaters was demonstrated. The use of developed catalytic tools, in particular TiO_2 coated textiles, in purification treatment of wastewaters represent a promising alternative to their integration within PHAs production process. The very good results encouraged the introduction of such catalytic components in analogous processes, improving environmental sustainability and saving water/energy consuming.

More generally, the developed strategy contributes to allocate the PHAs production process as very promising green biotechnology. The fully automatized prototype realized for PHA cultivation is a further step towards the achievement of a compact all in one system able to convert a specific waste into ready-to-use bioplastic. Further studies will be focused on to the implementation, tuning and testing of the system to produce PHA with different carbon sources obtained from different wastes. A particular attention will be paid to the possibility to combine ceramic active phase such as supported nano-TiO_2 and Ag as green and sustainable remediation against any possible bacterial contamination that bio-reactor can generate.

Acknowledgements This work has been funded by the Italian Ministry of Education, Universities and Research (MIUR) under the Flagship Project "Factories of the Future—Italy" (Progetto Bandiera "La Fabbrica del Futuro") [32], Sottoprogetto 2, research projects "innovative and sustainable Production of BioPolymers" (ProBioPol), "Prototyping ProBioPol results" (ProBioType), and "Prototyping ProBioPol Results II (UV/TiO_2 Photocatalytic Reactor)" (ProBioTypeII).

References

1. Telmo O (2013) Polymers and the environment. In: Dr. Yılmaz F (ed) Polymer science. InTech. https://doi.org/10.5772/51057. https://www.in
2. Kivanç M, Bahar H, Yilmaz M (2008) Production of poly(3-hydroxybutyrate) from molasses and peach pulp. J Biotechnol 136:S405–S406
3. Chanprateep S (2010) Current trends in biodegradable polyhydroxyalkanoates. J Biosci Bioeng 110:621–632. https://doi.org/10.1016/J.JBIOSC.2010.07.014
4. http://en.european-bioplastics.org/ n.d
5. Salehizadeh H, Van Loosdrecht MCM (2004) Production of polyhydroxyalkanoates by mixed culture: recent trends and biotechnological importance. Biotechnol Adv 22:261–279. https://doi.org/10.1016/j.biotechadv.2003.09.003
6. Serafim LS, Lemos PC, Torres C, Reis MAM, Ramos AM (2008) The influence of process parameters on the characteristics of polyhydroxyalkanoates produced by mixed cultures. Macromol Biosci 8:355–366. https://doi.org/10.1002/mabi.200700200

7. Jacquel N, Lo C-W, Wei Y-H, Wu H-S, Wang SS (2008) Isolation and purification of bacterial poly(3-hydroxyalkanoates). Biochem Eng J 39:15–27. https://doi.org/10.1016/J.BEJ.2007.11.029

8. Laycock B, Arcos-Hernandez MV, Langford A, Buchanan J, Halley PJ, Werker A, et al (2014) Thermal properties and crystallization behavior of fractionated blocky and random polyhydroxyalkanoate copolymers from mixed microbial cultures. J Appl Polym Sci 131. https://doi.org/10.1002/app.40836

9. Villano M, Valentino F, Barbetta A, Martino L, Scandola M (2014) Polyhydroxyalkanoates production with mixed microbial cultures: from culture selection to polymer recovery in a high-rate continuous process. N Biotechnol 31:289–296. https://doi.org/10.1016/J.NBT.2013.08.001

10. Whiffin VS, Cooney MJ, Cord-Ruwisch R (2004) Online detection of feed demand in high cell density cultures of Escherichia coli by measurement of changes in dissolved oxygen transients in complex media. Biotechnol Bioeng 85:422–433. https://doi.org/10.1002/bit.10802

11. Johnson K, Jiang Y, Kleerebezem R, Muyzer G, van Loosdrecht MCM (2009) Enrichment of a mixed bacterial culture with a high polyhydroxyalkanoate storage capacity. Biomacromol 10:670–676. https://doi.org/10.1021/bm8013796

12. Westerberg K (2007) Using the dissolved oxygen signal for automatic control in fed-batch production of PHA by a mixed culture

13. Jiang Y, Marang L, Tamis J, van Loosdrecht MCM, Dijkman H, Kleerebezem R (2012) Waste to resource: converting paper mill wastewater to bioplastic. Water Res 46:5517–5530. https://doi.org/10.1016/j.watres.2012.07.028

14. Popova S (2006) On-line state and parameters estimation based measurements of the glucose in mixed culture system. Biotechnol Biotechnol Equip 20:208–214. https://doi.org/10.1080/13102818.2006.10817402

15. Vargas A, Montaño L, Amaya R (2014) Enhanced polyhydroxyalkanoate production from organic wastes via process control. Bioresour Technol 156:248–255. https://doi.org/10.1016/j.biortech.2014.01.045

16. Oehmen A, Pinto FV, Silva V, Albuquerque MGE, Reis MAM (2014) The impact of pH control on the volumetric productivity of mixed culture PHA production from fermented molasses. Eng Life Sci 14:143–152. https://doi.org/10.1002/elsc.201200220

17. Kellerhals MB, Kessler B, Witholt B (1999) Closed-loop control of bacterial high-cell-density fed-batch cultures: production of mcl-PHAs by pseudomonas putida KT2442 under single-substrate and cofeeding conditions. Biotechnol Bioeng 65:306–315. https://doi.org/10.1002/(SICI)1097-0290(19991105)65:3%3c306:AID-BIT8%3e3.0.CO;2-0

18. Samorì C, Basaglia M, Casella S, Favaro L, Galletti P, Giorgini L et al (2015) Dimethyl carbonate and switchable anionic surfactants: two effective tools for the extraction of polyhydroxyalkanoates from microbial biomass. Green Chem 17:1047–1056. https://doi.org/10.1039/C4GC01821D

19. Patel M, Gapes DJ, Newman RH, Dare PH (2009) Physico-chemical properties of polyhydroxyalkanoate produced by mixed-culture nitrogen-fixing bacteria. Appl Microbiol Biotechnol 82:545–555. https://doi.org/10.1007/s00253-008-1836-0

20. Samorì C, Abbondanzi F, Galletti P, Giorgini L, Mazzocchetti L, Torri C et al (2015) Extraction of polyhydroxyalkanoates from mixed microbial cultures: impact on polymer quality and recovery. Bioresour Technol 189:195–202. https://doi.org/10.1016/j.biortech.2015.03.062

21. Dong S, Feng J, Fan M, Pi Y, Hu L, Han X et al (2015) Recent developments in heterogeneous photocatalytic water treatment using visible light-responsive photocatalysts: a review. RSC Adv 5:14610–14630. https://doi.org/10.1039/C4RA13734E

22. Bhatkhande DS, Pangarkar VG, Beenackers AACM (2002) Photocatalytic degradation for environmental applications—a review. J Chem Technol Biotechnol 77:102–116. https://doi.org/10.1002/jctb.532

23. Malato S, Fernández-Ibáñez P, Maldonado MI, Blanco J, Gernjak W (2009) Decontamination and disinfection of water by solar photocatalysis: recent overview and trends. Catal Today 147:1–59. https://doi.org/10.1016/j.cattod.2009.06.018

24. Kabra K, Chaudhary R, Sawhney RL (2004) Treatment of hazardous organic and inorganic compounds through aqueous-phase photocatalysis: a review. Ind Eng Chem Res 43:7683–7696. https://doi.org/10.1021/ie0498551
25. Chong MN, Jin B, Chow CWK, Saint C (2010) Recent developments in photocatalytic water treatment technology: a review. Water Res 44:2997–3027. https://doi.org/10.1016/J.WATRES.2010.02.039
26. Baldisserri C, Ortelli S, Blosi M, Costa AL (2018) Pilot—plant study for the photocatalytic/electrochemical degradation of Rhodamine B. J Environ Chem Eng 6:1794–1804. https://doi.org/10.1016/j.jece.2018.02.008
27. Jian Y (2009) US7514525B2 Recovery and purification of polyhydroxyalkanoates
28. Hecht SE, Niehoff RL, Narasimhan K, Neal CW, Forshey PA, Phan DV, et al (2010) US7763715B2 Extracting biopolymers from a biomass using ionic liquids
29. Hassan MA, Mohammadi M, Ab Razak NA, Chong ML, Abd Aziz S, Abdullah AA-A, et al (2011) WO2011108916A2 A method for recovering an intracellular PHA
30. Nielsen W, Mcgrath C (2013) WO2013016566 Methods of extracting polyhydroxyalkanoates from PHA-containing bacterial cells
31. Lafferty RM, Heinzle E (1978) US4101533 A: cyclic carbonic acid esters as solvents for poly-(β hydroxybutyric acid)
32. Terkaj W, Tolio T (2019) The Italian Flagship project: factories of the future. In: Tolio T, Copani G, Terkaj W (eds) Factories of the future. Springer

Chapter 7
Integrated Technological Solutions for Zero Waste Recycling of Printed Circuit Boards (PCBs)

Giacomo Copani, Marcello Colledani, Alessandro Brusaferri,
Antonio Pievatolo, Eugenio Amendola, Maurizio Avella and Monica Fabrizio

Abstract The demand for key metals for the production of high-tech products is constantly growing in Europe, leading to relevant problems both in terms of supply risks and costs. Waste from Electric and Electronic Equipment (WEEE) is growing very fast in Europe, with an annual increase rate between 3 and 5%. Printed Circuit Boards (PCBs), which are embedded in electric and electronics products, are very valuable waste products, since they are composed also of precious metals and key metals (about 25–30%). Recycling of PCBs is a very challenging task that has not been solved yet: recycling rates for traditional metals are around 30–35% and many critical key metals, as well as the non metal fraction, are not recycled. This work proposes a set of solutions to be adopted towards the automated zero-waste treatment of PCBs. They address selective disassembly of PCBs components, mechanical pre-treatments, chemical processes for the characterisation of metals material content of PCBs, as well as for the recycling of their non-metal fraction. New business models are finally proposed for the uptake of such solutions in a framework of integrated recycling chain.

G. Copani (✉) · M. Colledani · A. Brusaferri
CNR-STIIMA, Istituto di Sistemi e Tecnologie Industriali Intelligenti per il Manifatturiero Avanzato, Milan, Italy
e-mail: giacomo.copani@stiima.cnr.it

A. Pievatolo
CNR-IMATI, Istituto di Matematica Applicata e Tecnologie Informatiche, Milan, Italy

E. Amendola · M. Avella
CNR-IPCB, Istituto dei Polimeri, Compositi e Biomateriali, Pozzuoli, Italy

M. Fabrizio
CNR-ICMATE, Istituto di Chimica della Materia Condensata e di Tecnologie per l'Energia, Padova, Italy

T. Tolio et al. (eds.), *Factories of the Future,*
https://doi.org/10.1007/978-3-319-94358-9_7

7.1   Scientific and Industrial Motivations

Due to the globally raising production and consumption of high-tech products, the need of key-metals is constantly increasing. European countries face challenges in terms of supply risks, since they have limited key-metals reserves to be extracted, thus leading to possible negative significant impacts in the manufacturing sector. In 2010, the European Commission identified 14 critical metals for the future of Europe [1]. Extraction of such metals is mainly made in China, Russia, the Democratic Republic of Congo and Brazil, which makes their sourcing affected by political relationships. This situation generates considerable supply risk for Europe, considering that recycling rates are currently limited.

If properly recycled, Waste from Electric and Electronic Equipment (WEEE) represents a strategic source of key-metals. Besides post-use consumer electronics as mobile phones, personal computers, TVs, etc., also waste coming from other sectors using mechatronics product is rich of key-metals (such as automotive, aeronautics, military, industrial automation, etc.). For example, it can be estimated that the content of electronics in a modern vehicle accounts for the 40% of its entire manufacturing costs [2]. Due to rapid technological trends and to the increased diffusion of electronics, post-consumer WEEE is growing very fast in Europe, with an annual increase rate between 3 and 5%, which would lead to doubling volumes in the next decade [3]. All WEEE and mechatronics waste include PCBs, which are called *urban mineral resources* because, on average, they are composed for the 25–30% of metals such as copper, tin, nickel, gold and silver [4]. Moreover, they contain small quantities of critical metals such as indium, palladium, ruthenium, gallium, tantalum and platinum. Therefore, developing sustainable recycling technologies and systems for PCB key-metals recovery is of fundamental importance for the European recycling and high-tech product manufacturing sectors.

Because of its complexity, recycling of PCBs is a very challenging task that has not been solved yet from the point of view of systems technology and business model sustainability [5]. As a consequence, most of the PCB treatment takes place in China recurring to manual disassembly and processing, leading to negative consequences for human health and environment. Most PCB recycling approaches practiced today, based on pyrolysis and electrolysis, are high energy-demanding, release dioxin or acid drain water in the atmosphere and present significant limitation in the recovery performance. Indeed, only 30–35% of the PCBs metals can be recovered with achievable purity of 85–95%, depending on the type of material. The remaining materials cannot be economically recycled and recovered, thus they are landfilled. This last fraction includes especially key metals such as tantalum that is present in very low concentrations and is located in specific PCB components (e.g. 35% in capacitors). Despite the metal part of PCBs, the non-metal fraction, mainly composed of Fibre Reinforced Plastics (FRP), is currently not recycled at all [6].

Smart mechanical pre-treatments coupled with chemical recovery processes would be ideal techniques for addressing the recovery of materials from PCBs (metals and non-metals). In fact, compared to traditional pyro metallurgical processes, they involve very limited environmental impact, energy consumption and production of by-products, simultaneously allowing to recover currently landfilled fractions in a logic of zero waste. However, at present, metal recovery rate is still limited and a consistent fraction of precious metals is lost during shredding and separation operations [5]. This constitutes a barrier towards the industrial diffusion of mechanical plants for PCB recycling.

A sustainable PCBs treatment through novel technologies and processes will contribute not only to the solution of the environmental and economic challenge, but will also give the opportunity to increase employment through the establishment of a new advanced industrial recycling sector.

Moving from the outlined limitations and promising technological approaches, this work aims at the definition of a new systemic approach for the sustainable recovery of key-metals and of the non-metallic fraction of (PCBs) in Europe in a zero-waste logic, developing the methodological/technological solutions for supporting its effective use in industry.

The chapter is organised as follows. Section 7.2 presents the state of the art on flexible recycling systems. Section 7.3 outlines the research scope and approach, whereas Sect. 7.4 presents the results of the research. Industrial validation and testing activities are discussed in Sect. 7.5 and, finally, the conclusions together with indications for future research are presented in Sect. 7.6.

7.2 State of the Art

State-of-the-art PCB recycling process involves rough dismantling and recovery stages. Disassembly operations are required to remove dangerous components, such as batteries. In general, manual disassembly methods are adopted because of the high degree of variability of post-consumer product streams. Modular disassembly devices can in principle meet the requirements of continuously changing disassembly operations. Moreover, systems integrating manual and automated workstations (hybrid disassembly systems) represent viable solutions. Past research efforts addressed the implementation of semi-automatic solutions for electronic products disassembly. Nonetheless, an integrated solution, capable of processing heterogeneous products such as PCBs is still undeveloped [6]. Besides the limited automation level of the process, disassembly operations are nowadays poorly integrated with the downstream recovery treatments, thus making selective disassembly of highly concentrated components far from being achieved.

The recovery of materials from PCBs includes chemical and physical processes. Among chemical processes, pyrolysis and electrolysis are the most diffused. Pyrolysis enables to obtain friable solid residue from printed circuit boards made of organic, glass fibre and metallic fractions that can be easily separated [7]. The ashes in the residues mainly contain copper, iron, nickel and aluminium that can be further treated by electrolysis and pyro-metallurgical processes [8]. Biotechnology has been outlined as a suitable procedure in metallurgical processing. In particular, in the last years, bioleaching was exploited to recover precious metals and copper. However, limited research addressed the bioleaching of metals from WEEE.

The chemical composition of waste must be controlled to select the recovery technologies. Spectrometric techniques are excellent tools available for metal trace and ultratrace analysis of liquid and solid samples. Among them, Inductively Coupled Plasma-Mass Spectrometry (ICP-MS), Laser Ablation-Inductively Coupled Plasma-Mass Spectrometry (LA-ICP-MS) and Secondary Ion Mass Spectrometry (SIMS) can provide complete information about chemical composition of waste, but proper analysis procedures have to be developed.

Chemical processes are high energy-demanding and release in the atmosphere dioxin or acid drain water. The alternative is still manual re-usable component disassembly and incineration of the remaining PCB matrix; as a matter of fact, nearly the 80% of the PCBs collected in the world are processed in China adopting very poor ergonomic and safety conditions [9]. Key-metals, such as tantalum, gallium and indium, that are present in high concentrations only in specific components of PCBs, are currently not recovered and go to landfill [10]. Mechanical treatment represents a promising alternative to pyrolysis with low environmental pollution, limited equipment investments, low energy cost and diversified applications. This treatment consists in mechanical separation and size-reduction to obtain highly concentrated mixtures of metallic and non-metallic fractions. PCBs are an ideal candidate for mechanical shredding and separation but, given the high value of the roughly 30% metal fraction of a populated circuit board, mechanical shredding and separation systems should be highly efficient to maximize the recovered metals fraction. Detailed physical models estimating the distribution of the size and liberation degree of the output particles, as well as the processing energy required, can pinpoint ideal process operation [5]. Very little research has focused on this area, but stochastic process models, in particular Markov models, are a very promising avenue. Concerning the separation stage, physical models that predict particle trajectories have been developed for eddy current separation and corona electrostatic separation [11]. However, these models consider single trajectories of particles and fail to model two major motivations for the limited separation performance, i.e. particle interactions and impacts, and the presence of unliberated particles in the input material flow. The lack of validated models and control technologies makes comminution and separation processes still uncontrolled in industrial settings. Process parameters are manually set, based on experience coming from trial and error procedures. As a consequence, metal recovery rate is still limited and a consistent fraction of precious metals is lost during shredding and separation [12].

Chemical recycling methods of the Non-Metal Fraction (NMF) in PCBs include gasification, and depolymerisation process using supercritical fluids. The decomposition reactions of epoxy resin in supercritical water (SCW) or using supercritical methanol (SCM) have been studied in literature [13]. The goal of the research in this area is to control the chemical decomposition of organic fraction of PCB to obtain intermediate products suitable for further polymerization reaction. Another promising recycling option is the re-use of the polymer portion as filler in new compounds. Without treatment, these fillers can reduce mechanical properties of the polymers.

The use of mechano-chemical treatment to prepare scrap-based composites improves the compatibility in blends, since it favours the dispersion of nano-fillers following to the application of high compressive and shear forces.

Finally, from a business model perspective, literature is mainly explorative and takes a comparative country-perspective to illustrate how nations are managing the end-of-life of PCBs and electronic waste [14]. There are currently no clear evidences of consolidated successful business models for PCBs recycling in terms of reverse logistics practices, network supply chain management and companies' vertical integration, infrastructure, secondary markets and pricing strategies. Only some general indications in terms of supply chain actors can be found at waste management system level, but the whole business model with its supporting technologies is not analysed and quantitative performance assessments are lacking.

7.3 Problem Statement and Research Approach

From a technological, industrial and business point of view, the main difficulties for the establishment of successful practices for treating PCBs in Europe in a logic of circular economy [15] can be summarised as follows:

- Post-consumer PCBs are a very variable product due to different technologies, design, size and materials composition. This generates uncertainty in the type of waste to be treated. Thus, technology investments and business return are difficult to be forecasted and entrepreneurial risk is implicitly high.
- Limited knowledge is currently available on the elemental composition of PCB-mounted components and on the concentrations of key metals in them. Such a situation contributes to increase the above mentioned uncertainty.
- Flexible automation is not generally adopted in treatment processes of PCBs and process control is still poorly deployed in industrial practices: manual practices are the state of the art. This situation poses challenges for safety and health of human operators and, in addition, it makes the treatment non sustainable in Europe, due to the high labour cost compared to emerging countries.
- There is a poor link between the different recycling supply chain processes that are usually carried out by separated companies, in particular between mechanical pre-treatments and downstream recovery. This causes systemic inefficiencies and precious metal fraction losses [16].
- There is a lack of proven successful business models from the economic and environmental point of view. Such a lack discourages entrepreneurs to invest in this promising business application and limits the diffusion of circular economy practices based on PCBs recovery and recycling.

This work aims to address the previous challenges and to overcome the current fragmentation by developing a set of integrated technologies and business enablers in the phases of PCBs disassembly, shredding, separation and materials recycling. In particular, the following innovations are targeted:

- *New knowledge-based selective disassembly technologies for efficient pre-treatment of electronic waste*. The introduction of automation in operations that are currently manual will lead to a consequent increase of efficiency, safety and quality. Such technologies include: a visual inspection system for the localisation of components to be selectively disassembled on the PCBs; an automated knowledge-based component recognition system based on geometrical and functional analysis of components; an automated system for the characterisation of key-metals in PCBs, based on the elemental composition of the identified components; an automatic selective disassembly process for the extraction of valuable components with the goal of making the downstream mechanical treatment more efficient thanks to a higher concentration of valuable materials to be processed and to the reduction of parts heterogeneity (Sect. 7.4.4).
- *Advanced control models and automated technologies for mechanical shredding processes*. Knowledge-based models consist in statistical models for predicting the performance of shredding processes in order to optimise the performance of downstream separation processes. Based on these models, automated technologies allow optimal process control adapting parameters to the different type of waste to be processed (Sect. 7.4.1).
- *Environmentally-friendly solutions for the recovery and re-use of the metal and non-metal fractions from PCBs*. Innovative treatments are needed for recycling the metal and non-metal fraction from PCBs into innovative materials and products to be re-introduced in the market, under a zero-waste perspective. The performance of such solutions is assessed through evaluation and characterization of the physical and mechanical properties of the recycled products and the performance of developed technologies (Sect. 7.4.2).
- *New business models for zero-waste treatment of end-of-life PCBs*. Existing companies or new entrepreneurial actors will create new supply chains uptaking advanced PCBs recycling technologies to increase overall supply chain business performance (Sect. 7.4.3).

Through the development of the mentioned innovative solutions and business models, this work aims at the establishment of more sustainable PCBs recycling supply chains in Europe, as illustrated in Fig. 7.1. The final goal is to avoid the recourse to heavily impacting pyro-metallurgical processes that require big furnaces and economy of scale, through the establishment of recycling chains, mainly composed of SMEs, which implement highly efficient integrated processes of smart disassembly, mechanical pre-treatment and PCBs materials recovery and re-use.

The research activities reported in this work benefited from a pre-existing pilot plant for the End-Of-Life treatment of PCBs within the facilities of CNR-STIIMA (ex CNR-ITIA) [17]. This pilot plant supports disassembly, shredding and separation processes of mechatronics products embedding PCBs. This infrastructure allowed the testing and validation of developed models and technical innovations on real technologies at disposal of researchers.

Finally, being focused on getting the highest value from PCBs recycled components and materials, the proposed approach is thought to be sustainable even at

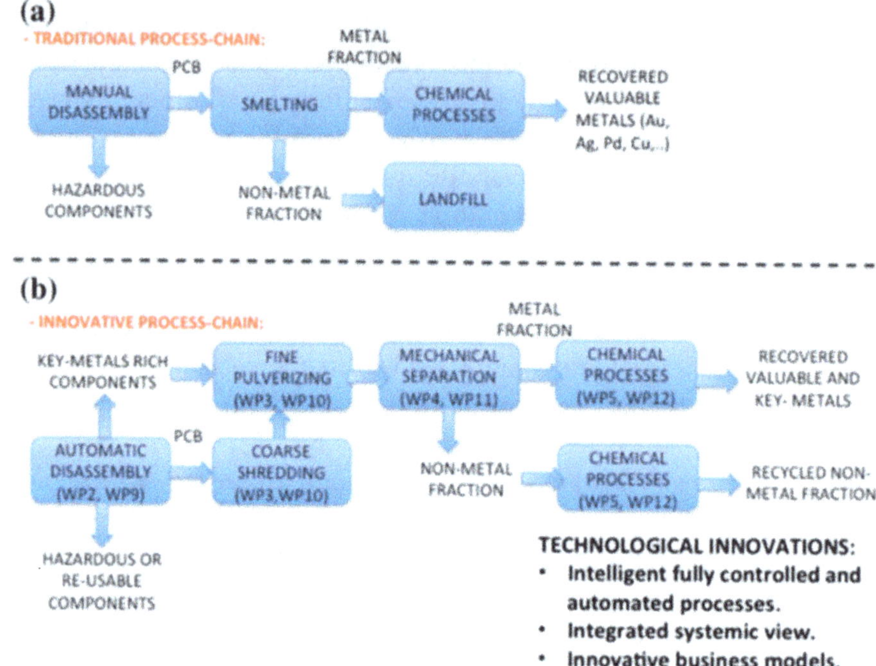

Fig. 7.1 Traditional (**a**) and innovative (**b**) PCBs recycling chains

relatively low WEEE collection levels, thus being targeted to SMEs operating in the recycling industry. This makes the results of the research particularly suited for uptake by European recycling industry, which is mainly composed of SMEs.

7.4 New Solutions for PCBs Recycling

A set of multi-disciplinary enablers was realised to address the current challenges limiting the recycling of PCBs. Technical solutions were developed to increase the performance of the disassembly and shredding process of PCBs, the assessment of the metal materials composition of shredded PCBs and the recycling of the non-metal fraction. New business models are proposed for the uptake of the developed technologies by supply chain actors. Results are described in the following subsections.

7.4.1 Shredding Process Modelling and Optimisation

In a shredding machine, the relationship between input and output mixture has a random nature. In the recycling industry, the operating shredding process parameters are traditionally set based on trial and error procedures and they remain constant regardless of the input mixture. Modelling these kinds of process would allow to set shredding parameters to obtain a desired target mixture out of specific input materials, with consequent advantages in time and material saving. For these reasons, a stochastic model has been developed taking into account physical machine parameters. In particular, design and controllable parameters have been considered. The former are the volume of the chamber and the number of the cutting tools (also called knives). The latter are the rotational speed of the rotor and the characteristic size of the output grate (the higher is this value, the largest is the maximum dimension of the particles exiting from the shredder).

The proposed model was built upon the Population Balance Model (PBM) framework [18]. This model study the behavior of particles inside the chamber of a comminution machine by defining three relationships: (1) the proportion of a specified particle type which is selected for breakage per unit of residence time; (2) the degree to which the selected particle type undergoes breakage; (3) the selection of particles which are to be removed from the process. The model predicts the evolution of the particle size and liberation classes during the shredding operations as a function of process parameters.

The transition of mass between size classes in a time interval is governed by a transition matrix P (named *Breakage and selection matrix*) of proportions that meets the mass conservation constraint. This matrix describes the probability to break a particle from one class to a smaller class. A stochastic dynamic was added to this framework, which is completely deterministic, by allowing the transition matrix itself to be random but centred around P. The parameters of the transition probability matrix P are functions of particle breakage and selection probabilities.

As P is unknown, it must be estimated (or trained) from experimental data. A correct quantification of the dispersion of the random matrices around P allowed to decide how many experimental runs were needed to achieve a sufficient precision. This study was first carried out for a batch process and then for a continuous process, which also includes a discharge matrix D, different for every grate size. This new matrix determines the fraction of mass in each size class able to leave the comminution chamber in a time interval. From these two matrices a PBM model has been developed, using Markov chains, based onto different assumptions named homogeneity and multiplication. In particular, the first one affirms that P does not depend on time while the second one that the size distribution of shredded particle only depends on the number of breakage intervals.

Based on the random PBM, a procedure for training a model of a specific shredder with a limited number of runs was designed. The trained model can predict the particle size distribution and the output flow in a stationary regime.

7.4.2 New Technologies to Recycle PCBs Materials

Recycling of PCBs materials poses multiple challenges considering the current state of the art. Recycling of the Metal Fractions (MF) is affected by uncertainty, since revenues depend by the content of metals of the different type of PBCs. Traditional materials characterisation is not a viable process, since it requires time, it should be carried out in certified laboratories and it is expensive. For evaluating the feasibility of MF recovery/re-use, an affordable procedure has been developed for measuring the MF content in PCBs powders obtained as output of mechanical pre-treatments. Literature data [13, 19] report mineralization techniques based on acid hydrofluoric (HF) attack. Considering the high danger of this acid, herein a HF-free mineralization method was developed and successfully applied to mechanically processed samples. The development of the HF-free mineralization is very attractive with a view to an industrial application of this method, because HF is a very dangerous acid for health and for the glass part of the spectrometer.

Herein, ICP-MS technique was mainly adopted because this technique is known as a unique tool to evaluate where the elements of interest are present and in which amount, and it is an efficient technique to evaluate their leachability. In fact, the combined use of mass spectrometry techniques (SIMS for surface analysis and ICP-MS, LA ICP-MS for bulk analysis) gave useful information about the composition of waste during the various steps of waste processing.

Printed circuit board scraps from different industries have been analysed to estimate the amount of lanthanides and precious metals. It is interesting to remark that ICP-MS results showed that the precious metal content is not correlated to lanthanide amount in the different matrices. For example, PCBs from automotive showed the higher Au and Pd amounts, but the higher Rare Earth content was found in home appliances grinded PCBs (in particular La, Ce and Nd).

Considering the Non Metal Fraction (NMF) of PCBs, new chemical compounds and new secondary applications should be defined to increase recycling practices. In this work, selective cleavage reactions have been defined for the production of small molecular weight fractions, to be used for the production of new reactive materials. PCBs motherboards are mainly made of an unsaturated polyester resin (UPR). The presence of reversible chemical linkages, such as esters, was exploited to split the thermosetting materials into smaller fragments that are suitable to further chemical processing. Although the direct polymerization reaction to produce UPR involves esterification by condensation of acids with alcohols, the corresponding equilibrium can be reversed to produce the monomeric components by hydrolysis as shown in Fig. 7.2a. An alternative to the hydrolytic cleavage involves the alcoholysis with consequent production of esters (Fig. 7.2b), which are conveniently soluble in common organic solvents.

While in theory all the aforementioned reactions occur spontaneously depending on temperature and pressure, their efficiency can be increased adding a catalyst. Common catalysts for trans-esterification reactions consist of transition metal salts like the widely used zinc acetate. Based on such premises, in this study the methanolysis of a model URP compound as a function of the catalyst nature, as well as of the reaction temperature and duration, was successfully investigated.

In terms of new applications for the recovered NMF of PCBs, recent researches and patents demonstrated the feasibility of innovative emulsion technologies for the production of thermoplastic composites with high proportions of recycled fillers of various size and nature (organic and inorganic) [20, 21]. By applying this methodology, powders of fibre-reinforced polymers such as NMF of PCBs can be emulsified

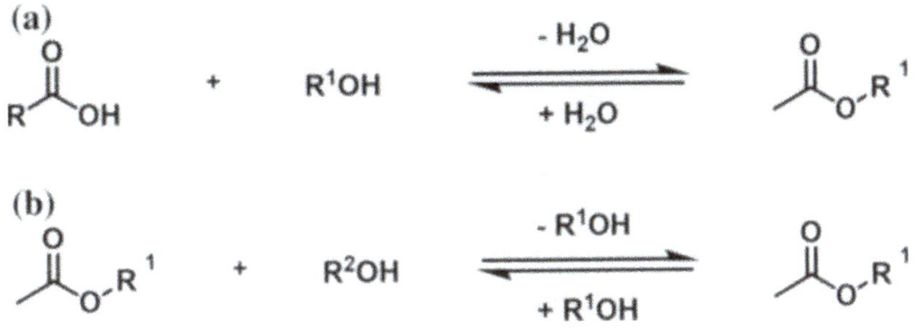

Fig. 7.2 Esterification-hydrolysis (**a**) and transesterification (**b**) reactions

and moulded in a thermoplastic matrix to obtain sheets or granules. In this study, this emulsification procedure was carried out at room temperature using a thermoplastic matrix made of expanded Polystyrene (EPS), which was preliminarily produced as a physical gel through a volatile organic solvent. Post-consumer packaging EPS is available in large quantities, since it is used as loose fill packaging for fragile products [22]. Moreover, its lightness makes its disposal highly difficult. The combination of these materials enabled to obtain low-cost recycled compounds with good performance thanks to the inclusion of a mixture of grinded PCBs and rubber tire into the EPS gel. In fact, EPS in the form of gel is able to combine a fluid state with the elasticity of the physical network. Such network results stable during subsequent machining operations, hence assuring good mechanical properties to the obtained composites.

7.4.3 New Business Models for PCBs Recycling

Multiple options are open for the uptake of smart PCBs pre-treatment processes and technologies (selective disassembly of PCBs components and optimised shredding and separation processes) by existing actors operating in the recycling chain or new companies. Existing literature was analysed [14] and a dedicated empirical analysis was carried in Italy to characterise typical recycling chains and to collect input for new business model design. Four treatment centres were visited and their management was interviewed to map the as-is situation and to ask their opinion on new technologies for efficient shredding and separation processes, which would substitute (or improve the performance) of current pyro-metallurgical processes. More precisely, the case study companies were an authorised consortium for the collection of WEEE, a company performing pyro-metallurgical processes on PCBs and two chemical companies recovering metals from PCBs ashes. Such a sample allowed to represent the whole recycling value chain, from PCBs collection to materials recovery. Based on this analysis, it can be argued that the main actors in the typical supply chain are:

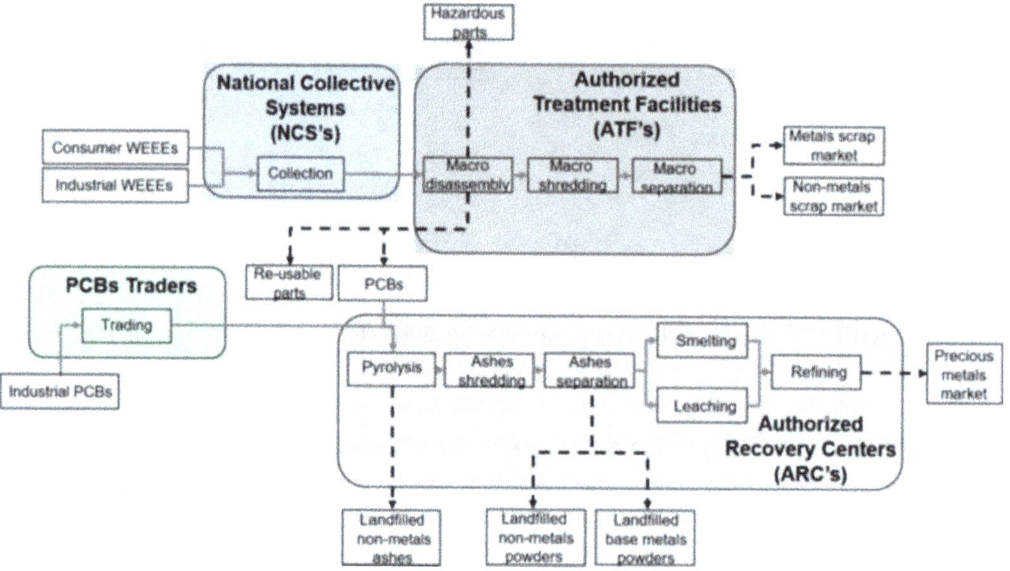

Fig. 7.3 Typical PCBs recycling chain

1. National Collective Systems that collect WEEE (NCSs). PCBs may also enter the recycling chain from specialised waste PCBs traders (Fig. 7.3).
2. Authorised Treatment Facilities (ATFs) that perform macro-disassembly of products and, eventually, macro-separation of scrap through rough shredding.
3. Authorised Recovery Centres (ARCs) that recover materials through chemical processes.

Three options were identified in this work for the uptake of novel PCBs recycling technologies, which would pave the way to new business models for supply chain actors. The first option is that new intelligent shredding and separation technologies, based on novel modelling and optimised control systems, are uptaken by new incoming actors, the *Independent De-manufacturing Centres* (IDCs). They will intercept PCBs flows coming from ATFs and scrap from PCBs traders, with the final aim to extract as much value as possible from them. Inside a hypothetical IDC, PCBs will be previously tested to identify valuable and reusable components that, in a following disassembly phase, will be separated and sold in the electronic components market. The remaining PCBs bare board will be sent to the shredding and separation phases. The resulting base metals and non-metals powders will be sold in their proper markets, while precious metals powders will be sent to ARCs for a further more detailed refining. This type of business model is the one seen as the most promising by the interviewed Italian actors in terms of possible implementation and profitability. From a scientific point of view, this type of business model could favour a high technology specialization and the development of knowledge around selective disassembly, shredding and separation processes. However, this option would offer lower proximity to the precious metals end-markets (managed by ARCs). It would also increase the exposure to PCBs collection fluctuations, with a high sourcing risk and contractual power of ATFs.

The second option implies that ATFs add to their traditional process the new advanced processes and technologies for the treatment of PCBs, transforming themselves in *Multi-treatment facilities* (MTFs). MTFs could exploit their current knowledge and technologies to treat various types of waste products, among which PCBs. PCSs treatment through smart processes might increase the profitability of ATFs due to the high content of valuable materials of this new category of waste, which is added to the traditional waste products that ATFs already process. Compared to the scenario of IDCs, the uptake of new technologies would imply a lower risk, since ATFs are recycling companies already operating in the market and experienced in treatment technologies.

Finally, the third option of business model would imply that the ARCs add new smart pre-treatment technologies prior to their traditional chemical recovery processes, becoming *Integrated Recovery Centres* (IRCs). This scenario would generate a radical innovation in the recycling chain, because it would allow the integration of mechanical and chemical treatment at the premises of the same company, with the consequent optimization of process parameters of both phases. This business model would overcome the current fragmentation of the recycling chain and the local optimization approaches which do not consider the recycling process as a whole and the opportunity to efficiently recover both MF and NMF of PCBs. However, such a business model appears very challenging, since ARCs have traditional competence on chemical recovery processes but are culturally distant from mechanical processes. Thus, they might be averse to new technologies or might experience difficulties in the exploitation of their benefits.

7.4.4 Prototypes

Three industrial prototypes were developed based on the results of this work. The first one is a station for PCBs automatic components selective disassembly for reuse, remanufacturing or recycling (Fig. 7.4). Such a prototype demonstrates how pre-selected PCBs components can be disassembled in a totally automated way to preserve their residual functionalities or to avoid precious materials loss during the disassembly process. In the prototype, disassembly of PCBs components is based on a thermal reflow process guaranteeing high disassembly accuracy. In such a process, a reflow heater is used to heat the board and the components to melt the solder paste which anchors components on the board. Proper time-temperature profiles were identified to avoid overheating and damages to PCBs (still functioning) components. Based on such profiles, a structured control solution was realised to coordinate the machine actuators (hot air nozzle and bottom heater) in terms of heating power. To this aim, two coordinated control loops have been implemented and thermocouples have been installed and connected on both control loops. A Modbus/TCP has been also included to support the information exchange to/from the process supervision, i.e. list of disassembly operations to be performed and related parameters.

Fig. 7.4 Automatic PCB components selective disassembly prototype station

A linear motors-based gantry architecture has been adopted to guarantee tools positioning accuracy as well as proper dynamic capabilities. Moreover, a force control table has been integrated to properly control the contact force impressed by the pneumatic disassembly tools onto the PCB during operation in order to avoid reusable chips damage. In fact, the robot is moved under position control during normal operation. The force control loop is activated when the tool is in proximity of the PCB to be operated (about 5 mm far from the PCB surface). An automatic tool exchange system has been integrated in the machine architecture to guarantee machine flexibility in processing different electronic chip case shapes. Furthermore, a system for automatic load and unload of the PCB pallet has been included within the machine operation area.

A computer vision system was developed and integrated in the station to support on-line automatic recognition of the components to be processed. Such a system is capable of identifying the components by matching the PCB image acquired by the camera and the information related to the components' geometry stored in a repository. Such a repository includes the main classes of PCBs components in terms of family, dimensions and rework operation constraints (such as maximum temperatures and thermal gradients). The repository was built on purpose by accessing a large number of components' materials specifications through public catalogues of vendors and represents a relevant knowledge source that will need to be constantly updated in the future according to new PCBs technology and components. Once identified the component to be disassembled, its coordinates in the machine operation plan are automatically adapted and delivered to the station controller to automatically reconfigure the operation point. The vision system is also used to compensate possible errors in manipulating the PCB component by the gripper.

The second developed prototype is a system for the in-line prediction of PCBs material content and value (depending on materials' price), based on the recognition of PCB components. The hardware is composed of a linear camera for PCB image acquisition, a transportation system and a lighting system. In the system, the PCBs to be analysed are manually loaded on a conveyor and the linear camera mounted at optimised distance from the conveyor scans the PCB for image acquisition. The lighting system (LED) provides a diffuse-like illumination modality avoiding spots or highlights. The software tool is composed of two modules: the *Image Processing Module* and the *Material Prediction Module*. The first module includes several segmentation algorithms that use the image of PCB to identify the specific areas of PCBs representing the electronic components mounted on it. The general procedure requires a thresholding operation for the segmentation of both the black plastic and the metal terminations. Afterwards, threshold images are post-processed to select the objects of interest. In the case of black plastic, a modified watershed algorithm is used for the definition of the mass centre and the approximate shape of black plastic bodies. For metallic parts, a logical operation is performed with an innovative proximity algorithm to assign to each plastic body the set of relative metallic terminations. Finally, the segmented regions are classified into the different categories of electronics packages that were previously characterized through an extensive analysis of open source documentation (materials declarations). A probabilistic knowledge-based method is used, since the real image differs from the ideal situation because of the presence of non-interesting objects. This method provides better results in term of classification rate with respect to other standard methods such as pattern matching or artificial neural network. This is justified considering that the spatial intensity distribution around an electronic package is very similar for most of the classes of electronic packages, thus increasing the classification error.

The second software module consists in statistical models able to predict the material contents of electronic components as a function of the electronic package type, size and geometry. The complete algorithm has been tested using several PCBs scraps coming from different application field. During the experiments, the PCBs have been transported through the conveyor belt under the linear camera and an image for each PCB has been taken. The image has been processed with the algorithm of segmentation and classification. Figure 7.5 shows the PCBs used in the testing, while Fig. 7.6 shows the final result of image processing. The classification error amounted to 15%.

Finally, the third prototype is a *Cyber-Physical System (CPS) for a self-optimizing shredding stage* to cope with the high variability of input products. In the prototype, coarse shredded particles undergo a Computerized Particle Analyser (CPA) for high speed granulometric analysis, which gives the dimensional distribution of the particles as output. The developed software tool acquires and stores this information,

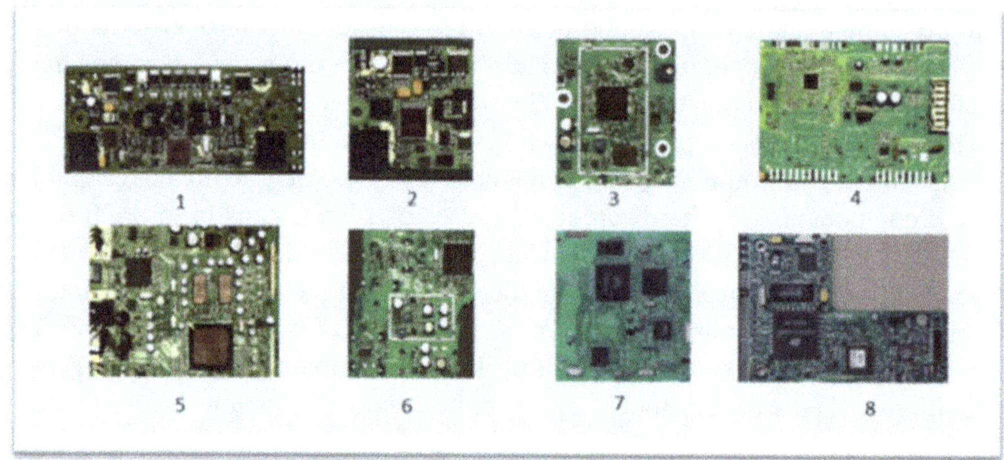

Fig. 7.5 PCBs used in prototype testing

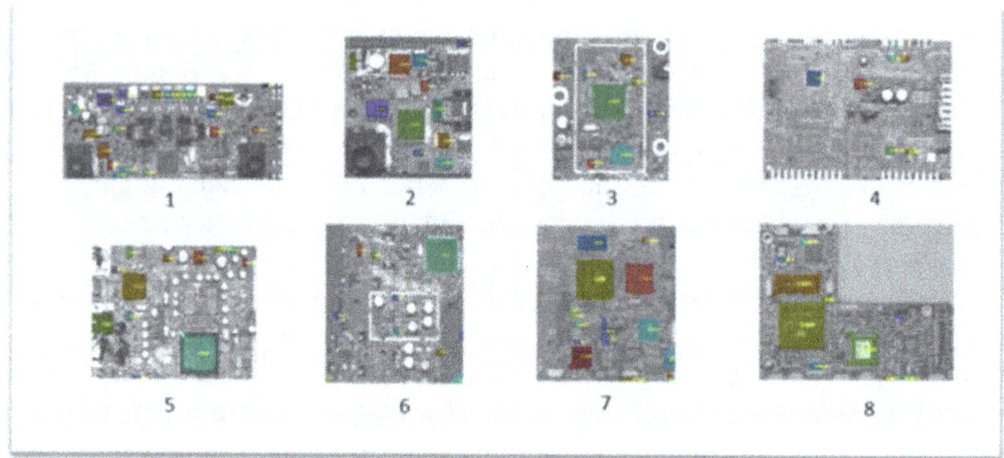

Fig. 7.6 Result of image processing

which is the input for a Population Balance Model (PBM) presented in Sect. 7.4.1. A human-machine interface was developed to acquire the target output distribution and throughput information from the operator. These data are processed by the optimization module that calculates and suggests (using least square method) the optimal process parameters to achieve targets. Meanwhile, particles are shredded by a fine comminution machine with optimal parameters and the output distribution is checked by the CPA before leaving the cyber-physical system. All the technologies have been integrated through a pneumatic transportation system. The overall prototype functioning is represented in Fig. 7.7.

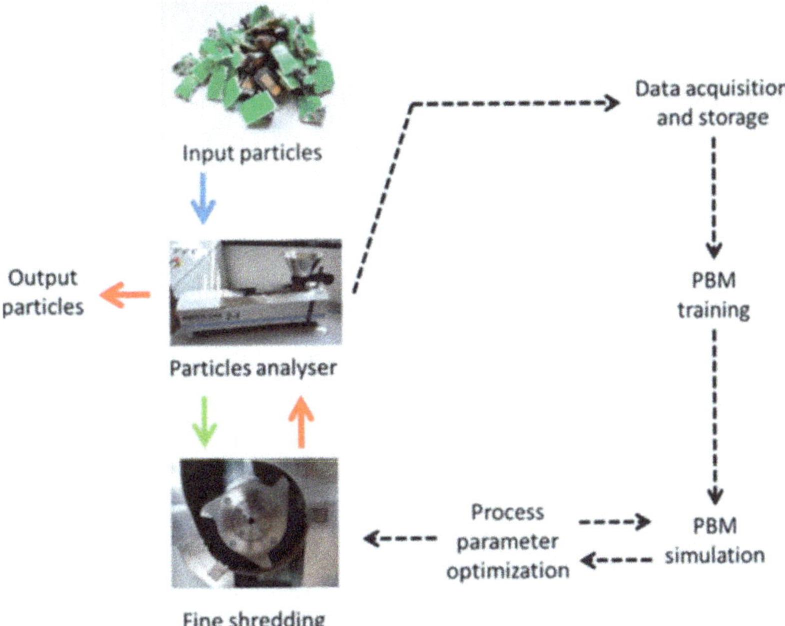

Fig. 7.7 CPS prototype functioning

7.5 Industrial Testing and Results

Technical solutions described in the previous section were tested and validated at laboratory scale. Results are described in the following subsections.

7.5.1 Shredding Model Validation and Parameters Optimisation

The experimental procedure mentioned in Sect. 7.4.1 was executed several times with input mixtures obtained from different types of PCBs (low, medium and high grade) coming from various sectors as automotive, white goods and personal computers, first for model training and then for model validation.

A relevant outcome for the rapid characterisation of the shredding machine, is that, within reasonable operational limits, the size distribution of output particles largely depends on the grate size and only weakly on the rotational speed. Therefore, this latter parameter can mainly be tuned to ensure an appropriate level of saturation of the comminution chamber for best performance depending on the input flow of material. Changing the speed will accelerate or slow down the process, affecting the throughput, but does not modify the output particles size distribution.

The validation experiments (both goodness-of-fit and out-of-sample validation) showed a good adhesion of predicted distribution to the real one. The weak dependence on rotational speed is a plausible assumption, as homogeneity and multiplication hypothesis. In the stationary state, the particles size distribution is well approximated by the distribution predicted by the model. As a consequence, by exploiting the PBM simulation model, it is necessary to identify the discharging grate with the most suitable size to optimize a comminution process obtaining an output particles size distribution in the stationary state as similar as possible to a target one (determined by the needs of downstream material processing).

Shredding prototype validation and verification activities have shown an accurate output size distribution, with a Kolmogorov-Smirnov metric for the comparison between the predicted and the measured distributions lower than 0.072, while the optimization time, defined as time to search for an optimal configuration of the controllable parameters, is almost instantaneous. Furthermore, the time to train the PBM breakage and selection function is below 120 s to obtain good results.

7.5.2   Validation of HF-Free Process for MF Assessment

The procedure developed for the assessment of the MF content in shredded PCBs was experimentally validated by mineralizing the same amount of powder from the same batch of shredded PCBs while using: (a) only HNO_3 acid; (b) HNO_3 acid and HCl acid in a ratio 1:3; (Aqua Regia); (c) HNO_3 and Aqua Regia mixture with a 1:1 ratio. The ICP-MS (X Series, Thermo) was calibrated by using a multi elements rare earth standard solution (Inorganic Ventures) and precious metal standard solution (Inorganic Ventures). Preliminary results showed that the most suited method to mineralize the elements of interest was the Aqua Regia attack because with this acid attack it is possible to digest both lanthanides and precious metals. Using this method, only small amounts of insoluble materials (silica and plastics) were detected. After preliminary mineralization assays, the samples were mineralized following this procedure: about 0.3 g of samples were placed in a Teflon tube and Aqua Regia was introduced and left overnight prior to microwave digestion. Samples were subjected to microwave (ETHOS D, Milestone) irradiation at 250 W for 5 min, 500 W for 5 min, 650 W for 5 min in sealed vials. Before the acid attack, some powders were ball milled to evaluate if a lower grain size could promote a faster mineralization. Results showed that no difference was observed in mineralization time.

The cleavage reaction presented in Sect. 7.4.2 was experimentally tested to assess the chemical process for the obtainment of intermediate products suitable for further polymerization reactions from the organic fraction of PCBs. The choice of catalysts was based on a study about a transesterification reaction of dimethyl terephthalate in ethylene glycol catalysed by different acetates metals. As a preliminary screening, zinc acetate was used to identify the optimal conditions for experiments run at low temperature, e.g. refluxing methanol (68 °C). It must be stressed how the system was active even at a low temperature, considering that this reaction is commonly carried out in supercritical conditions. A brief metal salts screening has also been carried out, but zinc acetate turned out to be the most active. Following these encouraging results, a higher temperature screening was conducted in a jacketed reactor thermostated at

Table 7.1 CPS prototype functioning

M_w (Da)	M_n (Da)	M_z (Da)	PDI	M_z/M_w	M_{z+1}/M_w
734	728	741	1.009	1.009	1.02

165 °C. Compared to the low temperature results, at a higher temperature the catalytic activity distribution resulted more narrow, with the cheap lead acetate being the best catalyst. For such reasons different lead salts were employed. Among the different counterions, the acetate salt still resulted to be the most active, as commonly reported in literature for zinc catalysed reactions.

Once defined the best catalytic conditions, the kinetic trend for this reaction was examined. Finally, the recovered depolymerised fraction was subjected to analysis by Gel Permeation Chromatography (GPS). The results of such analysis are reported in Table 7.1 and confirm, as expected, a monodispersed sample of low molecular weight which could be re-employed as a starting material for polymerisation.

Finally, the procedure to achieve new re-usable compounds from the NMF of PCBs was experimentally validated. EPS scraps were dissolved in an environmental low-impact solvent at room temperature obtaining a fluidified paste. Previously milled, PCB scraps with/without tire rubber powders (1:1 weight ratio with respect to EPS) were added. After mixing in a HAAKE Rheocord at room temperature, the obtained slurry was calendared between steel cylinders (distance 2 mm) and immersed in cold water for 48 h to remove the solvent. Finally, dried sheets were pelletized and moulded using a heated press at 190 °C for 12 min. The mechanical characterization performed on the samples has shown interesting properties of the prepared materials. In particular, flexural tests have shown that the compound containing only EPS+PCB is characterized by modulus and stress at the break values higher than the material containing also rubber tire scraps. The obtained mechanical properties are comparable to those of virgin PS and other technopolymers. On the contrary, samples containing also the rubber tire phase have shown higher resilience values evaluated by fracture tests, thus assessing that the presence of rubber component is able to significantly toughen the material.

7.6 Conclusions and Future Research

In this work, a set of technological and business enablers were proposed to address the challenge of PCBs recycling. Due to the complexity of this issue, a multi-disciplinary approach was taken to develop solutions impacting on various stages of the process, in a logic of supply chain integration. Such solutions support: the selective automated disassembly of valuable PCBs components for re-use or for the optimisation of the downstream shredding and separation processes; the optimisation of shredding through process modelling finalised to the definition of optimal parameters for the obtainment of a target mixture, depending on input waste material; the environmentally-friendly chemical characterisation of PCBs MF to analyse the content of precious materials in waste PCBs with high confidence; the recycling of PCBs NMF through a new cleavage chemical reaction to create re-usable compounds from organic powders and though a new application for the production of a composite material using compounds obtained from organic PCBs fraction, mixed with other fillers; the establishment of new recycling business models through the uptake of novel technological solutions by different supply chain actors.

Developed technical solutions were successfully implemented in industrial prototypes and tested in laboratory conditions. However, experimental conditions were limited because of the wide variety of possible PCBs and components technologies, as well as the business uncertainty that is typical of recycling business. In addition, the industrial applicability and sustainability of developed technologies and business models should be investigated more in detail. The following activities are outlined for future research and innovation:

- Evaluation of industrial feasibility of automatic selective disassembly process and identification of the product types that justify such an approach. This innovative process, in fact, is characterised by higher costs compared with low-cost manual disassembly. Thus, it is foreseen that such an application is justified only in the business cases where treated components are of high added-value. This can happen when they can be re-sold in the spare parts market or they are made of particularly valuable metals for which pre-concentration can guarantee higher efficiency of downstream separation processes and lower materials losses.
- Extend the validation of the shredding process model through a wider experimental campaign including different type of PCBs. Besides the validation and improvement of the model, other important activities deal with the estimation of the economic benefits of model-based process optimisation and the readiness of companies to accept such a sophisticated solution compared to the current way of operating shredding machines. From the research point of view, the integration between shredding and separation processes should be investigated in the future with a joint approach.
- Further testing of the robustness of the chemical methodologies developed to characterise PCBs MF and to allow recycling of PCBs NMF. In particular, for the characterisation process, the developed testing procedure should be applied to various samples of PBCs of different composition to test its performance. In addition, characterisation cost and lead time should be estimated and compared to industrial needs. Regarding PCBs NMF recycling, industrial and business sustainability should be assessed by identifying potential applications which makes it cost effective. Furthermore, Life Cycle Assessment analyses should be carried out to verify the effective environmental advantage of the recycling under the light of the additional operations that need to be introduced.
- Better characterisation of the proposed new business models through empirical research involving more companies at different levels of the supply chain and performing a quantitative assessment of economic and environmental performance at business model level, considering also risks associated with the uncertainty of

recycling business. Detailed return on investment simulations should be elaborated for the adoption of the solutions realised in this work.

Acknowledgements This work has been funded by the Italian Ministry of Education, University and Research (MIUR) under the Flagship Project "Factories of the Future—Italy" (Progetto Bandiera "La Fabbrica del Futuro" [23], research projects "Zero Waste PCBs—Integrated Technological Solutions for Zero Waste Recycling of Printed Circuit Boards (PCBs)", "PCB-ID—In-line automated device for the identification of components and the characterization of materials and value in waste PCBs" and "ShredIT—Self-Optimizing Shredding Station for De-manufacturing Plants"). Authors are grateful to Marco Diani and Francesco Baiguera for their support to the finalisation of this research and publication.

References

1. Report on Critical raw materials for the EU, Report of the Ad-hoc Working Group on defining critical raw materials (2014) https://ec.europa.eu/docsroom/documents/10010/attachments/1/translations/en/renditions/pdf. Accessed 30 July 2018
2. Copani G, Brusaferri A, Colledani M, Pedrocchi N, Sacco M, Tolio T (2012) Integrated de-manufacturing systems as new approach to end-of-life management of mechatronic devices. In: Proceedings of the 10th global conference on sustainable manufacturing, Istanbul, 31 Oct–2 Nov 2012
3. Colledani M, Copani G, Rosa P (2014) Zero Waste PCBs: a new integrated solution for key-metals recovery from PCBs. In: Proceedings SUM 2014, second symposium on urban mining, Bergamo, Italy, 19–21 May 2014
4. Borrotti M, Pievatolo A, Critelli I, Degiorgi A, Colledani M (2015) A computer-aided methodology for the optimization of electrostatic separation processes in recycling. Appl Stoch Models Bus Ind 32(1):133–148
5. Colledani M, Copani G, Tolio T (2014) De-manufacturing systems. Procedia CIRP 17:14–19
6. Kara S, Pornprasitpol P, Kaebernick H (2006) Selective disassembly sequencing: a methodology for the end-of-life products. CIRP Ann 55(1):37–40
7. Hall WJ, Williams PT (2007) Processing waste printed circuit boards for material recovery. Circuit World 33(4):43–50
8. Cui J, Zhang L, (2008) Metallurgical recovery of metals from electronic waste: a review. J Hazard Mater 158(2, 3):228–256
9. Huang K, Xu Z (2009) Recycling of waste printed circuit boards: a review of current technologies and treatment status in China. J Hazard Mater 164:399–408
10. Goosey M, Kellner R (2002) A scoping study end-of-life printed circuit boards. Intellect Shipley Europe Limited. www.cfsd.org.uk/seeba/TD/reports/PCB_Study.pdf
11. Li J, Xu Z, Zhou Y (2007) Application of corona discharge and electrostatic force to separate metals and non-metals from crushed particles of waste printed circuit boards. J Electrost 65:233–238
12. Li J, Xu Z (2010) Environmentally friendly automatic line for recovering metal from waste printed circuit boards. Environ Sci Technol 44:1418–1423
13. Kulkarni P, Chellam S, Flanagan JB, Jayanty RKM (2007) Microwave digestion-ICP-MS for elemental analysis in ambient airborne fine particulate matter: rare earth elements and validation using a filter borne fine particle certified reference material. Anal Chim Acta 599:170–176
14. Ongondo FO, Williams ID, Cherrett TJ (2011) How are WEEE doing? A global review of management of electrical and electronic wastes. Waste Manag 31:714–730
15. Tolio T, Copani G, Terkaj W (2019) Key research priorities for factories of the future—part I: missions. In: Tolio T, Copani G, Terkaj W (eds) Factories of the future. Springer

16. Colledani M, Tolio T (2013) Integrated process and system modelling for the design of material recycling systems. CIRP Ann 62(1):447–452
17. Tolio T, Copani G, Terkaj W (2019) Key research priorities for factories of the future—part II: pilot plants and funding mechanisms. In: Tolio T, Copani G, Terkaj W (eds) Factories of the future. Springer
18. Dehling HG, Gottschalk T, Hoffmann AC (2007). Stochastic modelling in process technology. Elsevier
19. Danadurai KSK, Chellam S, Lee CT, Fraser MP (2011) Trace elemental analysis of airborne particulate matter using dynamic reaction cell inductively coupled plasma-mass spectrometry: application to monitoring episodic industrial emission events. Anal Chim Acta 686:40–49
20. Cerruti P, Fedi F, Avolio R, Gentile G, Carfagna C, Persico P, Errico M, Malinconico M, Avella M (2014) Up-cycling end-of-use materials: highly filled thermoplastic composites obtained by loading waste carbon fiber composite into fluidified recycled polystyrene. Polym Compos 35(8):1621–1628
21. Avella M., Errico M, Fabozzi G, Lucchesi C, Lucchesi G, Malinconico M, Petrucci M, Ponzecchi E (2009) Process and plant for the production of composite thermoplastics and materials. International Patent WO/2009/072150
22. SUSRAC Report Summary, Cordis Europa. https://cordis.europa.eu/result/rcn/156514_en.html. Accessed 10 July 2018
23. Terkaj W, Tolio T (2019) The Italian flagship project: factories of the future. In: Tolio T, Copani G, Terkaj W (eds) Factories of the future. Springer

Part IV
Factory for the People

Chapter 8
Systemic Approach for the Definition of a Safer Human-Robot Interaction

Alessandro Pecora, Luca Maiolo, Antonio Minotti, Massimiliano Ruggeri, Luca Dariz, Matteo Giussani, Niccolò Iannacci, Loris Roveda, Nicola Pedrocchi and Federico Vicentini

Abstract Smart factories must speed up their processes to face new manufacturing challenges and, at the same time, demonstrate an extremely high degree of flexibility to reduce production costs and time. This kind of issues can be addressed by the cooperation between humans and robots in a mixed human-robot working environment. Robots have the compelling advantage of spatial precision and repeatability as well as the capability of applying defined forces. Humans, on the other hand, are especially skilled at complex manipulations and adapting to changing task requirements. In this complicate scenario of co-shared workplace and continuous human-robot interaction, safety strategies are a key requirement to avoid possible injuries to humans or fatal accidents. This chapter proposes a systemic approach to respond to these requirements. The approach merges and manages multiple sensing sources, redundant transmission protocols and software decision mechanisms, aiming to guarantee a continuous and reconfigurable co-share scenario that enables an operative interaction between human workers and robots in a controlled and safe environment. Furthermore, new technological solutions and innovative methodologies are presented for the definition of a safer workplace in human-robot interaction scenarios.

8.1 Scientific and Industrial Motivations

The need to face the production of complex families of product, with short life cycles, often characterized by strongly variable production patterns, requires factories to be able to quickly evolve and reconfigure [1]. These reconfigurable factories

A. Pecora (✉) · L. Maiolo · A. Minotti
CNR-IMM, Istituto per la Microelettronica e Microsistemi, Rome, Italy
e-mail: alessandro.pecora@cnr.it

M. Ruggeri · L. Dariz
CNR-IMAMOTER, Istituto per le Macchine Agricole e Movimento Terra, Cassana, FE, Italy

M. Giussani · N. Iannacci · L. Roveda · N. Pedrocchi · F. Vicentini
CNR-STIIMA, Istituto di Sistemi e Tecnologie Industriali Intelligenti per il Manifatturiero Avanzato, Milan, Italy

© The Author(s) 2019
T. Tolio et al. (eds.), *Factories of the Future*,
https://doi.org/10.1007/978-3-319-94358-9_8

173

must exhibit strong modularity and adaptability to exogenous and endogenous events [2, 3]. In accordance with production dynamics, the design of workplaces within factories must take into account the current social and demographic trends (e.g. increase in average life expectancy and increase in retirement age). Attention to issues such as safety, education and ergonomics can also be the key lever to improve productivity and profitability of the factory.

Traditionally, factory design has clearly distinguished between the roles and areas of expertise of human operators and automated and/or robotized production cells. Unfortunately, this operative model is nowadays obsolete since it cannot react and quickly adapt to the incoming opportunities arising from the global market. In this new scenario, it is crucial to provide innovative paradigms that ensure an efficient and safe work environment, as well as comfortable and stimulating for human workers. The new factories that make cooperation between humans and automated devices will be able to cope with difficult production scenarios characterized by products with short life cycles and high variability, thus requiring a fast adaptation of production capacity and the development of new knowledge [4].

The research presented in this chapter will specifically address the topic of designing and implementing human worker and robot cooperating systems. In particular, design methodologies, technologies and devices will be explored to guarantee a safe and fruitful human-robot interaction. To this aim, it is necessary to address themes related to the study of man-machine interactions, and hence the modelling of work areas and their perception by robots using advanced sensing systems, thus creating safe hybrid work environments. Indeed, the achievement of a strong interaction between a robot and a human worker, in a completely safe environment, relies on the development of a suitable framework including sensing technology, system architecture and proper cognitive models [5]. A sensor network must provide a sufficient amount of information to the robot about the surroundings and these data must be reliably transmitted and interpreted by the machine to adopt the right behaviours. Moreover, the suitable sensing technology must be shared and it has to be portable, light and easy to wear so that the worker can comfortably operate in the workplace [6].

As case study, the innovative approach will be proposed in a fenceless assembly/disassembly scenario considering peculiar issues in different workcells and along different workflows processes. Especially in disassembly processes, in fact, many diverse tasks must be executed according to local workflow, adapting the control settings for general and local motion/interaction planning with a high grade of versatility. Such a complex ensemble of abilities from both sides (robots and human workers) requires a transparent and seamless dynamic modification of the interaction at a system level in which human workers can suitably exchange tasks, enter/exit parts of the task execution, find a due—i.e. safe and comfortable—behaviour of the robots, thus avoiding any possibility of injuries or uneven unexpected robot behaviours. This contextual analysis represents a significant working environment since it joins a high level of adaptability and variability of processes, parts, manufacturing conditions, specific mechanical parameters and, more critically, unique human-robot association modes and assignments in assisted/cooperative mechanised manufacture [7].

In the next sections, the development of a full strategy to improve safety in typical fenceless scenarios will be described and examined starting from the concepts of dynamic supervision and adaptation of the interaction and combining multiple sensing sources with a novel WPAN (Wireless Personal Area Network) extended in the robot area. The proposed innovative approach will face also the development of algorithms and cognitive systems for the perception and representation of space-time environment of the factory and for the implementation of automatic processes of decision and action.

8.2 State of the Art

The first layer to build a systemic approach for the definition of a safer human machine environment relies on the possibility to maximize and diversify the different sensor sources deployed both on the robot and on the worker with the intent of providing a continuous feedback related to the position of the worker and his tasks. However, the presence of various materials (conductive, insulating, magnetic, etc.) in a specific workplace and the consequent issue related to anomalous absorbance in the electromagnetic spectrum still represents a critical aspect in the creation of a reliable set of data for the robot. In the years, many different sensing solutions have been proposed to overcome these issues and most of them rely on the use of multiples cameras and microphones [8]. Unfortunately, these technologies are limited by the line of sight of the associated camera or produce signals that are difficult to interpret for the robot. A further possible solution considers the specific operation distance between the worker and the robot (short and long-range). In case of long-range sensing technology, radio-frequency identification technology (RFID) has recently received a great attention due to manufacturing costs, power consumption, sensors wearability, response time, etc. [9]. Indeed, the tags embedded in worker smart garments (gloves, suite, etc.) can provide a reliable information of a tracked movement, without hindering the worker operation. Robots can use the signals coming from the RFID sensing technology to properly interpret the movements of the human worker and to consequently adapt their behaviour. However, also in RFID technology, different problems can arise from the signal detection at particular angle and from the specific distance between the tag and the reader involved in the localization process and the signal can be misinterpreted by the robot [10].

To properly manage and transmit the data produced by all the sensing networks, a robust wireless infrastructure needs to be adopted. In this case, the identification/recognition system identifies the operator presence and position in the robot area, using the combination of safe wireless channels and of unsafe low power wireless communication protocols (like the Ultra-low power Bluetooth end Zigbee). This process works both acquiring presence of communication link between the access points in the robot area and the units directly in the operator body, and applying an algorithm featuring the geo-referenced position of the operator based on the signal reception power from the operator unit. A deep analysis on communication physical

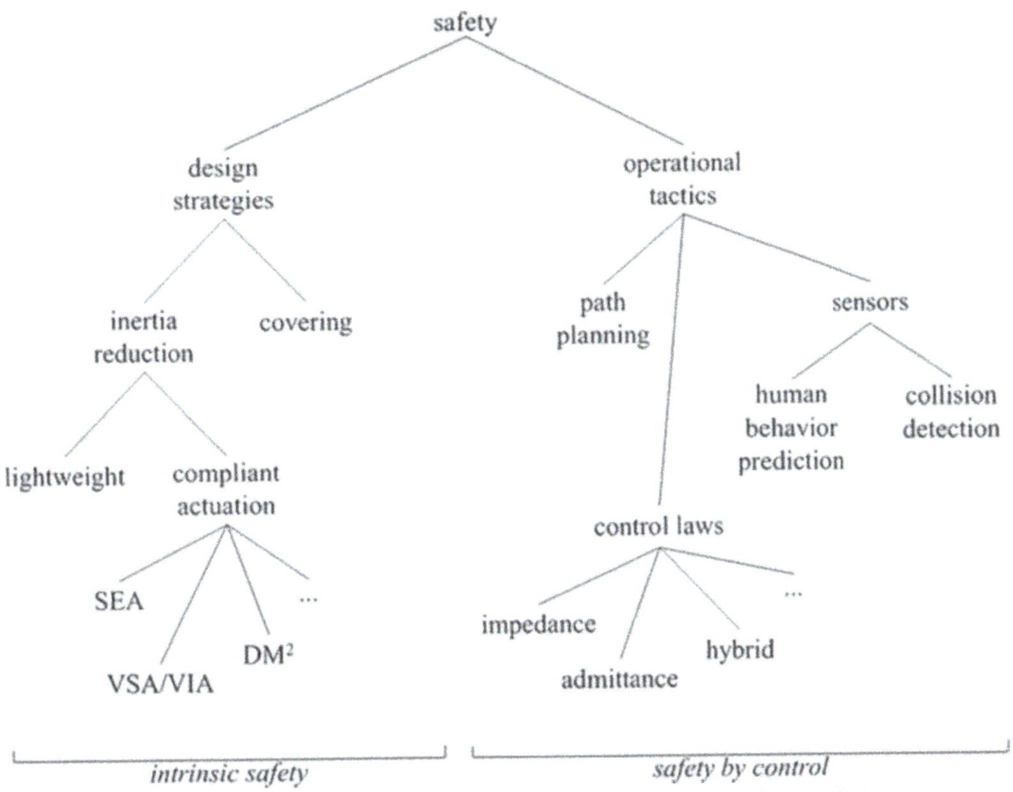

Fig. 8.1 Map of hardware features or design strategies that are used for embedding safety in robot systems

characteristics, power consumption, power management, easiness of reception power recognition systems, transceiver integration, availability and robustness in respect to industrial environment, must be taken into account to provide reliable data between the robot and the human workers.

Flexible/reconfigurable production, specifically when manual tasks can be assisted by robots, is remarkably supported by collaborative robotic methodologies that can guarantee a safe behaviour in unstructured dynamic environments. Lightweight solutions specifically designed for physical Human-Robot Interaction (pHRI) can be employed for these tasks. Since years, both academic [11–13] and manufacturing settlements [14, 15] have given careful consideration to broadening the coordinated effort space amongst robots and workers in industrial robotics as well. Industrial robots, from small to large size manipulators (e.g. anthropomorphic, Cartesian, parallel, etc.) still do not extensively display native safety features, either by design or by control (see Fig. 8.1). Nevertheless, such platforms are required to be integrated in safe setups at both machine and plant level.

On the normative side, most recent standards define human-robot cooperation modes and happen to apply on top of the consolidated standardization initiatives in machinery and robotics [16–21] that encapsulate the concept of *functional safety* altogether. Furthermore, safety is obviously an integrated concept, whose procedu-

Fig. 8.2 Dependability chart

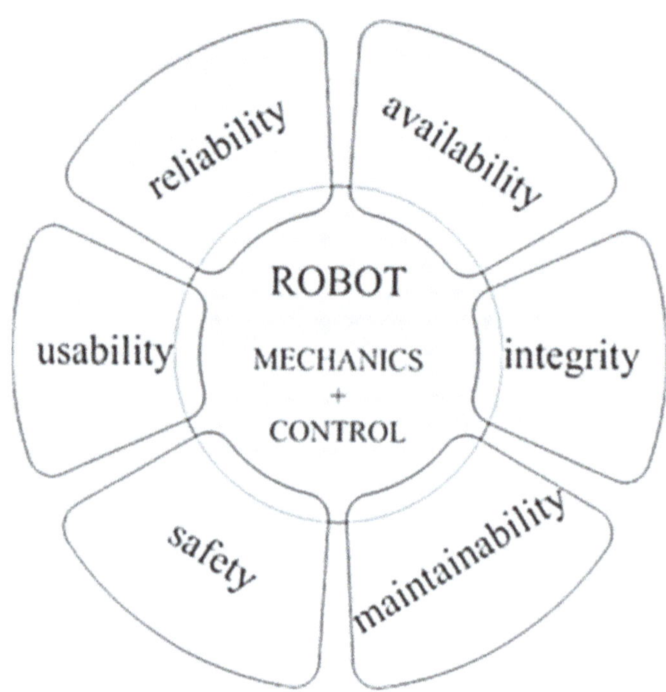

ral, logical and technological aspects are spread among hardware, algorithms and networks.

Safety options of devices are in all probability among the foremost self-addressed by technology suppliers due to a consolidated regulation [16, 18] and restricted needs in terms of integration may hinder the safety certification grade.

The notion of interaction in robotics can span different spatial conditions, from distant functional inter-action to close contact. Non-contact interaction is involved in the task cooperation and scenarios where the same workspace is concurrently visited. Contact interaction can occur either by planning or by accident. The former is primarily involved in most of the low-power tasks, like manipulation, handling, assembly, etc. where compliant controls represent a prominent class of settings. The latter has been considered as the major issue to be discussed and formalized in relationship with the normative requirements of safety, the very wide family of collision avoidance algorithms, etc.

At all interaction levels the cooperating robots should be transparent, i.e. they should resemble a *naturally* interacting companion, removing all cumbersome configurations or access modes, uneven motions or patterns, etc. This behaviour can be formulated through the concept of *dependability* that is considered as the property of displaying easiness, confidence and reliance to the user when performing interaction tasks. Under such conventionally abstract notion, dependability wraps a set of clearly defined child properties (see Fig. 8.2) [22] that altogether have a great impact over the robot control and resulting behaviour.

User confidence, as a matter of fact, stands on:

- usability, which can measure the feeling, comfort and easiness of learning and/or setting up the robot motion,
- reliability, which mainly addresses the continuity of service rarely throwing unexpected exceptions or rarely presenting failures,
- availability, which gives account of readiness of such service,
- maintainability and integrity, which are the ability to recover from system alteration and the absence of improper alterations, respectively,
- safety, which has been extendedly discussed before.

Most of dependability child properties can be fulfilled as a trade-off at whole system level. The dependability in the interaction, and its effects on the dynamic allocation of robot behaviour, relies on the cognitive aspects of the human robot interaction. Ongoing studies in the field of brain research have concentrated on psychological procedures of joint-activity among people to empower a successful cooperation [23]. Psychological investigations demonstrate that an effective collaboration in human teams requires members to arrange and execute their activities while anticipating the actions of the other colleagues, and not simply reacting to them [24]. Therefore, a proficient human/robot collaboration needs an effective communication [25–27]. The benefit of anticipatory actions in a human/robotic environment is proven in [28], showing a significant improvement of task efficiency compared to a reactive behaviour. A shared representation of human and robot abilities must be available to properly assign the tasks to the partners in the working group [29, 30]. Human-Robot collaboration using peer-to-peer approaches is a key research topic in robotics [31]. Laengle et al. presents a human-robot team employing a multi-agent control architecture [32], where the mobile system consists of an overhead-camera and a two-arm-manipulator. A proactive collaboration based on the recognition of intentions is proposed in [33]. Intention is a state of mind of the human that cannot be measured directly. However, human action is a result of intention, so Dynamic Bayesian Networks (DBNs) can be used to cope with these uncertainties [33]. The approach in [29] presents a scenario where a human and a robot system with two robotic arms cooperate to build a wooden model of an aircraft using a cognitive architecture consisting of high-level components input, interpretation, representation, reasoning, and output with several functional modules.

The execution of interactive tasks in an assembly cooperative workcell relies on measurements from distributed sensors that return system and users information.

8.3 Problem Statement and Proposed Approach

The proposed robot control architecture aims at providing safe collaborative behaviours of the robots that are sharing the workspace, or physically interacting with human operators. The technological limits for enabling a multi-model collaboration (i.e. contact-less or physical interaction) with robots, are mainly deriving from the lack of real-time, failsafe tracking of humans inside shared workspaces.

Additionally, all functions relevant to safety are required to comply with functional safety requirements.

The approach for solving the safe workspace sharing problem is then mainly based on the design of an architecture where all necessary sensing/control devices are integrated and synchronized, and that must be able to display failsafe properties.

The central infrastructure is a robotic Networked Control System (NCS, see Fig. 8.3) where tangible data are given by a heterogeneous arrangement of sources, including the sensors, diverse computational agents for the elaboration of functional application information and possible comparisons with stored data. The NCS provides an infrastructure for integrating general devices, including not safety-rated components. Fail-safe characteristics of the NCS architecture are based on several layers of cross-checking and packets validation data to decrease the probability to miss any detection of dangerous faults. Regarding functional safety, the target achievement is to reduce the Probability of Dangerous Failures per Hour (PFH$_d$) to negligible levels, in particular corresponding to ISO 13849 *PLd*, which is in general the normative requirement for industrial robots. Elements in a NCS are not required to be secure in standalone mode, but their data must be monitored and validated online.

This approach permits to make transparent and continuous changes in the robot behaviour, as long as such algorithms are always protected from data or protocol failures. In particular, the most relevant components in the NCS will be presented in the next section and can be anticipated as follows:

- The functional design and development of the architecture of unsafe devices to synchronize and process data coming from the localization sensing arrays and from robots. An outline of the architecture design is provided in Sect. 8.4.1.
- Design and development of ad hoc WAN (Wide-Area-Network) network and WPAN (Wireless Personal Area Network) networks with an enhanced level of redundancy of data. This set of protocols enables the interconnection of remote sensors into the architecture. Details are provided in Sect. 8.4.2.
- The input for localizing human operators, to be integrated in the safety architecture, deriving from general purpose sensors. The design and integration of innovative wearable smart sensing systems for long range (less than 2 m) and short range (less than 0.5 m) localization to be worn by the human operator are described in Sect. 8.4.3.

In case of long range sensor technology, the requirements based on low power consumption and precision of human positioning addressed the choice to the usage of distance estimation through Received Signal Strength (RSS) measuring. This measurement represents the most accessible transmission parameter to estimate the distance between the nodes.

Wearable infrared devices, based on photoresistors with flexible thin film germanium sensing layer, will be presented for short range sensors [34, 35]. The system is conceived by housing an infrared source or multiple infrared sources of the robot and integrating the photoresistors directly on a work glove. The system automatically bal-

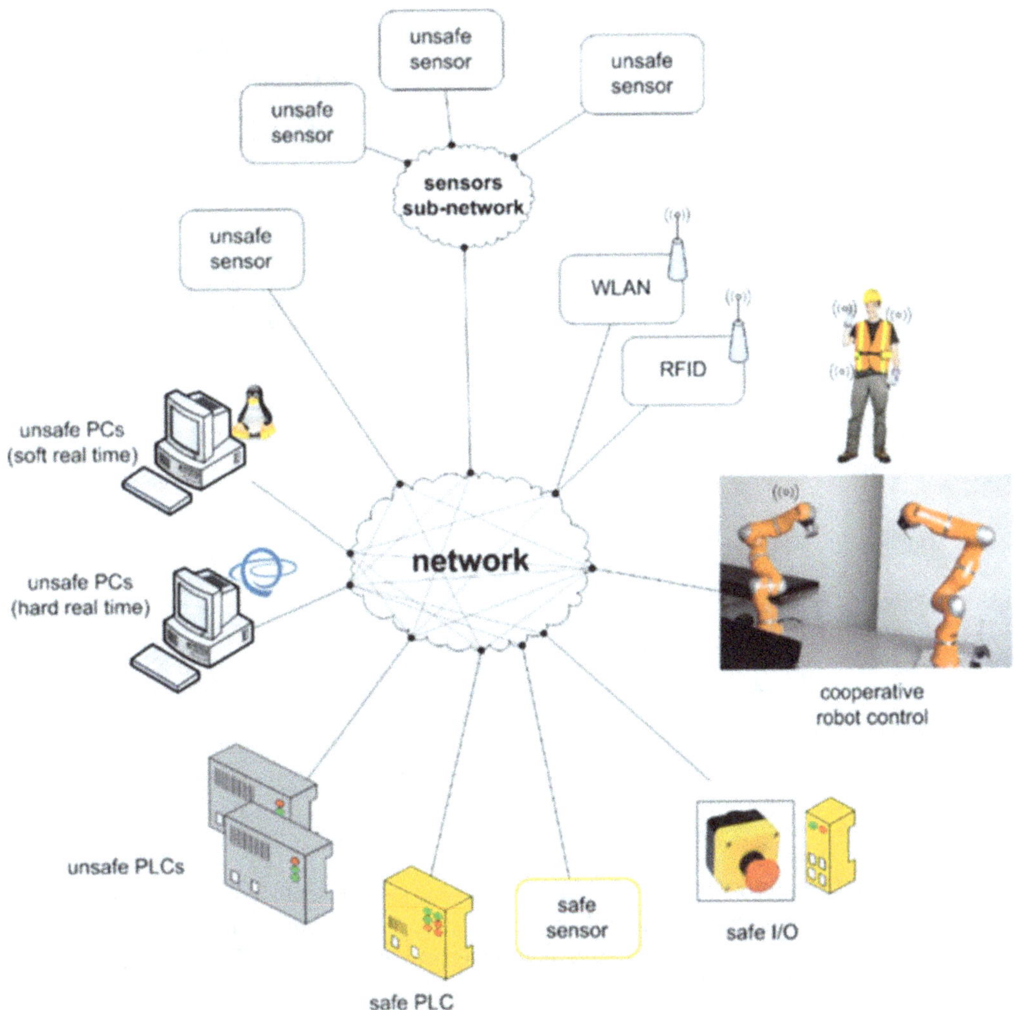

Fig. 8.3 Schema of a Robotic NCS. Safe-grade nodes (CPUs and sensors) are represented in yellow; CPUs can run different Operating Systems; sensors can be grouped in subnets

ances temperature variations and takes into account the infrared spatial distribution of emitting LED array.

8.4 The Experimental Solution

8.4.1 Architecture for Safe Distributed Robotic Systems

The core methodology for a safe distributed robot system consists in monitoring data channels and exchanged messages, to reduce the probability of failing in detecting data processed by sensor/monitoring nodes. Redundancy, diversity and monitoring are applied by a double independent elaboration of single channel sources. Then, the

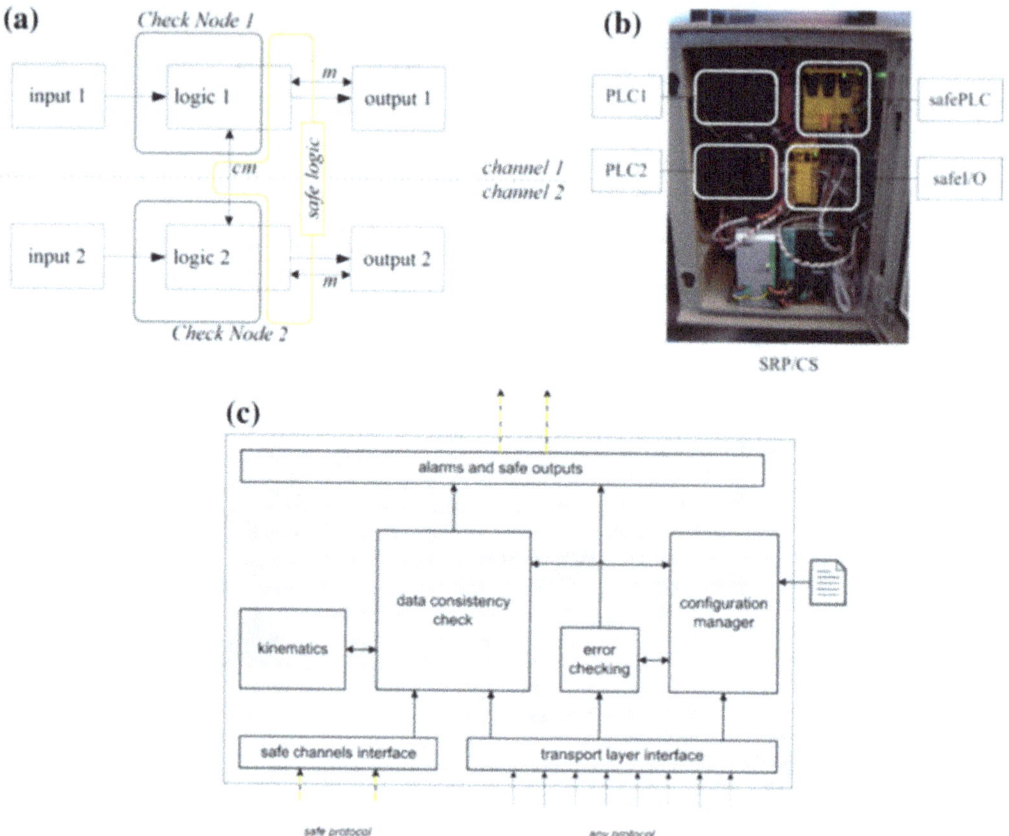

Fig. 8.4 **a** ISO13489-1:2006 model for Cat:3 architecture with deployed SRP/CS components (CPU1, CPU2 and safe CPU) outlining their logical domain. "m" are monitored safe state tasks execution. "cm" is the cross-monitoring of both channels. **b** Deployed SRP/CS featuring hardware components (PLC1, PLC2 and safePLC). **c** Architecture of the "logic" parts of the control system/safety related part (SRP/CS)

output data of comparative units (check nodes, programmed on standard PLC—see Fig. 8.4) are checked for consistency by a final safe node that plays the role of safety gate between the safety functions domain and all the general purpose CPUs or the unsafe sensors. Architecturally, the Safety-Related Part of a Control System (SRP/CS) is distributed in three logical components: two check-node CPUs, capable to support standard floating-point computation and to connect to user-defined libraries, and a component embedding a safe logic that is suitable to fulfil the preliminary conditions of dual structure and availability of monitoring coverage [36–38].

Components integrated in the NCS architecture can track devices and robots, using both wired and wireless communications infrastructures (see Fig. 8.5). In particular, the focus was on the verification of the communication interfaces (WLAN and UDP/IP) and analysis/validation of the specific parameters of the subsystems, including failures detection and protocol performance.

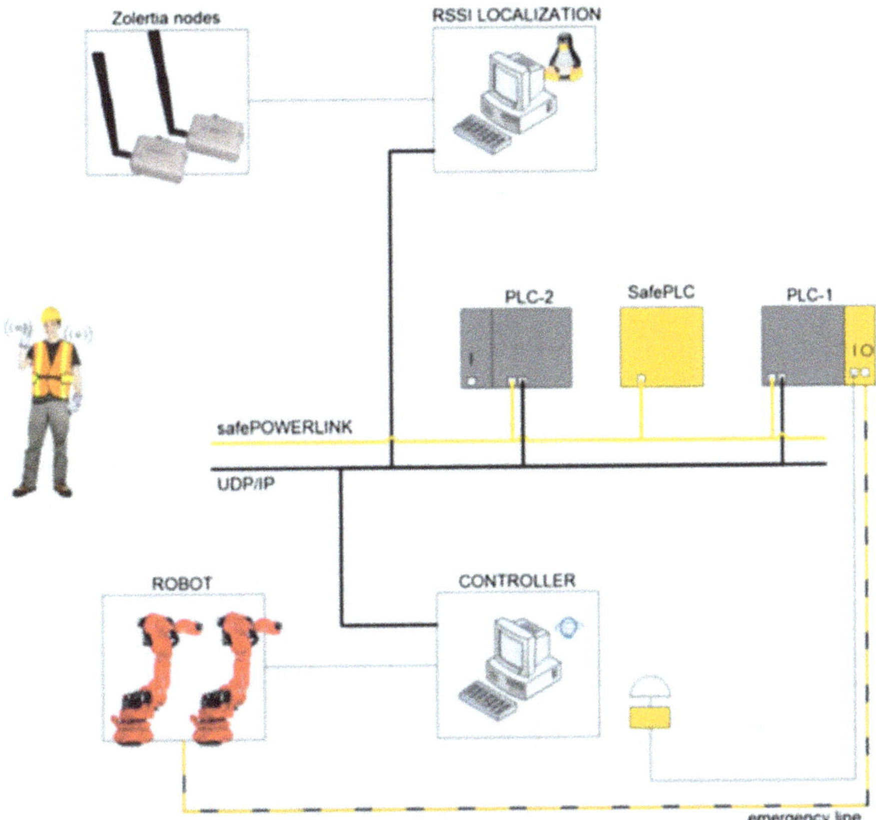

Fig. 8.5 Layout, devices and connections for the solution design [38]

8.4.1.1 Gateway Architecture for Wireless Node in the NCS

The Gateway Node is composed of a wireless node, based on the IEEE 802.15.4 standard, which handles the communication schedule of the real-time Wireless Sensor Network and forwards the data to an embedded PC through a USB link, as visible in Fig. 8.6. The embedded PC then forwards the data through a wired Ethernet network, encapsulating the data on UDP packets. The Gateway sends the raw data to a Localization Node, implementing the localization algorithm described in Sect. 8.4.3.3, which itself sends the estimated position to the Check Nodes, which provide the interface with the robot control network.

Two alternative scenarios have been considered as shown in Fig. 8.7.

With the first approach (Extended architecture in Fig. 8.7a), the Gateway Node must act as a bridge, i.e. it reads data from the real-time WSN, changes the safety header and forwards the packets to the Ethernet interface towards a separate Localization Node. Conversely, with the second approach (Reduced Architecture in Fig. 8.7b) the Gateway Node embeds the functionality of Localization and Check Node.

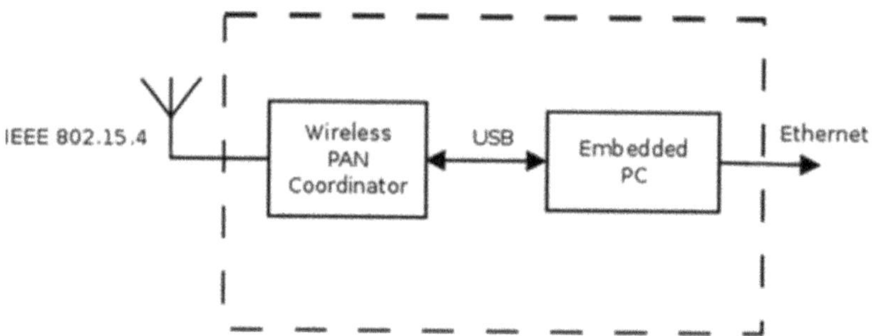

Fig. 8.6 Block diagram of the Gateway Node

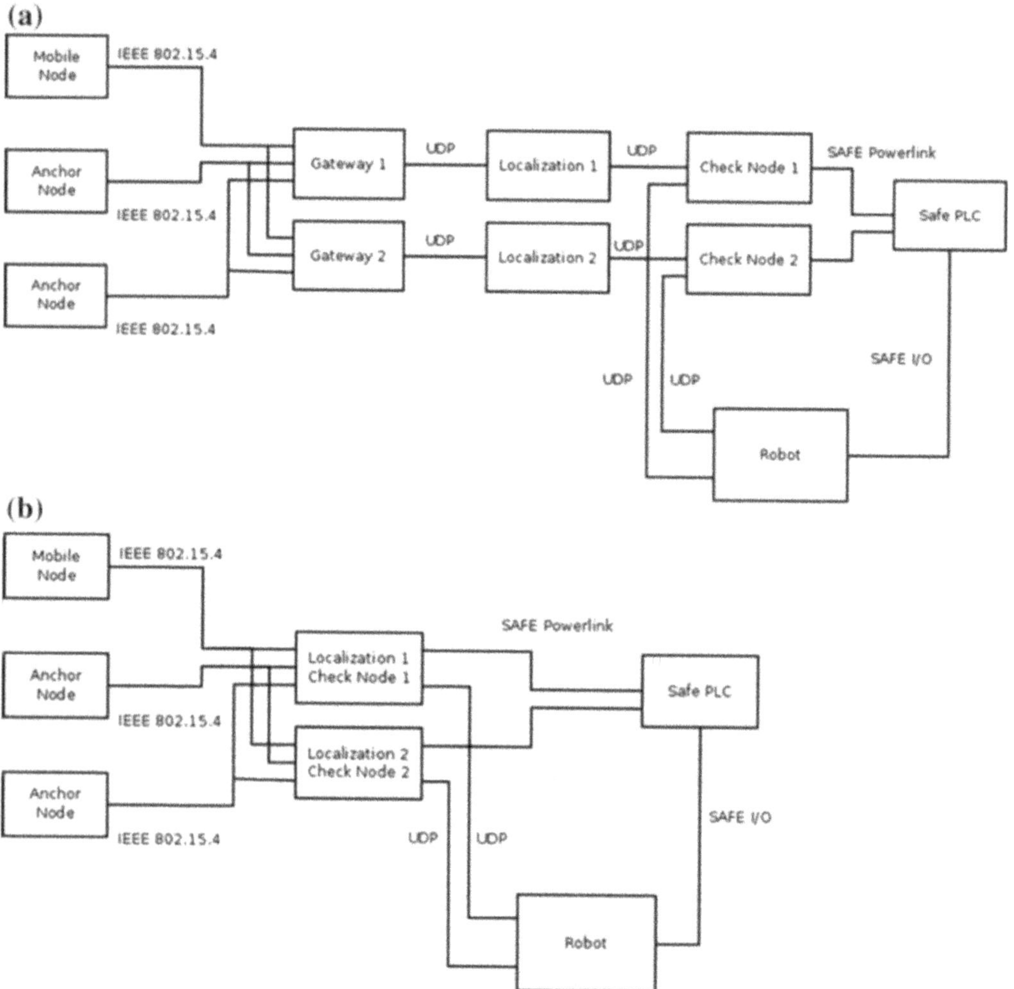

Fig. 8.7 **a** Extended architecture, **b** Reduced architecture

The choice between the two architectures depends on the computational power available on the Gateway Node; in the integration experiments the Gateway node was realised first with a low-power and low-performance embedded board (Zolertia Z1[1]), therefore corresponding to the Extended Architecture, and subsequently with a more powerful embedded board (Olimex P207[2]) that includes an Ethernet connection.

At the safety level the communication happens between the Wireless Sensor Nodes and the Localization Node through a safe communication channel. Data collected from the safe channel are used to compute the estimated position of the Mobile Node and to check errors that can occur, due to the intrinsic characteristics of the wireless channel. The localization algorithm should therefore not use expired or duplicated data, check data integrity and promptly forward to the Check Node an error status when it is not possible to reliably collect enough data from the Wireless Sensor network to perform the localization algorithm.

8.4.2 Software Stack for Real-Time Wireless Sensor Network in the NCS

The communication stack is based on IEEE 802.15.4e-LLDN (Low Latency Deterministic Network), with some modifications that enable to further reduce the worst-case communication latency. The code was initially designed to run on an embedded microcontroller without an operating system, subsequently it has been ported to a real-time operating system, which is needed to satisfy more strict real-time constraints. The software is composed of a high level Medium Access Control (MAC) and a low level MAC; the former controls the content of the packets, the configuration and the management of the network, while the latter handles the transmission and reception of the packets (see Fig. 8.8). The software is modular and can be easily ported to other platforms; as an example, it has been ported to an ARM Cortex-M3 and a discrete-time network simulator (NS3). The reference implementation is based on a commercial off-the-shelf wireless sensor network module, the Zolertia Z1, with a 16-bit RISC microcontroller (MSP430) running at 16 MHz.

The implementation presents some optimizations compared to the standard LLDN mode that reduce the maximum worst-case latency of the communication on the wireless channel and the time of system configuration; conversely the number of nodes that can be connected to the network with the same performance are also reduced [39–41].

The communication between the sensor nodes and the gateway node (Personal Area Network coordinator), according to IEEE 802.15.4e LLDN specifications, is synchronized through a temporal structure called superframe, which implements a Time Division Multiple Access method (TDMA).

---

[1] https://github.com/Zolertia/Resources/wiki/The-Z1-mote.
[2] https://www.olimex.com/Products/ARM/ST/STM32-P207/.

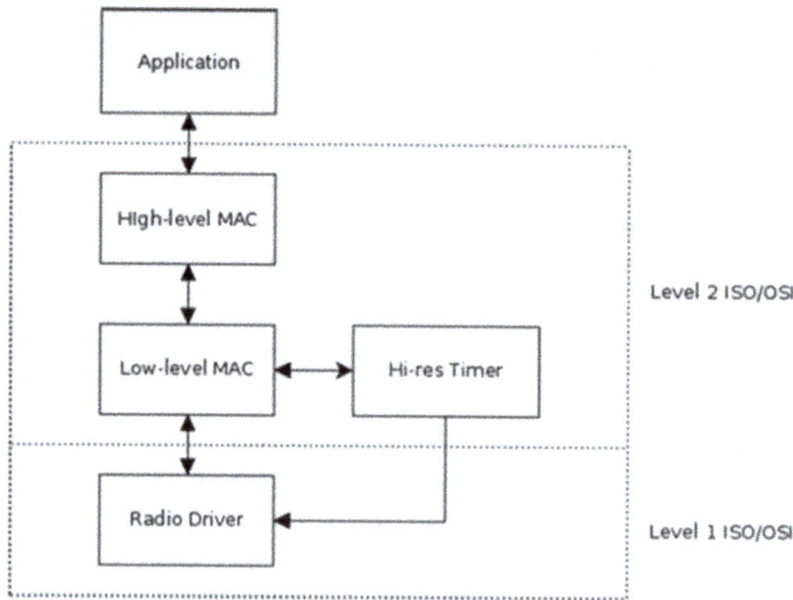

Fig. 8.8 Architecture of the wireless sensor network communication stack

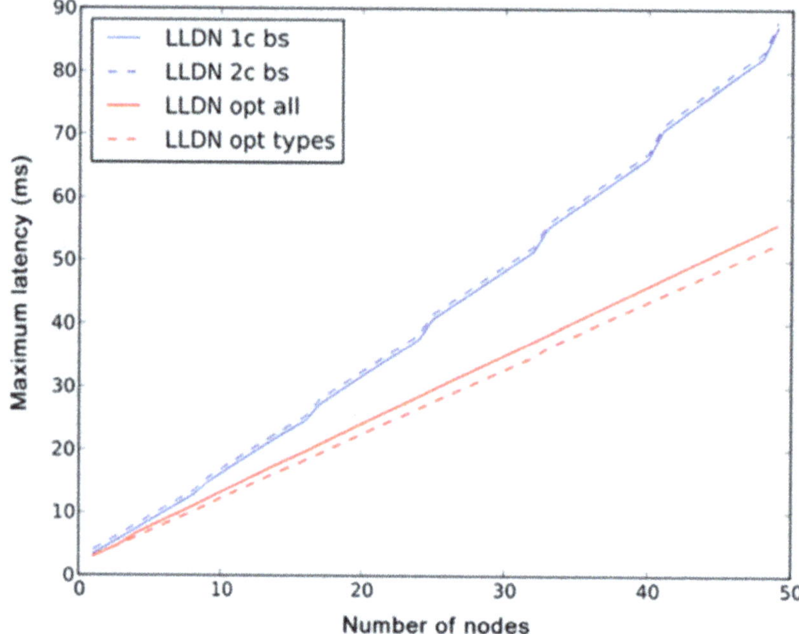

Fig. 8.9 Maximum latency achieved with sensor nodes and anchor nodes. It can be seen the net improvements in terms of time cycle in the regime phase of the network system thanks to the allocation of time slots

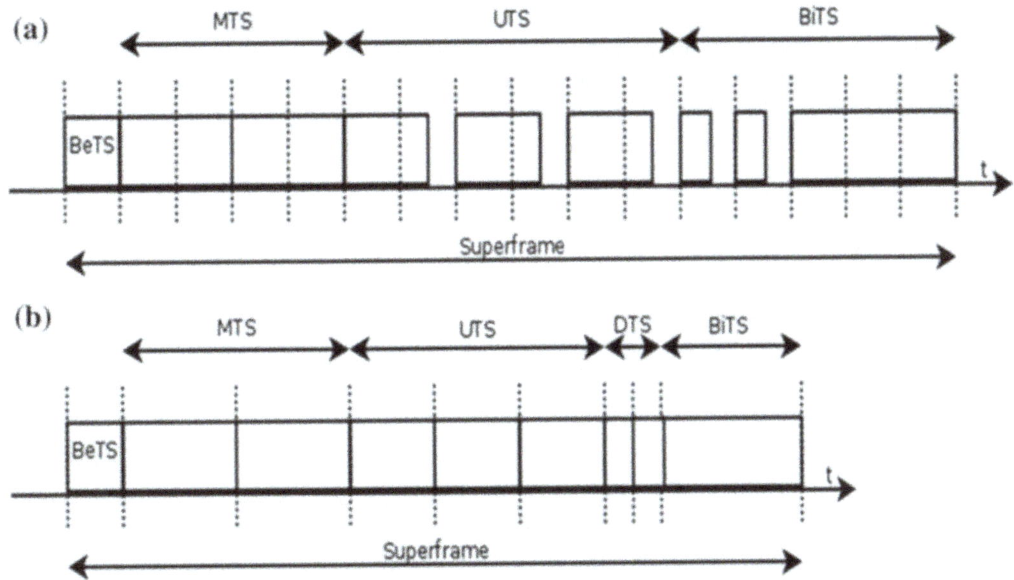

Fig. 8.10 **a** Standard IEEE 802.15.4 LLDN superframe format; **b** Proposed modified superframe format

Table 8.1 Superframe time for Standard and Modified LLDN for 10 and 20 nodes in a real-time WSN [38]

# Nodes	Superframe time [ms] Standard LLDN	Superframe time [ms] Modified LLDN
10	22.1	17.3
20	41.3	32.7

Figure 8.9 shows a comparison between two standard LLDN configurations (in blue) and two optimized configurations (in red), where a reduce communication latency is possible with the same number of nodes on the WSN. Figure 8.10 shows the comparison between a standard LLDN superframe structure and the superframe structure of the optimized LLDN protocol, which is the reason for the performance improvement showed in Fig. 8.9. Some numerical results are reported in Table 8.1.

The Configuration phase of the superframe structure has also been optimized, obtaining a significant reduction of configuration time of the 802.15.4-based wireless network. This has been achieved through an exclusive access of the nodes to the channel in the initial setup of the network, where in standard LLDN a shared access method is used. This required the modification of the packets used to discover all the nodes present on a WSN. The optimized configuration phase has been simulated using a Monte-Carlo approach for the standard LLDN, to model the shared access method, while for the optimized LLDN only a deterministic computation is required.

Figure 8.11 shows a comparison of the Configuration time between Standard and Modified LLDN Protocol.

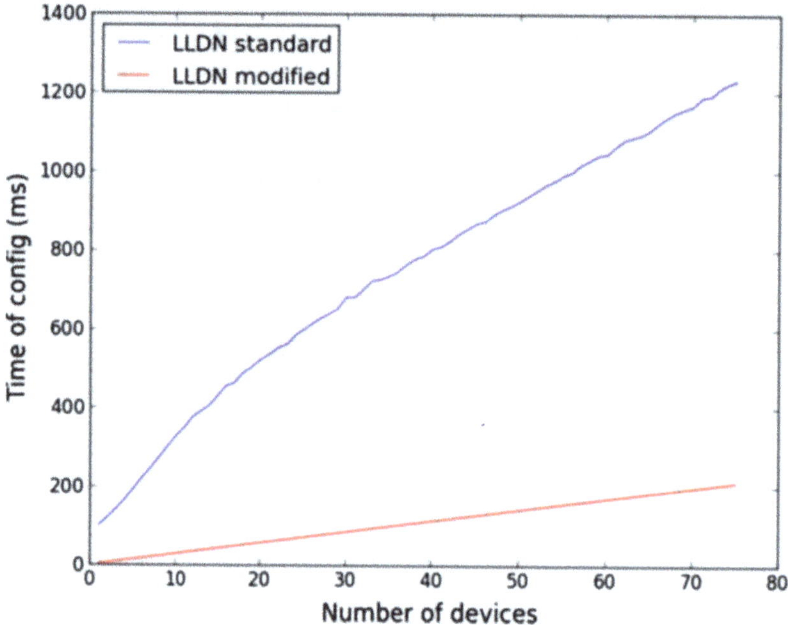

Fig. 8.11 Configuration Time for simulated Standard LLDN protocol, highlighting the time taken for procedure adhering to the standard (blue line) and calculated Modified LLDN (red line)

8.4.2.1 Safety Layer for Wireless Sensor Network

The safety layer implements the mechanisms needed to detect all the possible communication errors, thus enabling a reliable communication over a transmission channel that is intrinsically unreliable, such as the wireless channel. To check for transmission errors, a safety protocol has been designed on top of the layer 2 protocol used to access the wireless channel, by explicitly adding a timestamp to each message, a running number, an explicit acknowledgment and a time of validity, which combined together dramatically increase the reliability of the system. This enables the communication of safety-relevant data to and from a Wireless Sensor Network (WSN), according to the black channel principle. The counter measures adopted are represented in Fig. 8.12, showing how the possible transmission errors are detected. The safety layer is designed to minimize the added overhead, both in terms of packet length and complexity, so the resulting packet is well suited for the transmission over a Low-Rate WPAN with sufficient reliability; in particular, a compression of the timing information is necessary. The software prototype consists of a firmware part (source) and a PC part (sink), that are necessary to implement and monitor a reliable communication.

Type of error \ Measure	running number	timestamp	time expiration	reception ack	sender/receiver id	data integrity assurance	redundancy with cross check	different data integrity assurance systems
unintended repetition	X O	X					X	
loss	X O			X			X	
insertion	X O			X	X		X	
incorrect sequence	X O	X					X	
falsification				X O		X	X	
delay		X O	X O					
coupling of SR/non-SR information				X O	X			X

X - The measure protects against the type of error
O - Implemented in the safety layer

Fig. 8.12 Sources of error messages on black channels and implemented countermeasures

8.4.3 Sensing System

As the input side of the safe infrastructure, a sensing layer has been adopted with the intent of providing in real time a continuous feedback between the worker and robot. In particular, the sensing layer is composed of two wearable prototypes: (1) a long-range detection system integrated into a safety jacket, (2) a short-range sensing platform mounted on work gloves.

In the first case, a set of five inertial modules provides the position of the work in real time and a central ZigBee unit guarantees the data transmission. All the electronics and wires are embedded in an internal pocket architecture, thus creating a soft housing for the system. Also the power supply, in form of a battery, is connected to the central ZigBee unit on the chest.

For the second prototype, standard work gloves have been modified in order to house flexible photoresistors based on germanium thin film technology. A sensor for each glove has been integrated together with the electronics to investigate the reliability of the system and the limits of this technology. The devices are conceived to interact with an infrared source placed in the central part of the robot, with the intent of monitoring a distance ranging from 30 cm to the contact.

8.4.3.1 Smart Working Suit

A specific working suit has been developed to capture the worker posture inside a production line. The suit has been endowed with several Magnetic Angular Rate and Gravity (MARG) sensors and a RF transmission module. The fusion of data coming from the MARG sensors enables to predict the relative position of the sensors

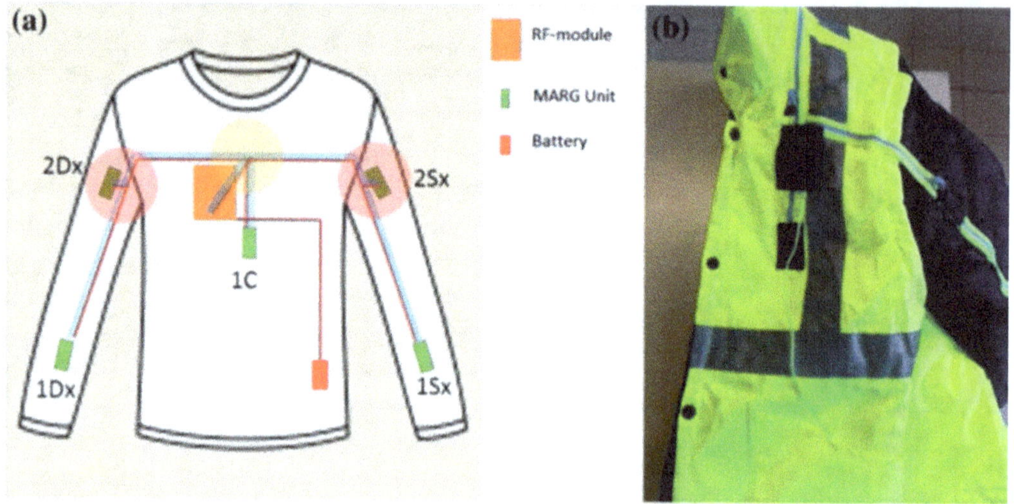

Fig. 8.13 **a** Schematic representation of the working suit. In the figure, MARG units, RF- module, battery and the wired connection are highlighted. **b** Photo of the assembled prototype

themselves and thus of the rigid body firmly bound to it, while the RF module is used both to transmit the MARG sensor data to the PC and as localization sensor. The MARG sensor units were placed on the arms, forearms and chest, to capture the motion of the human joints. The sensor units are connected by means of a shared serial bus directly integrated into the woven suit (see Fig. 8.13).

8.4.3.2 Short Range Sensor

A set of flexible infrared sensors has been fabricated to be housed on a glove to provide a valuable feedback for interactions between the worker and the robot in close proximity.

The prototype sensors have been developed using the RF-PECVD (Radio Frequency-Plasma Enhanced Chemical Vapour Deposition) technique. RF-PECVD enables to deposit materials on surface by means of vapour deposition of its gaseous state, in this case heating germane gas (GH_4) and blends of Germane-Silane (SiH_4) at 250–300 °C. Changing the relative concentration of the two gases, it is possible to obtain devices with different characteristics, thus maximizing the sensing performance of the devices.

Finally, a simple read-out electronic has been designed and developed to perform a first stage of signal conditioning in the proximity of the sensor. In particular, an interface circuit for the resistive optical sensor in Germanium, has been designed and realized in through the use of technology based on thin film transistors (TFTs) of polycrystalline silicon at low temperature (LT-PS) deposited on the flexible substrate of polyimide (PI).

(a)

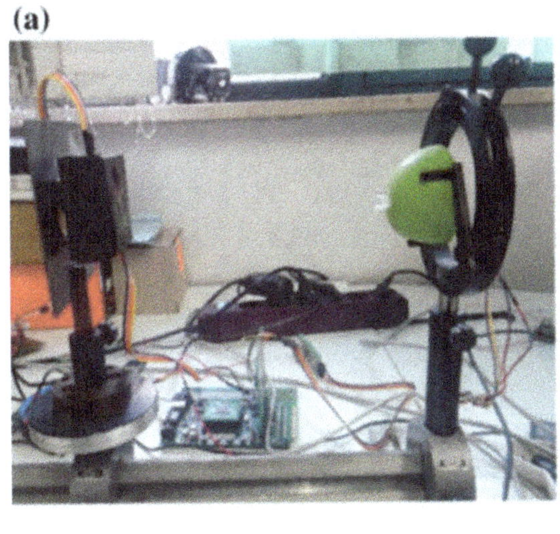

(b)

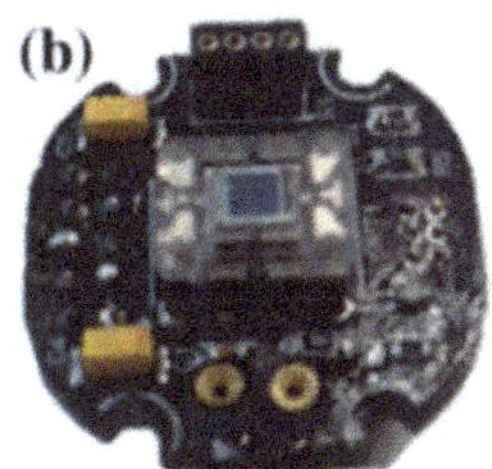

(c)

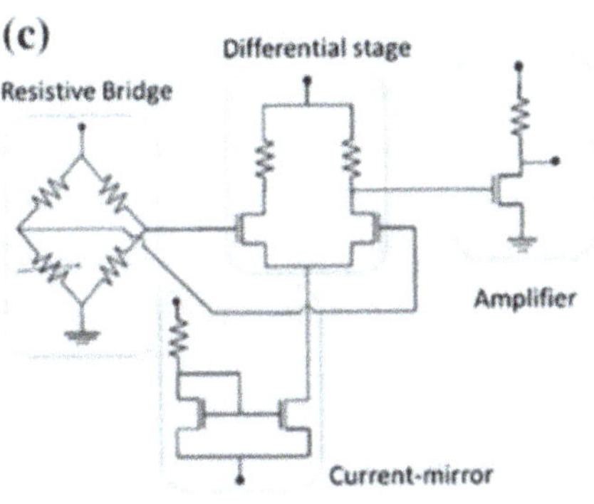

(d)

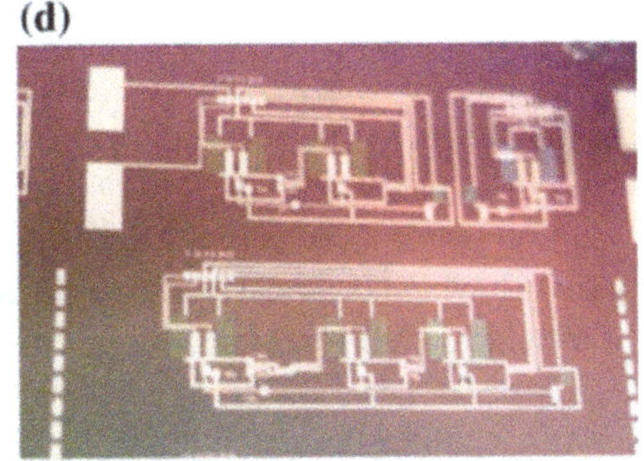

Fig. 8.14 **a** Graduated linear optical demonstrator mounted on a bench used to align first arrangement of IR sources and receivers. **b** Commercial IR-sensor. **c** Diagram of read-out electronic for the IR resistive sensors. **d** Photograph of the detail of the block of differential stages and current mirrors and Si-Ge IR-Sensor prototype

The circuit architecture is divided into several functional blocks: the first of these is a Wheatstone bridge to ¼, inside of which the optical sensor is inserted. The second functional block is constituted by the differential amplifier, fed by a current mirror. As the last step, a common-source amplifier allows to decouple the output of the differential stage from that of the circuit.

The entire circuit design was based on a process of n-MOS LTPS, (see Fig. 8.14c, d) [42]. To properly measure the sensors, preliminary tests have been carried out on a graduated linear optical bench. In the first step, a commercial IR sensor has been used to optimize the calibration system and then to realize a benchmark system for the prototype sensors (see Fig. 8.14a, b) [43, 44].

8.4.3.3 Long Range Sensor

According to the requirement of detecting operator position in a real industrial scenario with an average surface of about 25 square, a system equipped with protocol

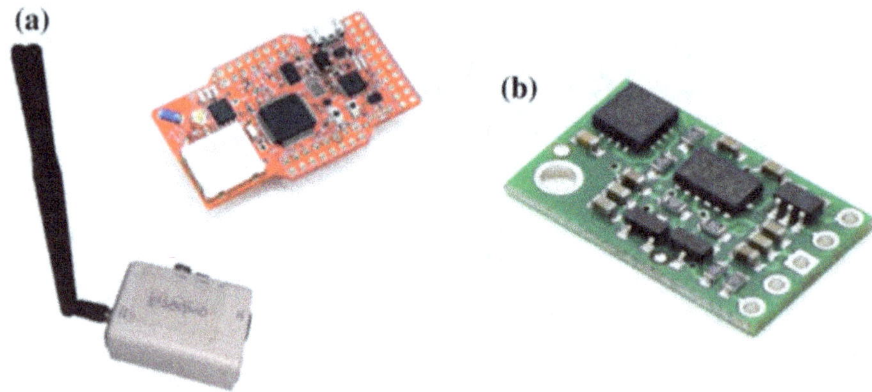

Fig. 8.15 a Zigbee Zolertia z1 node, **b** MARG Polulu inertial module (Size12.8 mm × 20.3 mm. Consumption 10 mA to 3.3 V)

ZigBee like has been adopted to evaluate in real time the interaction between the robot and the human worker. This system can exploit wireless localization by using different algorithms (typically Anchor based and Range based). In particular, the position detection platform consists of fixed reference nodes (anchor nodes), at known position, with which it is possible to evaluate the dynamic distance of the human worker by measuring the Received Signal Strength between the different radio frequency modules.

This localization technique is based on a triangulation method: in this case the detection of the distances between the object and at least three non-collinear reference anchors, permit to evaluate the relative position by using a suitable Lateration algorithm.

The developed localization algorithm consists of three phases: (1) ranging, in which distances are computed, (2) positioning, where nodes positions are estimated and (3) refinement, where the error is reduced through iterative methods. The solution chosen to implement the wireless sensor network for the localization, makes use of the commercial device Zolertia z1 (Fig. 8.15), that is a low-power module for sensor networks of small dimensions (34.5 × 56.8 mm) and it is based on a TI MSP430 microcontroller, programmable in C language. The system is equipped with ZigBee module with integrated ceramic antenna. The card is completely expandable on account of the quantity of ports accessible (USB, I2C, SPI, 2xUARTs), so that inertial sensors and different other devices can be effectively externally connected.

The localization hardware system is very small in size and can be easily integrated in work clothes without being invasive. This system also lends itself to a more advanced integration in clothing, due to the relative simplicity of its electronic circuits, its open source type and completeness of the technical specifications that can be further miniaturized and made on thin, flexible PCBs.

Five inertial modules were connected to the mobile node to improve the assessment of the movement and position of the operator. These are placed on the upper limb (arm and forearm) and on the chest, for a total of five modules. The system is powered by 2 AA batteries 3.3 V connected to the mobile node placed at the chest height. An inner pocket has been prepared to house the batteries and the connection wiring with the mobile node [45, 46].

The five inertial modules are connected to the mobile node with which they share the communication protocol and the power supply. The wires are five for each module, i.e. three wires (SDA and SCL I/O) and two power cords (3.3 V and GND).

The proposed system is able to recognize the position of a person in the space with an uncertainty of about 30 cm. This value is perfectly compatible with the other recognition system based on short range sensors. This uncertainty strongly depends on the number of the fixed antennas located in the experimental space and it can be further reduced by adding heavier acquisition protocols. The position has been successfully acquired by real-time monitoring the movements of the person in a room equipped with different furniture and equipment made of metal and wood.

8.4.4 Contact-Less Modes for Safe Workspace Sharing

ISO/TS 15066:2016 defines the notion of Separation Monitoring that is intended to specify the concept of ISO 13855 minimum safety distance in the case of moving robots in the same workspace of human operators. Task-dependent safe distances are instead proportional to the current velocities of both robots and human operators. As a result, trajectory-dependent safety areas can be intuitively designed according to a given time-varying robot task, provided the availability of a safety-rated sensor system that measures relative positions and speeds. If the wireless sensors nodes (Sect. 8.4.3) are integrated in the overall architecture (Sect. 8.4.1) and use the proposed communication protocols (Sect. 8.4.2), then they support the dynamic computation of the minimum safety distance to maintain for the protection of the operator.

Safe emergency states or the violation of separation conditions can be triggered far less frequently than regular safeguarded cases, because the separation area is optimized (i.e. reduced) along the current task movement direction. The concrete result is that the entire virtual envelope of the protected space around the interacting bodies allows some proximity conditions that are useful for adaptability and efficiency in production.

Figure 8.16 shows that the human operator H, while moving at a given velocity, is safe as long as it is not inside the red region that envelopes all the safe distances that separate H from all the parts of the robot R. The shape of the safety region changes over time depending on the mutual H-R configurations.

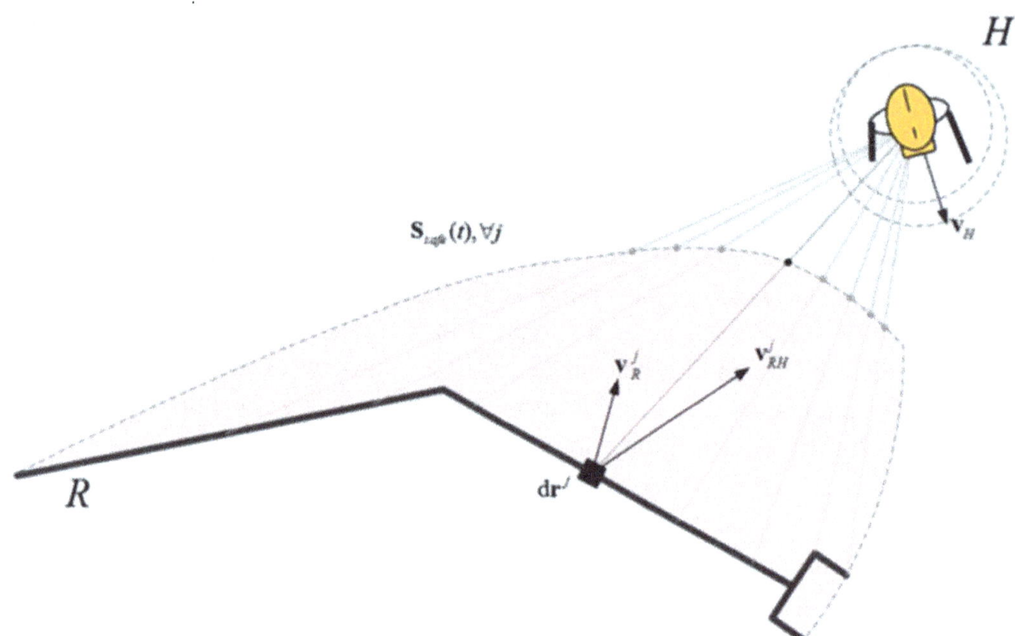

Fig. 8.16 Visualization of dynamic safe minimum distance. The red region minimizes the portion of the collaborative workspace that cannot be occupied by the user at a given time

8.5 Conclusions and Future Research

A multiple sensor network based on wearable and conformable technology represents a unique tool in advanced Wireless Sensor Network environment. Specifically, the implementation of WSN to track workers in joint human/robot workspaces has been presented as a source of environmental information that is merged with additional technologies to estimate risky conditions in collaborative tasks. With the aim to provide a logical settlement for safety-related operations, architectural solutions have been presented to deploy information, from unsafe sources to safe Inputs/Outputs. Proper network protocols, for easy and low-cost integration of general purpose sensors, are employed as black channels to connect safe processing units that are in charge of providing the required protection level. It has been shows that the modified WSN protocol can increase significantly the channel real time for upgrading and/or preserving the overall Safety Integrity Level. The enhanced protocol bridges some gaps in the service availability of 802.15.4 protocol, thus fulfilling relevant solutions for safe applications in industry. While a full risk assessment procedure has to consider the analysis of faults and their severity, functional safety of communication protocols are considered beneficial because of the implementation of monitored failure modes, instead of unrecoverable dangerous failure modes, in standard communication networks [38].

These results will boost the interest of small industries and stakeholders for applications and technologies to be deployed in different scenarios from automotive assembly line to fenceless interactions between humans and robots in general.

Even in different applications especially in biomedical market, wearable technology and prototypes could exploit the solutions proposed in this work to overcome several issue related to device durability, comfort, and contact force detection [47].

Acknowledgements This work has been funded by the Italian Ministry of Education, Universities and Research (MIUR) under the Flagship Project "Factories of the Future—Italy" (Progetto Bandiera "La Fabbrica del Futuro") [48], Sottoprogetto 1, research projects "FACTOry Technologies for HUMans Safety" (FACTOTHUMS) and "Improving human-robot cooperation and safety in shared automated workplaces of the automotive industry". The latter project was selected by the Canada-Italy Concurrent Call on Automotive Manufacturing R&D and its twin Canadian project was funded by Automotive Partnership Canada to assess the usability of the technological and architectural solutions in a pre-industrial automotive scenario [43, 44].

References

1. Burke R, Mussomeli A, Laaper S, Hartigan M, Sniderman B (2017) The smart factory—responsive, adaptive, connected manufacturing. Deloitte University Press
2. Tolio T, Ceglarek D, Elmaraghy HA, Fischer A, Hu SJ, Laperriere L, Newman ST, Vancza J (2010) SPECIES-co-evolution of products, processes and production systems. CIRP Ann Manuf Technol 59(2):672–693
3. Terkaj W, Tolio T, Valente A (2009) Designing manufacturing flexibility in dynamic production contexts. In: Tolio T (ed) Design of flexible production systems. Springer, pp 1–18
4. Müller R, Vette M, Scholer M (2016) Robot Workmate: a trustworthy coworker for the continuous automotive assembly line and its implementation. Procedia CIRP 44:263–268
5. Burghart C, Mikut R, Stiefelhagen R, Asfour T, Holzapfel H, Steinhaus P, Dillmann R (2005) A cognitive architecture for a humanoid robot: a first approach. In: Proceedings of 5th IEEERAS international conference in humanoid robots, Tsukuba, Japan
6. Robla-Gómez S, Becerra VM, Llata JR, González-Sarabia E, Torre-Ferrero C, Pérez-Oria J (2017) Working together: a review on safe human-robot collaboration in industrial environments. IEEE Access 5:26754–26773
7. Tolio T, Copani G, Terkaj W (2019) Key research priorities for factories of the future—part I: missions. In: Tolio T, Copani G, Terkaj W (eds) Factories of the future. Springer
8. Athalye A, Savić V, Bolić M, Djurić PM (2011) A radio frequency identification system for accurate indoor localization. In: Proceedings of IEEE international conference on acoustics, speech and signal processing (ICASSP), Prague, Czech Republic
9. Bouet M, dos Santos AL (2008) RFID tags: positioning principles and localization techniques. In: Proceedings of IFIP wireless days 2008, Dubai, United Arab Emirates
10. Hagelauer A, Ussmueller T, Weigel R (2012) SAW and CMOS RFID transponder-based wireless systems and their applications. In: Proceedings of frequency control symposium (FCS), 2012 IEEE international, Baltimore, MD, USA
11. De Santis A, Siciliano B, De Luca A, Bicchi A (2008) An atlas of physical human–robot interaction. Mech Mach Theory 43(3):253–270
12. Pires J (2009) New challenges for industrial robotic cell programming. Ind Robot Int J 36(1)
13. Albu-Schaeffer A, Bicchi A, Boccadamo G, Chatila R, De Luca A, De Santis A, Giralt G, Hirzinger G, Lippiello V, Schiavi R, Siciliano B, Tonietti G, Villani L (2005) Physical human-robot interaction in anthropic domains: safety and dependability. In: Proceedings of 4th IARP/IEEE-Euron workshop on technical challenges for dependable robots in human environments
14. Behnisch K (2008) White paper—safe collaboration with ABB robots—electronic position switch and SafeMove. WHP–EPS-2006

15. Matthias B, Kock S, Jerregard H, Källman M, Lundberg I (2011) Safety of collaborative industrial robots: certification possibilities for a collaborative assembly robot concept. In: Proceedings of IEEE international symposium on assembly and manufacturing (ISAM), Tampere, Finland
16. EN-ISO10218-1 (2011) Robots for industrial environments-safety requirements-part 1, Geneva
17. EN-ISO2018-2 (2011) Robots and robotic devices-safety requirements for industrial robots-part 2, Geneva
18. EN-ISO12100 (2010) Safety of machinery-general principles for design-risk assessment and risk reduction. ISO, Geneva
19. EN-ISO14121 (2007) Safety of machinery–risk assessment. ISO, Geneva
20. EN-ISO13849 (2006) Safety of machinery–safety-related parts of control systems. ISO, Geneva
21. EN-IEC61508 (2006) Functional safety of electrical/electronic/programmable electronic safety-related systems. ISO, Geneva
22. Vicentini F, Pedrocchi N, Molinari Tosatti L (2014) SafeNet: a methodology for integrating general-purpose unsafe devices in safe-robot rehabilitation systems. Comput Methods Programs Biomed 116(2):156–168
23. Lenz C, Nair S, Rickert AKM (2008) Joint-action for humans and industrial robots for assembly tasks. In: Proceedings of the 17th IEEE international symposium on robot and human interactive communication, Munich, Germany
24. Knoblich G, Jordan JS (2003) Action coordination in groups and individuals: learning anticipatory control. J Exp Psychol Learn Mem Cogn 29(5):1006–1016
25. Laengle T, Hoeniger T, Zhu L (1997) Cooperation in human-robot teams. In: Proceedings of the IEEE international symposium on industrial electronics, Guimaraes, Portugal
26. Breazeal C, Brooks A, Gray J, Hoffman G, Kidd C, Lee H, Lieberman J, Lockerd A, Chilongo D (2004) Tutelage and collaboration for humanoid robots. Int J Humanoid Rob 1(2):315–348
27. Sidner CL, Dzikovska M (2005) A first experiment in engagement for human-robot interaction in hosting activities. In: Advances in natural multimodal dialogue systems. Springer
28. Hoffman G, Breazeal C (2007) Effects of anticipatory action on human-robot teamwork: efficiency, fluency, and perception of team. In: 2nd ACM/IEEE international conference on human-robot interaction (HRI), Arlington, VA, USA
29. Rickert M, Foster M, Giuliani M, By T, Panin G, Knoll A (2007) Integrating language, vision and action for human robot dialog systems. In: Proceedings of international conference on human-computer interaction, Beijing, PRC
30. Foster M, By T, Rickert M, Knoll A (2006) Human-robot dialogue for joint construction tasks. In: Proceedings of the international conference on multimodal interfaces. ACM Press
31. Tang F, Parker LE (2006) Peer-to-peer human-robot teaming through reconfigurable schemas. In: AAAI spring symposium on "To boldly go where no human-robot team has gone before", Stanford University
32. Längle T, Lueth T, Rembold U, Wörn H (1997) A distributed control architecture for autonomous mobile robots—implementation of the Karlsruhe multi-agent robot architecture (KAMARA). Adv Robot 12(4):411–431
33. Schrempf O, Hanebeck U, Schmid A, Worn H (2005) A novel approach to proactive human-robot cooperation. In: Proceedings of IEEE international workshop on robot and human interactive communication, 2005, ROMAN 2005, pp 555–560
34. Chen E, Shih CY (2011) Polymer infrared proximity sensor array. IEEE Trans Electron Devices 58:1215–1220
35. Pecora A, Maiolo L, Maita F, Minotti A (2012) Flexible PVDF-TrFE pyroelectric sensor driven by polysilicon thin film transistor fabricated on ultra-thin polyimide substrate. Sens Actuators A Phys 185:39–43
36. Vicentini F, Pedrocchi N, Giussani M, Molinari Tosatti L (2014) Dynamic safety in collaborative robot workspaces through a network of devices fulfilling functional safety requirements. In: Proceedings of ISR/Robotik 2014; 41st international symposium on robotics, Munich, Germany

37. Vicentini F, Pedrocchi N, Molinari Tosatti L (2013) SafeNet of unsafe devices—extending the robot safety in collaborative workspaces. In: Proceedings of the 10th international conference on informatics in control, automation and robotics
38. Vicentini F, Ruggeri M, Dariz L, Pecora A, Maiolo L, Polese D, Pazzini L (2014) Wireless sensor networks and safe protocols for user tracking in human-robot cooperative workspaces. In: Proceedings of IEEE 23rd international symposium on industrial electronics (ISIE), Istanbul, Turkey
39. Dariz L, Ruggeri M, Malaguti G (2013) A proposal for enhancement towards bidirectional quasi-deterministic communications using IEEE 802.15.4. In: Proceedings of telecommunications forum (TELFOR), Belgrade, Serbia
40. Dariz L, Malaguti G, Ruggeri M (2014) Performance analysis of IEEE 802.15.4 real-time enhancement. In: Proceedings of IEEE 23rd international symposium on industrial electronics (ISIE), Istanbul, Turkey
41. Dariz L, Ruggeri M, Ferraresi C(2015) A comparison between configuration strategies for IEEE 802.15.4 low-latency networks. In: Proceedings of IEEE international conference on industrial technology (ICIT), Sevilla, Spain
42. Pecora A, Maiolo L, Cuscunà M, Simeone D, Fortunato G (2008) Low-temperature polysilicon thin film transistors on polyimide substrates for electronics on plastic. Solid State Electron 52:348–352
43. Ferrone A, Maiolo L, Minotti A, Pecora A, Iacovo D, Colace L, Grayli S, Leach G, Bahreyni B (2016) Flexible near infrared photoresistors based on recrystallized amorphous germanium thin films. In: Proceedings of 2016 IEEE SENSORS, Orlando, Florida, USA
44. Grayli S, Ferrone A, Maiolo L, Iacovo AD, Pecora A, Colace L, Leach G, Bahreyni B (2017) Infrared photo-resistors based on recrystallized amorphous germanium films on flexible substrates. Sens Actuators A Phys 263:341–348
45. Polese D, Pazzini L, Minotti A, Maiolo L, Pecora A (2014) Compensation of the antenna polarization misalignment in the RSSI Estimation. In: Proceedings of SENSORNETS 2014, Lisbon, Portugal
46. Polese D, Pazzini L, Minotti A, Maiolo L, Pecora A (2015) Tunable transmission power to improve 2D RSSI based localization algorithm. In: Proceedings of SENSORNETS 2015, 4th international conference on sensor networks
47. Ferrone A, Maita F, Maiolo L, Arquilla M, Castiello A, Pecora A, Jiang X, Menon C, Ferrone A, Colace L (2016) Wearable band for hand gesture recognition based on strain sensors. In: IEEE international conference on biomedical robotics and biomechatronics (BioRob), Singapore
48. Terkaj W, Tolio T (2019) The Italian flagship project: factories of the future. In: Tolio T, Copani G, Terkaj W (eds) Factories of the future. Springer

Chapter 9
Haptic Teleoperation of UAV Equipped with Gamma-Ray Spectrometer for Detection and Identification of Radio-Active Materials in Industrial Plants

Jacopo Aleotti, Giorgio Micconi, Stefano Caselli, Giacomo Benassi, Nicola Zambelli, Manuele Bettelli, Davide Calestani and Andrea Zappettini

Abstract Large scale factories such as steel, wood, construction, recycling plants and landfills involve the procurement of raw material which may include radiating parts, that must be monitored, because potentially dangerous for workers. Manufacturing operations are carried out in unstructured environments, where fully autonomous unmanned aerial vehicle (UAV) inspection is hardly applicable. In this work we report on the development of a haptic teleoperated UAV for localization of radiation sources in industrial plants. Radiation sources can be localized and identified thanks to a novel CZT-based custom gamma-ray detector integrated on the UAV, providing light, compact, spectroscopic, and low power operation. UAV operation with a human in the loop allows an expert operator to focus on selected candidate areas, thereby optimizing short flight mission in face of the constrained acquisition times required by nuclear inspection. To cope with the reduced situational awareness of the remote operator, force feedback is exploited as an additional sensory channel. The developed prototype has been demonstrated both in relevant and operational environments.

J. Aleotti · G. Micconi · S. Caselli
Università di Parma, Dipartimento di Ingegneria e Architettura, Parma, Italy

G. Benassi · N. Zambelli
due2lab s.r.l., Parma, Italy

M. Bettelli · D. Calestani · A. Zappettini (✉)
CNR-IMEM, Istituto dei Materiali per l'Elettronica ed il Magnetismo, Parma, Italy
e-mail: andrea.zappettini@imem.cnr.it

© The Author(s) 2019
T. Tolio et al. (eds.), *Factories of the Future*,
https://doi.org/10.1007/978-3-319-94358-9_9

9.1 Scientific and Industrial Motivations, Goals and Objectives

The presence of radioactive material is a real risk in different contexts such as steel, wood, construction, recycling industries, mines, and landfills. For example, not only secondary material of uranium mines results in a legacy radiological contamination [1], but also gold mining may produce the accumulation of radioactive materials in certain stages of the process [2]. Also, radioactive contamination was found in wood and its products [3], in particular in wood pellet [4]. Another source of contamination comes from the illicit dropout of nuclear sources that are molten in scrap metal [5]. Indeed, not in all large-scale factories or landscapes the control of the materials at the entrance is compulsory, and, even when such control nominally exists, the monitoring procedures have become only recently operative.

At present, if a large industrial area has to be monitored for the presence of radioactive material, then the search is mainly led by a human operator carrying proper instrumentation, with the unavoidable exposure of the operator to the radiation. Moreover, stacks of materials to be monitored in most cases are too large or too high for a direct monitoring by a human operator so that the human operator is inhibited to map all the zones of the stack.

In this context, the exploitation of an unmanned aerial vehicle (UAV) equipped with a radiation detector permits, at the same time, to guarantee the safety of the operator that can guide the UAV from a secure zone, and the monitoring of an area that can be reached only by an UAV.

The research field on radiation detection using robotic systems comprises two main topics, i.e. localization and mapping. In particular, localization addresses the problem of finding a point source of radiation using the perceived radiation distribution as a guide in the search behaviour of the robot. Radiation mapping is focused on the problem of generating a complete map of an area. Dealing with illicit or unforeseen radioactive materials in unknown locations involves major safety hazards, hence procedures and techniques for reducing human exposure time and radiation risks can provide benefits of utmost importance, while also serving as a deterrent measure.

Teleoperation of aerial vehicles is complex due to the lack of situation awareness of the operator. Indeed, when only visual feedback is available to the operator, such limited feedback is often inadequate to support critical remote operations. Indeed, the information provided by direct sight or through an on-board camera may not be sufficient due to the limited field of view. Moreover, time delays in the aerial environment can make teleoperation a difficult task. For example, an obstacle outside the field of view cannot be detected by the operator, which may lead to a collision.

Hence, it is crucial to investigate alternative strategies for aerial teleoperation, which provide additional information sources to the operator by exploiting signals from sensors. Such information is helpful to increase the perception of the environment constraints, which can increase the situation awareness of the operator and improve efficiency and task performance.

Using haptic teleoperation with force feedback is a promising solution for aerial teleoperation. In haptic teleoperation the control system feeds external forces to a haptic device.

In this paper a unique platform is presented for studying the potential of non-conventional human-robot interaction techniques in the context of aerial radiation localization. The study was concerned with the general goal of developing technologies for improving the interaction between operator and machines in challenging industrial contexts by the exploitation of a haptic user interface for remote control of an Unmanned Aerial vehicle (UAV). By using a single bilateral, haptic interface the user is able to provide 3D motion commands to the UAV and receive at the same time an informative force feedback via the interface. The advantages obtained with this approach in the context of radiation detection will be discussed in the following of this paper. Moreover, a novel CdZnTe-based custom radiation detector is integrated on board of the UAV, so that radiation sources can be localized and identified without a direct exposure of the human operator. By means of CdZnTe crystals, solid-state radiation detectors can be realized such that their limited weight and power consumption enable sensor integration even in UAVs offering a small payload. Moreover, spectroscopic detectors could be realized in a similar technology, with the advantage of enabling localization and identification of the radiation source at once. By combining haptic guidance and CdZnTe-based custom detector, novel teleoperation technologies are developed to reduce the risks connected to dangerous processes in large-scale industrial environments.

Since the final goal was to make the prototype available for future exploitation by third parties (public agencies for environment monitoring, other territorial agencies, private companies), the UAV equipped with the radiation sensor and the haptic guidance system has undergone the protocol required to achieve the Italian certification for standard operations. Also, operators have been required to obtain the licence for piloting the UAV. Achieving the national certification for the UAV, as well as proper driving licences, has been an important milestone towards potential market applications of the system.

The prototype has been demonstrated in relevant and operational environments. These experiments have been carried out with the cooperation of the local agency for environmental control, ARPAE Emilia Romagna, and Emiltest s.r.l. in Gossolengo, Piacenza, Italy. The aim of these experiments was to demonstrate the effective sensitivity of the system to weak nuclear sources and the localization capability of the prototype. A second set of experiments has been carried out in operational environments. Flights were performed in a waste disposal landfill in Novellara, Reggio Emilia, Italy, monitoring the specific radiation levels and checking for any hidden radiating source.

9.2 State of the Art

The use of UAVs for environmental nuclear radiation monitoring was investigated in several works [1, 6–13] either by exploiting standard remote controllers or by specifying pre-programmed flight missions. None of previous works investigated haptic teleoperation for nuclear radiation detection. Indeed, so far haptic teleoperation of UAVs has been investigated exclusively to assist the operator for collision avoidance as reported in [14].

A UAV system was proposed in [6] for air quality control, water pollution monitoring and radiation leakage detection using a cell biosensor system. Okuyama et al. [7] reported the experimentation with an autonomous helicopter to measure radiation data and provide real-time data transmission to a ground station, including images. Boudergui et al. [8] carried out a preliminary evaluation of a teleoperated UAV equipped with a CdZnTe-based sensor and a gamma camera for nuclear and radiological risk characterization in indoor environments. Operating an UAV in outdoor environment poses significantly different challenges in terms of situational awareness and flight mission optimization. In [9] and [10] fixed wing UAVs capable of flying at high altitude and high speed were developed for radiation detection. Experiments were performed around the Fukushima Daiichi nuclear power plant in [11–13] where the radioactive caesium deposition on the ground was successfully measured. In other works [15–17] experiments in simulated environments were reported. In [15] a simulation of an autonomous helicopter was developed for radiation detection and imaging. In [16] a multiple UAVs radiation contour mapping task was simulated with formation flight control. In [17] a method for multiple source localization and radiation contour mapping using a UAV was proposed and evaluated with numerical simulations.

Several gamma ray detectors are available on the market [18]. The most exploited ones are Geiger-Müller detectors. However, these gas detectors show little sensitivity, are fragile (they are basically gas-filled glass tubes), and do not provide spectroscopic performances. Scintillators are also quite diffused. The light produced by the scintillating material can be picked up by photodiodes instead of fragile phototubes. However, the energy resolution of scintillators is not satisfactory, especially if low energy gamma rays are detected. On the market, several semiconductor-based detectors are available. Silicon detectors show good spectroscopic performances, however they provide limited stopping power for energy larger than few tenths of KeV. Germanium detectors show excellent spectroscopic performance, however they require cryogenic cooling, and thus they are not easily integrated on a UAV.

Due to the fact that the detector must work at room temperature with excellent energy resolution in a wide energy range, the best candidate material is CdZnTe. Also, the high electrical resistivity of CdZnTe crystals ensures low power consumption.

CdZnTe detectors were used in many fields and exploited for medical applications [19], security [20], environmental control [21], and astrophysics [19].

For these reasons, CZT detectors can easily integrated on a UAV. In [8] a teleoperated quad-rotor had previously proposed that performs nuclear and radiological risk characterization using a CdZnTe sensor and a gamma camera for security purposes. The system was proposed mainly for security applications, thus in small indoor contexts. Moreover, the system was only proposed, and its actual performance was not shown.

9.3 Problem Statement and Proposed Approach

The exploitation of a UAV represents an optimal solution for the localization and identification of nuclear material. If only direct visual feedback is available to the operator of the UAV, teleoperation at a distance is a complex task due to the lack of situation awareness. Therefore, it is important to provide additional information to the pilot by exploiting the UAV on-board sensors. This work proposes a teleoperation system where a three degree of freedom haptic device is used for sending motion commands to the UAV as well as for receiving force feedback to guide the exploration of the environment. In particular, force feedback provides real-time information that reduces operator's perceptual load in localizing the hazardous materials. The force vector is computed in real-time from the CdZnTe gamma ray sensor measurements to generate an attractive zone surrounding the detected radioactive source. In this way the pilot is helped to keep the UAV close to the uncovered radiation source. In the proposed teleoperation approach the UAV flies at a constant height while the operator sends directional commands in the horizontal plane and receives a 2D force feedback on the xy space of the haptic device. The usual flying procedures such as taking off, landing, and emergency operations are carried out using a standard remote controller. The teleoperation system includes a ground station that receives both UAV telemetry data and data from the CdZnTe gamma ray sent through a wireless link. The approach is represented in Fig. 9.1.

Another important aspect is the use of a CdZnTe-based gamma ray spectrometer. Since the detector must localize and identify nuclear sources, and must be mounted and carried by the UAV, the detector characteristics must fulfil the following requirements: (i) overall weight (detector, read out electronics, and batteries) lower than 0.5 kg, (ii) operate at room temperature, (iii) low power consumption (battery operated for 1 h), (iv) large energy range: 10 keV–1.5 MeV, (v) good energy resolution (<3% @662 keV), (vi) high sensitivity (>1000 cps for a ^{57}Co source of 1 mSv/year at 2 m from the detector), (vii) large field of view. The extended autonomy of the gamma-ray sensor (at least 1 h) will enable multiple UAV missions by simply swapping the battery of the UAV, without requiring unmounting and reconfiguration of the sensor unit.

The detector must work in a wide energy range (10 keV–1.5 MeV) to reveal the main potential nuclear materials. High energy gamma can be efficiently stopped only by thick detectors. This means that in any case, a several millimetre thick CdZnTe detector must be employed. Nowadays, on the market, CdZnTe detectors as large as $20 \times 20 \times 6$ mm^3 are available, thus it makes sense to evaluate whether the volume of such detector is large enough for the scope of the proposed application.

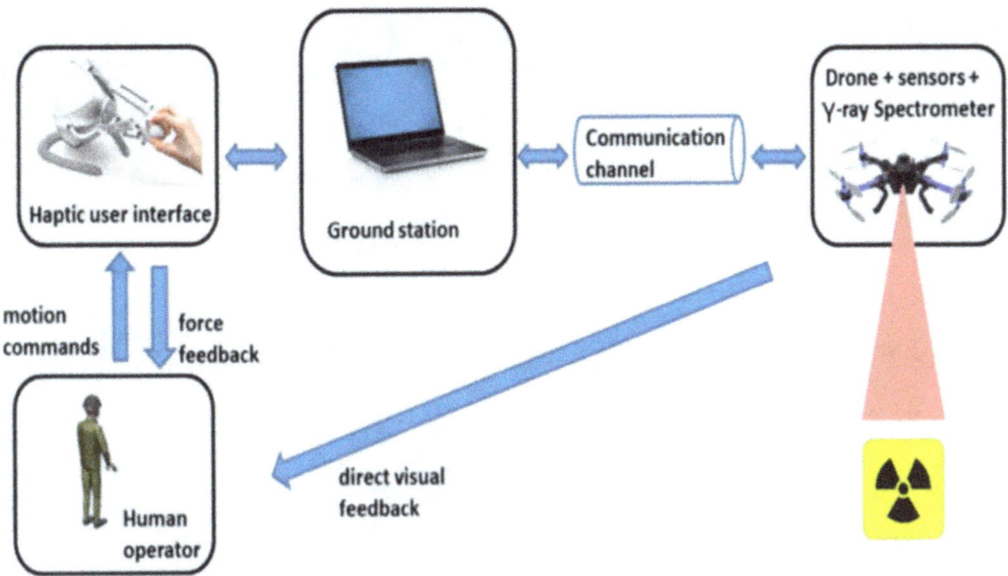

Fig. 9.1 Overall architecture of XDrone system

Table 9.1 Counts per second measured by a $20 \times 20 \times 6$ mm^3 CdZnTe detector at 1 m from nuclear sources potentially dangerous for a worker

Nuclear source	Dose (mSv/year)	Source activity (MBq)	Counts per second
^{241}Americium	1	160	5080
^{57}Cobalt	1	54	1644
^{137}Caesium	1	8.2	636

In order to evaluate the required sensitivity it should be recalled that the aim is to detect nuclear sources in industrial plants or material stocking plants whose intensity can be dangerous for the health of operators or workers. Italian regulations [22], in accordance with the recommendations of the International Commission on Radiological Protection [23], set the limit of exposition doses for workers to 1 mSv/year. It is therefore reasonable for the UAV to look for nuclear sources whose activity is enough to expose a worker standing at one meter distance to such a dose. In order to preserve the integrity of the UAV, it is better to keep the UAV at a height of about one meter from the ground, even though in most situations the UAV can fly at lower height.

Table 9.1 reports for different nuclear sources the counts per second measured by a $20 \times 20 \times 6$ mm^3 detector at a distance of one meter from a nuclear source. Table 9.1 shows that such detector is able to obtain a number of counts per second that is enough to localize a nuclear source posing a threat to worker safety.

Fig. 9.2 UAV equipped with gamma ray detector

9.4 Developed Technologies, Methodologies and Tools

The adopted UAV (Fig. 9.2) is an octocopter in coaxial configuration whose size is about 550 mm (without propellers). The UAV gross payload is 4 kg and the maximum flight time is about 15 min. The aerial vehicle is equipped with MEMS accelerometer, gyro, magnetometer, and GPS sensors. The autopilot system is a 168 MHz ARM CortexM4F microcontroller running an Arducopter firmware. The gamma-ray detector, enclosed in a box, is attached to a two-axes brushless gimbal unit. An embedded system reads sensor data from the gamma-ray detector is supplied with a Li-Ion battery pack ensuring long-lasting use.

The UAV was also equipped with a remotely Flight Terminator and with a parachute. Finally, the UAV was equipped with a video camera pointing to the ground to evaluate offline the error between the actual location of the radiation substance and the location estimated by the operator through the haptic feedback.

The haptic device is a 3 degrees of freedom Novint Falcon (top left inset in Fig. 9.1), which features a position resolution of 0.0635 mm, a range of motion of about 10 cm^3 and a maximum force feedback of 10 N.

The teleoperation approach is based on an impedance control mode where the user specifies the horizontal heading direction of the UAV by moving the tool point of the haptic device. In particular, a movement of the haptic device causes the change of the UAV waypoint [24], and so the direction of the UAV movement.

If the tool point of the haptic device is within a fixed small range from the centre of the haptic reference frame the waypoint is not updated and the UAV hovers. The UAV altitude set point is provided by the standard controller and can be changed at any time.

A force feedback is provided to the operator as a 2D vector on the horizontal plane of the haptic workspace, acting as a basin of attraction with quadratic profile, causing

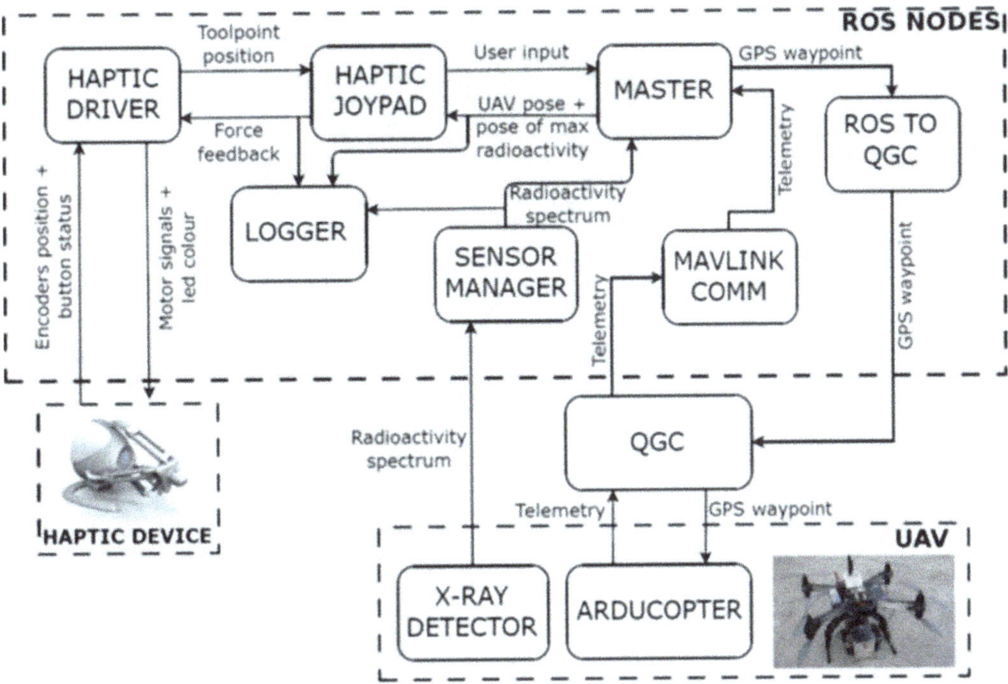

Fig. 9.3 Software architecture of the XDrone system

the UAV to be attracted by the radiation source. Indeed, as the pilot moves away from radiation source, a force is generated towards the maximum of the detected radiation whose position is updated over time.

The number of counts is integrated over a time frame of 2 s. Outside the zone of attraction, no force feedback is calculated. Force feedback can be reset by simply acting on the haptic device, so that multiple concentrated radiation sources can be localized one by one.

Transmission of measured radioactivity data from the UAV to the ground station is performed through a dedicated long-range 5 GHz WiFi connection.

Figure 9.3 shows the software architecture, which is based on the robot operating system (ROS) middleware, including the haptic interface, the UAV ArduPilot system and several ROS nodes. A modified version of the QGroundControl software platform was used on the ground station for data transmission with the on-board ArduPilot system controlling the UAV. In particular, the MAVLink communication node translates MAVLink messages to ROS messages. The Master node, which subscribes to ROS topics, manages information regarding the UAV state and the radioactivity sensor data. The Haptic joypad node reads the current tool point position from the haptic driver node, computes a heading direction, and publishes it as a ROS message. Moreover, it computes force feedback for the haptic device.

CdZnTe detectors as large as $20 \times 20 \times 6$ mm^3 are available on the market, but they are usually pixels detectors with a number of read-out channels that is not compatible with the available power consumption and the maximum payload. Since the detector is 6 mm thick and a good energy resolution has to be achieved, a single

Fig. 9.4 Anode contact configuration of the CZT drift strip detectors

carrier device has to be considered, in such a way to avoid the hole contribution to the signal. Drift strip devices are realized for this purpose and have been used by some of the authors to obtain large CdZnTe detectors with excellent energy resolution and imaging capabilities [25]. More recently, we have realized a $5 \times 5 \times 20$ mm^3 drift strip detector to monitor the 470 keV gamma emissions produced by the 10B(n, α)7Li nuclear reaction that is exploited for tumour treatment. The detector showed very good spectral resolution and efficiency [26]. Thus, we realized four $5 \times 6 \times 20$ mm^3 strip detectors like the one shown in Fig. 9.4 to obtain good energy resolution and, at the same time, provide the volume requested to achieve the required sensitivity as reported in Table 9.1 [27]. The central grid is actually the collecting grid and is coupled to the charge sensitivity pre-amplifier. The other strip grids are biased in such a way to drift the electrons toward the central strip, thus obtaining a focusing effect.

For sensor manufacturing, standard pixel detectors from REDLEN were acquired, characterized with certified spectroscopic properties. Four $5 \times 6 \times 20$ mm^3 samples were obtained by carefully sawing. All the six surfaces of each detector were polished. In fact, it is known that also the preparation of lateral surfaces is important in order to reduce leakage currents. Polishing consists of two main steps, the first one with abrasive paper and the second one involving alumina powder suspension. The strip grid configuration shown in Fig. 9.4 was obtained using a photolithographic process and depositing gold contacts by the electroless technique with alcoholic solutions [28]. This type of contacts ensures good mechanical stability and blocking current-voltage characteristics, thus limiting dark current. This is very important to get good spectroscopy detectors. The detectors were bonded on Teflon-based substrates in order to ensure optimal electrical insulation using silver epoxy and gold wirings [29].

A battery provides the required power for the detector and the read out electronics. The system consumption is relatively low, so that a small battery is enough to power the detectors for 8 h operations. A low noise DC-DC converter was used to bias the detectors with the required 600 V. The read out electronic chain was constituted for each of the four detectors by a charge sensitive preamplifier, a shaping amplifier, and a peak detector. Signals from the four detectors were collected by a microcontroller converting the analogue signal into a digital signal and operating as a multichannel analyser to generate the energy spectra. These data are then sent to the ground station by the Intel-Galileo system described before.

The detectors and the electronics were housed in a 3D printed plastic box (see Fig. 9.2). The plastic box was covered with aluminium tape to create a noise shielding. The total weight of the box, with the detector and the electronics was about 300 g, much lower than the UAV maximum payload.

9.5 Testing and Validation of Results

The first test of the prototype has been carried out in laboratory and concerned the capability of the detector integrated on the UAV to collect spectra of a number of nuclear sources (Sect. 9.5.1). Then, the prototype has been tested in a relevant environment (Sect. 9.5.2). The aim of these experiments was to demonstrate the effective sensitivity of the system to weak nuclear sources and the localization capability of the prototype. Finally, a set of experiments has been carried out in operational environments (Sect. 9.5.3). Flights were performed in a waste disposal landfill checking for any hidden radiating source.

9.5.1 Laboratory Test

The detector integrated on the UAV was firstly tested in the lab. The CdZnTe detector was tested using laboratory nuclear sources ^{57}Co and ^{137}Cs. The response of the detector at ^{137}Cs is shown in Fig. 9.5 (red line). The resolution of the 662 keV line is 1.7%. For comparison it is shown also the spectrum obtained by a $5 \times 6 \times 20$ mm^3 detector realized with simple planar contacts. The spectra obtained with the strip grid contacts shows better resolved peaks. Similar results were obtained using a ^{57}Co nuclear source. An energy resolution of 4% at 122 keV was obtained.

It is worth mentioning that the drift strip detector also showed a two times larger sensitivity with respect to the other.

Preliminary experiments for validation of the haptic interface have been performed using a small, easily manoeuvrable UAV not equipped with the actual gamma detector. The presence of a nuclear radiating source on the ground was simulated. The

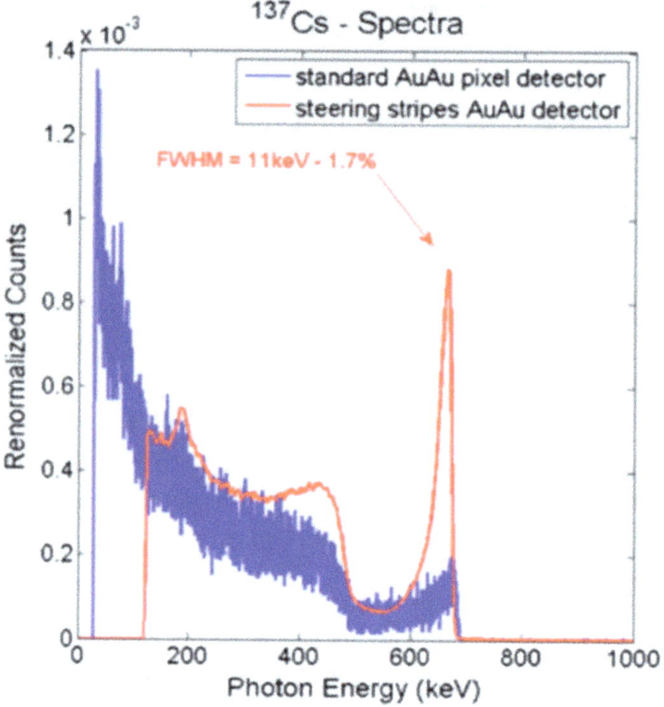

Fig. 9.5 Energy spectra obtained with a ^{137}Cs nuclear source using two CdZnTe $5 \times 6 \times 20$ mm^3 detectors with two planar contacts (blue line) and the strip grid contact geometry (red line)

operator, indeed, was unaware of the actual location of the simulated sources. These preliminary experiments allowed to establish the effectiveness of the approach as well as to refine the haptic interaction modality [29] and verify the stability of UAV teleoperation at the operating speed required for radiation monitoring purposes.

Experiments showed that the haptic-based teleoperation system helps the operator in focusing the search in the areas close to the radiating sources, thanks to the better support received through the haptic feedback channel.

9.5.2 Test of the Prototype in a Relevant Environment

The UAV prototype was validated in field experiments in two outdoor environments under the supervision of the regional environmental protection agency. The first location was an industrial plant in Gossolengo, Italy. A nuclear source (^{192}Ir) was located in the service area. This source is routinely used for non-destructive testing of large metallic elements soldering. Because of the large intensity of the nuclear source, the UAV can detect the gamma radiation in a large area. The nuclear source was inserted into a vertical lead container (to prevent lateral radiation emission) placed on the ground at the centre of a target with concentric circles (0.5 m radius step size). Preliminary experiments have been performed with prior knowledge of the position of the nuclear source that was placed in direct sight of the operator. These experiments were carried out to actually test the functionality of the whole prototype and of the haptic interface in particular. As the UAV entered inside the

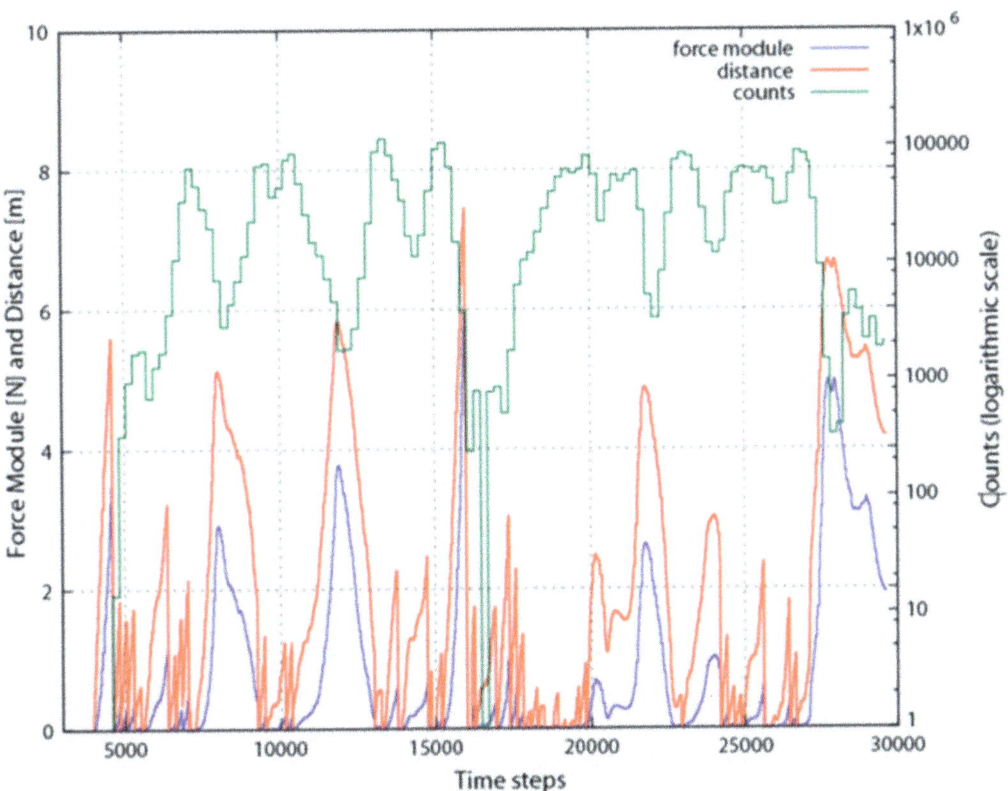

Fig. 9.6 Data obtained during the UAV flight in the presence of a ^{192}Ir nuclear source

radiation basin, the detector started to reveal gamma events and the pilot could feel a force on the haptic interface. Thus, the pilot flew the UAV back and forth several times in order to collect data on the whole system response.

The results of one of the flights are summarized in Fig. 9.6 [30]. The red line describes the distance of the UAV from the nuclear source, as calculated taking into account the data of the GPS. The green line accounts for the number of events measured by the CdZnTe detector in a log scale. Since the radiation field approximately follows the inverse square dependence, the minima of the distance of the nuclear sources correspond to the maxima of the detector counts. The pale blue line represents the module of the force computed in real-time by the XDrone software subsystem and experienced by the pilot at the haptic interface. When the UAV goes away from the nuclear source, the pilot feels an increasing force that pushes him back toward the nuclear source [30].

During the flight, the spectroscopic detector also collects the energy spectrum shown in Fig. 9.7. The spectrum contains the main peaks of the ^{192}Ir nuclear source, thus enabling the possibility of identification of the nuclear source.

Next, a set of experiments has been performed where the position of the nuclear source was unknown to the operator (Fig. 9.8). The position of the nuclear source was changed in each experiment. The UAV maintained a height from the ground ranging from 1.5 to 3 m and the average flight time was about 5 min for each mission.

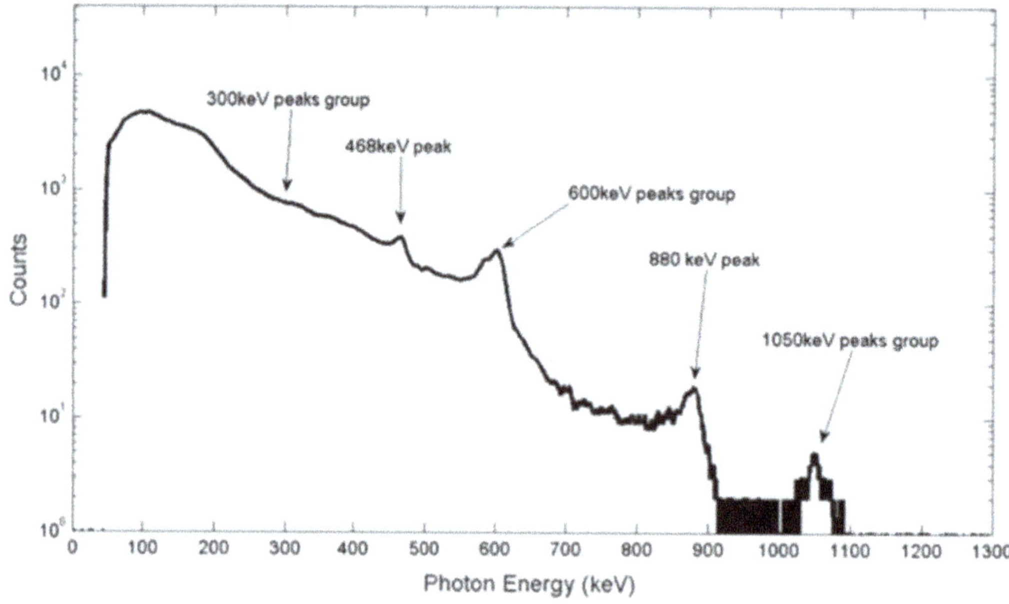

Fig. 9.7 Data obtained by the detector in the presence of the ^{192}Ir nuclear source

Figure 9.8 shows images of the experiments. Figure 9.8a shows in the foreground the haptic device handled by the pilot. In the background the nuclear source is shielded by an iron cylinder. Figure 9.8b was captured by the on board camera and shows the nuclear source in the centre of the target with concentric circles. Videos of these experiments are available.[1]

9.5.3   Test of the Prototype in Operational Environment

Further experiments were carried at a location that was chosen to represent a possible operational environment for the exploitation of the prototype. With this in mind, we performed flight experiments in an inter-municipal landfill in Novellara (RE), Italy, used for the disposal of waste materials. The facility is classified as a type B landfill which in principle excludes the presence hazardous waste. However, a recurrent environmental issue is whether radioactive material has been illicitly hidden in landfills, and hence landfill monitoring is indeed a potential operational environment for the drone. Clearly, in these environments the challenge in searching for nuclear sources it is not only given by the threat of the exposure to gamma radiation, but also by the harsh environment itself.

   The environment of the landfill in Novellara was not flat, hence multiple exploration trials have been performed at different heights by changing the UAV altitude set-point.

---

[1] http://www.bo.cnr.it/imem-old/xdrone/videos.php.

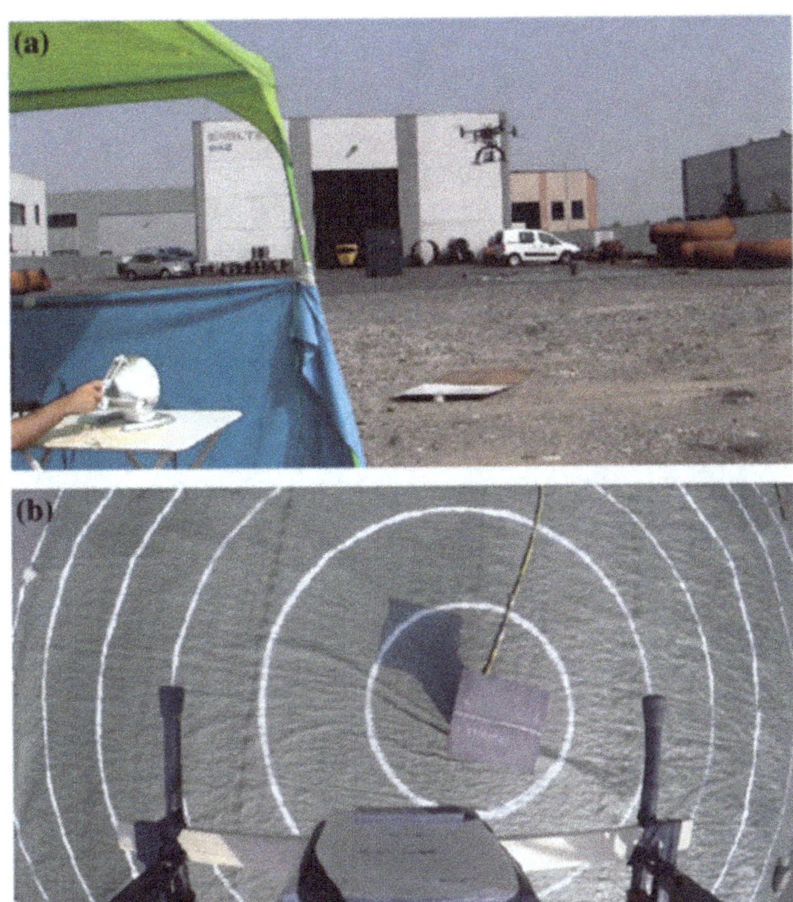

Fig. 9.8 Localization of the nuclear source; **a** in foreground the haptic device handled by the pilot; **b** board camera image of the nuclear source

The outcome of the tests confirmed the absence of nuclear waste material. The registered background gamma radiation was actually lower (about a half) of what was measured in the surrounding fields. This has to be attributed to shielding effect of the garbage with respect to the natural radiation from the ground. Images of the experiments are shown in Fig. 9.9.

The haptic user interface was also evaluated in field experiments in comparison to a standard mode of operation where the operator could only see sensor feedback from the detector by looking at a monitor screen. Results, reported in [29], indicated that without force feedback the operator had to spend about 23% of the time looking at the screen. On the other hand, the operator remained steadily focused on the flying UAV when the haptic feedback was turned on. This demonstrated that the use of force feedback can help to reduce the operator workload and minimize the risk of dangerous events like collisions and crashes.

Fig. 9.9   Field experiments in a landfill used for the disposal of waste materials

9.6   Conclusions and Future Research

The developed prototype proved to be an effective tool for localization and identification of nuclear material dispersed in the environment or illicitly stocked in industrial plants. The effectiveness of the approach was demonstrated in the lab, in a relevant environment in the presence of an intense ^{192}Ir nuclear source located in a service area, and in an operational environment consisting of an inter-municipal waste disposal landfill.

The prototype integrates a UAV with an airborne innovative CdZnTe-based gamma ray detector, able to detect with high sensitivity all the main nuclear contaminants. Thanks to the spectroscopic grade of the detector, identification of the localized nuclear source was also demonstrated in the project.

A haptic interface was adopted to drive the UAV, exploiting the real time counting rate experienced by the detector and an innovative software tool, fully compatible with the standard piloting software of the UAV. The pilot experienced on the haptic interface an attractive force that drove him towards the nuclear source. This teleoperation solution not only reduces the localization time, but also enables the operator to drive the UAV without missing the continuous view of the vehicle.

The results of this research have been presented at several conferences and the prototype has been shown at several national and international events An International patent has been filed [31] disclosing the main achievements of this research. Moreover, the realized prototype is now ready for exploitation by third parties (e.g. ARPA Emilia-Romagna, other territorial agencies, private companies), under the responsibility and control of IMEM-CNR and University of Parma. A web site was created[2] illustrating the main features of the prototype, and the experiments carried out. Through the website it is also possible to access the procedure to be followed by third parties and stakeholders in order to take advantage of this new prototype.

Acknowledgements This work has been funded by the Italian Ministry of Education, Universities and Research (MIUR) under the Flagship Project "Factories of the Future—Italy" (Progetto Bandiera "La Fabbrica del Futuro") [32], Sottoprogetto 2, research projects "XDrone" and "X-Drone2".

We thank ARPA Emilia Romagna and in particular Roberto Sogni for providing supervision and support for the experimental evaluation. We also thank "Emiltest srl Controlli Non Distruttivi" in Gossolengo, Piacenza, Italy and in particular Filippo Panciroli for granting access to their facility. We also thank Giancarla Rossetti for helping us with the authorization procedures for moving the nuclear source. Finally, we want to thank S.a.ba.r. S.p.A. for hosting our experiments on the landfill.

References

1. Martin PG, Payton OD, Fardoulis JS et al (2015) The use of unmanned aerial systems for the mapping of legacy uranium mines. J Environ Radioact 143:135–140
2. Wendel G (1998) Radioactivity in mines and mine water-sources and mechanisms. J South Afr Inst Min Metall 98(2):87–92
3. Hus M, Kosutic K, Lulic S (2001) Radioactive contamination of wood and its products. J Environ Radioact 55:179–186
4. Calabrese M, Quarantotto M, Cantaluppi C et al (2015) Quality characteristics and radioactive contamination of wood pellet imported in Italy. Open J Appl Sci 5:183–190
5. Ortiz P, Friedrich V, Wheatley J et al (1999) Lost and found danger—orphan radiation sources raise global concern. IAES Bull 41:18–25
6. Lu Y, Macias D, Dean ZS, Kreger NR et al (2015) A UAV-mounted whole cell biosensor system for environmental monitoring applications. IEEE Trans Nanobiosci 14(8):811–817
7. Okuyama S, Torii T, Suzuki A et al (2008) A remote radiation monitoring system using an autonomous unmanned helicopter for nuclear emergencies. J Nucl Sci Technol 45(sup5):414–416
8. Boudergui K et al (2011) Development of a drone equipped with optimized sensors for nuclear and radiological risk characterization. In: 2nd international conference on advancements in nuclear instrumentation measurement methods and their applications (ANIMMA). IEEE, pp 1–9

---

[2]http://www.bo.cnr.it/imem-old/xdrone/.

9. Kurvinen K, Smolander P, Pöllänen R et al (2005) Design of a radiation surveillance unit for an unmanned aerial vehicle. J Environ Radioact 67(2):340–344

10. Pöllänen R, Toivonen H, Peräjärvi K et al (2009) Radiation surveillance using an unmanned aerial vehicle. Appl Radiat Isot 67(2):340–344

11. MacFarlane J, Payton O, Keatley A et al (2014) Lightweight aerial vehicles for monitoring assessment and mapping of radiation anomalies. J Environ Radioact 136:127–130

12. Sanada Y, Torii T (2015) Aerial radiation monitoring around the Fukushima Daiichi nuclear power plant using an unmanned helicopter. J Environ Radioact 139:294–299

13. Martin P, Payton O, Fardoulis J et al (2016) Low altitude unmanned aerial vehicle for characterising remediation effectiveness following the FDNPP accident. J Environ Radioact 151(Part 1):58–63

14. Carloni R, Lippiello V, D'Auria M et al (2013) Robot vision: obstacle-avoidance techniques for unmanned aerial vehicles. IEEE Robot Autom Mag 20(4):22–31

15. Towler J, Krawiec B, Kochersberger K et al (2012) Radiation mapping in post-disaster environments using an autonomous helicopter. Remote Sens 4(7)

16. Han J, Xu Y, Di L et al (2013) Low-cost multi-UAV technologies for contour mapping of nuclear radiation field. J Intell Rob Syst 70(1–4):401–410

17. Newaz AAR, Jeong S, Lee H et al (2016) UAV-based multiple source localization and contour mapping of radiation fields. Robot Auton Syst 85:12–25

18. Knoll Glenn F (2000) Radiation detection and measurement, 3rd edn. Wiley, USA

19. Del Sordo S, Abbene L, Caroli E et al (2009) Progress in the development of CdTe and CdZnTe semiconductor radiation detectors for astrophysical and medical applications. Sensors 9:3491–3526

20. Camarda GS et al (2007) CdZnTe room-temperature semiconductor gamma-ray detector for national-security applications. In: Systems, applications and technology conference, LISAT 2007. IEEE Long Island, INSPEC, Accession Number: 9830508

21. Kowatari M, Kubota T, Shibahara Y et al (2015) Application of a CZT detector to in situ environmental radioactivity measurement in the Fukushima area. Radiat Prot Dosimetry 167:348–352

22. D.Lgs. 230/95 e s.m.i. http://www.repertoriosalute.it/wp-content/uploads/2016/11/dlgs230-95.pdf

23. International Commission on Radiological Protection. http://www.icrp.org

24. Micconi G et al (2015) Haptic guided UAV for detection of radiation sources in outdoor environments. In: Proceedings of the 3rd RED-UAS 2015 workshop on research, education and development of unmanned aerial systems, Cancun, Mexico, 23–25 Nov 2015

25. Kuvvetli I et al (2014) A 3D CZT high resolution detector for x- and gamma-ray astronomy. In: SPIE—the international society for optical engineering. 2014. 9154, art. n.:91540X

26. Bettelli M et al (2016) CdZnTe detector prototype for boron imaging by SPECT during BNCT treatment: simulations and measurements in a neutron field. In: 2016 IEEE NSS/MIC/RTSD, Strasbourg, Nov 2017, R01-7

27. Aleotti J et al (2017) Detection of nuclear sources by UAV teleoperation using a visuo-haptic augmented reality interface. Sensors 17:2234

28. Benassi G, Nasi L, Bettelli M et al (2017) Strong mechanical adhesion of gold electroless contacts on CdZnTe deposited by alcoholic solutions. J Instrum 12:P02018

29. Micconi G et al (2016) Evaluation of a haptic interface for UAV teleoperation in detection of radiation sources. In: 18th Mediterranean electrotechnical conference: intelligent and efficient technologies and services for the citizen, MELECON 2016

30. Aleotti J, Micconi G, Caselli S, Benassi G, Zambelli N, Calestani D, Zanichelli M, Bettelli M, Zappettini A (2015) Unmanned aerial vehicle equipped with spectroscopic CdZnTe detector for detection and identification of radiological and nuclear material. In: 2015 IEEE nuclear science symposium and medical imaging conference (NSS/MIC), pp 1–5
31. Aleotti J et al (2017) A system and a relative method for detecting polluting substances using a remotely piloted vehicle from a haptic command device. Patent WO 2017055962 A1
32. Terkaj W, Tolio T (2019) The Italian flagship project: factories of the future. In: Tolio T, Copani G, Terkaj W (eds) Factories of the future. Springer

Part V
Factory for Customised and Personalised Products

Chapter 10
Proposing a Tool for Supply Chain Configuration: An Application to Customised Production

Laura Macchion, Irene Marchiori, Andrea Vinelli and Rosanna Fornasiero

Abstract The full implementation of collaborative production networks is crucial for companies willing to respond to consumer demand strongly focused on product customisation. This chapter proposes an approach to evaluate the performance of different Supply Chain (SC) configurations in a customised production context. The model is based on discrete-event simulation and is applied to the case of supply chain in the fashion sector to support the comparison between mass and customised production. A prototype web-based interface is also developed and proposed to facilitate the use of the model not only for experts in simulation but for any user in the SC management field.

10.1 Scientific and Industrial Motivations, Goals and Objectives

Empirical evidence indicates the growing application of product customization strategies thanks to the possibility to consider the customers' opinion along the production process and realize goods more in line with customer's preferences [1]. It is expected that product customization will increase in the next 10 years by about 30% [2]. Fashion industry, and in particular footwear, appears to have good possibilities in implementing product customization, since Nike and Adidas are working on materials and technologies to support this business model and also small and medium companies are becoming more and more interested in customizing their offer [3, 4]. Moreover, a customized production can be a way to restore the competitiveness of European companies, enabling them to differentiate their offer from mass-productions realized in low-wage countries. However, to realize personalized productions the application

L. Macchion · A. Vinelli
Dipartimento di Tecnica e Gestione dei Sistemi Industriali, Università di Padova, Padua, Italy

I. Marchiori · R. Fornasiero (✉)
CNR-STIIMA, Istituto di Sistemi e Tecnologie Industriali Intelligenti per il Manifatturiero Avanzato, Milan, Italy
e-mail: rosanna.fornasiero@stiima.cnr.it

T. Tolio et al. (eds.), *Factories of the Future*,
https://doi.org/10.1007/978-3-319-94358-9_10

of advanced production technologies and innovative management practices [5]. This is particularly relevant within the footwear industry in which customer requirements can vary widely in terms for example of fit, colours, functional and aesthetic features [4] along different dimensions like seasons, target groups, and geographical areas.

There are several studies proposing innovative strategies for customization at the company level, but it is still needed to identify the supply-chain implications of customized productions especially in sectors like footwear where production is highly parcelled to many different suppliers and subcontractors, each of them specialised in specific operations. The complexity of the footwear production is given mainly by the high variability of the product and the short life of each model requiring fast changing production systems and flexible production networks [4, 5].

The potential benefits of personalization are based on better adherence to customers' requirements and the resulting customers' satisfaction. Several studies suggest that this strategy may be particularly suitable in highly competitive sectors like fashion industry, where product differentiation is a key factor to compete in the market and achieve a competitive advantage [6]. A company may decide to offer only a few customizable products, or could adopt a fully made-to-measure strategy involving all aspects of its production [7–11].

This chapter proposes an approach to evaluate Supply Chain (SC) performance of customized productions characterized by small-customized orders within the footwear industry. Thanks to a prototype web-based interface, a simulation model evaluates different type of performance indicators like inventory volume, order delivery, product quality for customized (i.e. small production) and traditional collections (i.e. large productions), and quality and delivery time of suppliers [4]. The model is based on deep analysis of historical data from 8 companies to study the common features of their SCs and to define a model that can represent the most critical and important steps in production.

The chapter is organized as follows. Section 10.2 gives an overview of the literature related to SC management for customized production. Section 10.3 presents the proposed approach, whereas Sect. 10.4 presents the details about the simulation model and the scenarios to be evaluated. Finally, Sect. 10.5 discusses the conclusions and future developments.

10.2 State of the Art

Customization strategy should enable a company to provide the right product variety without compromising rapid response to market or impinging upon its need for efficient production [9, 10]. In this way, customization is like mass production of customized goods while needing to find a trade-off between efficiency and flexibility [6, 9, 10, 12, 13]. This issue has been extensively studied from the company's internal point of view by identifying product design aspects (such as modularization and postponement [14, 15]) that can combine the different market demand needs with the production capability of a company [8]. In particular, for what concerns footwear industry, the development of new customized collections has become a significant challenge, where the desire for style, quality and comfort is very high and is influenced by the customer individual expectations and needs. Several footwear companies have decided to shift from the production of standard products to customization of some product collections for final consumers or for their direct first tier customers (i.e. retailers having direct contact with the final market that are able to specify the type of product customization that will be offered and appreciated within their shops). Therefore, footwear industry has started to create collections that match the stylistic and aesthetic trends of the market, e.g. flash collections characterized by a low number of products compared to traditional collections, produced in compressed lead times and characterized by features based on the specific market requirements [16]. However, an analysis that considers the implication that customization has on the entire SC is still missing and this works aims at filling this gap. Customization remains a real challenge within the footwear sector because of the complexity of the entire production process, which encompasses the assembly of several shoe parts while using different materials and manufacturing technologies and needing the involvement of many SC partners.

In customized contexts, the supply network configuration must be supported by the systemic monitoring of performance indicators [17], in particular of the suppliers' performance, as a way to assure the timing of the delivering process [18]. Therefore, how to physically locate production and manage supplier relationships have become an increasingly important strategic decision as shown both in review [19], in explorative studies [20–22] and quantitative modelling [16, 23–26] to optimize sourcing and production activities.

These strategic decisions become even more critical in case of customized production [27]. An evaluation of inter-company and intra-company performance is required to improve SC capability in order to optimize the use of both resources and suppliers. In particular, the analysis of suppliers' performance leads to understand how the production nodes can be linked in a network, and how they can collaborate to propose new practices that aim at filling the gap between the current performance (in terms of quality and delivery time) and the expected performance targets [28, 29]. However, even though many studies have discussed the need to monitor the performance of each supplier, relatively few attempts have been made to systematically identify how the supplier performance can affect the performance of the entire SC network [30].

10.3 Modelling a Supply Network for Customized Production

Several different aspects need to be taken into consideration to address the product customization issue by adopting a SC point of view. The proposed approach aims to evaluate network configurations over time to understand the organizational decision-making process, investigate the relations between the actors in the SC and analyse the consistency between the SC coordination model and the decision-making policies [31]. A discrete-event simulation (DES) model was developed to evaluate the performance of supply networks for customised productions in terms of inventory volume and order delivery time for small or large productions. Moreover, the model allows evaluating how the suppliers' performance in terms of quality and delivery time can influence the overall ability of the network to fulfil the customized requests [4, 32].

As explained in [4], discrete-event simulation is chosen because various problems related to SC configurations have already been solved with this technique. For instance, Cigolini et al. [33] employ discrete-event simulation to analyse SC configurations in terms of stocks and stock-out by evaluating if intermediaries influence product cost and quality. Ramanathan [34] implements a simulation method to evaluate the importance of collaboration in a SC, while Bottani and Montanari [35] propose a simulation model to support demand forecast with information sharing mechanisms in a SC. In terms of applying a simulation method in the fashion industry, Zülch et al. [36] employed simulation to examine different scenarios for customised orders in the garment sector, although that model made a comparison at the company level instead of considering the entire SC. This work takes inspiration from the previous approaches in literature to design a simulation model to support the comparison of different SC configurations.

The proposed SC configuration approach was developed for the fashion sector and the structure of the network is based on the experience gained during the analysis of several case studies within the footwear sector [27, 37, 38]. Figure 10.1 represents the network flow where different suppliers collaborate with the focal company (i.e. the shoe producer) by realizing specific components of final products and then by delivering the products to the warehouse of the shoe producer. The shoe producer assigns specific orders to suppliers based on their manufacturing skills after the verification of the specific needs of consumers in terms of stylistic preferences and ordered quantities. Finally, when each product component arrives at the components' warehouse, the shoe producer finalizes the production activities (mainly assembling) and sends final products to warehouse where orders are organized and shipped to customers. As a proxy for the implementation of product customization, the model considers the dimension of the customer order, which can range from the request of a single customized product to a single customized batch and compares these customization possibilities with a production realized with large batches of non-customized products.

The developed model supports the analysis of the impact of changing external parameters (e.g. customer demand, environmental regulations) on business, in particular in relation to the introduction of new product opportunities into their production network. The model mainly targets supply managers who have the possibility to create comparative SC configuration scenarios based on the variations in the suppliers' performance. Some of the variables that are available for the user are the following:

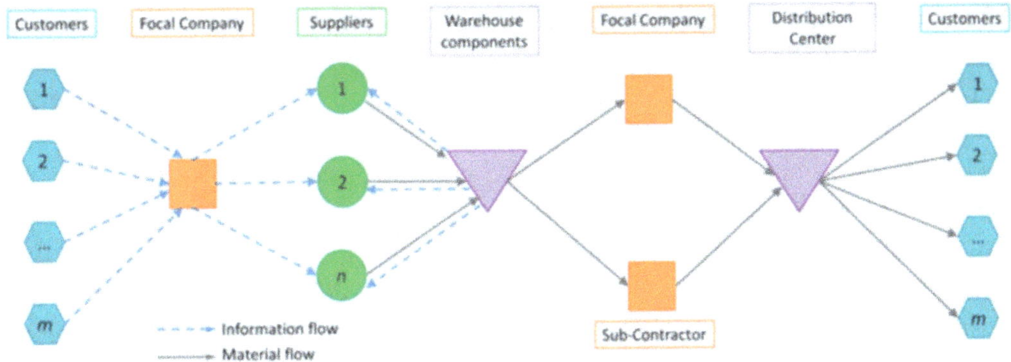

Fig. 10.1 Simplified representation of the network flow

- Activation of suppliers for the specific scenario to be considered.
- Updating their performance according to the data analysed in ERP.
- Implementing different re-order strategy (EOI-Economic Order Interval versus EOQ-Economic Order Quantity).
- Changing the customer order quantity and typology.
- Changing the mix of products to be produced in the considered period and related quantities.
- Changing BOM (Bill of Material).
- Changing initial stock of components in the warehouse.

10.4 Outcomes

This section is organised in three parts: Sect. 10.4.1 describes the developed simulation model, Sect. 10.4.2 presents the software architecture that is behind the web-based user interface and that enables the use of the model also to people not expert in simulation. Finally, Sect. 10.4.3 gives an insight on the experiments carried out to compare different scenarios.

10.4.1 Simulation Model

The optimization of a SC configuration requires first to characterize the *as-is* situation in terms of available suppliers, type of supplied components, type of managed order (small or large quantities) and performance indicators (quality, delivery time for each order).

In this study, we adopt the representation of the supply chain where there are two main flows, goods and order flows. The generic node i ($I = 1, …, N$) receives orders from the node $i - 1$ and goods from the node $i + 1$, through transport activities. For each node i, a lead time $L(i)$ includes transport, ordering and warehousing activities. We model the flow of multiple goods, with different BOM and with different order dimension. Customer demand is a stochastic variable, with normal distribution $N(\mu, \sigma)$.

The proposed Simulation Model evaluates the daily production for a period defined by the user (usually season production of six months or one year). The production mix considers the use of a fixed number of components chosen among the most important for customisation for the specific product. Moreover, the model assumes that a share of the products are produced only by the focal company or by a sub-contractor while the remaining by either the focal company or the subcontractor based on the shortest queue in the production process. The sub-contractor production time includes the extra time required to cover both the time it takes to deliver the components to the focal company and the time it takes to ship the final products. It is assumed that there are two different re-ordering policies (ROP):

- *ROP*1, i.e. the suppliers send the ordered quantity every time it is requested by the focal company.
- *ROP*2, i.e. the components are sent at a fixed time or when the accumulated quantity of re-orders reaches a predetermined value.

The materials are sent to the longest production queue for products that can be produced either by the focal company or by a subcontractor.

One centralised warehouse is located at the focal company site where all suppliers send components for all the product models. When components arrive at the centralised warehouse, they undergo a quality check with an associated processing time. In case of defects, the system sends back the request of a new order to the suppliers to cover the discarded amount. In the *ROP*2 policy, since the inventories cover the mean consumption of the supplier lead-time, the re-stocking order will be sent to the suppliers only when the stock reaches the re-order point. This can shorten the delivery time, but requires larger inventories [39]. When the production is completed, all the products are sent to the distribution centre located at the focal company site, which is then responsible for delivering the required orders to the retailers.

Since the proposed Simulation Model adopts a SC perspective, the model is based on the key performance indicators (KPIs) of each supplier in terms of order delivery time and order quality, as suggested in literature [13]:

- $T(i)$, i.e. delivery time of supplier i evaluated as the average time to deliver an order. This KPI is particularly relevant in a customization context because the need to provide the customized product in a short period calls for high flexibility.
- $Q(i)$, i.e. quality of supplier i evaluated as the average percentage of defective pieces for each delivered order. This KPI is also particularly critical in a customization context because consumers willing to pay a higher premium price for customized products do not tolerate defective pieces; defective products create delivery delays due to reworking.

After the definition of the *as-is* situation, different *to-be* scenarios for each network configuration can be defined to analyse the performance of both traditional and customised products. Starting from the collected data, a set of experiments allows to vary from the *as-is* analysis depending on the company's choice, for instance, to

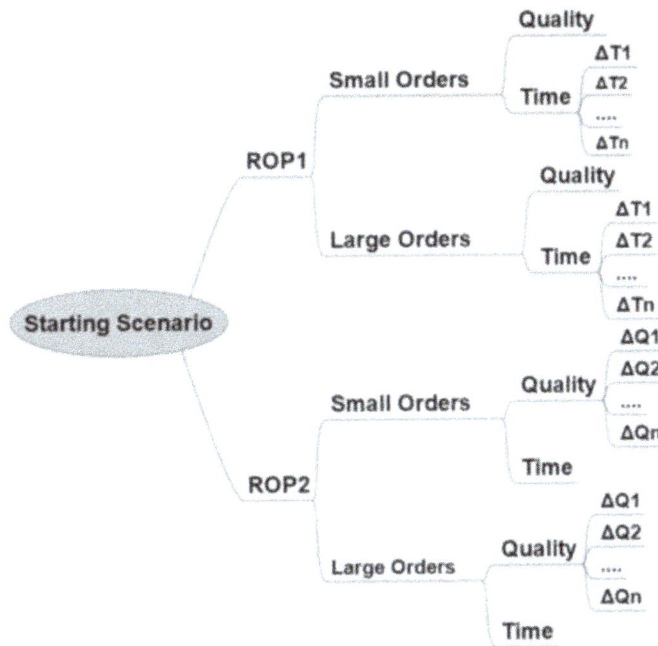

Fig. 10.2 Scenarios according to the chosen variables

collaborate with new suppliers or to outsource specific activities to different suppliers. The simulation runs for each of the *to-be* scenarios along the dimensions represented in Fig. 10.2 (i.e. re-ordering policy order dimension, supplier KPIs) to create different supply chain configurations.

The overall network performance is analysed for each configuration, based on variations in the suppliers' performance in terms of quality and delivery time. Thus, according to the supply network, different SC configurations are automatically analysed and compared in order to identify the best SC solution. Each network configuration is evaluated based on the following supply chain performance indicators:

- Order Lead-Time (OLT), i.e. the average time window from the issue of the order to the retailer of the focal company until the delivery of products to the customer.
- Inventory Volume (IV), i.e. average amount of inventory in the supply chain.

More details about the development of the Simulation Model can be found in [4].

10.4.2 Software Architecture

The software architecture represented in Fig. 10.3 was designed and implemented to take advantage the Simulation Model presented in the previous section. The architecture consists of the following components: Simulation Model, Simulation Engine, Web Interface, and external data sources.

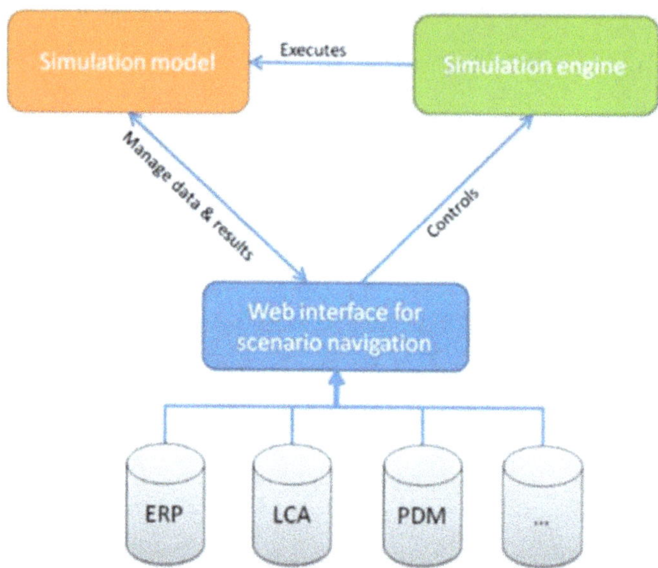

Fig. 10.3 The software architecture

The web interface supports supply chain managers in the use of the simulation model. Indeed, this interface provides an analysis of the *as-is* situation of the SC configuration in terms of type of supplied components, type of orders (small or large quantities, customized orders) and performance indicators (quality, delivery time for each order). Then, using data from external data sources (e.g. ERP, PDM), various *to-be* network configurations can be generated by taking into consideration different combinations for suppliers' performance. For each network configuration, the user can see the impact on the overall network performance.

The generated simulation models are executed by a simulation engine (i.e. SIMIO[1] in this case) and the web interface enables to navigate through the scenarios in a use friendly way. The user can access the models, so that input data can be gathered and the results can be easily visualized and analysed.

The web interface was developed on the Bloomy Decision platform[2] that is an executive decision platform providing a framework to support *what-if* analysis based on advanced analytics (predictive and prescriptive). The choice on Bloomy Decision is based on the fact that it can tackle complex problems in multi-decision making contexts and exploits data and computation distribution, while managing information with hybrid data structures (SQL, NO-SQL, and Hadoop HDFS techniques).

Moreover, this platform enables multiple users to share input data, models, and calculus engine results. The platform offers the ability to manage simulation and optimization problems, by enabling the user to setup a simulation or optimization model in a guided and structured manner. The web interface enables to control:

- Simulation input data and parameters.

---

[1] https://www.simio.com.

[2] www.act-OperationsResearch.com.

- Creation of simulation scenarios.
- Execution of simulation runs exploiting the SIMIO simulation engine.

The software architecture is based on the web application that manages the data flow feeding the simulation engine (SIMIO) to start the simulation process based on the predefined model. Moreover, the platform enables *what-if* analysis and comparative visualization of simulation scenario results.

After the simulation run, the user can compare different scenarios (up to 5) with different views summarising the results. In particular, the web application provides views on simulation results and scenario comparison. The Simulation view enables to select a scenario and visualize the related results via a graph and a set of tables. For example, in the case of the simulation model described in Sect. 10.4.1, it is possible to control the amount of items assembled by the focal company or by the sub-contractor, the average order lead time, the level of pollution and the final stock of the components. The user can select a scenario and compare it while exploring a table reporting all the evaluated scenarios. The simulation engine dynamically creates a set of bar charts for each control variable, where each graph compares different scenarios in the considered dimensions.

10.4.3 Experiments

The simulation model was tested with data collected from footwear companies by means of interviews and extraction of data from their ERP to have the possibility to make a cross-case analysis on most important practices related to fashion supply management. Data from different sources were aggregated to consider an average value of most important performance indicators.

Table 10.1 summarizes some important data related to the selected companies. It is clear that these companies need to manage orders with large variability (see *Order dimension* column in Table 10.1), requiring them high flexibility in the production process and supply chain.

The initial value of the inventory level is set to an average of 20 orders (both for large and small orders) when the ROP1 policy is adopted and an average of 40 orders when the ROP2 policy is used. Large orders are characterised by a normal distribution $N(800, 20)$, whereas small orders by a normal distribution $N(40, 5)$. The model simulates the daily production for a period of six months. The model assumes that 30% of the products are produced only by the focal company, 30% only by a sub-contractor and the remaining 40% by either the focal company or the subcontractor.

After setting the *as-is* supply chain scenario with the starting figures, the simulation runs are based on the scenarios presented in Fig. 10.2. For each ROP strategy and for each order dimension, the variations in the supplier performance (quality and delivery time) lead to different configurations. Therefore, scenarios are characterized by the improvement in the suppliers performance in terms of either $T(i)$

or $Q(i)$ (i.e. improvement in the delivery time or percentage of defects). Table 10.2 reports the results for 24 scenarios in terms of change in the supply chain IV and OLT. Each scenarios $Jxy(s)$ is characterized by x that indicates the order dimension (S=small, L=large), y that indicates the supplier performance improvement (Q=quality, T=time) and s that specifies the magnitude of the improvement ranging from 3 to 18%.

Figure 10.4 puts together all the scenarios from Table 10.2 and shows the two indicators (IV and OLT) considered for SC performance. The results show that, assuming the ROP1 strategy is applied, most of the scenarios with small orders (scenarios $JSQ(s)$ and $JST(s)$) have a large impact on both IV and OLT compared with the scenarios with large orders (scenarios $JLQ(s)$ and $JLT(s)$). Moreover, the improvement in the supplier average delivery time $T(i)$ [scenarios $JST(si$ and $JLT(s))$] has a greater impact on IV than the improvement in the quality $Q(i)$ [scenarios $JSQ(s)$ and $JLQ(s)$].

Similar results can be obtained for the ROP2 strategy and a more detailed analysis can be found in [4]. Furthermore, as addressed in [40], the simulation model can been integrated with Life-cycle assessment (LCA) analysis to create scenarios where it is

Table 10.1 Master data of companies

Company	Turnover € 2015[a]	Employees	Seasonal volume[b]	Seasonal n. models	Order dimension[c]	Price range
A	53	230	220	600	10–300	medium/high
B	10	20	150	180–250	80–600	high
C	10	40	37	150–180	50–1500	medium
D	8	40	32	200–300	50–600	medium/high
E	17	100	60	300	10–1000	medium/high
F	6	53	25	40–60	300–5000	medium/high

[a]in million euro
[b]in thousands of products
[c]no. of products per order

Table 10.2 Results of the scenarios for $ROP1$

s	Δ improvement (%)	Δ Inventory Volume (IV) (%)				Δ Order Lead Time (OLT) (%)			
		JSQ(s)	JLQ(s)	JST(s)	JLT(s)	JSQ(s)	JLQ(s)	JST(s)	JLT(s)
1	3	1.4	0.3	5.1	2.3	2.3	1.4	6.1	1.4
2	6	2.5	0.5	6.7	3.1	3.5	1.6	6.3	4.4
3	9	3.1	0.7	7.9	3.9	7.6	1.8	7.6	4.9
4	12	3.4	0.8	8.8	4.5	7.9	2.4	8.5	5.4
5	15	3.6	1.0	9.7	4.9	8.1	3.3	9.2	5.5
6	18	3.7	1.0	10.2	5.1	8.4	4.0	13.7	5.9

Fig. 10.4 Impact on supply chain performance

possible to evaluate the impact on pollution generated by the SC due to changes in the performance of the suppliers.

10.5 Conclusions and Future Research

This chapter has proposed an approach to analyse which factors have a greater impact on SC performance, to help top management to design the best configuration with upstream partners. Increasing the number of partners in SCs can complicate the decision making and hence the performance of model slows down.

The simulation model enables SC managers to analyse different SC scenarios by changing input parameters (e.g. lead times, inventory policies, demand volumes) and checking the impact of these modifications on several performance indicators. The model can be accessed via a web interface to create and compare scenarios using charts and figures. In this way, also people not expert in simulation can work with the simulation model. Thanks to the software architecture, the simulation model can be linked to the ERP, LCA and PDM systems of a company to collect input data necessary for the simulation runs.

The model has been tested with data collected from the footwear sector, demonstrating that the improvement in the supplier performance has a positive impact on the overall SC performance. The degree of impact is related to the flexibility of suppliers in reacting to fast changing requests measured in terms of order dimension and product variety.

The proposed approach and model can be easily applied to different business contexts where product customization represents a new business strategy to improve the performance in terms of inventory volume and order lead-time. In fact, although the quantitative results of the study are specific to the case of the footwear industry, the presented simulation model can be applied to other companies to compare different production scenarios and evaluate the performance of the SC.

The proposed model could also be improved by including other performance indicators and the variation of exogenous factors, such as demand variability. Changes in the management of orders (e.g. varying when and how customer orders become production orders) could also be considered as a way to improve the scenarios to be considered. Finally, the study could be further deepened by adopting a long-term perspective that investigates the performance trends over a number of years to study SC configurations with a longitudinal perspective.

Further developments can be done to improve the web-interface, to extend the integration of the tool with other databases and to assure a multi-criteria comparison between the scenarios.

During the development of the model, several interactions have been organized with experts at Politecnico Calzaturiero, an organisation representing footwear companies in Veneto Region, to disseminate customisation paradigm and to validate the software prototype.

Finally, the exploitation of the scientific results led to the submission and approval of two European research projects titled DISRUPT (FOF-11, 2016) and NEXT-NET (NMBP-37-CSA). The former deals with the management of the SC, whereas the latter is related to definition of scenarios for SC of the future. Indeed, one important focus will be on new SC models for customization and how technologies can support the process towards customisation.

Acknowledgements This work has been funded by the Italian Ministry of Education, Universities and Research (MIUR) under the Flagship Project "Factories of the Future—Italy" (Progetto Bandiera "La Fabbrica del Futuro") [41], Sottoprogetto 2, research projects "Innovative Methodologies, ADvanced processes and systEms for Fashion and Wellbeing in Footwear" (Made4Foot) and "Tool for Customer-driven Supply Networks configuration" (CD-NET). Part of this work is a compendium of what has already been published by the authors in [4, 40].

References

1. Forrester Research (2010) The customer experience index. https://www.forrester.com/report/The+Customer+Experience+Index+2010/-/E-RES55833. Accessed January 2016
2. Bain & Company (2013) Making it personal: rules for success in product customization. http://www.bain.com/publications/articles/making-it-personal-rules-for-success-in-product-customization.aspx
3. GeekWire Summit Conference (2015). http://www.geekwire.com/events/geekwire-summit-2015/
4. Macchion L, Fornasiero R, Vinelli A (2017) Supply chain configurations: a model to evaluate performance in customised productions. Int J Prod Res 55(5):1386–1399. https://doi.org/10.1080/00207543.2016.1221161
5. Tolio T, Copani G, Terkaj W (2019) Key research priorities for factories of the future—part I: missions. In: Tolio T, Copani G, Terkaj W (eds) Factories of the future. Springer

6. Fiore A, Lee SE, Kunz G (2004) Individual differences, motivations, and willingness to use a mass customization option for fashion products. Eur J Mark 38(7):835–849. https://doi.org/10.1108/03090560410539276

7. Anderson-Connell LJ, Ulrich PV, Brannon EL (2002) A consumer-driven model for mass customization in the apparel market. J Fashion Market Manag 6(3):240–258. https://doi.org/10.1108/13612020210441346

8. Fogliatto FS, Da Silveira GJC, Borenstein D (2012) The mass customization decade. An updated review of the literature. Int J Prod Econ 138(1):14–25. https://doi.org/10.1016/j.ijpe.2012.03.002

9. Jiao J, Ma Q, Tseng MM (2003) Towards high value-added products and services mass customization and beyond. Technovation 23(10):809–821. https://doi.org/10.1016/S0166-4972(02)00023-8

10. Dong B, Jia H, Li Z, Dong K (2012) Implementing mass customization in garmet industry. Syst Eng Procedia 3:372–380. https://doi.org/10.1016/j.sepro.2011.10.059

11. Macchion L, Danese P, Fornasiero R, Vinelli A (2017) Personalisation management in supply networks: an empirical study within the footwear industry. Int J Manuf Technol Manag 31(4):362–386. https://doi.org/10.1504/ijmtm.2017.086138

12. Pine BJII, Victor B (1993) Making mass customization work. Harvard Bus Rev 71(5):108–118

13. Salvador F, Rungtusanatham M, Forza C (2004) Supply-chain configurations for mass customization. Prod Plan Control 15(4):381–397. https://doi.org/10.1080/0953728042000238818

14. Shao XF, Ji JH (2008) Evaluation of postponement strategies in mass-customization with service guarantees. Int J Prod Res 46(1):153–171. https://doi.org/10.1080/00207540600844027

15. Harrison A, Skipworth H (2008) Implications of form postponement to manufacturing: across case comparison. Int J Prod Res 46(1):173–195. https://doi.org/10.1080/00207540600844076

16. Macchion L, Moretto A, Caniato F, Caridi M, Danese P, Vinelli A (2015) Production and supply network strategies within the fashion industry. Int J Prod Econ 163(C):173–188. https://doi.org/10.1016/j.ijpe.2014.09.006

17. Liu GJ, Shah R, Schroeder RG (2012) The relationships among functional integration, mass customisation, and firm performance. Int J Prod Res 50(3):677–690. https://doi.org/10.1080/00207543.2010.537390

18. Shepherd C, Günter H (2011) Measuring supply chain performance: current research and future directions. In: Behavioral operations in planning and scheduling. Springer, Berlin, Heidelberg, pp 105–121. https://doi.org/10.1007/978-3-642-13382-4_6

19. Şen A (2008) The U.S. fashion industry: a supply chain review. Int J Prod Econ 114(2):571–593. https://doi.org/10.1016/j.ijpe.2007.05.022

20. Brun A, Caniato F, Caridi M, Castelli C, Miragliotta G, Ronchi S, Sianesi A, Spina G (2008) Logistics and supply chain management in the luxury fashion retail: empirical investigation of Italian firms. Int J Prod Econ 114(2):554–570. https://doi.org/10.1016/j.ijpe.2008.02.003

21. Belso-Martínez JA (2010) International outsourcing and partner location in the Spanish footwear sector an analysis based in industrial district SMEs. Eur Urban Reg Stud 17(1):65–82. https://doi.org/10.1177/0969776409350789

22. Kinkel S (2012) Trends in production relocation and backshoring activities: changing patterns in the course of the global economic crisis. Int J Oper Prod Manag 32(6):696–720. https://doi.org/10.1108/01443571211230934

23. Jin B (2004) Achieving an optimal global versus domestic sourcing balance under demand uncertainty. Int J Oper Prod Manag 24(12):1292–1305. https://doi.org/10.1108/01443570410569056

24. Taplin IM (2006) Restructuring and reconfiguration. The EU textile and clothing industry adapts to change. Eur Bus Rev 18(3):256–261. https://doi.org/10.1108/09555340610663719

25. Christopher M, Peck H, Towill DR (2006) A taxonomy for selecting global supply chain strategies. Int J Logist Manag 17(2):277–287. https://doi.org/10.1108/09574090610689998

26. MacCarthy BL, Jayarathne PG (2013) Supply network structures in the international clothing industry: differences across retailer types. Int J Oper Prod Manag 33(7):858–886. https://doi.org/10.1108/ijopm-12-2011-0478

27. Shamsuzzoha A, Kankaanpaa T, Carneiro LM, Almeida R, Chiodi A, Fornasiero R (2013) Dynamic and collaborative business networks in the fashion industry. Int J Comput Integr Manuf 26(1–2):125–139. https://doi.org/10.1080/0951192X.2012.681916

28. Askoy A, Öztürk N (2011) Supplier selection and performance evaluation in just-in-time production environments. Expert Syst Appl 38(5):6351–6359. https://doi.org/10.1016/j.eswa.2010.11.104

29. Dey PK, Bhattacharya A, Ho W (2015) Strategic supplier performance evaluation: a case-based action research of a UK manufacturing organisation. Int J Prod Econ 166:192–214. https://doi.org/10.1016/j.ijpe.2014.09.021

30. Akyuz GA, Erkan TE (2010) Supply chain performance measurement: a literature review. Int J Prod Res 48(17):5137–5155. https://doi.org/10.1080/00207540903089536

31. Tako AA, Robinson S (2012) The application of discrete event simulation and system dynamics in the logistics and supply chain context. Decis Support Syst 52(4):172–186. https://doi.org/10.1016/j.dss.2011.11.015

32. Umeda S, Zhang F (2006) Supply chain simulation: generic models and application examples. Prod Plan Control 17(2):155–166. https://doi.org/10.1080/09537280500224028

33. Cigolini R, Pero M, Rossi T, Sianesi A (2014) Linking supply chain configuration to supply chain performance: a discrete event simulation model. Simul Model Pract Theory 40:1–11. https://doi.org/10.1016/j.simpat.2013.08.002

34. Ramanathan U (2014) Performance of supply chain collaboration—a simulation study. Expert Syst Appl 41(1):210–220. https://doi.org/10.1016/j.eswa.2013.07.022

35. Bottani E, Montanari R (2010) Supply chain design and cost analysis through simulation. Int J Prod Res 48(10):2859–2886. https://doi.org/10.1080/00207540902960299

36. Zülch G, Koruca HI, Börkircher M (2011) Simulation-supported change process for product customization—a case study in a garment company. Comput Ind 62(6):568–577. https://doi.org/10.1016/j.compind.2011.04.006

37. Zangiacomi A, Fornasiero R, Franchini V, Vinelli A (2017) Supply chain capabilities for customization: a case study. Prod Plan Control 28(6–8):587–598. https://doi.org/10.1080/09537287.2017.1309718

38. Macchion L, Fornasiero R, Danese P, Vinelli A (2016) The effects of personalization on collaborative production networks location. In: Afsarmanesh H, Camarinha-Matos L, Lucas Soares A (eds) Collaboration in a hyperconnected world. PRO-VE 2016. IFIP advances in information and communication technology, vol 480. Springer, Cham, pp 433–440. https://doi.org/10.1007/978-3-319-45390-3_37

39. Fornasiero R, Macchion L, Vinelli A (2015) Supply chain configuration towards customization: a comparison between small and large series production. IFAC-PapersOnLine 48(3):1428–1433

40. Brondi C, Fornasiero R, Collatina D (2017) Sustainability assessments for mass customization supply chains. In: Modrak V (ed) Mass customized manufacturing: theoretical concepts and practical approaches. CRC Press. ISBN 9781498755450

41. Terkaj W, Tolio T (2019) The Italian flagship project: factories of the future. In: Tolio T, Copani G, Terkaj W (eds) Factories of the future. Springer

Chapter 11
Hospital Factory for Manufacturing Customised, Patient-Specific 3D Anatomo-Functional Models and Prostheses

Ettore Lanzarone, Stefania Marconi, Michele Conti, Ferdinando Auricchio, Irene Fassi, Francesco Modica, Claudia Pagano and Golboo Pourabdollahian

Abstract The fabrication of personalised prostheses tailored on each patient is one of the major needs and key issues for the future of several surgical specialties. Moreover, the production of patient-specific anatomo-functional models for preoperative planning is an important requirement in the presence of tailored prostheses, as also the surgical treatment must be optimised for each patient. The presence of a prototyping service inside the hospital would be a benefit for the clinical activity, as its location would allow a closer interaction with clinicians, leading to significant time and cost reductions. However, at present, these services are extremely rare worldwide. Based on these considerations, we investigate enhanced methods and technologies for implementing such a service. Moreover, we analyse the sustainability of the service and, thanks to the development of two prototypes, we show the feasibility of the production inside the hospital.

E. Lanzarone (✉)
CNR-IMATI, Istituto di Matematica Applicata e Tecnologie Informatiche
"Enrico Magenes", Milan, Italy
e-mail: ettore.lanzarone@cnr.it

S. Marconi · M. Conti · F. Auricchio
Dipartimento di Ingegneria Civile e Architettura, Università di Pavia, Pavia, Italy

I. Fassi · C. Pagano · G. Pourabdollahian
CNR-STIIMA, Istituto di Sistemi e Tecnologie Industriali Intelligenti per il
Manifatturiero Avanzato, Milan, Italy

F. Modica
CNR-STIIMA, Istituto di Sistemi e Tecnologie Industriali Intelligenti per il
Manifatturiero Avanzato, Bari, Italy

© The Author(s) 2019
T. Tolio et al. (eds.), *Factories of the Future*,
https://doi.org/10.1007/978-3-319-94358-9_11

233

11.1 Scientific and Industrial Motivations, Goals and Objectives

The significant increase of life expectancy over the last decades [1], which was made possible thanks to the progress of medical sciences, generated as a counterpart a higher demand for health care services, as elderly people generally need more intensive medical assistance [2, 3]. In addition, the modern possibilities to treat patients affected by severe diseases are generating new classes of chronic patients who need specific and highly qualified treatments [4]. This new demand for more intensive, advanced and personalised care services is clearly in contrast with the limited budget of national and regional health care systems. Consequently, new models are necessary to guarantee adequate care treatments to each patient.

In parallel, new production technologies are boosting the manufacturing of patient-specific solutions. On the one hand, personalised prostheses tailored on the specific patient are becoming crucial for the development of several surgical specialties; on the other hand, patient-specific anatomic models for preoperative planning are essential in the presence of customised prostheses, as also the surgical treatment must be optimised.

However, in the common practice, most of medical products are produced in standard sizes and shapes, and then stocked in hospitals, where the product to implant is chosen as the one closest to patient's anatomy. Thus, it is often necessary to manually adapt the product to the anatomic characteristics of the patient, with the risk of damaging the product and not reaching the optimal size and shape for the patient. This may also determine longer surgery times, with higher costs per patient.

Just recently, we may notice a significant growth of rapid prototyping services dedicated to the medical field, which are mostly provided by external companies that established a medical division.[1] However, the presence of a prototyping service inside the hospital would be a benefit for the clinical activity. The location inside the hospital would allow a closer interaction with clinicians during the model development, leading to significant time and cost reductions, and to a higher effectiveness of the products.

At present, to the best of our knowledge, there are only four services located inside the hospital worldwide: 3D Print Lab@USB (Basel, Switzerland)[2]; RIH 3D Lab (Rhode Island, Providence, RI, US)[3]; Austin Health 3D Medical Printing Laboratory (Austin, Melbourne, Victoria, Australia)[4]; 3D and Quantitative Imaging Laboratory (Stanford University School of Medicine, Stanford, CA, US).[5] However, though they

---

[1]Materialise—www.materialise.com/en/industries/healthcare.
Zare—www.zare-prototyping.eu/en/medical-division.

[2]3D Print Lab@USB—www.unispital-basel.ch/das-universitaetsspital/bereiche/medizinische-querschnittsfunktionen/kliniken-institute-abteilungen/departement-radiologie/kliniken-institute/klinik-fuer-radiologie-und-nuklearmedizin/3d-print-lab.

[3]RIH 3D Lab—www.brown.edu/Research/3DLab.

[4]Austin Health 3D Medical Printing Laboratory—www.austin.org.au/page?ID=1839.

[5]3D and Quantitative Imaging Laboratory—http://3dqlab.stanford.edu/.

are located inside the hospital, these laboratories have several limitations in terms of available technologies, as they are equipped with low cost machines and rely on outsourcing for the complex cases that require high-resolution 3D printers. Moreover, their work is confined to the orthopaedic and maxillo-facial specialties, which are the easiest to manage.

In this context, the project *Hospital Factory for Manufacturing Customised, Patient-Specific 3D Anatomo-Functional Models and Prostheses* (Fab@Hospital) proposed an innovative paradigm, i.e., the production of personalised medical products (e.g., prostheses) in an environment closely integrated with the hospital, through new design approaches and technologies, to guarantee a direct interaction between patients, medical personnel, and product manufacturers. This may improve the quality of life for patients, the performance of the health care system, and the competitiveness of manufacturers.

The Fab@Hospital paradigm consists of:

- advanced mathematical tools and modelling technologies tailored to manufacture personalised products;
- location of a *hospital factory* inside or near the hospital;
- production of personalised products (e.g., prostheses) at the *hospital factory* in a short time, thanks to the combination of innovative technologies and processes.

Besides the products, the hospital factory would produce personalised anatomic models (e.g., reconstructions of vascular districts), which may help the surgeons in studying the treatment in advance by simulating different strategies.

To meet these goals, the following scientific and technological objectives were identified:

1. Define new mathematical methods and approaches to build accurate anatomo-functional models from medical images.
2. Define improvements to the existing technologies for personalised products, e.g., additive manufacturing and micro Electrical Discharge Machining (EDM).
3. Propose innovative process combinations to reduce the production costs, thus supporting a wide diffusion of personalised medical products.
4. Demonstrate the applicability of the new technological approaches and process combinations through the development of two prototypes.
5. Propose new business models for the production of personalised medical products inside or near the hospital.

The rest of the chapter is organised as follows. Section 11.2 overviews literature related to the addressed problems, which are stated in Sect. 11.3. The developed technologies and methodologies are detailed in Sect. 11.4, while the outcomes of the work (in terms of two prototypes, several mathematical tools and a business model structure) are shown in Sect. 11.5.

11.2 State of the Art

Some medical specialties have started to benefit from rapid prototyping in the last few years, especially for preoperative planning purposes [5, 6]. In fact, clinicians may obtain more information from physical objects than from computer virtual models only. Moreover, their educational value to train new surgeons is also recognised [7].

Among the others, maxillofacial and orthopaedic specialties currently employ physical models to test different solutions, e.g., for the implant of bone fixation plates.

Vascular surgery has seen the development of a dedicated rapid prototyping sector [8]. In fact, the benefit of a physical vascular model for planning the implant of stents or vascular prostheses is linked not only to its morphological characteristics, but also to the mechanical properties of the reproduced vascular district. As the surgeon tests the placement and the release of the prosthesis, and he/she evaluates the prosthesis-vessel interaction, the vessel is required to have a behaviour as consistent as possible with the real pathophysiology; thus, compliant models with controlled elasticity are needed. However, vascular applications are mainly at the research level, and very few companies deal with patient-specific silicone vascular models.

In all cases, the current production scheme in factories not linked to the hospitals has some drawbacks:

- *Anatomo-functional properties*. Clinicians do not have the expertise to retrieve functional properties from the common medical imaging. At the same time, manufacturers do not have the expertise to translate these properties into suitable production specifications. Consequently, sending the production request to an external factory without a discussion between clinicians and manufacturers about the specifications may reduce the effectiveness of the personalised medical product.
- *Production times*. The direct interaction between clinicians and manufacturers would avoid loss of information and the need to correct the product, thus speeding up the process. Moreover, the production inside or near the hospital would significantly reduce transportation times.
- *Costs*. Even though the prices of the anatomic models for visualisation purposes are lowering, thanks to the progressive spread of prototyping technologies, moving the production inside the hospital would significantly reduce the costs. Moreover, compliant functional models with proper elasticity are still extremely expensive (thousands of euros even for very small vascular districts) and the production inside the hospital would make them more affordable.

Our literature analysis focuses on the four topics addressed in this work, i.e., patient-specific anatomo-functional cardiovascular models, patient-specific fixation plates, mathematical tools to support their design, and business models to make their production more efficient.

- **Cardiovascular models**. In vitro analyses of the vascular fluid-dynamics may help clinicians in understanding the impact of specific pathologies and devices. In this context, additive manufacturing is playing a crucial role, allowing the production of highly complex geometries at lower costs and in lower times than with standard subtractive technologies. Thus, additive manufacturing is rapidly spreading in the medical field to produce patient-specific anatomic models. In particular, 3D printed anatomic models have been used to test different operative approaches [9–12], or to improve the design process of endovascular devices. In fact, in vitro fluid-dynamic analyses are useful to identify the interaction between the device (e.g., valves, endo-prostheses, stents) and the human vascular system [13–16].

- **Fixation plates and screws**. Patient-specific fixation plates and screws are not widely adopted, due to the difficulties in the small-scale production of 3D complex components, usually made of stainless steel (ASTM F-55 and F-56), pure titanium and its alloys (ASTM F-136), and cobalt-chromium-tungsten-nickel alloy (ASTM F-90). In this context, micro-EDM could be a suitable technology for producing customised plates, due to the ability to perform complex and high-precision machining on electro-conductive materials [17]. Being a thermal process, it can be used with considerable success also for the machining of extremely hard and strong materials [18], including conductive ceramics [19, 20]. Also techno-polymers, like the PolyEther Ether Ketone (PEEK), have been recently introduced for the manufacturing of fixation plates. In this case, 3D printing technologies such as Fusion Deposition Modelling (FDM) and Stereo Lithography Apparatus (SLA) can be applied.
- **Mathematical models**. Tools for reconstructing the geometrical features of some districts from medical imagining are nowadays widespread and widely used in clinics. However, as for the mechanical properties, commercial tools do not generally include this possibility, which still represents an open research issue.
- **Business models**. There are very few works investigating the business and managerial sides of manufacturing customised medical products. The existing studies, which have been conducted only recently, mainly address the problem from an economic perspective. Some works investigated the economic implications of 3D printing in general [24], while others focused on evaluating the cost structure and developing cost models for additive manufacturing [25–27]. Lindemann et al. [28] developed a business model for evaluating the cost of additive manufacturing. Schröder et al. [29] investigated the manufacturing of customised medical devices from a business model perspective with a specific focus on value chain; they emphasised that interoperability is a significant driver for the efficient manufacturing of customised medical devices, as customisation is not a single-stakeholder process but a multi-actor process that includes suppliers, surgeons and patients.

11.3 Problem Statement and Proposed Approach

Our work includes two main applications. On the one hand, we employ additive manufacturing (3D printing) to produce anatomo-functional models for the cardiovascular specialty; on the other hand, we exploit micro-EDM and additive manufacturing to produce fixation plates for the orthopaedic specialty. Moreover, our work also involves two supporting activities, i.e., the development of mathematical models and tools for the design of patient-specific products, and the development of appropriate business models to suggest the best management strategies and to prove the benefits of the Fab@Hospital paradigm (i.e., the production inside or near the hospital). All these activities are detailed in the next subsections.

11.3.1 Additive Manufacturing for Cardiovascular Models

The goal is to manufacture deformable vascular models, with realistic mechanical and geometrical properties, to be employed for in vitro analyses. In particular, we focus on benchmark aortic models to test innovative endovascular devices and new surgical procedures.

The proposed production approach is based on additive manufacturing in combination with a moulding technique, to produce silicone models featuring both mechanical and geometrical properties of the vessel.

Ideally, the highest adherence to reality is possible through a full patient-specific approach, in which all mechanical and geometrical information, along with flows and pressures, are acquired from the specific patient. In the absence of all information, a less specific approach can be pursued by combining patient-specific information and general knowledge common to several individuals (retrieved from the literature or measured on a significant number of patients).

11.3.2 Micro-EDM and Additive Manufacturing for Fixation Plates

Although customised implants are occasionally used for surgery, they are not oriented to traumatic pathologies. In fact, the treatment of traumatic pathologies requires

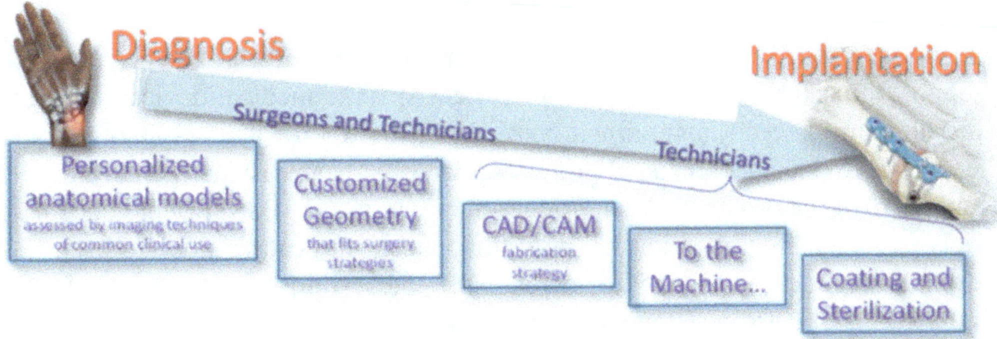

Fig. 11.1 Procedure for fabricating customised fixation plates

manufacturing the fixation plate in less than seven days, thus requiring a strong interaction between clinicians and manufacturers (see Fig. 11.1 for the fabrication procedure).

The rapid production and the clinicians-manufactures interaction could be easily achieved if the prototyping station were inside the hospital. However, to make such prototyping station available, the fabrication technology for customised devices should fulfil several constraints, due to the confined space and the controlled environment. Additive manufacturing and micro-EDM fulfil these constraints. On the one hand, complex 3D shapes can be manufactured with micro-EDM on every electro-conductive material (e.g., titanium and surgical steel); on the other hand, FDM and SLA are suitable for the fabrication of polymeric objects with complex shape at low cost and low environmental impact.

11.3.3 Prediction Models for Patient-Specific Functional Properties

Stochastic tools may support the estimation of patient-specific mechanical properties in several districts, based on non-invasive measurements and patient's characteristics. This is of particular importance in case of soft tissues (e.g., for the cardiovascular specialty). Thus, we focus on two cardiovascular applications: (i) the estimation of the aortic stiffness and its spatial variations; (ii) the estimation of the ultimate mechanical properties and of the stress-strain characteristics in patients with ascending aorta aneurysm. Moreover, due to the lack of effective tools to support Finite Element Analysis (FEA) under uncertain parameters and Structural Topology Optimization (STO), which are common problems when dealing with patient-specific problems, we propose an approach for efficiently solving FEA problems in the presence of stochastic parameters or within iterative optimization algorithms.

11.3.4 Business Models

As mentioned above, there are no appropriate business models that can be employed to support the additive manufacturing of patient-specific medical devices. The gap in the literature is even larger when considering the role of the hospitals in manufacturing individualised medical products, as hospitals are usually perceived only as end-users of the products. Also Product-Service System (PSS) oriented business models pay very little attention to apply the concept in health care, and even less to extend the practices of PSS to increase the integration between hospitals and manufacturers of customised medical products.

Thus, our goal is to develop a reference structure for business models that can support the Fab@Hospital paradigm.

11.4 Developed Technologies, Methodologies and Tools

In the following, we present the technologies, the methodologies and the tools developed to address the four problems presented in Sect. 11.3. Moreover, as for additive manufacturing and micro-EDM, we describe the features of the associated prototypes. The first prototype (described in Sect. 11.4.1) is a preoperative model for the cardiovascular specialty, while the second prototype (described in Sect. 11.4.2) consists of a set of fixation plates for the orthopaedic specialty.

11.4.1 Additive Manufacturing for Cardiovascular Models

We relied on a moulding technique to create an aorta benchmark model, where the mould is produced by means of 3D printing technology. We built up a flexible and completely parametric model, which is able to adapt in a consistent way to the modifications of the structural parameters. In particular, for the prototype, we considered geometrical and mechanical parameters retrieved from the literature; in addition, in the absence of literature data, some geometrical data were obtained from several Computed Tomography (CT) images.

The mould is composed of three parts: two outer shells and an inner part that creates the inner lumen of the vessel (Fig. 11.2). The thickness of the model, given by the distance between inner and outer moulds, was selected according the desired

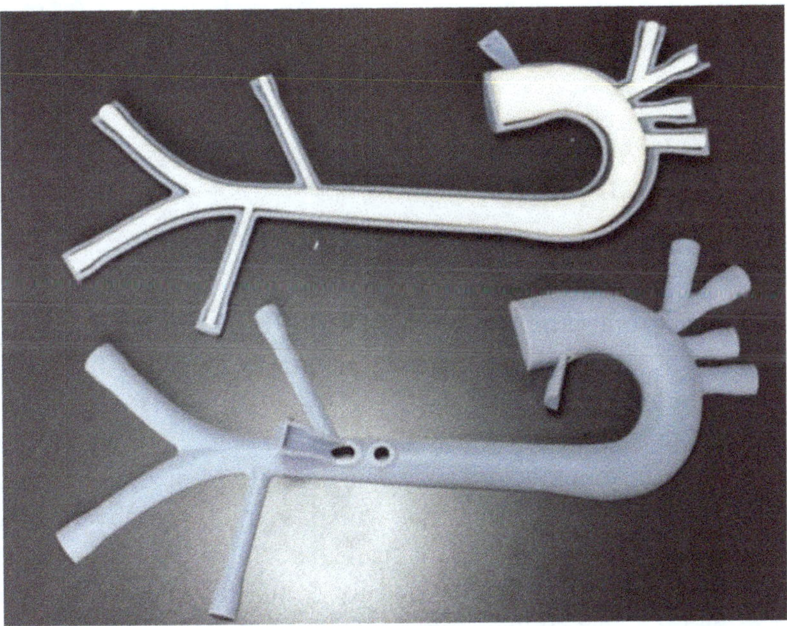

Fig. 11.2 3D printed mould of the aorta model, composed of an inner lumen (white) and two outer shells (blue)

punctual compliance. Indeed, starting from the inner lumen geometry, we computed the thickness in order to get the desired compliance.

The mould was manufactured using the 3D printing technology. We employed the Objet 30 Pro printer (Stratasys, MN, US), which is based on the *Material Jetting* technology, where layers of photopolymer resin are deployed on a building tray and cured by means of ultraviolet light. We employed the commercial photopolymer VeroWhitePlus RGD835 and the support material FullCure 705, which is necessary to support the parts of the model that do not lay directly on the tray or the underlying layer. After printing, a fine post-processing was performed to ensure a smooth finishing of all surfaces of the mould that will be in contact with the silicone, and the mould was finally assembled.

To create the aorta model, we employed the two-part Sylgard 184 Silicone Elastomer (Dow Corning, MI, US) in a ratio of 10 parts of silicone base to 1 part of curing agent by weight. The silicone mixture was poured into the mould, placed into a vacuum chamber to eliminate air, and then left curing at room temperature (about $23\,°C$) for 48 h. Such two-part ratio and curing temperature were chosen in order to tune the mechanical properties of the silicone [30]; the selected combination leads to a final elastic modulus of 1.32 ± 0.07 MPa. After curing, the mould was removed to get the final silicone model.

The resulting prototype, which can be considered a benchmark deformable aortic model, is shown in Fig. 11.3. It is endowed with terminations that can be connected to in vitro circuits by means of pipe junctions. Moreover, the use of transparent silicone allows inner visibility, which is fundamental for several experiments, e.g., endograft deployment or particle tracking. Transparency is also fundamental to exploit the model to train beginner surgeons and clinicians; allowing inner visibility, the student is able to look through the vessel at the movements of the endoscopic instruments.

11.4.2 Micro-EDM and Additive Manufacturing for Fixation Plates

Three manufacturing technologies (namely, micro-EDM, FDM and SLA) were tested for manufacturing several prototypes made of different materials (metals and polymers). The two additive manufacturing technologies were studied using two geometries and different polymers.

The two most appropriate micro-EDM approaches (i.e., milling-EDM and wire-EDM) were tested in terms of material removal rate, to minimise the machining time. Two materials were investigated: titanium, which is largely employed in the medical sector due to biocompatibility and mechanical characteristics, and the ceramic composite Si3N4-TiN, which is currently used for dental implants. Several tests were carried out to find the best strategy and parameters to achieve the typical features of fixation plates, such as holes and bores. Holes are necessary on plates both for temporary and permanent fixture because, according to the anatomy of the patient

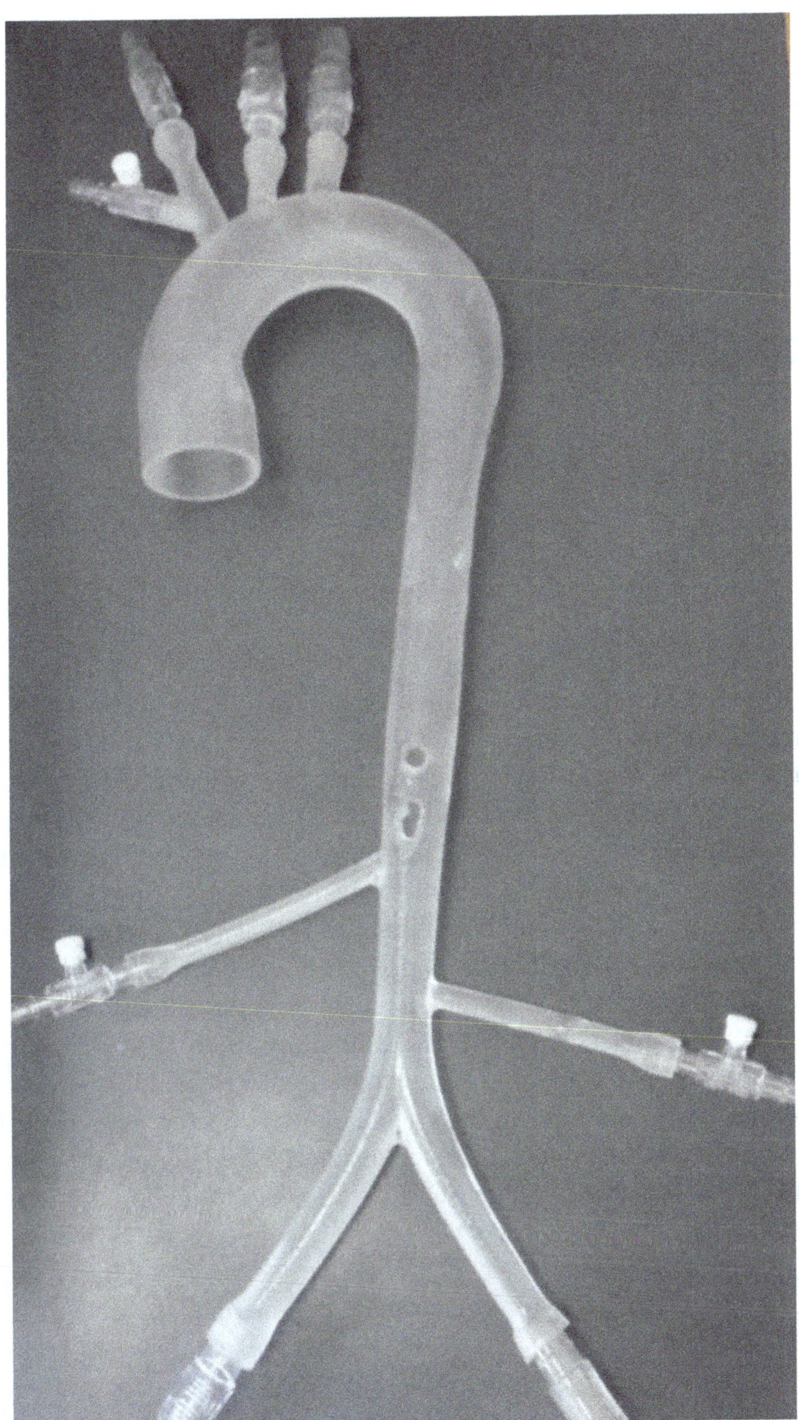

Fig. 11.3 Silicone aorta model after mould removal

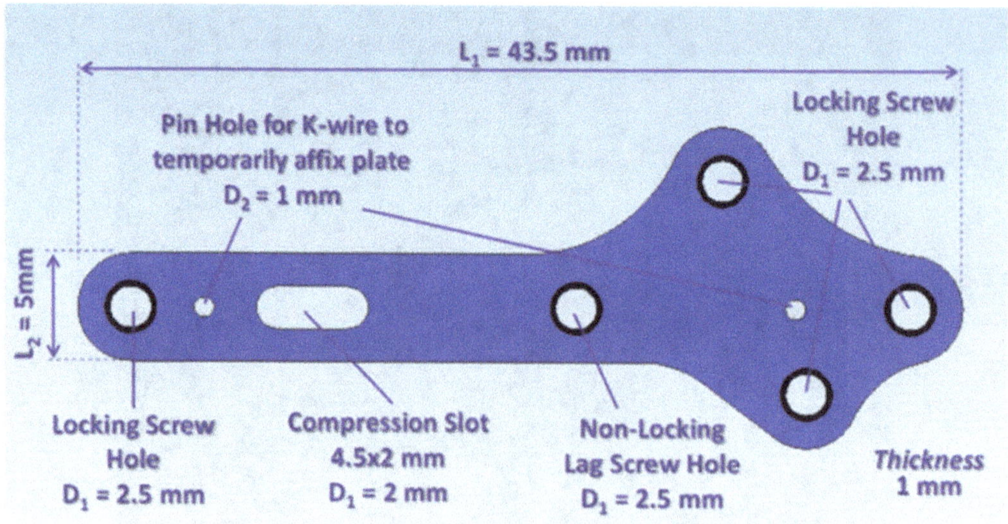

Fig. 11.4 Fracture plate for osteotomy and arthrodesis in the foot

and the fracture, different fixation points are required; bores lighten the plate and help assembling during the surgery.

The FDM process was tested with engineering polymers and composites. Tests were carried out to correlate the FDM process parameters with the highest resolution achievable. They were conducted on both ad hoc designed samples and on a few designs of commercial fixation plates, using the printers Sharebot Next Generation (Sharebot, Italy) and S2 (Gimax 3D, Italy). Moreover, to tailor the mechanical properties of the biocompatible polymers, carbon nanotubes (CNT) were added as filler in several polymeric matrixes, e.g., Polymethyl Methacrylate (PMMA), Polyoxymethylene (POM) and Polyamide (PA). The mechanical characterisation of these composites was carried out using FDM 3D printing and SLA dog-bone tensile specimens. SLA technology, with the equipment Form 1+ (Formalabs, MA, US) was finally selected as alternative additive process, in order to compare its performance with the FDM in terms of production time, mechanical strength and surface quality.

As for the prototype, the micro-EDM was tested on the prototype shown in Fig. 11.4, which is inspired by a fixation plate commercialised by Vilex (McMinnville, TN, USA). The plate includes four holes for the screws, to lock the plate to the bone, a non-locking lag screw hole, one compression slot and two smaller holes for temporary fixing. Samples were manufactured to prove the customisation methodology, the machining performance and the process ability.

A commercial fixation plate for the *Rolando fracture* (i.e., a fracture to the base of the thumb, see Fig. 11.5a) was selected as test specimen in order to assess additive manufacturing capabilities. The prototype was fabricated in several polymers (both commercial and internally developed) using FDM and SLA. As expected, the two technologies have different potentialities. FDM allowed the use of a greater variety

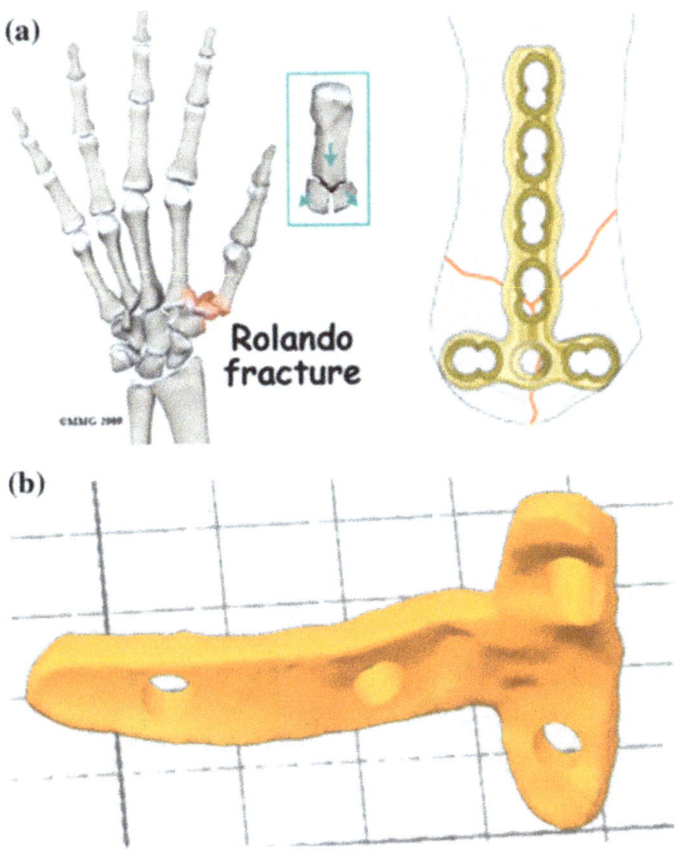

Fig. 11.5 Rolando fracture and commercial fixation plate (**a**); proposed patient-specific plate (**b**)

of polymers and composites with tailored properties, while the selection of materials for SLA was much reduced.

Finally, a patient-specific fixation plate for the *Rolando fracture* was designed and fabricated with different polymers (Fig. 11.5b). Figure 11.6 shows some of the produced plates, made using FDM in PA-CNT4% and SLA. SLA samples present very smooth surfaces due to the higher resolution of the technology.

11.4.3 Prediction Models to Identify Patient-Specific Functional Properties

As mentioned in Sect. 11.3.3, we focused our activity on three applications, for which an adequate solution in the literature was still lacking.

The first one deals with a Bayesian estimation approach to assess the stiffness and its spatial variations in a given aortic region, based on CT Angiography (CTA) images acquired over a cardiac cycle [21]. The arterial stiffness was derived by linking the kinematic information from the CTA images with pressure waveforms, generated by a lumped parameter model of the circulation [22]. The proposed approach includes

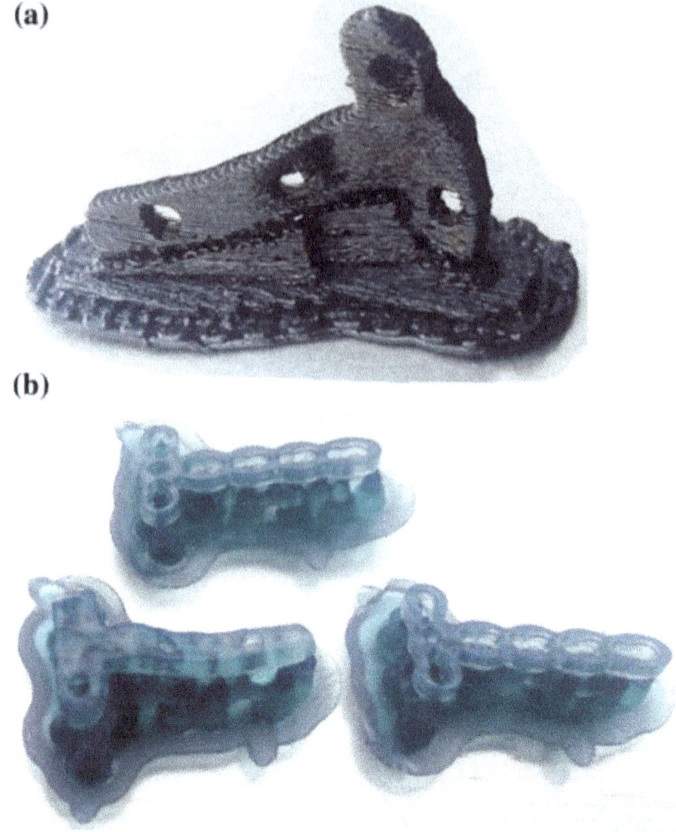

Fig. 11.6 Rolando fracture fixation plates made using FDM (**a**) and SLA (**b**)

the uncertainty of the input variables and exploits the entire diameter and pressure waveforms over the cardiac cycle.

The second application deals with the ascending aorta aneurysm, which is a severe life-threatening condition with asymptomatic rupture risk. We developed an approach to estimate the patient-specific ultimate mechanical properties and the stress-strain characteristics based on non-invasive data [23]. Through a regression model, we built the response surfaces for the ultimate stress and strain, and for the coefficients of the stress-strain characteristics, all in function of patient data commonly available in the clinical practice.

Moreover, due to the lack of effective tools to support Finite Element Analysis (FEA) under uncertain parameters and Structural Topology Optimization (STO), which are common problems when dealing with patient-specific problems, we propose an approach for efficiently solving FEA problems in the presence of stochastic parameters or within iterative optimization algorithms.

Finally, a relevant issue for studying the mechanical behaviour of biological tissues and structures with computational tools (e.g., FEA) is the uncertainty associated with the model parameters. We addressed the problem of solving the FEA in the presence of uncertain parameters by exploiting the functional principal component analysis to get acceptable computational efforts [31]. Indeed, the approach allows to construct an optimal basis of the solutions space and to project the full FEA problem into a smaller space spanned by this basis. The same approach was also used to reduce the computational effort of iterative optimization algorithms for STO [32].

11.4.4 Business Models

A general structure to configure potential business models for customised manufacturing in health care was developed. The proposed structure is based on the Product-Service System (PSS) concept, considering the morphological box defined by Lay et al. [33]. At the same time, it entails some modifications on the role of hospital and machinery supplier.

In particular, the proposed model consists of a set of building blocks, i.e., characteristic features that define the main aspects and decision points to be set (Fig. 11.7). For each feature, a number of options were defined, which describe the potential alternatives that can be selected to configure the business model. The features consider six relevant perspectives for a PSS-oriented business model: (i) *Location*, which refers to the physical production location of customized medical device; (ii) *Operational personnel*, which refers to the workforce allocation for production; (iii) *Equipment ownership*, which describes the property right to use the manufacturing equipment and machinery; (iv) *Maintenance*, which describes the party responsible for carrying out the maintenance of the equipment; (v) *Payment mode*, which define whether the payment is made in a traditional or an alternative way; (vi) *Target segment*, which clarifies whether the fabricated devices are produced only to serve the internal use of the hospital, or also to be offered and sold to other potential external customers [43]. Business models are thus configured by selecting different options for each characteristic feature; obviously, each configuration defines a particular strategy for the customer and the supplier.

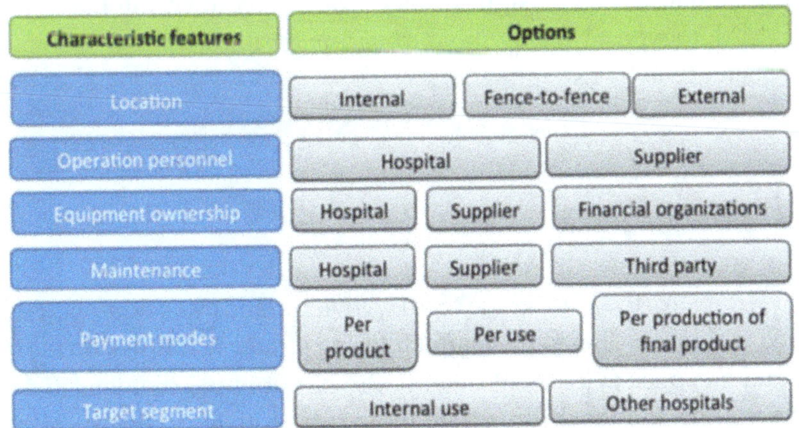

Fig. 11.7 Structure of the proposed PSS-oriented business model [35]

11.5 Outcomes

We present the outcomes and the results separately for each addressed problem. In particular, we refer to the prototypes, the mathematical models and the business model.

11.5.1 Aorta Silicone Model

The model was first analysed from a macroscopic point of view: surfaces are smooth, homogenous, even if slightly damaged only in the subtlest parts. Also the transparency is guaranteed.

For a quantitative analysis, the silicone model was tested to assess the actual compliance of the mock aorta. A series of CT scans of the model were performed at different inner pressures, using a 64 slice Definition AS CT (Siemens, Germany). To do so, all branches were closed and an inner pressure was imposed by means of a sphygmomanometer, ranging from 40 to 220 mmHg. Then, all acquisitions were post-processed, performing the segmentation of the air in the aorta lumen with the open source software ITK-Snap. At the end of the segmentation process, a set of labels was obtained, one for each slice of the CT scan, which were interpolated to create the inner volume rendering. Through the inner volume at different pressures, we identified the compliance of the model, expressed as the slope of the volume-pressure curve, here equal to $0.2008\,cm^3$/mmHg. Results show a linear volume-pressure relation that properly replicates the physiology of the aorta, even though the slope of the curve is slightly lower than expected, meaning that the model is more rigid than a physiologic aorta [34].

This small mismatch can be explained by the amount of factors that may impair the properties of the silicone mixture. Actually, a fine tuning of the mechanical properties can be performed by acting on the ratio between base and curing agent, i.e., the elastic modulus can be significantly reduced by lowering the amount of the curing agent. Moreover, lowering the curing temperature leads to the same result, even if the decrease is less significant.

11.5.2 Fixation Plates

The SX200 HP micro-EDM machine (Sarix, Switzerland) was equipped with different electrode tools according to the manufacturing strategy. A tungsten carbide cylindrical rod and a copper cylindrical tube, both with nominal diameter 0.4 mm, were used for the milling and the drilling operations, respectively. The overall machining time to complete a fracture plate in Titanium was equal to 375 min.

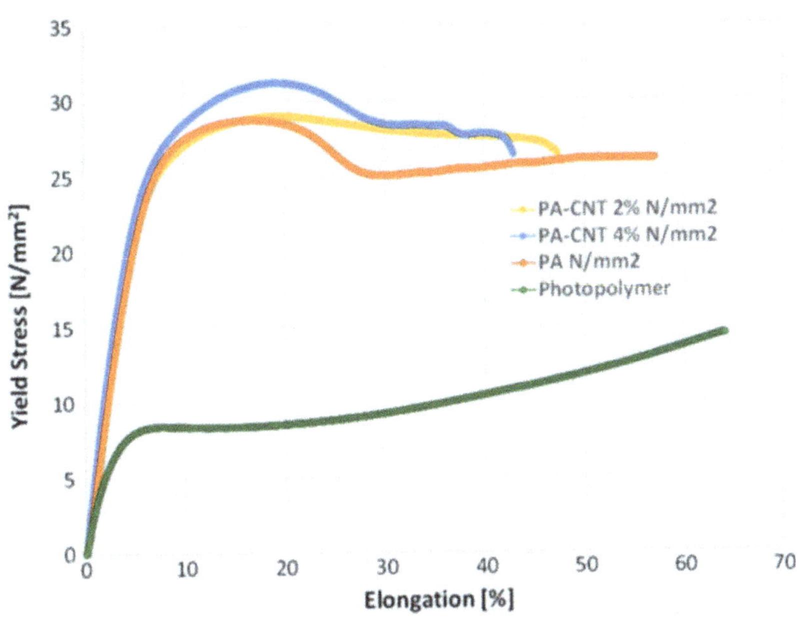

Fig. 11.8 Tensile tests for PA, PA with different CNT concentrations, and photopolymer from SLA

The same plate was made in a ceramic composite (Si3N4-TiN) using micro-EDM. Due to the higher material removal rate, the total machining time was reduced to 245 min. Unfortunately, the machining time is a severe drawback for this technology, which might not be suitable for more complex devices with a higher number of holes and bores.

Concerning the additive manufacturing, Fig. 11.8 shows the results of the tensile tests performed on different polymers by using dog-bone tensile specimens manufactured using FDM 3D printing and SLA. It can be noticed the poor mechanical performance of the photopolymer with respect to PA. The PA-CNT composites slightly increase the maximum yield stress, while they do not seem to influence the Young modulus.

11.5.3 Prediction Models

For the Bayesian estimation of the stiffness and its spatial variations, efficiency and accuracy of the proposed method were tested on some simulated cases and on a real patient [21]. The proposed approach showed to be powerful and to catch regional stiffness variations in human aorta using non-invasive data. The obtained estimates can also be used for producing patient-specific prostheses and preoperative tools that respect the estimated mechanical properties.

As for the estimation of patient-specific ultimate mechanical properties and stress-strain characteristics, we applied the approach to a dataset of 59 patients. The approach was validated, as accurate response surfaces were obtained for both ultimate properties and stress-strain coefficients [23]: prediction errors are acceptable, even though a larger patient dataset would be required to stabilise the surfaces, making it possible an effective application in the clinical practice.

Finally, considering the reduced basis approach to solve FEA problems in the presence of uncertain parameters [31] or for STO [32], results are promising. We assessed the applicability of the proposed approach on several test cases, obtaining satisfactory results. On the one hand, solving the problem in the reduced space spanned the functional principal components is computationally effective; on the other hand, very good approximations are obtained by upper bounding the error between the full FEA solution and the reduced one.

11.5.4 Business Models

The generated configurations of business model were defined by combining different options for each characteristic feature. Three main configurations were developed [35]:

- *Product-oriented business model*, in which the hospital buys the production machinery from a supplier with additional services. In this scenario, the hospital is the manufacturer and the production can take place either inside or near the hospital.
- *Use-oriented business model*, in which the hospital does not acquire any production machinery and the supplier retains the ownership of all equipments. In this configuration, the hospital rents or leases the equipments, and installs them in an internal production lab. While the ownership is not transferred to the hospital, the equipment is run by operating personnel of the hospital.
- *Result-oriented business model*, in which the hospital takes a step forward toward collaboration and integration with the supplier. While the fabrication place remains inside or near the hospital and the production takes place under the supervision of the hospital, the supplier is responsible for running the production. The supplier owns the equipments, provides additional maintenance services, and is responsible for running the production through its own personnel. The hospital provides the physical space for the production, and pays for the production of each final product.

Figure 11.9 shows the three different configurations.

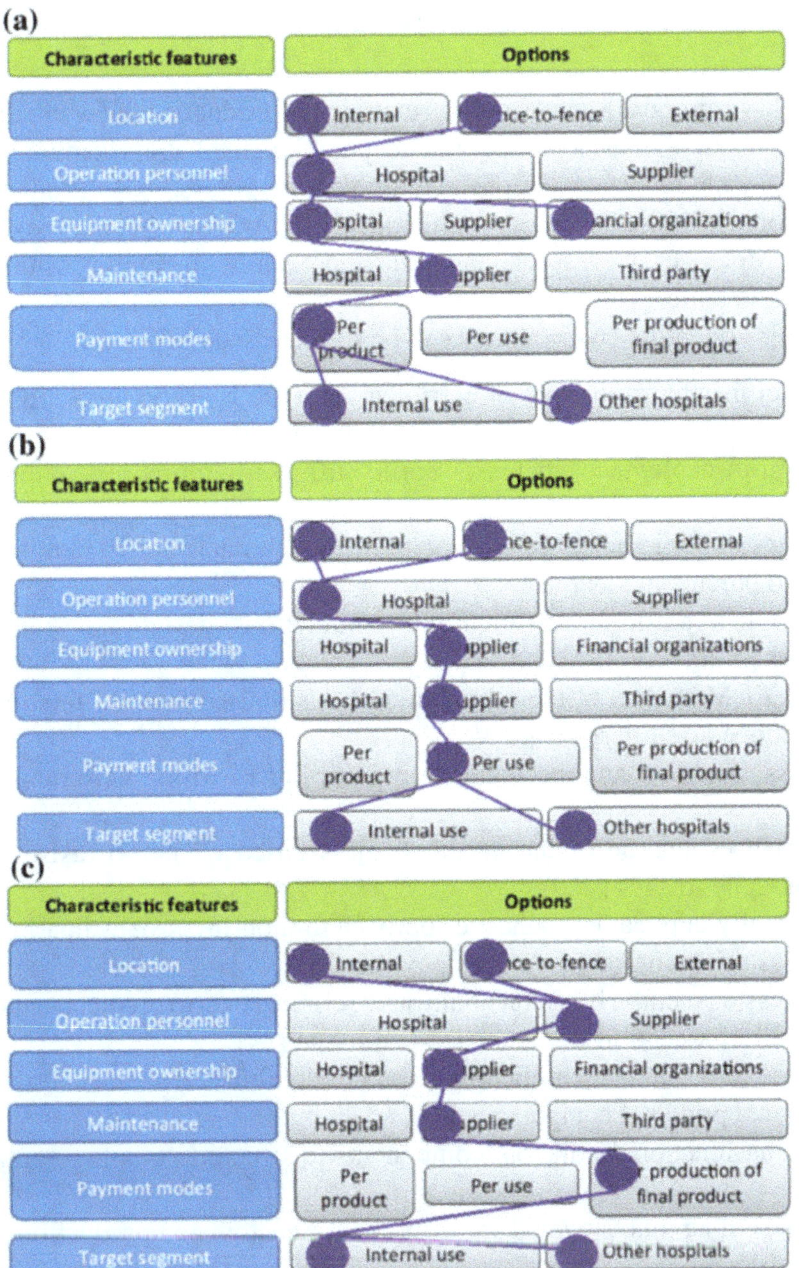

Fig. 11.9 Three possible business model configurations [35]: product-oriented business model (**a**); use-oriented-oriented business model (**b**); result-oriented business model (**c**)

11.6 Conclusions and Future Research

In this work, we propose a new paradigm to bring the production of personalised products (e.g., prostheses) inside or near the hospital, i.e., the Fab@Hospital paradigm.

Through some relevant examples, we proved the possibility to produce patient-specific products in small factories with production processes that may easily involve clinicians. Moreover, we also validated the approach by interacting with clinicians of several specialties.

The major scientific contributions can be summarised as follows:

1. A novel insight on the role of entry tears in type B aortic dissection: pressure measurements in an in vitro 3D printed model [36].
2. Stent graft deployment increases aortic stiffness in an ex-vivo porcine model [37].
3. Changes in aortic pulse wave velocity of four thoracic aortic stent grafts in an ex vivo porcine model [38].
4. A compliant aortic model for in vitro simulations: design and manufacturing process [34].
5. Micro-EDM studies of the fabrication of customised internal fixation devices for orthopedic surgery [39].
6. Process capability and mechanical properties of FDM in micro manufacturing [40].
7. Bayesian estimation of the aortic stiffness based on non-invasive computed tomography images [41].
8. A clinically-applicable stochastic approach for non-invasive estimation of aortic stiffness using computed tomography data [21].
9. A regression method based on noninvasive clinical data to predict the mechanical behavior of ascending aorta aneurysmal tissue [23].
10. Efficient uncertainty quantification in stochastic finite element analysis based on functional principal components [31].
11. Applying functional principal components to structural topology optimization [32].
12. Proposal of an innovative business model for customized production in healthcare [42].
13. Development of a PSS-oriented business model for customized production in healthcare [35].
14. A new perspective of product-service business models for customized manufacturing in healthcare [43].

Items 1–4 refer to the additive manufacturing for cardiovascular models; items 5–6 to the fixation plates made using micro-EDM and additive manufacturing; items 7–11 to prediction and mathematical tools; items 12–14 to the business models.

Future work will consider the implementation of the Fab@Hospital paradigm in a small hospital factory, to simulate the entire production process from the clinical request up to the final product.

Acknowledgements This work has been funded by the Italian Ministry of Education, Universities and Research (MIUR) under the Flagship Project "Factories of the Future—Italy" (Progetto Bandiera "La Fabbrica del Futuro") [44], Sottoprogetto 2, research projects "Hospital Factory for Manufacturing Customized, Patient Specific 3D Anatomo-Functional Model and Prostheses" (Fab@Hospital) and "Fab@Hospital for bone plate fabrication and patient anatomy reconstruction using rapid prototyping technologies" (F@H for 3D plates).

Authors thank the industrial and clinical interest group of project Fab@Hospital for the involvement in the activities: OVERMACH S.p.A. (Giacomo Cacciani); Sarix SA (Franck Leleu); I.R.C.C.S. Policlinico San Donato (Santi Trimarchi); I.R.C.C.S. Policlinico San Matteo (Andrea Pietrabissa); Azienda USL della Valle d'Aosta (Raffaella Pagano).

References

1. Vaupel JW (2010) Biodemography of human ageing. Nature 464:536–542
2. Angus DC, Kelley MA, Schmitz RJ, White A, Popovich J Jr (2000) Caring for the critically ill patient. Current and projected workforce requirements for care of the critically ill and patients with pulmonary disease: can we meet the requirements of an aging population? JAMA 284:2762–2770
3. Wunsch H, Angus DC, Harrison DA, Collange O, Fowler R, Hoste EA, De Keizer NF, Kersten A, Linde-Zwirble WT, Sandiumenge A, Rowan KM (2008) Variation in critical care services across North America and Western Europe. Crit Care Med 36:2787–2793
4. Megari K (2013) Quality of life in chronic disease patients. Health Psychol Res 1:e27
5. Singare S, Lian Q, Wang WP, Wang J, Liu Y, Li D, Lu B (2009) Rapid prototyping assisted surgery planning and custom implant design. Rapid Prototyp J 15:19–23
6. Pietrabissa A, Marconi S, Peri A, Pugliese L, Cavazzi E, Vinci A, Botti M, Auricchio F (2016) From CT scanning to 3D printing technology for the preoperative planning in laparoscopic splenectomy. Surg Endosc 30:366–371
7. Marconi S, Pugliese L, Botti M, Peri A, Cavazzi E, Latteri S, Auricchio F, Pietrabissa A (2017) Value of 3D-printing for the comprehension of surgical anatomy. Surg Endosc 31:4102–4110
8. Torres IO, De Luccia N (2017) A simulator for training in endovascular aneurysm repair: the use of three dimensional printers. Eur J Vasc Endovasc Surg 54:247–253
9. Mottl-Link S, Hübler M, Kühne T, Rietdorf U, Krueger JJ, Schnackenburg B, De Simone R, Berger F, Juraszek A, Meinzer HP, Karck M, Hetzer R, Wolf I (2008) Physical models aiding in complex congenital heart surgery. Ann Thorac Surg 86:273–277
10. Giannatsis J, Dedoussis V (2009) Additive fabrication technologies applied to medicine and health care: a review. Int J Adv Manuf Tech 40:116–127
11. Bisdas T, Teebken OE (2011) Future perspectives for the role of 3D rapid prototyping aortic biomodels in vascular medicine. VASA 40:427–428
12. Torres K, Staśkiewicz G, Śnieżyński M, Drop D, Maciejewski R (2011) Application of rapid prototyping techniques for modelling of anatomical structures in medical training and education. Folia Morphol (Warsz) 70:1–4
13. Camp TA, Stewart KC, Figliola RS, McQuinn T (2007) In vitro study of flow regulation for pulmonary insufficiency. J Biomech Eng 129:284–288
14. Yoffe B, Vaysbeyn I, Urin Y, Waysbeyn I, Zubkova O, Chernyavskiy V, Ben-Dor D (2007) Experimental study of a novel suture-less aortic anastomotic device. Eur J Vasc Endovasc Surg 34:79–86
15. Kalejs M, von Segesser LK (2009) Rapid prototyping of compliant human aortic roots for assessment of valved stents. Interact Cardiovasc Thorac Surg 8:182–186
16. Shiraishi I, Yamagishi M, Hamaoka K, Fukuzawa M, Yagihara T (2010) Simulative operation on congenital heart disease using rubber-like urethane stereolithographic biomodels based on 3D datasets of multislice computed tomography. Eur J Cardiothorac Surg 37:302–306

17. Liu K, Lauwers B, Reynaerts D (2010) Process capabilities of micro-EDM and its applications. Int J Adv Manuf Tech 47:11–19
18. Reissig L, Völkl R, Mills MJ, Glatzel U (2004) Investigation of near surface structure in order to determine process-temperatures during different machining process of Ti6Al4V. Scripta Mater 50:121–126
19. Liu CC, Huang JL (2003) Effect of the electrical discharge machining on strength and reliability of TiN/Si3N4 composites. Ceram Int 29:679–687
20. Liu K, Paris J, Ferraris E, Lauwers B, Reynaerts D (2007) Process investigation of precision Micro-machining of Si3N4-TiN ceramic composites by electrical discharge machining (EDM). In: Proceedings of the 15th international symposium on electromachining (ISEM), pp 221–226
21. Auricchio F, Conti M, Ferrara A, Lanzarone E (2015) A clinically-applicable stochastic approach for non-invasive estimation of aortic stiffness using computed tomography data. IEEE Trans Bio-Med Eng 62:176–187
22. Lanzarone E, Liani P, Baselli G, Costantino ML (2007) Model of arterial tree and peripheral control for the study of physiological and assisted circulation. Med Eng Phys 29:542–555
23. Auricchio F, Ferrara A, Lanzarone E, Morganti S, Totaro P (2017) A regression method based on noninvasive clinical data to predict the mechanical behavior of ascending aorta aneurysmal tissue. IEEE Trans Bio-Med Eng 64:2607–2617
24. Weller C, Kleer R, Piller FT (2015) Economic implications of 3D printing: market structure models in light of additive manufacturing revisited. Int J Prod Econ 164:43–56
25. Hopkinson N, Dickens PM (2003) Analysis of rapid manufacturing—using layer manufacturing processes for production. Proc Inst Mech Eng C-J Mec 217:31–39
26. Ruffo M, Tuck C, Hague R (2006) Cost estimation for rapid manufacturing—laser sintering production for low to medium volumes. Proc Inst Mech Eng B-J Eng 220:1417–1427
27. Ingole DS, Kuthe AM, Thakare SB, Talankar AS (2009) Rapid prototyping—a technology transfer approach for development of rapid tooling. Rapid Prototyp J 15:280–290
28. Lindemann C, Jahnke U, Moi M, Koch R (2012) Analyzing product lifecycle costs for a better understanding of cost drivers in additive manufacturing. In: Proceedings of solid freeform fabrication symposium, pp 177–188
29. Schröder M, Falk B, Schmitt R (2015) Evaluation of cost structures of additive manufacturing processes using a new business model. Proc CIRP 30:311–316
30. Johnston ID, McCluskey DK, Tan CKL, Tracey MC (2014) Mechanical characterization of bulk Sylgard 184 for microfluidics and microengineering. J Micromech Microeng 24:035017
31. Bianchini I, Argiento R, Auricchio F, Lanzarone E (2015) Efficient uncertainty quantification in stochastic finite element analysis based on functional principal components. Comput Mech 56:533–549
32. Alaimo G, Auricchio F, Bianchini I, Lanzarone E (2018) Applying functional principal components to structural topology optimization. Int J Numer Meth Eng 115:189–208
33. Lay G, Schroeter M, Biege S (2009) Service-based business concepts: a typology for business-to-business markets. Eur Manag J 27:442–455
34. Marconi S, Lanzarone E, Van Bogerijen G, Conti M, Secchi F, Trimarchi S, Auricchio F (2018) A compliant aortic model for in vitro simulations: design and manufacturing process. Med Eng Phys 59:21–29
35. Pourabdollahian G, Copani G (2015) Development of a PSS-oriented business model for customized production in healthcare. Proc CIRP 30:492–497
36. Marconi S, Lanzarone E, De Beaufort HWL, Conti M, Trimarchi S, Auricchio F (2017) A novel insight on the role of entry tears in type B aortic dissection: pressure measurements in an in vitro 3D printed model. Int J Artif Organs 40:563–574
37. De Beaufort HWL, Conti M, Kamman AV, Nauta FJH, Lanzarone E, Moll FL, Van Herwaarden JA, Auricchio F, Trimarchi S (2017) Stent graft deployment increases aortic stiffness in an ex-vivo porcine model. Ann Vasc Surg 43:302–308
38. De Beaufort HWL, Coda M, Conti M, Van Bakel TMJ, Nauta FJH, Lanzarone E, Moll FL, Van Herwaarden JA, Auricchio F, Trimarchi S (2017) Changes in aortic pulse wave velocity of four thoracic aortic stent grafts in an ex vivo porcine model. PLoS One 12:e0186080

39. Modica F, Pagano C, Marrocco V, Fassi I (2015) Micro-EDM studies of the fabrication of customised internal fixation devices for orthopedic surgery. In Proceedings of the ASME international design engineering technical conferences and computers and information in engineering conference IDETC/CIE
40. Pagano C, Basile V, Fassi I (2016) Process capability and mechanical properties of FDM in micro manufacturing. In: Proceedings of the ICOMM international conference on micromanufacturing, vol 41
41. Lanzarone E, Auricchio F, Conti M, Ferrara A (2014) Bayesian estimation of the aortic stiffness based on non-invasive computed tomography images. Bayesian statistics from methods to models and applications—proceedings of BAYSM. Springer proceedings in mathematics & statistics, pp 133–142
42. Pourabdollahian G, Copani G (2014) Proposal of an innovative business model for customized production in healthcare. Mod Econ 5:1147–1160
43. Pourabdollahian G, Copani G (2016) A new perspective of product-service business models for customized manufacturing in healthcare. In: Service business model innovation in healthcare and hospital management, pp 87–109
44. Terkaj W, Tolio T (2019) The Italian flagship project: factories of the future. In: Tolio T, Copani G, Terkaj W (eds) Factories of the future. Springer

Chapter 12
Polymer Nanostructuring by Two-Photon Absorption

**Tommaso Zandrini, Raffaella Suriano, Carmela De Marco,
Roberto Osellame, Stefano Turri and Francesca Bragheri**

Abstract Two-photon polymerization (2PP) is an innovative technology that in
recent years showed a tremendous potential for three-dimensional structuring of
photopolymers at the submicron scale. It is based on the nonlinear absorption of
ultrashort laser pulses in transparent photosensitive materials. 2PP has been so far
exploited in various fields, including photonics, microfluidics, regenerative medicine
and MEMS prototyping. The versatility of this technology relies also on the photo-
materials; indeed, polymers are easy to process, low cost and they allow the tailoring
of their chemical and mechanical properties. 2PP nanotechnology is here exploited
to produce micro and nanostructures that can be easily customized both in the geometry and in polymer functionalization. In particular, atomic force microscopy tips are
fabricated on top of commercial cantilevers to demonstrate the technology feasibility and customizability. Moreover nanoporous membranes that can be fabricated by
2PP as a single custom product or as a mould for mass production through replica
moulding are realized to evaluate the scalability of the fabrication process.

12.1 Scientific and Industrial Motivations

In recent years, the development of industries capable of supplying increasingly customized products according to the needs and desires of customers is taking hold. The
concept of *on-demand fabrication* is therefore attracting much attention in several
fields. Laser fabrication is in general a flexible and rapid prototyping process for the
creation of a variety of microstructures in a broad range of materials, but it allows only
serial fabrication processes. This aspect, which is highly desirable for customization,
poses some limitations for mass production and even for mass customization.

T. Zandrini · R. Osellame · F. Bragheri (✉)
CNR-IFN, Istituto di Fotonica e Nanotecnologie, Milan, Italy
e-mail: francesca.bragheri@ifn.cnr.it

R. Suriano · C. De Marco · S. Turri
Politecnico di Milano, Dipartimento di Chimica, Materiali e Ingegneria Chimica "G.Natta",
Milan, Italy

Two-photon polymerization (2PP) by femtosecond laser pulses is a fabrication technology that showed great potential in the fabrication of three-dimensional (3D) polymeric sub-micron structures with resolution down to 100 nm [1–3]. 2PP process is based on a photochemical polymerization reaction triggered by the two photon absorption that occurs in the focal point of a tightly focused femtosecond laser pulse. By moving the focal spot in a UV photopolymerizable transparent photoresist, complex patterns of polymerized material can be realized; the non-irradiated polymer can be removed by a development step that leaves free-standing 3D microstructures of polymerized material. Two-photon polymerization is a highly reconfigurable lithographic technique that allows a high control of the fabrication parameters and shows many advantages such as the submicron resolution and the ability to directly write three-dimensional structures without constraints related to their shape. This technology is thus very promising in the frame of prototyping since it allows an unprecedented freedom in the product customization and provides an important tool to create nanostructures otherwise not feasible. The areas of application are numerous and range from the biomedical sciences to telecommunications, micromechanics and microfluidics [5–12]. The limitations linked to its serial nature should be overcome to promote this fabrication technology from the laboratories to the industries. In some cases this can be done by exploiting the potential of 2PP to fabricate the master mould to be used in replica moulding (REM), i.e. a promising technology for duplicating the shape, size and pattern of features present on a master [13].

The main goal of this work is therefore to tackle the potentialities for an industrialization of the two-photon polymerization process by demonstrating its capability to produce micro and nanostructures that can be easily customized both in their geometry and in material functionalization and the scalability of products whenever replica moulding is coupled to this technology.

2PP is a highly flexible rapid prototyping technology allowing a high control of the fabrication parameters. Photopolymers are low cost, easy to process and tailorable in their chemical and mechanical properties. The combination of these aspects allows satisfying the need for products and services that can be adapted to the customer specific requirements.

The intermediate objectives to successfully achieve the main goal are then the development of photopolymerizable polymeric materials with peculiar chemical properties and of the laser processing procedures that allow the nanostructuring of these materials by 2PP with specific resolution ranges. In particular, a relevant case study has been identified to validate these approaches in the manufacturing of atomic force microscopy (AFM) tips and of nanoporous membranes. Because of the wide diffusion of this technology in many research fields, the customization of the tips both in terms of the geometry of the structure and in terms of their mechanical and chemical properties has potentially high impact for an on-demand product customization, perfectly matching the trend that industry is experiencing in the products personalization.

In the following sections the problem tackled by our research is introduced by evaluating the state of the art of the proposed technology applied on the envisaged applications. Afterwards the achieved results are reported first by showing the choice of the polymer through tests on its polymerizability, and then the fabrication of AFM tips is shown both on glass coverslips and in commercial AFM cantilevers. The fabricated AFM tips have been validated by nanoindentation measurements and the obtained results on biologically relevant soft materials are reported. At last preliminary experiments on nanoporous membrane fabrication by replica moulding are introduced by showing the process exploited to fabricate both the master by 2PP and the membranes by replica moulding.

12.2   State of the Art

Many technologies are available to fabricate microstructures. The leading technology for AFM tips is photolithography, usually on hard materials such as silicon and its compounds. Unfortunately, the mechanical properties of these materials cannot be tailored with the freedom which is typical of polymers; moreover it is very difficult to produce curved features, and they require a complex and expensive multi-step process. Polymers can be processed by photolithography or by electron beam lithography with very high lateral resolution, but only in a single layer fashion, preventing the realization of complex geometries. Stereolithography has the ability of producing more complex structures but with a lower resolution, especially in the vertical direction, and always with a layer-by-layer approach.

An innovative technology for 3D structuring of photopolymers at the submicroscale able to overcome these limitation is two-photon polymerization (2PP), based on the nonlinear absorption of ultrashort pulses in transparent photosensitive materials [1–4]. Because of the nonlinear nature of the process, a lateral resolution beyond the diffraction limit can be realized by controlling the laser pulse energy deposition. Moreover, the absorption along the beam propagation direction is tightly confined in the focal region. As a result, the technique provides much better structural resolution and quality than the aforementioned techniques. Microstructuring of photosensitive materials by 2PP is effective for the fabrication of 3D structures with a resolution of 100 nm or better, achievable through the combination of computer-controlled positioning systems and, typically, near-infrared laser oscillators.

The potential applications of 2PP technology cover diverse research fields, ranging from the fabrication of photonic elements [5, 6] and of scaffolds for regenerative medicine [7–9], to the prototyping of MEMS [10] and the fabrication of microfluidic components [11, 12]. Another promising application is the fabrication of polymeric tips for Atomic Force Microscopy (AFM). AFM tips are currently manufactured with standard microfabrication technologies, by using hard inorganic materials (such as silicon nitride, silicon, gold) and requiring expensive and time-consuming facilities. The main advantages of polymeric AFM tips realization by 2PP relies on the combination of a cheap manufacturing process and the appealing capacity of customization. AFM tips can be tailored in terms of both geometry and materials; in particular the customization of mechanical and chemical properties may be realized by exploiting polymers with specific functionalities for a targeted application. The first AFM tips structured by 2PP were demonstrated in 2005 by Kim and Muramatsu [14]. Since then only a few related studies were reported and all of them employed the same commercial polymer, which is composed of acrylic and epoxy monomers [15]. The main advantages of acrylic photoresists are their wide commercial availability, the optical transparency, and the ease of processing [16]. Nevertheless, acrylic polymers present poor chemical resistance to organic solvents that hinders the realization of 3D structures to be used in applications requiring chemically resistant elements. Perfluoropolyether polyurethanes (PFPE) are well known macromolecular families that can be used to obtain hydrophobic and chemically resistant materials [17–19]. The synthesis and processing by 2PP of a new PFPE-based have been recently demonstrated by the partners of the project [20]. 2PP fabricated hydrophobic AFM tips were already proposed [14], but the contact angle of the photopolymer was about $70°$, too small to be considered a low wettability surface. Adhesion forces between AFM tips and hydrophilic surfaces can be reduced by exploiting hydrophobic self-assembling coatings; anyway these are easily removed by the tip interaction with the surface [14]. The availability of AFM tips directly fabricated with a hydrophobic material, as for tips realized by 2PP structuring of low surface tension polymers, would be advantageous for applications as topographic imaging of biomacromolecules or hydrogels; in fact, the adhesive forces between the standard AFM tips and their hydrophilic surfaces may affect the imaging quality. Moreover, AFM emerged as an essential tool to measure the elastic modulus of soft materials otherwise difficult to be estimated [21]. For example, Cross et al. [22], using the AFM, showed that cancer cells are 70% less stiff than normal cells, but the model used to estimate the elastic modulus approximated the real shape of the AFM tip by a cone or a paraboloid. Reproducing the exact cone or paraboloid shape by 2PP may reduce the errors, which affect this type of measurements.

2PP technology has demonstrated an excellent potential for the fabrication of complex polymeric structures at the microscale, but its serial nature usually hinders a mass production of components. For this reason a parallel replication method of 2PP structures is very desirable.

2D nanoporous membranes have many applications, ranging from optics and electronics, to filtration and purification or biosensing and single-molecule detection. Despite the extensive research carried out in fabrication of nanoporous materials, there are still challenges to create nanoporous systems, mainly related to the poor flexibility in terms of pore shape and uniformity intrinsic to standard lithography and ion-track etching [23]. Replica moulding enables to duplicate the shape, size and pattern of features of a master, usually fabricated on silicon by photolithography, to a polymeric replica by means of an intermediate polymer mould [13]. Productivity, precision and fidelity of REM replication process are largely dependent on the characteristics of polymeric moulding material. To date, poly(dimethylsiloxane) (PDMS) elastomers have been largely used; however, PDMS swells in contact with many solvents and chemicals, and it has some limitations when used to produce very low aspect ratio structures [24]. Photocurable PFPEs have been recently proposed as higher performance moulding materials [25]. In particular, PFPE moulds were recently used to replicate the submicron structures patterned on poly(methyl methacrylate) (PMMA) and polystyrene (PS) substrates by femtosecond laser ablation [26]. 2PP technology would then be a suitable rapid prototyping technology to realize customized masters to be used for nanoporous membranes replica.

12.3 Problem Statement and Proposed Approach

The aims of the work are the validation of the two-photon polymerization technology as a new methodology to fabricate customized micro and nanostructures to be used in various applications and the preliminary demonstration of customized mass production through 2PP and replica moulding technologies. Given the ample range of possible applications for the proposed technology, the attention was focused on the fabrication of atomic force microscopy (AFM) tips, as an archetypical and relevant example of the potentiality of the 2PP technology. AFM and scanning probe microscopy in general are widely diffused technologies exploited in many fields of research for morphological and mechanical material characterization with exceptional spatial accuracy; on the basis of the specific application, different geometries or functionalization of the tip are required. For example, hydrophobic tips are desirable when biological elements are under investigations so as to minimize adhesion forces between the tip and the sample under study; high aspect ratio tips would be beneficial in the analysis of materials with high asperities, while spherical tips would be more appropriate for soft material studies. The second objective, which is of potential interest for the industrialization of the process, was the feasibility study of customized mass production by 2PP. To overcome the limitation of 2PP due to its serial nature REM could be applied to 2PP fabricated master structures. In this way, the high prototyping capability and reconfigurability of 2PP is exploited to realize custom moulds that can be then massively reproduced by means of replica moulding technique. As interesting example for the feasibility of the demonstration, nanoporous membranes have been chosen. Membranes with nanometer-scale features may have several applications, e.g. in electronics, optics, catalysis, selective molecule separation, filtration and purification, biosensing and single-molecule detection.

To achieve these goals, the attention has been first concentrated on the optimization of the 2PP fabrication process for different photoresists, either commercially available or specifically developed. In particular, new photoresists have been developed with tailored chemical and mechanical properties to satisfy the desired hydrophobicity and resistance to solvents. After a preliminary material characterization, attention has been devoted to study and identify the laser processing windows that allow obtaining specific resolution ranges in the fabricated nanostructures for the chosen materials. The obtained results permitted to fabricate prototype tips both on glass coverslip substrates and on commercial cantilevers to be tested for AFM indentation measurements. Soft hydrophilic materials important in biomedical applications such as, for example, regenerative medicine, have been then chosen as test materials to validate the prototypes produced. The attention has been then moved to the choice of the materials to be used both for the realization of the master by 2PP and for the membrane replication by REM. Different conical structures have been fabricated by 2PP as master, which has been used to replicate membranes with different porosities.

The strategy to fulfil the proposed objectives followed a three-step work-plan: (i) development of new functionalized photoresist and definition of the operating parameters of the laser for 2PP fabrication; (ii) demonstration of the realization of hydrophobically functionalized AFM tips by fabricating the tips on commercial cantilevers and by experimentally validating them through AFM nanoindentation measurements; (iii) realization by replica moulding of nanostructured devices such as porous membranes for biosensing applications, whose master has been fabricated by 2PP technology.

12.4 Developed Technologies, Methodologies and Tools

The development of the technology consisted in three steps: first the polymers, offered by industries or synthetized on purpose, were tested to evaluate their photopolymerizability by 2PP. Then the laser writing process was optimized to obtain the desired tip shape. Finally the tips have been fabricated on top of commercial cantilevers to be used in AFM systems.

The achievement of the proposed objectives allowed the development of new materials and the optimization of the fabrication methods for both the proposed case studies, i.e. AFM tips and nanoporous membranes. In fact in order to realize AFM tips, different hydrophobic polymers have been produced and exploited to fabricate microstructures by 2PP. As shown in the following subsections, the optimization of the processes allowed identifying one of these polymers as the best to realize hydrophobic AFM tips in terms of shape and mechanical stability. The chosen polymer has been also exploited in replica moulding process to realize nanoporous membrane with the goal of demonstrating the feasibility of the process.

12.4.1 Realization of AFM Tips

With the aim of obtaining hydrophobic photoresists to fabricate AFM tips with reduced tip-sample interactions, fluoropolymers were chosen among the various classes of polymers. The use of fluoropolymers, in fact, guarantees to achieve the properties of hydrophobicity and solvent-resistance, due to the very low polarizability of C–F bonds. More specifically, perfluoropolyethers (briefly PFPEs) are a class of polymeric oils having the following chain $-(CF_2O)_q(CF_2CF_2O)_p-$ which can be chemically functionalized to obtain photopolymers suitable for standard UV curing processes.

Different PFPE-based photoresist to be effectively structured by 2PP were considered. Their validation in terms of chemical and physical properties is reported in

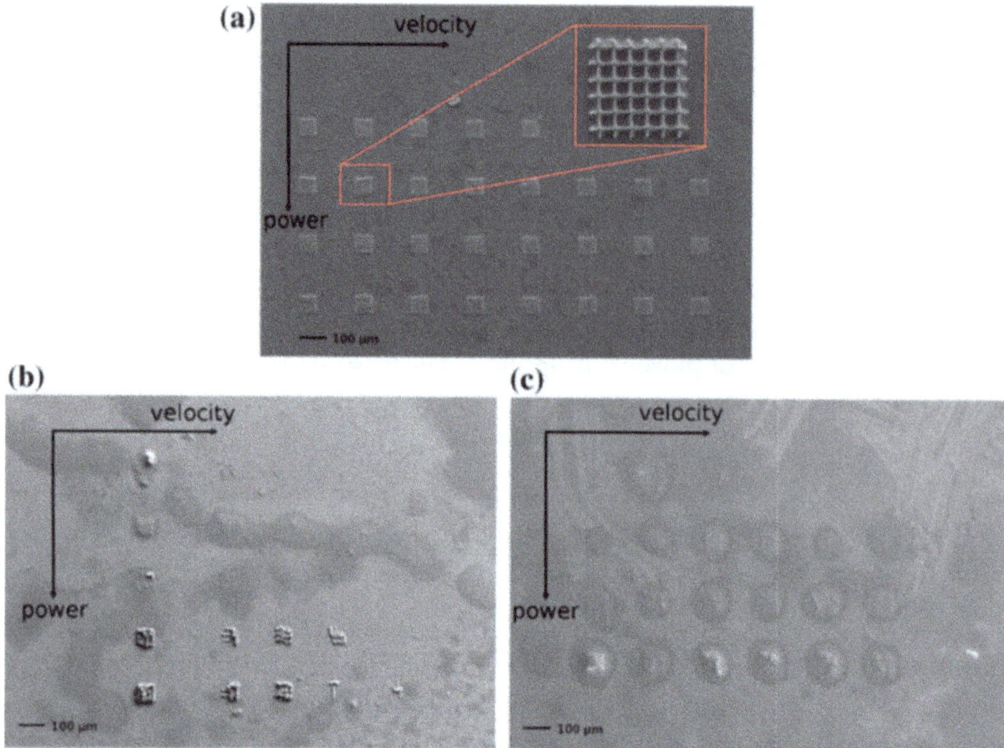

Fig. 12.1 Power-velocity arrays of multilayer grids, power and scanning velocity ranges respectively from 5 mW to 30 mW and from 5 to 250 µm/s. **a** PFPE-PETA, **b** PFPE-TUMA and **c** PFPE-DUTA structures

Sect. 12.5, while here the development of the 2PP technology for these photoresist is described. The testing has been carried out by polymerizing simple structures made by layers of parallel lines, each layer perpendicular to the previous one, and varying laser power and irradiation velocity to identify the optimal processing parameters. The deposited energy is crucial in the 2PP process, since when the energy is not sufficient to generate a critical amount of free radicals in the materials, polymerization will not occur, while if it is too much the material will be damaged, becoming strongly absorbent and spreading the damage to a large surrounding area. These energy values allowed defining a polymerization threshold and a damage threshold for each material.

The laser source employed for 2PP was a Toptica FemtoFiber PRO Er-doped fibre laser, whose second harmonic at 780 nm produces pulses of less than 100 fs at a repetition rate of 80 MHz. Figure 12.1 shows the outcomes of the preliminary tests where the laser power was varied between 5 and 30 mW, while the scan speed between 5 and 250 µm/s. These ranges are shown in the figure along the vertical and the horizontal direction, respectively. The values of laser power and irradiation velocity that gave well-formed lines define an initial fabrication window, included between the polymerization threshold and the damage threshold, from which a finer study of the stability of more complex structures could start.

As visible from Fig. 12.1a, the structures fabricated in PFPE-PETA are generally well formed, which means that with this material a wide fabrication window is available. This is a critical aspect, since it allows a robust fabrication process, where small variations in the laser intensity and in the fabrication environment will not spoil the final result. On the other hand, in PFPE-TUMA and PFPE-DUTA, respectively in Fig. 12.1b, c, the multilayer grids are either non polymerized or damaged, which means that it was not possible to identify a good combination of laser power and fabrication speed for these materials with the employed laser source. The other PFPE based materials tested showed no evidence of polymerization. Given these results, PFPE-PETA has been identified as the only suitable material for 2PP, among the ones considered in this study.

A second aspect that should be taken into account in the development of the optimized fabrication method concerns the geometry of the desired structure. The shape chosen for the AFM tip was a cylindrical base with a hemisphere on top; the sphere, with a well-defined radius of curvature, is necessary for nanoindentation measurements. Indeed it allows a simple modelling of the interaction between the tip and the material, and it is thus required by the mathematical models used to extract the mechanical properties of the sample from the curves acquired with the AFM. The base, instead, prevents any attraction or repulsion between the AFM cantilever and the sample surface. Conical tips have also been proposed for this task, but the smaller curvature radius of the tip was more difficult to measure via scanning electron microscope (SEM) images, and less reproducible, since it varied much with the laser irradiation parameters, so this geometry has been discarded.

To irradiate the volume of such structures with the femtosecond laser, a layer-by-layer irradiation pattern has been chosen, as it is common in many 3D-printing techniques for compact volumes. The tip has been divided in horizontal slices, spaced less than the single polymerized laser line, and each slice was scanned with parallel lines. To study the mechanical stability of these structures, before proceeding with the fabrication on AFM cantilevers, some tips were realized on top of a coverglass with PFPE-PETA. This was necessary to identify the optimal distances between slices and between lines, which give well defined shapes and avoid energy accumulation, due to the excessive superposition of nearby lines. Once completed this first step, stability and repeatability of the tips was studied, by varying laser power and irradiation speed within the previously defined fabrication window, and realizing for each combination nine identical structures.

A parameter scan with power ranging from 10 to 30 mW, and speed from 10 to 100 μm/s is shown in Fig. 12.2. It can be observed how the tips shown in the inset are well formed and reproducible, while the ones with much higher power in the top line are swollen, and the ones in the bottom line, with the lowest power, are collapsed.

Fig. 12.2 Optimization of the fabrication parameters. Groups of nine identical cylinders with an hemisphere on top have been fabricated for each power-velocity combination. A detail of well-defined structures written at 20 mW, 30 and 50 μm/s is shown in the inset [27]

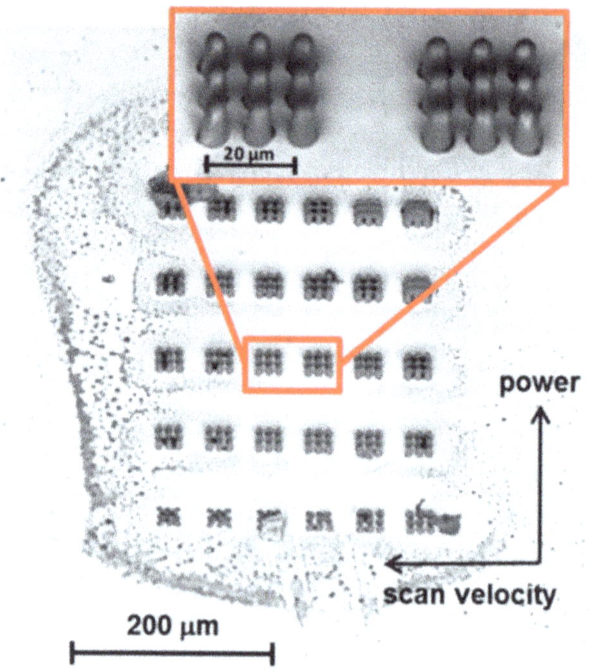

12.4.2 Realization of Nanoporous Membranes

The realization of nanoporous membranes could be carried out according to two different strategies, i.e. (i) direct fabrication of the membrane by polymerizing the photoresist and (ii) fabrication by 2PP of a master for REM. In the first case the fabrication of the membrane would allow a very high precision in the positioning and shape of the pores, enabling also the formation of very small pores. Since the polymerization process would occur around the pore, large regions should be polymerized and almost no limits due to the resolution of the process would occur. In the second case the fabrication time is shorter, even if the pore dimension is limited to the smallest feature that can be fabricated by 2PP, i.e. by the resolution of the 2PP process.

Taking advantage from the results obtained in the realization of AFM tips, the second strategy has been chosen. The activity has been focused on the fabrication by 2PP on a glass substrate of a stable matrix of cones to be used as master for the membrane replica. The conical tips were made of SZ2080 material, which was proven to be suitable for producing well-defined structures via 2PP. First, the optimal fabrication parameters to achieve cones with very sharp tips have been found, since this would determine the minimum size of the pore achievable after replica moulding and the results are reported in Fig. 12.3a. Secondly, the stability of the fabricated structures has been evaluated; indeed, the cones should have high adhesion to the substrate in order not to detach during the membrane stripping process. No treatment of the glass could be performed; in fact, silanization would increase the photopolymer adhesion, but also the membrane adhesion, hindering the stripping process. Two

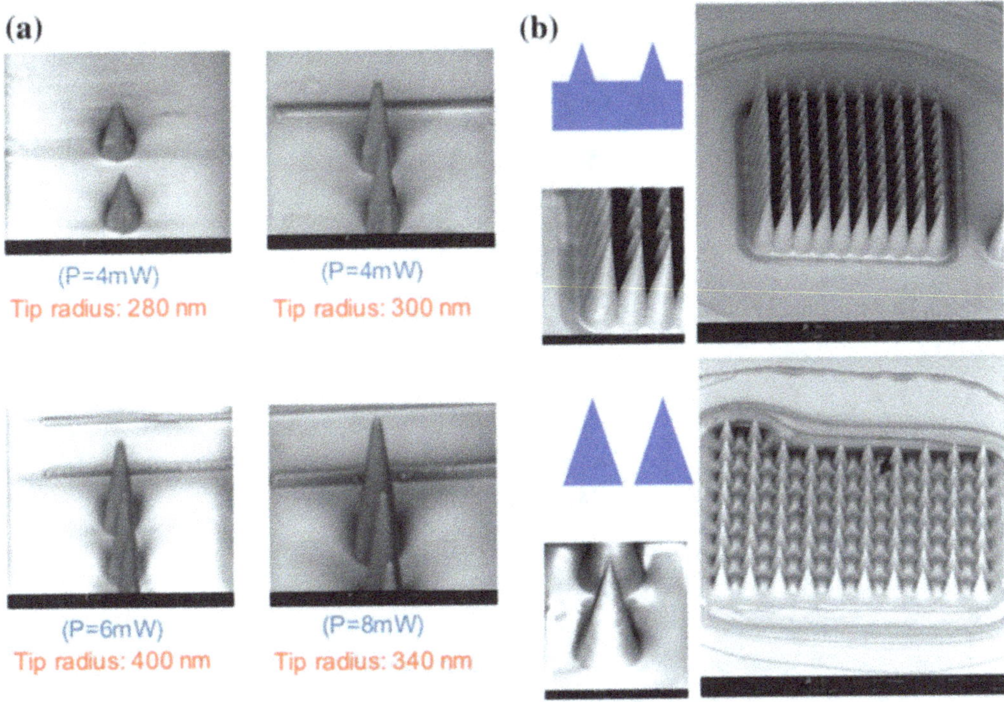

Fig. 12.3 Development of tips matrices to be used as master for nanoporous membranes replica. **a** Optimization of the laser parameters for fabrication of cones with sharp tips. In all the shown cases the distance between irradiated adjacent planes is 0.3 nm and the scan velocity is equal to 0.07 mm/s **b** Optimization of the matrix of cone stability and geometry

different structures of cones with a large basement or larger bases to increase the stability and the adhesion of the cones have been therefore tested. The results reported in Fig. 12.3b show that both strategies allowed obtaining a stable matrix of cones; anyway, cones with large bases and without a basement have been later used since this allowed realizing tips of cones without defects and larger coverslip areas covered by cones.

In order to produce high fidelity membranes as negative replicas of the master, different PFPE-based photoresists have been tested, taking advantage of their low surface tension, which makes de-moulding operations more efficient. Among these, we have decided to employ a UV-curable PFPE-dimethacrylate (PFPE-DMA 4000), synthesized with a reaction between 2-isocyanate ethylmethacrylate and a PFPE macrodiol (average molecular weight of nearly 4000 g/mol). The product of this reaction was cured after adding a photoinitiator and the photo-crosslinked material exhibited a high hydrophobicity with a water contact angle of $113 \pm 0.7°$ and a shear modulus of 0.55 MPa, suitable to have soft membranes that can be easily peeled off from the mould. Replicated membranes were then obtained by means of the following 4-step process: (1) drop-casting of the PFPE photocurable resin on the mould; (2) spin-coating; (3) UV curing; and (4) de-moulding of crosslinked nano-structured layers.

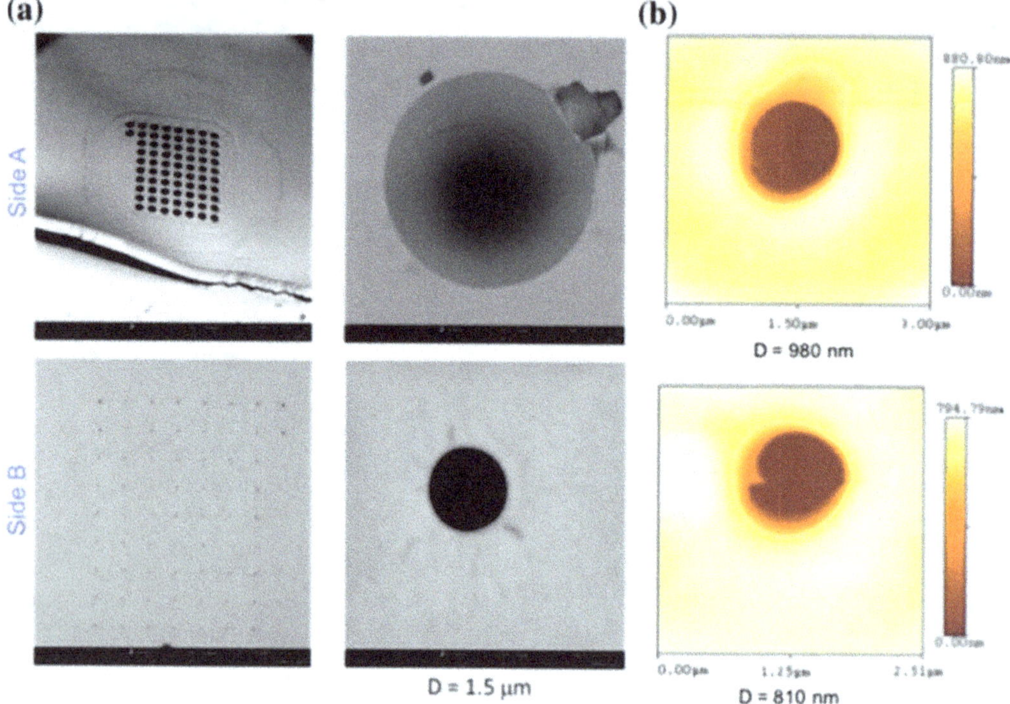

Fig. 12.4 Pores of PFPE membrane obtained by REM characterized by **a** SEM and by **b** AFM analysis

This process enabled to obtain membranes with a high fidelity to the master, due to a successful peeling of cured layers from the master, to the low surface tension and elastic modulus of PFPE-DMA. The step of spin-coating in the replica moulding process was needed to obtain thin layers of the viscous PFPE photoresist, but it also allowed controlling the thickness of the resist cured on the 2PP mould and to tailor the dimensions of the membrane pores. The membranes were finally characterized by SEM and AFM analysis. The results shown in Fig. 12.4 revealed the presence of pores with an average diameter of around 1 µm, thus demonstrating the successful exploitation of the proposed fabrication tool.

12.4.3 Developed Prototypes

The fabrication technology has been developed by optimizing the fabrication parameters of the tips on coverglass. The prototype has been later obtained by polymerizing a PFPE-PETA tip on top of a tipless commercial cantilever (AppNano HYDRA2R-200NTL). Usually, gold-coated cantilevers are employed for AFM measurements to increase laser intensity and thus enhance the detection of AFM probe movements. However, to avoid power absorption by the cantilever during laser irradiation, which would have led to material damage, an uncoated cantilever has been used. In order to

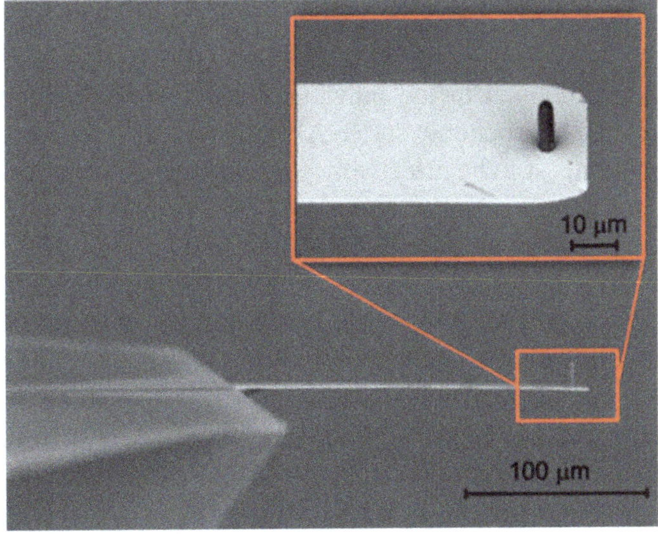

Fig. 12.5 SEM image of a PFPE-PETA AFM tip fabricated on top of a commercial cantilever [27]. The tip has a diameter of 3 μm and is 12.5 μm high

increase the adhesion between the resist and the Silicon Nitride cantilever, the latter has been silanized with 3-acryloxypropyl trimethoxysilane in vapour phase.

A sandwich support has been built to hold the cantilever and to allow the laser beam to reach the surface of the cantilever from above. A microscope slide has been covered with a layer of PDMS that could support the cantilever holder and two PDMS spacers. The spacers were holding a glass coverslip, so that the distance between the coverslip and the cantilever surface was more than the tip height, but less than the working distance of the objective used to focus the laser. The photoresist was inserted into the sandwich from the sides. The coverslip was necessary to ensure that the laser was always impinging on flat surfaces, and to avoid the transmission of the motion of the translation stages to the resist and to the cantilever.

The fabrication of the tip shown in Fig. 12.5 has been finally performed at 100 μm/s with a laser power of 15 mW. The distance between lines was set to 0.3 μm, and the spacing between planes to 0.1 μm. A one second stop between each plane has been introduced into the fabrication program to give the cantilever the time to stop vibrating after the vertical translation.

The cylindrical tip obtained after the fabrication on the cantilever is shown in Fig. 12.5, where it is possible to notice the hemispherical tip on top of it in the inset. The base diameter was measured to be 3 μm, and the tip was 12 μm high.

The uncoated cantilever caused the AFM feedback laser to give a very low signal during the prototype validation, hence a gold coating has been added by sputtering on the back of the cantilever after the tip fabrication. To prevent the gold from reaching the cantilever bottom surface and the tip, a mask has been realized via laser irradiation and chemical etching in fused silica. This enabled a precise insertion of the cantilever between two vertical walls, exposing only the desired surface to the gold target of the sputter coater.

12.5 Testing and Validation of Results

As mentioned in the previous section, new photoresists have been developed to realize hydrophobic stable AFM tips by 2PP. In particular, an optimal PFPE-based photoresist to be effectively structured by 2PP has been identified and examined in terms of chemical and physical properties. The exploitation of this photopolymer to successfully produce hydrophobic AFM tips by 2PP on commercial cantilevers has been demonstrated. In fact, AFM tips were validated by performing AFM indentations and by measuring the elastic modulus of medically relevant materials to be used as extracellular matrices in biological systems. Moreover, a clear understanding of tip-sample interactions exhibited by soft matter and hydrogels on the nanoscale were achieved, by comparing hydrophobic commercial probes and PFPE-based AFM tips.

12.5.1 Validation of Chosen Photoresist

A preliminary study was performed to investigate the chemical and physical properties of 6 PFPE photoresists, including commercially available grades. The PFPE-based resists used are different in terms of molecular weights and photo-reactive functionalities (Table 12.1). The exact chemical structure of PFPE photoresist is undisclosed apart from the PFPE-PETA resist, which was already synthesized and presented in a previous work by De Marco et al. [20]. All the PFPE-based resists showed a similar hydrophobicity, with a static contact angle versus water of 109–113° (Table 12.1) and a very low surface energy, suggesting the enrichment of a fluorine-based component on the resist surfaces.

In order to characterize the photo-reactivity of PFPE resists, the heats of crosslinking reaction were measured by a differential scanning calorimeter coupled with a UV source (UV-DSC), recording sample enthalpy changes when exposed to UV light. The results show that the heats of crosslinking for resists with methacrylate groups are lower than those with acrylates, probably due to the steric hindrance of methyl groups that decrease the reactivity of methacrylates during the polymerization [28, 29].

Among the photoresists with acrylate groups, a higher heat of crosslinking was obtained for PFPE-PETA with a value of 166 J/g (Table 12.1). This reveals that the resist with 6 acrylate groups is more reactive than others, suggesting why only PFPE-PETA provides reproducible and stable structures fabricated by 2PP, as shown in the previous paragraph. This finding also implies that it may be necessary to overcome a threshold value of heat of crosslinking to fabricate structures by 2PP, shedding light on mechanisms that makes photoresists suitable for 2PP structuring process. PFPE-PETA resist stands out as a suitable candidate for the microfabrication of hydrophobic and rigid AFM tips by considering also the relatively high shear modulus of 210 MPa measured for PFPE-PETA by dynamic mechanical analysis (DMA).

Table 12.1 PFPE -OH equivalent weights, total number and type of photo-reactive functionalities, enthalpies of crosslinking measured by differential scanning calorimetry coupled with a UV source (UV-DSC), water contact angles and surface energies for all the PFPE resists

Acronym	PFPE HO eq. weight	Photo-reactive groups per molecule	Enthalpy of crosslinking (J/g)	Water contact angle (°)	Surface energy (mN/m^2)
PFPE-DA	630	2 acrylates	111	n.a.[a]	n.a.[a]
PFPE-DUMA	630	2 methacrylates	55	110.0 ± 0.7	16.1 ± 0.5
PFPE-DUTA	1082	4 acrylates	116	112.9 ± 0.5	24.6 ± 3.9
PFPE-TUA	501	4 acrylates	111	112.6 ± 0.6	24.4 ± 4.0
PFPE-TUMA	636	4 methacrylates	69	112.8 ± 0.4	17.4 ± 1.4
PFPE-PETA	520	6 acrylates	166	108.9 ± 0.8	20.9 ± 0.5

[a]PFPE-DA resist solution was not homogeneous, providing a very rough surface where it was difficult to measure contact angles

12.5.2 *Validation of Tailored Atomic Force Microscopy (AFM) Tips*

The capability of 2PP fabricated tips to perform nanoindentation tests was validated by performing AFM measurements in air on two bio-medically relevant compliant materials, poly(dimethyl siloxane)-based elastomers and poly(ethylene glycol) hydrogels. The results obtained with 2PP fabricated tips were compared with those obtained using a commercially available probe with a spherical SiO_2 tip silanized with 1H, 1H, 2H, 2H perfluorooctyltriethoxysilane. As already shown in [30, 31], the functionalization of commercial probes with a fluorosilane is essential to decrease tip-sample interaction and perform AFM nanoindentations otherwise unfeasible on soft and hydrophilic materials in air.

Figure 12.6 shows force versus z-piezo displacement and force versus indentation curves obtained performing AFM indentation tests on PDMS 10:1 using a 2PP PFPE-PETA tip. A value of adhesion force of 6 nN was calculated as the difference between the minimum value of the force and the final value in the retraction curves measured with 2PP PFPE-PETA tips. This value of adhesion force is lower than the value obtained from curves measured using a fluorosilane-modified commercial tip (30 nN). This indicates that, using tips fabricated by 2PP structuring of PFPE-PETA photoresist, reduced tip-sample adhesive interactions occur in comparison with a fluorinated silicon tip, when samples like PDMS are tested. Moreover, an increase of maximum indentation ranges was achieved when using 2PP fabricated tips: from 35–45 nm obtained with a silanized commercial probe, to 60–75 nm using PEFE-PETA tips. This result demonstrates that PEFE-PETA photoresist enables the mea-

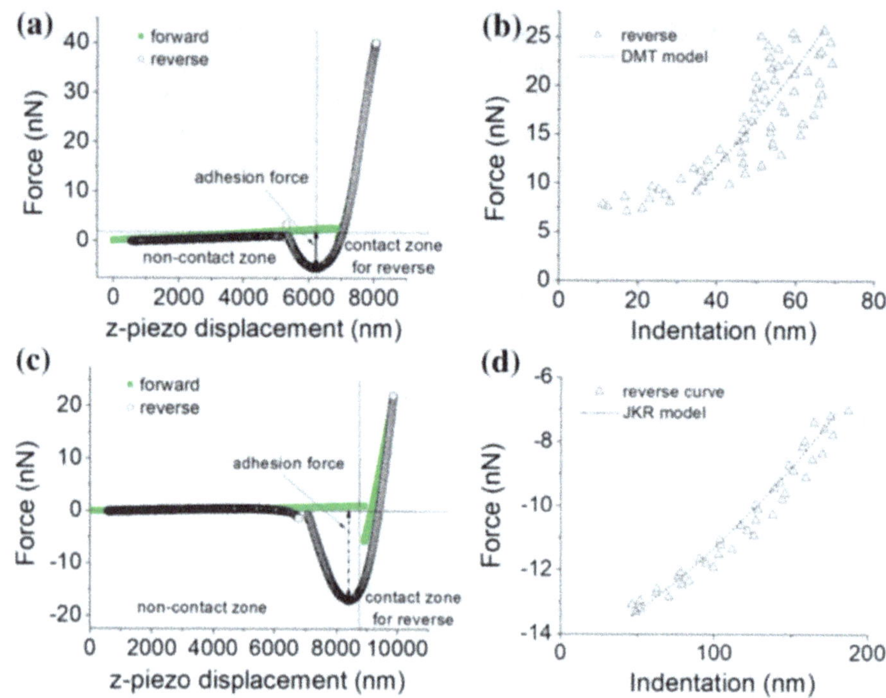

Fig. 12.6 Force versus displacement curves of **a** a PDMS 10:1 and **c** a PEGDA hydrogel samples measured with a AFM tip fabricated by 2PP with PFPE-PETA photoresist on a commercial cantilever. Force versus indentation curves obtained from reverse phase with superimposed **b** DMT and **d** JKR model curves in the dataset selected for the fitting for PDMS and for PEGDA, respectively [27]

Table 12.2 Maximum elastic indentation (δ), adhesion force (F_{ad}), Young's modulus (E_s) and reference Young's modulus values (E) for PDMS samples crosslinked with a prepolymer base/curing agent weight ratio of 10:1 (PDMS 10:1) and PEGDA hydrogels [27]

	Indentation δ (nm)	Adhesion force F_{ad} (nN)	Young's modulus E_s (kPa)	Reference value E (kPa)
PDMS 10:1	67 ± 6	5.9 ± 0.7	190 ± 191	1240 ± 50^a
PEGDA hydrogels	195 ± 7	16.4 ± 0.5	31 ± 5	22 ± 2^b

[a]Values measured by dynamic mechanical analysis
[b]Values derived from rheological measurements

surement of a large indentation range, leading to a better quality of elastic modulus prediction from force-distance curves.

Young's modulus of PDMS was determined by employing Derjaguin-Müller-Toporov (DMT) model, suitable for soft samples with a low surface energy [32]. A value of elastic modulus of 1.2 ± 0.2 MPa was obtained and found to be in good agreement with reference values measured by conventional techniques such as DMA (Table 12.2) [27].

Nanoindentation tests using 2PP fabricated tips were also performed on crosslinked poly(ethylene glycol) hydrogels prepared by UV polymerization of a poly(ethylene glycol) diacrylate (PEGDA, 10% wt.) and water (90% wt.), because these materials are increasingly studied and selected to act as artificial extracellular matrices. Figure 12.6c, d show force versus displacement and force versus indentation curves measured on PEGDA hydrogels. PEGDA hydrogels exhibit an adhesion force higher than PDMS samples 10:1 in the unloading curves (see Table 12.2). By using a commercial tip with a fluorosilane coating, an adhesion force of 36.7 ± 0.1 nN was obtained on PEGDA hydrogels. The higher hydrophobicity of 2PP fabricated tips led to a higher value of adhesion force with respect to the 2PP PFPE-PETA tips (16.4 ± 0.5 nN). Therefore the 2PP PFPE-PETA tips are more effective than fluoro-silanized tips at reducing adhesive interaction. Using the 2PP fabricated tips, a larger indentation range of 195 ± 5 nm was measured when compared to fluoro-silanized SiO_2 probes (120 ± 18 nm), thus resulting in a more accurate calculation of the elastic modulus also for PEGDA hydrogels. The JKR model was selected to calculate Young's modulus of hydrogels, because this model is more appropriate than DMT model for soft samples with a high surface energy [33]. Only unloading curves were considered for the evaluation of elastic moduli to neglect any dissipative phenomena, thus leading to an average value of 31 ± 5 kPa. The obtained value of modulus matches the average value measured using standard tips (32 ± 6 kPa) and it is also very similar to the average value of modulus estimated from rheological measurements (22 kPa). These findings confirm the ability of 2PP tips to perform nanoindentation tests in air on soft samples with very high water content and to measure their Young's modulus [27].

12.6   Conclusions and Future Research

The results of this work demonstrated that the 3D prototyping capability of 2PP technology is suitable for the realization of specific products following the customer needs. In fact, the results showed that this technology is suitable to fabricate AFM tips with customized geometries, mechanical and chemical properties. This can be easily achieved thanks to the 2PP capability of efficiently nanostructuring different photopolymers. Preliminary results on the realization of porous membrane by replica moulding of a 2PP fabricated master showed the possibility to scale up the fabrication process for an example customizable product.

In particular, the obtained results cover different aspects ranging from the development of a new photoresist [33] to the optimization of the laser fabrication protocols for an efficient structuring of the developed photoresist, to the fabrication of AFM tips on commercial cantilevers and to their experimental validation [27]. A PFPE-based photoresist has been indeed developed to fabricate hydrophobic and rigid AFM tips on commercial cantilever. The chosen tip geometry is suitable for nanoindentation measurements on soft materials and the realized prototypes have been used to characterize the elastic modulus of biologically relevant materials as PDMS and PEGDA hydrogels. The tips demonstrated to have low adhesion interactions with the sample surface thanks to the high hydrophobicity of the chosen polymer. This feature permitted to characterize the sample by measuring large indentation ranges, thus enabling one to retrieve the elastic modulus from the force-distance curves with higher precision. The obtained results, in comparison with fluorosilane-functionalized commercial tips, demonstrated that the hydrophobic PFPE tips fabricated by 2PP can be used for nanoindentation test. Indeed, these showed a better performance in combination with the advantage of being tailor-made for samples under investigation in terms of tips shape and dimension.

The positive results give birth to new opportunities to further explore the customization offered by two-photon polymerization and replica moulding. The functionalization of the tip allows testing biological and functional surfaces and the development of AFM micromechanical testing models and protocols for soft and biological surfaces. Moreover the development of specialty photoresists with tailored surface properties would open the possibility to use 2PP fabricated tips for chemical sensing. Further work can be also focused on the membranes in order to achieve smaller pore and validate them for some applications, e.g. integrating them in microfluidic chips for water filtering analysis.

Acknowledgements This work has been funded by the Italian Ministry of Education, Universities and Research (MIUR) under the Flagship Project "Factories of the Future—Italy" (Progetto Bandiera "La Fabbrica del Futuro") [34], Sottoprogetto 2, research project "POLYmer nanostructuring by two-PHoton ABsorption" (POLYPHAB). The authors would like to thank Solvay for kindly suppling PFPE-based photoresists.

References

1. Farsari M, Chichkov BN (2009) Two-photon fabrication. Nat Photonics 3:450–452
2. Kawata S, Sun HB, Tanaka T, Takada K (2001) Finer features for functional microdevices. Nature 412:697–698
3. Maruo S, Fourkas JT (2008) Recent progress in multiphoton microfabrication. Laser Photon Rev 2:100–111
4. Juodkazis S, Mizeikis V, Misawa H (2009) Three-dimensional microfabrication of materials by femtosecond lasers for photonics applications. J Appl Phys 106:5
5. Osipov V, Pavelyev V, Kachalov D, Žukauskas A, Chichkov BN (2010) Realization of binary radial diffractive optical elements by two-photon polymerization technique. Opt Express 18:25808–25814
6. Ovsianikov A, Ostendorf A, Chichkov BN (2007) Three-dimensional photofabrication with femtosecond lasers for applications in photonics and biomedicine. Appl Surf Sci 253:6599–6602
7. Tayalia P, Mendonca CR, Baldacchini T, Mooney DJ, Mazur E (2008) 3D cell-migration studies using two-photon engineered polymer scaffolds. Adv Mater 20:4494–4498
8. Ovsianikov A, Deiwick A, Van Vlierberghe S, Dubruel P, Möller L, Drager G, Chichkov BN (2011) Laser fabrication of three-dimensional CAD scaffolds from photosensitive gelatin for applications in tissue engineering. Biomacromol 12:851–858

9. Raimondi M, Eaton SM, Laganà M, Aprile V, Nava MM, Cerullo G, Osellame R (2013) Three-dimensional structural niches engineered via two-photon laser polymerization promote stem cell homing. Acta Biomaterialia 9:4579–84

10. Park SH, Yang DY, Lee KS (2009) Two-photon stereolithography for realizing ultraprecise three-dimensional nano/microdevices. Laser Photon Rev 3:1–11

11. Tian Y et al (2010) High performance magnetically controllable microturbines. Lab Chip 10:2902–2905

12. Amato L, Gu Y, Bellini N, Eaton SM, Cerullo G, Osellame R (2012) Integrated three-dimensional filter separates nanoscale from microscale elements in a microfluidic chip. Lab Chip 12:1135–1142

13. Xia YN, Whitesides GM (1998) Soft lithography. Angew Chem Int Edit 37:551–575

14. Kim JM, Muramatsu H (2005) Two-photon photopolymerized tips for adhesion-free scanning-probe microscopy. Nano Lett 5:309–314

15. Okada T, Yamamoto Y, Sano M, Muramatsu H (2009) Fabrication of various tip-size AFM probes for evaluating single-molecular retraction force between actin and anti-actin. Ultramicroscopy 109:1299–1303

16. Baldacchini T, LaFratta CN, Farrer RA, Teich MC, Saleh BEA, Naughton MJ, Fourkas JT (2004) Acrylic-based resin with favorable properties for three-dimensional two-photon polymerization. J Appl Phys 95:6072–6076

17. Tonelli C, Trombetta T, Scicchitano M, Castiglioni G (1995) New Perfluoropolyether soft segment containing Polyurethanes. J Appl Polym Sci 58:1407–14007

18. Turri S, Radice S, Canteri R, Speranza G, Anderle M (2000) Surface study of perfluoropolyether-urethane cross-linked polymers. Surf Interface Anal 29:873–886

19. Temtchenko T, Turri S, Novelli S, Delucchi M (2001) New developments in perfluoropolyether resins technology: high solid and durable polyurethanes for heavy duty and clear OEM coatings. Prog Org Coat 43:75–84

20. De Marco C, Gaidukeviciute A, Kiyan R, Eaton SM, Levi M, Osellame R, Chichkov BN, Turri S (2013) A new Perfluoropolyether-Based hydrophobic and chemically resistant photoresist structured by two-photon polymerization. Langmuir 29:426–431

21. Butt HJ, Cappella B, Kappl M (2005) Force measurements with the atomic force microscope: technique, interpretation and applications. Surf Sci Rep 59:1–152

22. Cross SE, Jin YS, Rao J, Gimzewski JK (2009) Applicability of AFM in cancer detection. Nat Nano 4:72–73

23. Stroeve P, Ileri N (2011) Biotechnical and other applications of nanoporous membranes. Trends Biotechnol 29:259–266

24. Rolland JP, Hagberg EC, Denison GM, Carter KR, DeSimone JM (2004) High-resolution soft lithography: enabling materials for nanotechnologies. Angew Chem Int Edit 43:5796–5799

25. Rolland JP, Van Dam RM, Schorzman DA, Sr Q, DeSimone JM (2004) Solvent resistant photocurable "liquid teflon" for microfluidic device fabrication. J Am Chem Soc 126:8349–8349

26. De Marco C, Eaton SM, Levi M, Cerullo G, Turri S, Osellame R (2011) High-Fidelity Solvent-Resistant replica molding of hydrophobic polymer surfaces produced by femtosecond laser nanofabrication. Langmuir 27:8391–8395

27. Suriano R, Zandrini T, De Marco C, Osellame R, Turri S, Bragheri F (2016) Nanomechanical probing of soft matter through hydrophobic AFM tips fabricated by two-photon polymerization. Nanotechnology 27:155702

28. Evans AG, Tyrrall E (1947) Heats of polymerization of acrylic acid and derivatives. J Polymer Sci 2:387–396

29. Roberts DE (1950) Heats of polymerization. A summary of published values and their relation to structure. J Res NBS 44:221–232

30. Credi C, Biella S, De Marco C, Levi M, Suriano R, Turri S (2014) Fine tuning and measurement of mechanical properties of crosslinked hyaluronic acid hydrogels as biomimetic scaffold coating in regenerative medicine. J Mech Behav Biomed Mater 29:309–316

31. Suriano R, Oldani V, Bianchi CL, Turri S (2015) AFM nanomechanical properties and durability of new hybrid fluorinated sol-gel coatings. Surf Coat Technol 264:87–96

32. Derjaguin BV, Muller VM, Toporov YP (1975) Effect of contact deformations on the adhesion of particles. J Colloid Interface Sci 53:314–326
33. Johnson KL, Kendall K, Roberts AD (1971) Surface energy and the contact of elastic solids. Proc R Soc Lond Ser A Math Phys Sci 324:301–313
34. Terkaj W, Tolio T (2019) The Italian flagship project: factories of the future. In: Tolio T, Copani G, Terkaj W (eds) Factories of the future. Springer

Chapter 13
Use of Nanostructured Coating to Improve Heat Exchanger Efficiency

Antonino Bonanno, Mariarosa Raimondo and Michele Pinelli

Abstract This work investigates the potential of surface nano-coatings on the heat transfer surface of heat exchangers, in order to improve their overall efficiency in terms of pressure drop and heat power exchanged. The work started from the consideration that, due to the increasingly strict international standards, the machines will be provided with optimized engines with new and improved devices to reduce the exhaust emissions (liquid cooled EGR valves, new catalytic converters and DPF systems) which reduce inevitably the space in the engine compartment. In this paper it is studied the feasibility of the coating processes and the adaptability of deposition techniques to the industrial production process of cross flow heat exchanger fins. The innovative exchanger performance, investigated using dedicated test rigs, shows that the results are influenced by the coating technology.

13.1 Scientific and Industrial Motivations

Heat exchanger efficiency is one of the main concerns in trying to improve the overall efficiency of mobile off-road machines and industrial plants. Compact radiators play a fundamental role in temperature control of internal combustion engines (ICE) and hybrid drivelines. An improvement of this component could affect the global efficiency of the machine. Typically, the problem is studied from a geometrical point of view, using different production lines to make fins geometries dedicated to improve the heat exchange or to reduce the pressure drop. Heat exchanger optimization is needed to obtain smaller, lighter, but at the same time more efficient radiators.

A. Bonanno (✉)
CNR-IMAMOTER, Istituto per le Macchine Agricole e Movimento Terra, Ferrara, Italy
e-mail: a.bonanno@imamoter.cnr.it

M. Raimondo
CNR-ISTEC, Istituto di Scienza e Tecnologia dei Materiali Ceramici, Faenza, Ravenna, Italy

M. Pinelli
Università di Ferrara, Ferrara, Italy

© The Author(s) 2019
T. Tolio et al. (eds.), *Factories of the Future*,
https://doi.org/10.1007/978-3-319-94358-9_13

In this paper, we propose to focus the attention on the possibility to change the performance not only via geometrical modifications, but also considering new production processes to functionalize the heat exchanger surfaces, thus enabling to customize the performance based on the specific requests. Indeed, the laminar layer height, which influences the ability to exchange heat and the pressure drop could be modified by changing the surface functionalization while keeping the same geometry. Therefore, the use of functionalized coating could improve the heat exchanger performance, giving the possibility to develop new product families able to satisfy the market requests, especially for tailored (high value—high technological content) products. The use of functionalized coating (superhydrophobic and oleophobic) in heat exchanger has not been tested until now. Recently some studies dedicated to the applicability of nanostructured coating on hydraulic piston pumps have been presented [1]. The structure of a nanoscale surfaces enables to modify its wetting and the fluid interaction. Thus, it is possible to reach high levels of repulsion against water (super-hydrophobicity) or oils/alcohols (oleophobicity).

This work studies the potential of a novel nanostructured coating, featuring super-hydrophobic and oleophobic behaviour, in providing better performance to the heat exchanger for fixed and mobile circuit. Different production technologies were investigated with the aim to further develop the currently most easily adaptable methodology, i.e. hot brazing.

Moreover, a software tool was developed to optimize the heat exchanger geometry performance, thus ensuring the best results in term of heat exchanger performance regardless of its overall dimensions. The software tool is able to split the exchanger into sub-domains having homogeneous boundary conditions and then simulates the performance, both on the cold and hot side, in terms of the WHTC (Wall Heat Transfer Coefficient) and pressure drop, using CFD (Computational fluid dynamics) analysis. The basic idea is to simulate the entire radiator using sub-domains, considered as a combination of elements with series and parallel layout, as a function of the regions to be simulated.

The chapter is organized as follows. Section 13.2 presents the state of the art regarding the ability to develop a nanostructured coating able to change the fluid surface interaction, specifically to reach a contact angle greater than 110 °C. Section 13.3 proposes the overall approach to functionalize the whole heat exchanger without modifying the actual construction methodology. Section 13.4 presents the developed methodologies, tools, and prototypes. The testing and analysis of thermomechanical and corrosion analysis of functional layer is reported in Sect. 13.5. Finally, the conclusions are drawn in Sect. 13.6.

13.2 State of the Art

Although the heat exchangers are widely used both in mobile and fixed applications, to the best of our knowledge there are few examples of scientific works focused to the same approach proposed in this paper. The effect of contact angle (or surface wettability) on the convective heat transfer coefficient has been studied and a correlation between the pressure drop reduction and heat transfer coefficient has been found considering a hydrophilic surface [2].

In a steam condensation application, the use of a hydrophobic coating has improved the overall heat coefficient of more than 10 times [3]. In air conditioning applications, the pressure drop and the heat transfer have been studied considering a hydrophilic coating and a new correlation between heat, mass and momentum transfer has been proposed to describe the obtained results [4]. Furthermore, the use of a hydrophilic coating has enhanced the heat transfer performance and has reduced the pressure drop in a dehumidifying application [5]. The coating ability to promote fluid motion after the application of external forces was studied in the laminar field of motion [6] and in turbulent conditions [7, 8]. A relation between slip length and contact angle was studied in [9].

Wettability is the ability of a liquid to maintain contact with a solid surface. The wettability of the surface has a great influence on heat exchange, because it influences the type of motion that is established between surface and fluid. The impact of wettability effect on heat transfer was studied in [10]. The effect of contact angle (or surface wettability) on the convective heat transfer coefficient in microchannels was studied in [11], while a correlation between the contact angle and the wetting properties is proposed in [12]. In none of the previous studies, however, the possible connection between the reduction of the friction between the fluid and the surface, by means of the nanostructured coating, and the increase in the heat exchange performance was investigated. About the oleophobic performance obtained using the technology proposed in this paper, the main literature can be found in [1, 13–17].

13.3 Problem Statement and Proposed Approach

Cross-flow heat exchangers are widely used in automotive and earthmoving machinery applications because they can transfer a great quantity of thermal power in a relative small volume. The need to cool process fluids is always present and becomes critical when there is the need to concentrate a big amount of power in increasingly small dimensions. This is the case, for example, of new hybrid solutions, where to the thermal power generated by the ICE is added the thermal power developed by the driveline and the electric motor. This work is based on the hypothesis that, if heat transfer coefficient increases and friction power losses do not increase too much or even decrease, then it is then possible to transfer the same power by smaller dimensions or to transfer more thermal power by the same overall dimensions. The problem was addressed through five phases.

The first part of the work was dedicated to reduce the wettability of the surface thanks to the design and synthesis of coatings of various types, both organic and hybrid (organic-inorganic). In particular, we focused on the morphology of the nanostructure and on the surface chemistry that both have to be controlled and modulated (Sect. 13.4.1). In this context, the synthesis of ceramic oxides nanoparticles (Al_2O_3,

ZrO_2, etc.) was studied to form a stable colloids or nanosuspensions in different media (i.e. alcohols or water). In this phase it was mandatory to have a high control degree on phases, particle size and composition. Also specific synthetic approaches were developed (i.e. sol-gel), involving particle nucleation directly from precursors present into the fluid, in the meantime avoiding particle precipitation and ensuring a better control over physical variables of suspensions.

The second phase was the adaptation of laboratory deposition techniques to the specific production process followed to make the brazed heat exchanger (Sect. 13.4.2). In this context, coating deposition by simple immersion techniques, such as dip-coating or spin-coating in controlled conditions, revealed to be suitable to produce homogeneous functional surfaces (turbolators), to easily control the layer thickness in the order of nanometers and to promote the formation of micro- and nanostructures. The main problem is related to the fact that the turbolators could be easily functionalized using the dip-coating, but they lost their superhydrophobic/oleophobic performance after the brazing phase, necessary to assembly turbolators and fins in order to produce a complete heat exchanger. For this reason other production techniques (e.g. pumping, suction) were experimented with the aim to investigate which of them could interfere as less as possible with the currently used production technique.

The third phase was the development of a software tool able to simulate the heat exchanger, considering the different possible combinations of turbolators and fins surfaces (Sect. 13.4.3). The heat exchanger was decomposed in main portions in which the flow has some peculiar characteristics (mainly from the point of view of boundary conditions applied). The final aim is to have a customized, yet flexible, tool able to predict the heat exchanger performance under different environmental conditions and geometry dimensions, considering not only the possibility to optimize the radiator by the fin geometry but also using the surface functionalization.

The assessment of functional performance of the obtained samples is the fourth phase of the work (Sect. 13.4.4). The setup of new measurement stations, in order to have practical estimates of the behaviour of different surface coating, was also taken into consideration during this part of the work. For instance, a prototype with reduced dimension was used to monitor the pressure drop and the exchanged power considering different nano-coating functionalized surface. The reasons grounding the choice of the test cycles, foreseen for the functional performance tests, derive from the experience gained in functional characterization of the components at the experimental stage and from the investigation of the relevant laboratory tests required by the applicable international standards. A key factor at this stage was the assessment of changes or damages to the microstructure of the functionalized layers after the test cycles.

The fifth and final phase was dedicated to the verification of new coating ability, in the expected operative conditions, to assure the efficiency improvement (Sect. 13.5). A complete tribological, corrosion and mechanical characterization of functional layers was performed.

13.4 Developed Technologies, Methodologies and Tools

13.4.1 Development of Functional Layer

Super-hydrophobicity can be achieved, as exposed in the work by Bonanno et al. [1], by inducing a flower-like structure on a micro or nano-scale so that liquid droplets are not able to adhere to the surface because of air entrapped among the surface roughness. Attempting to repeat the Cassie-Baxter model in practice the target was to obtain a structure like the one in Fig. 13.1.

A super-hydrophobic or super-hydrophilic behaviour of the liquid droplets in contact to the surfaces in question can be obtained by changing a phase in the process:

- Super-hydrophilic behaviour

  - Al_2O_3 nanoparticles sol deposition
  - Heat treating

- Super-hydrophobic behaviour

  - Al_2O_3 nanoparticles sol deposition (inorganic component)
  - Heat treating
  - Fluoridation with fluoroalkilsylane (FAS—organic component) F8263.

A lubricant film can be used in a porous solid [18] to convey better mechanical resistance to the coating layer, high pressure resistance, self-healing, sand low friction. If the lubricant is immiscible, then the interface between the solid and the

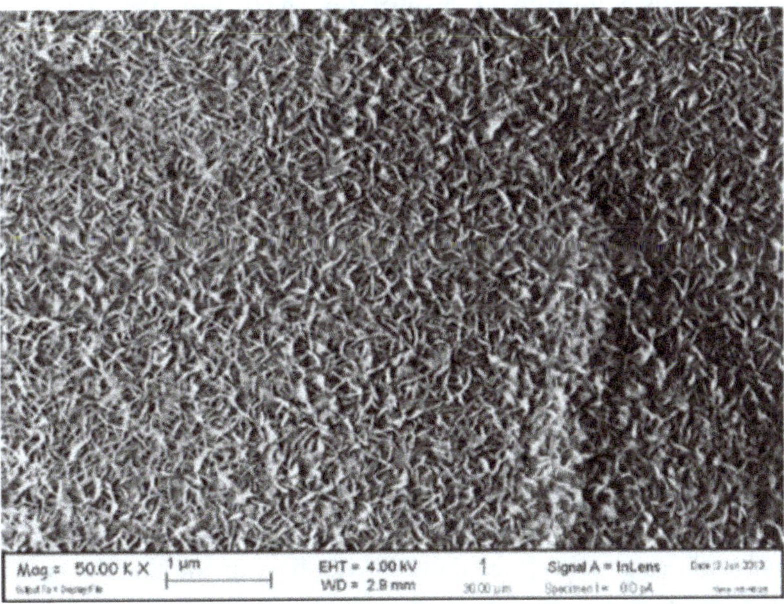

Fig. 13.1 SEM image of a Cassie-Baxter structure

immiscible liquid is no more solid but liquid, meaning that the lubricant interposed between solid and liquid reduces the friction.

13.4.2 Deposition Techniques

A surface must be optimised to keep repellency stable over time. In fact, keeping the free energy of the solid/lubricant/liquid system as low as possible, it is possible to make the liquid reside on the surface of the lubricant without replacing it, thus eliminating the fixing points and obtaining extremely low sliding angles (super-repellent surfaces). This is true also for liquids with very low surface tensions [19]. A qualitative analysis showed the formation of the pinning effect, therefore the components were further functionalized with fluoridated lubricant FC-43 in order to obtain a stable repellency. The functionalization of singular components of a heat exchanger implicates a following bonding with further heat treatments that could damage the coating, and so threaten the relative hydrophobic effect. The general coating technique is the same described in the work by Bonanno et al. [1].

Three different coating techniques were developed and tested in this work:

- Functionalization by emptying. The functionalization by dip-coating (immersion in the solution) implies a strict dependence of the sol container on the heat exchanger dimensions. Besides the volume of sol to be produced must be sometimes much greater than the volume to be filled in the heat exchanger. The emptying method works taking advantage of the same principle as the dip-coating one, namely the relative motion between solid surface and coating liquid. In this case liquid moves to the solid which is steady whereas by dip-coating the solid moves and liquid is steady.
- Emptying by suction. This method consists in sucking the inner liquid, previously filled in the heat exchanger, using a controlled volume variation container like a syringe. Since the optimal liquid evacuation velocity is 2 mm/s for both alumina suspension solution and FAS solution, an appropriate piston velocity must be assessed. This method presents the non-negligible advantage of using the exact volume of liquid necessary to fill the heat exchanger, furthermore, since both nanoparticles solution and FAS can be used more than once to coat the solid surface, several heat exchangers can be treated with the same solutions.
- Emptying by pumping. This method is based on emptying the treated heat exchanger by pumping the chemical solutions out of it by using a peristaltic pump which is volumetric (flow rate can be easily controlled by varying rotational speed) and suitable for low flow rate values (0–100 ml/min). The pump, activated by the electrical supply, sucks the solution from the heat exchanger and sends it to the bottle that works as a recovery container. After having emptied the heat exchanger the solution can be used again in another one.

Fig. 13.2 Heat exchanger
sub-domains

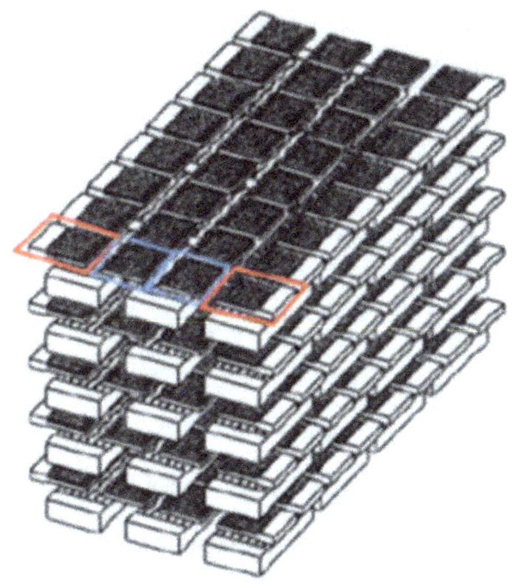

13.4.3 Simulation of Heat Exchanger Performance

As previously described, a dedicated software tool was developed to simulate the heat
exchanger performance. By positioning on the XZ plane, the heat exchanger can be
divided into a number of basic elements, which have certain geometric characteristics
and fluid dynamics features. In particular, it can be distinguished an entrance area,
an inner section and an output section of the cooling fluid. The same can be said for
the hot fluid. Preliminary analyses have highlighted a substantial independence of
the results from the direction of the fluid: heat transfer coefficients and pressure drop
were slightly affected by the direction in which the transition plenum-interstices takes
place. This consideration leads to match subdomains that present the same geometric
properties regardless of the fluid dynamic condition in the subdomain. The result
was to group the nine possible subdomains into four classes, in this way, only four
subdomains have been analysed as representative from each class. Summarizing, the
adopted methodology does not simulate the coupling of one or more rows of hot side
fins with one or more cold side fins. Rather a small element of every single fin was
analysed, by imposing on the single domain the boundary conditions that most likely
will occur during operation. Then, through a revision of the results that is performed
by means of a spreadsheet, the subdomains are reconstructed in a virtual exchanger
so as to have the characteristic of the global heat exchanger. The heat exchanger
is thermo-fluid-dynamically reconstructed (Fig. 13.2) by spacing layers (in parallel)
consisting of an element of *Type 2* (shown in red) input, which has *n* elements in series
Type 5 (highlighted in blue) and then conclude with an element of *Type 2* output. In
this way, the types of simulation domains are cut without significantly reducing the
accuracy of the entire analysis. Initially, it was assumed that the size of the element
could have an influence on the transfer coefficient due to the actual distribution of
temperature on the wall of the fins. However, by performing a preliminary analysis
on the influence of the size of the computational domain on the value of the heat
transfer coefficient, there was a substantial independence from such a value.

Table 13.1 Simulation conditions

Fluid	Tested geometries	Temperatures (°C)	Subdomains	Velocities (m/s)
Air	A0, A1, A2, A3, A4	20–40–60	Type 2–Type 5	2–4–6
Water-glycol	P0–P2	60–80–120	Type 2–Type 5	0.2–0.4–0.6
Oil	P0–P2	60–80–120	Type 2–Type 5	0.2–0.4– 0.6
Hot air	P1–P3	120–160–200	Type 2–Type 5	4–6–8

For each designed geometry, we have identified two types of domain:

- *Type 2*, characterized by the presence of the inlet plenum that is used to simulate the input section;
- *Type 5*, in which the plenum is absent because it represents the internal geometry of the heat exchanger.

A total amount of 5 geometries for fins (A0, A1, A2, A3, A4), 2 geometries for water-glycol and oil channels (P0 and P2) and 2 geometries for the air channels (P1 and P3) were analysed to assess the specific heat transfer coefficient and pressure drop. The simulations performed were 180, as reported in Table 13.1.

13.4.4 Test Rig and Prototypes

During the work several prototypes were realized and a dedicated test rig was built to properly test them. The test rig is shown in Fig. 13.3 and is composed of a service pipeline that is connected to a proportional valve, i.e. a gate valve controlling the flow rate through the heat exchanger. Before the beginning of the tests, a heater controls the temperature of the inlet oil by heating the whole mass of oil contained in the tank, so that the oil has a high heat capacity to keep at a constant temperature during the tests. The air flow (the cold fluid) is driven by a fan, actuated by a hydraulic motor, whose suction side is connected to the heat exchanger by a divergent duct. The oil temperatures (inlet and outlet), the oil flow, the air temperature, the oil and air pressures were all collected by a dedicated acquisition software, in order to have all the measured values coming from different transducers. In particular, two flow meters were installed to measure a wide range of volume flow rate: one has a full scale of 10 l/min and the other of 300 l/min that can be activated separately by ball valves placed upstream them.

All the measures were printed in a text file that can be imported into a post-processing software to examine and derive other physical quantities not directly measured. A control panel of the acquisition software was developed from the ground up (both hardware and software) and used to monitor the working conditions of the test rig.

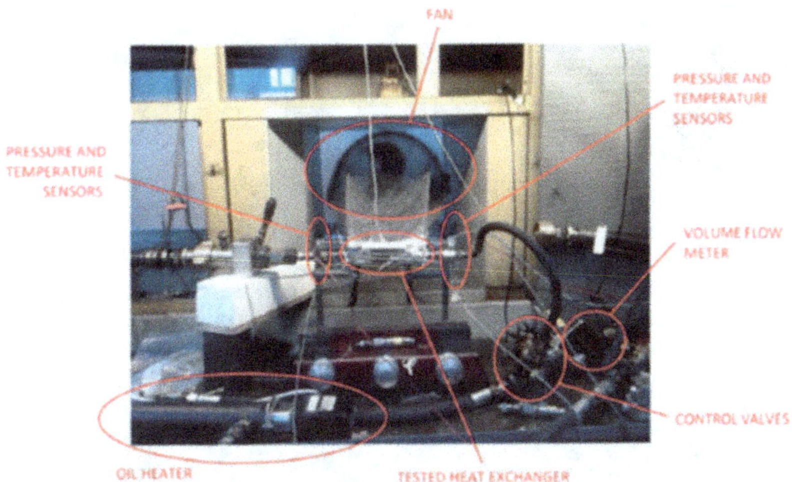

Fig. 13.3 Test rig for heat exchanger prototypes

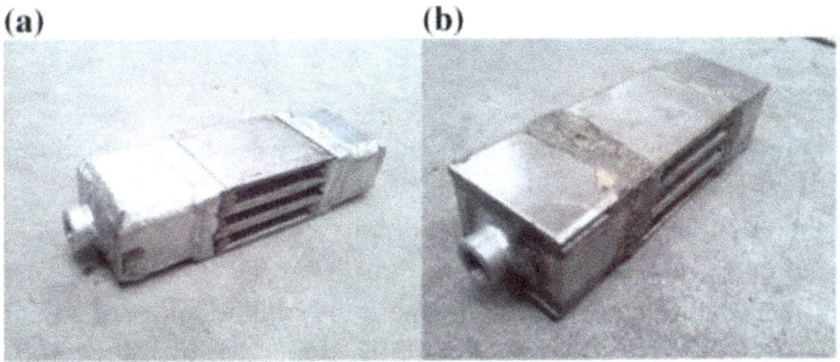

Fig. 13.4 *Type A* heat exchanger **a** standard and **b** functionalized (right)

The first set of prototypes was developed by using the brazing technology to assemble the fin packs (Fig. 13.4) at about 600 °C, temperature at which the nanostructured coating would be irreparably damaged. To avoid this, the samples were first assembled and brazed, and then coated, according to one of the technologies presented in Sect. 13.4.

Four different set of cross flow heat exchangers were realized as previously presented and tested:

- *Type A*: 63 mm by 100 mm by 46 mm;
- *Type B*: 94 mm by 93 mm by 200 mm;
- *Type C*: 160 mm by 160 mm by 45 mm;
- *Bonded*: 63 mm by 100 mm by 46 mm.

The choice of testing three different shapes and dimensions of heat exchangers is justified by the necessity to verify if scale effect would have amplified their performance differences.

In the *Type A* the lateral bowls were glued after the functionalization (by empting) of the fins pack as showed in Fig. 13.4 on the right. In the *Types B* and *C* the lateral

Fig. 13.5 *Bonded* heat exchanger

bowls were pre-assembled and the functionalization was obtained by pumping or by suction.

Due to the fact that the heat exchangers fins pack are always pre-assembled, it was impossible to evaluate if the coating process worked correctly, deposing and structuring the correct nano-structured superhydrophobic and oleophobic layer. A completely different set of prototypes were developed to solve this problem; these were assembled by bonding (Fig. 13.5). In this case the fins were singularly coated and the assembled together by bonding. All the prototypes showed until now (*Type A*, *Type B*, *Type C* and *Bonded*) were tested using oil, which is the fluid normally used with this type of heat exchangers.

Finally, a fifth set of prototypes (Fig. 13.6) was prepared in order to evaluate the performance of a completely flat surface and using water as working fluid. This last set was developed to investigate the theoretical hypothesis for which the greater is the contact angle the lower is the friction between the fluid and the surface [20–22].

Because the nanostructured coating gives the best performance, in terms of contact angle as well as contact angle hysteresis and energetic interaction between fluid and surface, with water, we decided to develop a completely dedicated experimental set up and test. The testing rig works in a closed loop, when the prototype is selected, the water previously heated flows into the prototype heat exchanger and then returns to the pump. It is also possible to select the coated or the non-coated channel, because they are in parallel. This last test rig was used only with the *channel* prototype, whereas the test rig shown in Fig. 13.2 was used with all other prototypes tested using oil as working fluid.

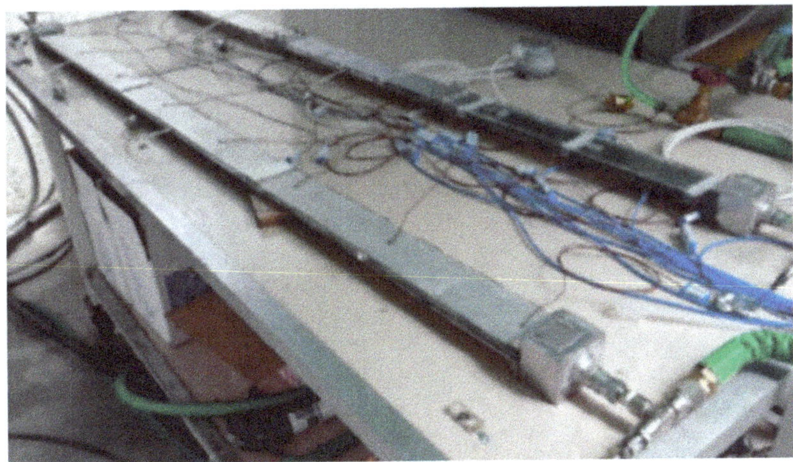

Fig. 13.6 Channel prototype

13.5 Testing and Validation of Results

Wettability of the samples treated with the techniques described in the previous section were measured using diiodomethane (CH_2I_2 with a surface tension of 50.8 mN/m) because of the impossibility to measure the water contact angle due to a complete water proofing of the surface of heat exchanger fins (Fig. 13.7). The results showed a contact angle of $133.1° \pm 7.1°$ using the hybrid methodology and $123.8° \pm 6.6°$ using the lubricant. It can be noticed the good hydrophobicity achieved with the hybrid method against the lubricant one on the liquid side fins. It was also tested the possibility to achieve a direct functionalization through fluoridation only (agent F 8263). The assessment was done through water contact angle (WCA), CH_2I_2 contact angle, and surface energy (SE) measurement. The WCA was $136.2° \pm 34.7°$, the CH_2I_2 contact angle was $70.9° \pm 10.9°$ whereas the SE was 7.28 ± 23.91 mN/m. From the results obtained, it comes out the nano-structuring necessity to achieve super-hydrophobicity.

The heat resistance assessment of the coated layers was performed at three different temperatures:

- Heating at 200 °C for 2 h—the results showed that the superhydrophobicity/oleophobicity is maintained;
- Heating at 300 °C for 2 h—still hydrophobic surfaces (spherical droplets) but lower slipping;
- Heating at 400 °C for 2 h—destruction of the organic component of the coating, with subsequent super-hydrophilicity.

The results confirm that the brazing production methodology currently employed is not compatible with the nanostructured superhydrophobic layer.

The test rig (see Sect. 13.4.4) was designed to supply prototype heat exchangers with suitable thermal and hydraulic power to be dissipated. After measuring the three physical quantities (volume flow rate, static temperature and static pressure) thanks to

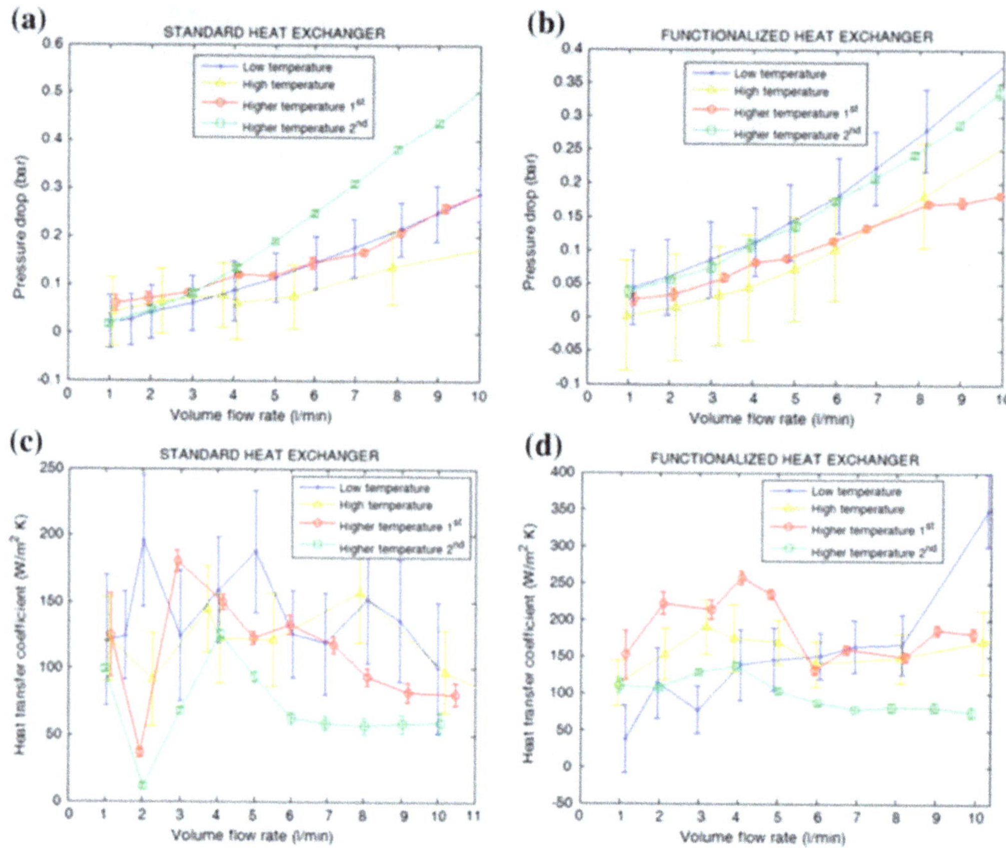

Fig. 13.7 Experimental results: **a** standard heat exchanger pressure drop at different temperatures; **b** functionalized heat exchanger pressure drop at different temperatures; **c** standard heat exchanger heat transfer coefficient at different temperatures; **d** functionalized heat exchanger heat transfer coefficient at different temperatures

the sensors in the test rig and knowing the thermo-physical properties of the process fluid (hydraulic oil), it is possible to extrapolate the overall Darcy viscous friction factor f according to Eq. (13.1) and the overall heat transfer coefficient U according to Eq. (13.2).

$$f = \frac{2 * S_f^2 * \Delta p_m}{\rho(t) * Q^2} \tag{13.1}$$

$$U = \frac{\rho(t) * Q * c_p * \Delta T_m}{S_h * \Delta T_m * F} \tag{13.2}$$

S_f is the flow through surface at the inlet of the finned part of the heat exchanger, Δp_m is the static pressure difference measured upstream and downstream the heat exchanger, $\rho(t)$ is the mass density of the fluid as a function of the instant temperature of the fluid, Q is the volume flow rate of the fluid, c_p is the specific heat capacity at constant pressure of the fluid, ΔT_m is the measured temperature difference between upstream and downstream the heat exchanger, S_h is the heat transfer surface of the finned part of the heat exchanger, ΔT_{ml} is the logarithmic mean temperature difference, F is a correction factor [23] depending on the shape of the heat exchanger and the way the fluids interact each other thermally and dynamically.

Four tests were carried out in two periods (or stages):

1. Test of *Type A* heat exchangers at three different inlet oil temperature conditions: 40, 65, 70 °C (two tests have been performed at 70 °C). For each working condition, volume flow rate was varied between 1 and 10 l/min with a step of 1 l/min.
2. Test of *Type A*, *Type B*, *Type C* and *Bonded* in the updated test rig at three different inlet oil temperature conditions: 50, 70, 85 °C.

The second stage corresponds to an evolution of the first one inasmuch the test rig has been improved and more attention paid to the data acquisition system. The geometrical correction factor F can be considered equal to one with good accuracy in each working condition.

The idea of a channel (basically a rectangular cross section pipe as showed in Fig. 13.6 has been conceived to study the behaviour of water in contact to straight walls that interact with liquid only through viscous friction.

The next subsections present the results of the main experiments.

13.5.1 Type A Heat Exchanger

Figure 13.7a and b show how differences in pressure losses behaviour occur as inlet oil temperature values change. At 40 °C the pressure drop distribution seems to be regular and relative to a smooth curve in the standard heat exchanger and in the functionalized one; at 65 and 70 °C the standard heat exchanger pressure losses curves present regions of flow instability whereas the functionalized curves exhibit it only at 70 °C inlet oil temperature. This behaviour could be explained considering laminar to turbulent flow transitions. Heat transfer coefficient is characterized by a great measurement uncertainty for volume flow rate values less than 7 l/min so it is not possible to extrapolate a precise trend in that range. After 7 l/min the trend is to reach an asymptote which in the functionalized heat exchanger seems to be greater of about 50% than in the standard one. By observing the curves in Fig. 13.7c and d it can be said that heat transfer coefficient values, for volume flow rate values less than 7 l/min, are basically diffused nearby a mean value of 116.46 ± 38.73 (W/m² K) for standard heat exchanger and 176.47 ± 59.45 (W/m² K) for the functionalized one.

13.5.2 Type B, Type C and Bonded Heat Exchangers

Contrary to *Type A*, no remarkable improvements of the functionalized heat exchanger on the standard one have been recorded, both in terms of thermal and

Table 13.2 Channel test results (Test 1)

Flow rate (l/min)	Standard channel exchanged thermal power (W)		Functionalized channel exchanged thermal power (W)	
	Mean	Std. Dev.	Mean	Std. Dev.
1	127	±5	192	±5
1.6	165	±4	212	±5
2.1	178	±4	231	±11
3.1	182	±7	323	±9

pressure loss performance. Probably the results are related to the different coating technology that was used.

13.5.3   Test on Channel

The tests on channel were performed using water instead of oil. Between the first test and the others some weeks passed, due to a temporary unavailability of test rig. During this stand-by period, the channel remained full of water. The results obtained in the first test could be summarised as reported in Table 13.2.

In none of subsequent tests similar results were obtained. The thermal power exchanged with functionalized channel was quite similar to those obtained using standard channel. The reason of this behaviour could be explained thanks the corrosion test results. Corrosion behaviour of AA1050 aluminium alloy samples with super-hydrophobic coating has been assessed during salt fog and water/ethylene glycol mixture exposure at 80 °C, both in presence and in absence of chloride ions.

The survey evidenced that the super-hydrophobic layer is able to offer just a partial protection to the metallic sublayer when exposed to an aggressive environment like the salt fog one. Even after short exposure time, the layer loses its super-hydrophobic features because of local corrosion process arise on the layer defects that evolve towards deep penetrating forms. The same behaviour was noted using a less aggressive environment, like water/ethylene glycol mixture exposure at 80 °C, in presence of 200 ppm of chloride ions. Therefore, due to the fact that the tests with channels were performed using normal potable water, the different behaviour noted could be explained with a degradation of super-hydrophobic layer due to corrosion.

13.5.4   Customized Software to Design Tailored Heat Exchanger

We performed 180 CFD simulation to characterize the different fins geometry (both for oil, water and air), considering different temperatures and flow velocities. Once

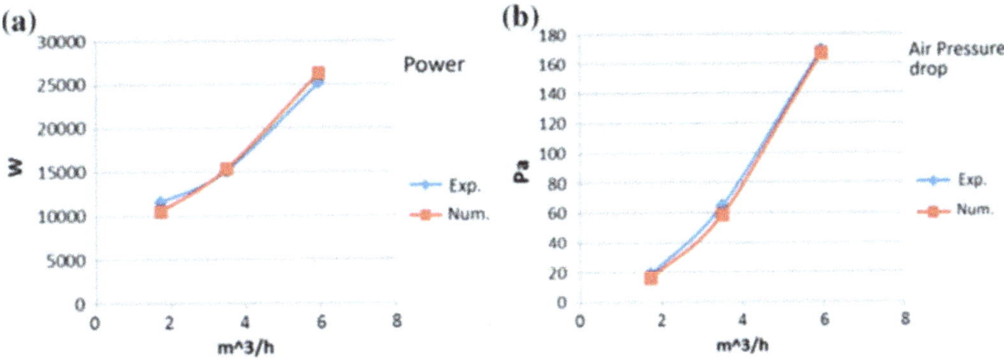

Fig. 13.8 Comparison between numerical (red square) and experimental (blue diamond) results. **a** Exchanged power; **b** Air pressure drop

the generic exchanger is reconstructed as a set of *Type 2* and *Type 5* sub-domains, it is necessary to use the results of the fluid dynamic analyses, in order to define the behaviour of each sub-domain in terms of WHTC (wall heat transfer coefficient) and pressure losses. The obtained data were interpolated, using cubic splines, to describe the subdomain behaviour in working conditions different from those simulated. This calculation was repeated for every sub-domain (*Type 2* and *Type 5*) of the exchanger as a function of the mean passage velocity and temperature. An identical approach is applied to determinate the pressure losses. The same procedure was also utilized for generating the curves of the density variation of the generic fluid that flows into the generic sub-domain as a function of the temperature. The results obtained using this approach show a good agreement between the calculated and tested performance, as showed in the Fig. 13.8. The mean error between the numerical and experimental results is ±6% for the exchanged power (Fig. 13.8a) and 3% for the pressure drop (Fig. 13.8b).

13.6 Conclusions and Future Research

This work investigated the potential of surface nano-coating of heat exchanger in order to improve the thermal efficiency of the component. We have designed and synthesized a functional layer, obtaining a good granulometric distribution of the suspension. We have demonstrated the feasibility of the coating processes and the adaptability of deposition techniques to the cross flow heat exchanger fins. Using dip-coating technique we were able to coat many finned surfaces. The results showed a contact angle of $133.1° \pm 7.1°$ using the hybrid methodology and $123.8° \pm 6.6°$ using the lubricant. It can be noticed the good hydrophobicity achieved with the hybrid method against the lubricant one on the liquid side fins. This means that one of the objectives of this project has been reached: to change completely the surface behaviour. Different coating methodologies were tested (pumping, empting, etc.) and none of them showed clearly the possibility to follow the actual production methodology (by brazing) of cross flow heat exchanger, which should be probably modified.

The influence of the new superhydrophobic and oleophobic surface on the heat exchanger behaviour was studied thanks to a dedicated test rig. The results show that, for the *Type A* heat exchanger, the heat transfer coefficient values, for volume flow rate values less than 7 l/min, are basically diffused nearby a mean value of 116.46 ± 38.73 (W/m^2 K) for standard heat exchanger and 176.47 ± 59.45 (W/m^2 K) for the functionalized one. The same results were not found in the other heat exchanger (*Type B, Type C, Bonded*) probably due to the different coating methodology used. A last test was performed using water instead of oil and considering a different heat exchanger geometry (channel). In this last test the first set of results obtained show a good improvement of heat exchange coefficient ($\approx$+20%), but the test performed after some weeks on the same component did not confirm this trend. The results become clearer after the corrosion test, were the survey evidenced that the super-hydrophobic layer is able to offer just a partial protection to the metallic sublayer when exposed to an aggressive environment like the salt fog one. Even after short exposure time, the layer loses its super-hydrophobic features because of local corrosion process arise on the layer defects that evolve towards deep penetrating forms. Finally, a dedicated modular software was developed. The results obtained comparing the experimental results with the numerical one show good agreement considering both the heat exchanger power and pressure drop. Further results related to this work are presented in the papers [24–30].

Finally, the exploitation of the scientific results led to the submission and approval of a research project titled HEAT (nano coated Heat ExchAnger with improved Thermal performances) funded under the call POR-FESR 2014–2020 of Regione Emilia Romagna. This project is coordinated by ISTEC-CNR in cooperation with FIRA spa and COMEX EUROPE srl.

Acknowledgements This work has been funded by the Italian Ministry of Education, Universities and Research (MIUR) under the Flagship Project "Factories of the Future - Italy" (Progetto Bandiera "La Fabbrica del Futuro") [31], Sottoprogetto 2, research projects "Customized Heat exchanger with Improved Nano-coated surface for earth moving machines Applications" (CHINA) and "Thermally Improved Heat Exchangers prototypes with Superhydrophobic Internal Surfaces: new assembly procedures and materials" (THESIS).

The authors express their gratitude to the following researcher, for the precious and effective collaboration, without which all this work would not be possible: Eng. Giuseppe Rizzo, Eng. Giorgio Paolo Massarotti, Eng. Luca Pastorello, Dr. Maria Giulia Faga, Dr. Giovanna Gautier, Dr. Federico Veronesi, Dr. Magda Blosi, Dr. Aurora Caldarelli, Dr. Giulio Boveri, Mrs. Guia Guarini.

The authors would like to thank for his indispensable technical and methodological contribution, FIRA S.p.A., which developed the heat exchangers and gave the necessary support to perform all the tests and to develop the different prototypes.

This work would not have been possible without the foresight of Prof. Roberto Paoluzzi, in memory of whom this paper was written.

References

1. Bonanno A, Raimondo MR, Zapperi S (2019) Surface nano-structured coating for improved performance of axial piston pumps. In: Tolio T, Copani G, Terkaj W (eds) Factories of the future. Springer
2. Rosengarten G, Cooper-White J, Metcalfe G (2006) Experimental and analytical study of the effect of contact angle on liquid convective heat transfer in microchannels. Int J Heat Mass Trans 49(21)
3. Lara JR, Holtzapple MT (2011) Experimental investigation of dropwise condensation on hydrophobic heat exchanger. Part II: effect of coatings and surface geometry. Desalination 280:363–369
4. Sheng Y, Quing-Ding W (2006) Experimental study on physical mechanism of drag reduction of hydrophobic materials in laminar flow. Chin Phys Lett 23(6):1634
5. Xiaokui M, Ding G, Zhang Y, Wang K (2009) Airside characteristics of heat, mass transfer and pressure drop for heat exchangers of tube-in hydrophilic coating wavy fin under dehumidifying conditions. Int J Heat Mass Transf 52:4358–4370
6. Jiaguo Y, Xiujian Z, Qingnan Z, Gao W (2001) Preparation and characterization of super-hydrophilic porous TiO2 coating films. Mater Chem Phys 68:253–259
7. Rothstein PJ (2010) Slip on Superhydrophobic Surfaces. Annu Rev Fluid Mech 42:89–109. https://doi.org/10.1146/annurev-fluid-121108-145558
8. Martell MB, Rothstein JP, Perot JB (2010) An analysis of superhydrophobic turbulent drag reduction mechanisms using direct numerical simulation. Phys Fluids (22), 065102. American Institute of Physics, https://doi.org/10.1063/1.3432514
9. Bsushan B, Wang Y, Maali A (2009) Boundary slip study on hydrophilic, hydrophobic, and superhydrophobic surfaces with dynamic atomic force microscopy. Langmuir 25(14):8117–8121. https://doi.org/10.1021/la900612s
10. Choi C, Moohwan K (2011) Wettability effects on heat transfer, two phase flow, phase change and numerical modeling. Ahsan A (ed) InTech. https://doi.org/10.5772/19512
11. Rosengarten G, Cooper-White J, Metcalfe G (2010) Experimental and analytical study of the effect of contact angle on liquid convective heat transfer in microchannels. Int J Heat Mass Transf 49:4161–4170
12. Liu K, Yao X, Jiang L (2010) Recent development in bio-inspired special wettability. Chem Soc Rev 39:3240–3255. https://doi.org/10.1039/b917112f
13. Tuteja A, Choi W, Mabry JM, McKinley GH, Cohen RE (2007) Designing super-oleophobic surfaces. Science 318:1618–1622. https://doi.org/10.1126/science.1148326
14. Bhushan B (2011) Biomimetic inspired surfaces for drag reduction and oleophobicity/philicity. Beilstein J Nanotech 2:66–68. https://doi.org/10.3762/bjnano.2.9
15. Rimondo M, Blosi M, Bezzi F, Mingazzini C (2011) Metodo per il Trattamento di Superfici Ceramiche per Conferire alle Stesse una Elevata Idrofobicità e Oleofobicità. Patent RM2011A000104
16. Raimondo M, Blosi M, Bezzi F, Mingazzini C (2012) Metodo di Trattamento di Superfici Metalliche per Conferire alle Stesse una Elevata Idrofobicità ed Oleofobicità. Patent RM2012A000291
17. Raimondo M (2012) Making super-hydrophobic building materials: static and dynamic behavior of nanostructured surface. In: IV ICC4 international ceramic conference, Chicago (USA), July 14–19 2012
18. Wong TS, Kang SH, Tang SK, Smythe EJ, Hatton BD, Grinthal A, Aizenberg J (2011) Bioinspired self-repairing slippery surfaces with pressure-stable omniphobicity. Nature 477(7365):443–447. https://doi.org/10.1038/nature10447
19. Vogel N, Belisle R, Hatton B, Wong T, Aizenberg J (2013) Transparency and damage tolerance of patternable omniphobic lubricated surfaces based on inverse colloidal monolayers. Nat Commun 4:2176. https://doi.org/10.1038/ncomms3176
20. Ingebrigtsen T, Toxvaerd S (2007) Contact angles of Lennard-Jones liquids and droplets on planar surfaces. J Phys Chem C 111:8518

21. Hong SD, Ha MY, Balachandar S (2009) Static and dynamic contact angles of water droplet on a solid surface using molecular dynamics simulation. J Coll Interf Sci 339:187
22. Park JY, Ha MY, Choi HJ, Hong SD, Yoon HS (2011) A study on the contact angles of a water droplet on smooth and rough solid surfaces. J Mech Sci Technol 25:323
23. Kakaç S, Liu H (1998) Heat exchangers selection, rating and thermal design. CRC Press LLC
24. Pastorello L, Bonanno A (2015) Application of Nano-structured coatings to heat transfer surface of heat exchangers. In: The fourteenth scandinavian international conference on fluid power, May 20–22, 2015, Tampere, Finland
25. Paoluzzi R, Bonanno A, Ferrari C, Martelli M (2014) Procedure for hydraulic oil heat exchanger performance improvement through integrated CFD analysis. Int J Fluid Power 15(3):169–180
26. Raimondo M, Blosi M, Caldarelli A, Guarini G, Veronesi F (2014) Wetting behavior and remarkable durability of amphiphobic aluminum alloys surfaces in a wide range of environmental conditions. Chem Eng J 258:101–109. https://doi.org/10.1016/j.cej.2014.07.076
27. Caldarelli A, Raimondo M, Veronesi F, Boveri G, Guarini G (2015) Sol–gel route for the building up of superhydrophobic nanostructured hybrid-coatings on copper surfaces. Surf Coat Technol 276:408–415. https://doi.org/10.1016/j.surfcoat.2015.06.037
28. Motta A, Cannelli O, Boccia A, Zanoni R, Raimondo M, Caldarelli A, Veronesi F (2015) A mechanistic explanation of the peculiar amphiphobic properties of hybrid organic−inorganic coatings by combining XPS characterization and DFT modeling. ACS Appl Mater Interfaces 7:19941–19947. https://doi.org/10.1021/acsami.5b04376
29. Malavasi Veronesi F, Caldarelli A, Zani M, Raimondo M, Marengo M (2016) Is a knowledge of surface topology and contact angles enough to define the drop impact outcome? Langmuir 32:6255–6262. https://doi.org/10.1021/acs.langmuir.6b01117
30. Raimondo M, Veronesi F, Boveri G, Guarini G, Motta A, Zanoni R (2017) Superhydrophobic properties induced by sol-gel routes on copper surfaces. Appl Surf Sci 422:1022–1029. https://doi.org/10.1016/j.apsusc.2017.05.257
31. Terkaj W, Tolio T (2019) The Italian flagship project: factories of the future. In: Tolio T, Copani G, Terkaj W (eds) Factories of the future. Springer

Part VI
Advanced-Performance Factory

Chapter 14
Surface Nano-structured Coating for Improved Performance of Axial Piston Pumps

Antonino Bonanno, Mariarosa Raimondo and Stefano Zapperi

Abstract The work starts from the consideration that most of the power losses in a hydraulic pump is due to frictional losses made by the relative motion between moving parts. This fact is particularly true at low operating velocities, when the hydraulic lift effect must be able to maintain a minimum clearance in meatus to limit the volumetric losses. The potential of structured coatings at nanoscale, with super-hydrophobic and oleophobic characteristics, has never been exploited before in an industrial application. The work studies the potential application of nano-coating on piston slippers surface in a real industrial case. The aim is to develop a new industrial solution to increase the energetic efficiency of hydraulic pump used in earthmoving machines. The proposed solution is investigated using a dedicated test bench, designed to reproduce real working conditions of the pump. The results show a reduction of friction coefficient while changing working pressure and rotation velocity.

14.1 Scientific and Industrial Motivations

Hydraulic pump efficiency is one of the main concerns in trying to improve the overall efficiency of mobile off-road machines and industrial plants. Axial piston pumps and motors are mainly used in heavy-duty applications; the improvement, albeit marginal, in overall efficiency may significantly impact the global efficiency of the machine. The overall efficiency of energy conversion from fuel to unit of production is as low as 18% on average, and the largest part of the losses are in the internal combustion engine (~65%). The efficiency of axial piston pumps can be

A. Bonanno (✉)
CNR-IMAMOTER, Istituto per le Macchine Agricole e Movimento Terra, Ferrara, Italy
e-mail: a.bonanno@imamoter.cnr.it

M. Raimondo
CNR-ISTEC, Istituto di Scienza e Tecnologia dei Materiali Ceramici, Faenza, Ravenna, Italy

S. Zapperi
CNR-IENI, Istituto per l'energetica e le Interfasi, Milan, Italy

improved by means of a surface treatment technology mimicking what in the nature is well known as the *Lotus effect*. The application of this technology to a real product, with measurable objectives, would represent a result that can be quickly exported to a new generation of industrial products.

In general, structuring of nanoscale surfaces has an impact on their wetting and therefore fluid interaction to obtain high levels of repulsion against water (superhydrophobicity) or oils (oleophobicity) to which they come into contact. Many living organisms in nature have superhydrophobicity or oleophobicity (Lotus leaves, Nephentes plant, etc.), providing a valid chemical-structural model for the production of synthetic surfaces of industrial interest. The literature presents many examples of methodologies, or techniques, for modulating liquid-surface interaction through surface roughness control, which induces the so-called, Wenzel state (partial wettability) to Cassie-Baxter's level that can guarantee high values of contact angle (CA) between liquid and surface and, hence, high repulsion [1]. It is known that a nanometric surface roughness (i.e. nanostructures with appropriate morphology-like pillars, protuberances, etc.) causes that the liquid in contact with the surface is not able to penetrate these discontinuities and is forced to remain on top of them. In this case, the CA between liquid and surface reaches very high values (greater than 150°), ensuring very low wettability. CA values also depend on other physical parameters, including surface liquid tension and surface energy of the material, which is in turn related to surface chemistry suitably combined with geometric factors. The technology proposed in [2, 3] is able to produce very high CA surfaces (up to 170°) and extremely low surface energy (<1mN/m) on different types of substrates (ceramics, metals, copper alloys and aluminium).

The aim of this work is to develop a new generation of axial piston pumps by investigating the potential of a novel nanostructured coating, featuring superhydrophobic and oleophobic behaviour that can provide better performance to the interface between lubricated moving parts of the hydraulic pumps and motors. This requires to study at microscopic level the influence of surface roughness on the static or dynamic wetting behaviour of fluids by means of numerical simulations. A key issue is to find the optimal properties in terms of nano-patterning for friction minimization. Therefore it is needed to investigate how and to what extent the superhydrophobicity and oleophobicity of the coatings depends on their specific structure at atomic scale level. In summary, this work set the following goals:

- Development of a nanostructured coating able to drastically change the fluid surface interaction, specifically reaching a contact angle greater than 150° for water and 110° for hydraulic oil, reducing the CAH (contact angle hysteresis) towards water to values lower than 5°, and reducing the surface energy to values lower than 10 mN/m.
- Reduction of friction coefficient (10/15% at least) in parts of a hydraulic pump subjected to relative motion.
- Characterization of friction and wear behaviour of the developed coat.
- Development of a computational model of friction and lubrication of confined layer.

- Development of a hydraulic pump prototype able to work using the nanocoated components.

The chapter is organized as follows. Section 14.2 gives an overview of the state of the art. Section 14.3 presents the steps of the proposed approach to functionalize the surfaces of moving elements. Section 14.4 give the details of the developed methodologies and tools, whereas the experiments and the analysis of the results is addressed in Sect. 14.5. The conclusions are drawn in Sect. 14.6.

14.2 State of the Art

The main causes for efficiency reduction in hydraulic components are the lack of adequate drain flow to maintain lubrication between moving parts and the lack of hydrodynamic lift to slippers and delivery port plate [4]. Considering a performance curve of a typical open circuit, variable displacement, axial piston pump, the critical areas can be easily identified (for a given rotational speed) at low delivery pressure. On the other hand, if the rotational speed is changed while keeping fixed the delivery pressure, then it can be demonstrated that low speed conditions are the most critical. Some attempts have been made in the past to overcome these limitations by using ceramic materials [5] or special metallic coatings [6], but state of the art commercial solutions still rely on surface finishing and geometric design. The potential benefits for hydraulic pumps and motors, deriving from the application of the approach proposed in this work, have been studied only recently.

The improvement of the energy performance (friction reduction) of the slipper in axial piston pumps with inclined plate has been addressed by several authors [4, 5, 7] using experimental measures or through the use of computational fluid dynamics (CFD) codes. Most of these used Reynolds' simplified lubrication or a complete 3D Navier-Stokes approach [8]. All models are based on the existence of an anti-slip bond between fluid and solid. On the contrary, this paper removes such assumptions thanks to the potential of a new surface geometrical configuration that is able to generate a lubrication micro-channel with a slipping length of the same order of magnitude of the geometric interface [1, 4, 5]. The movement of the fluid, after the application of an external force, is influenced by the dynamic behavior of the sliding surfaces. The resistance and lifting forces generated by the fluid are also a function of the dynamic interaction between surface and fluid. There are very few scientific articles presenting and describing these phenomena. Some [5] consider the field of laminar motion, others [7, 8] turbulent conditions. However, among the reported studies, no one applies the concept to real application cases and no existing document links the effect of reduction of friction resistance, by means of coating, to the efficiency of the components.

It is well known from the literature that structural modifications can be induced by controlling the roughness degree of surfaces, so let wetting phenomena moving from an homogeneous state (termed as *Wenzel state*) to a discontinuous one (termed as *Cassie-Baxter state*), this latter characterizing low wetting conditions [9]. When roughness increases and nanostructures (i.e. pillars, bumps, etc.) develop, fluids droplets falling on a surface are forced to stay on the top of asperities, giving rise to high, stable hydrophobicity which, in some cases where surface chemistry is adequately controlled, can also mean high oleophobicity. The air trapped on the rough surface, in fact, makes the fluid droplet bead up at very high contact angle, which in the case of superhydrophobic surfaces can reach values greater than 150° or greater than 120° for oleophobic ones. The modification of chemistry, and hence of surface energy, is also necessary to improve repellence to fluids, especially when the target is the repellence towards oils. Surface energy and surface tension, in fact, are the other two fundamental parameters for phenomena occurring at interfaces of a solid-liquid-air system. Due to the quite low values of surface tensions of oils or alkane (much lower for example than water), the design of oleophobic materials requires to lower, as much as possible, the solid surface energy. Oleophobic performance described in the literature [10, 11] correspond to surface energy values lower than 10 mN/m. In this context, the preliminary results concerning the technology applied in this work, revealed to be more efficient, producing materials with surface energy lower than 1 mN/m (substrates: ceramics, metals or alloys) [12–14].

The new superhydrophobic and oleophobic nano-coated surfaces need to be studied, also, from a computation modelling point of view. The nanoscale properties of boundary lubrication were modelled using large scale molecular dynamics (MD) simulations, in order to obtain a complete multiscale description of the frictional dissipation. MD simulations are really important to estimate the friction and lubrication phenomena, considering the new coated surfaces in a mathematical model of the pump. The numerical boundary lubrication model was obtained performing molecular dynamics (MD) simulations of the frictional properties of lubricant molecules in confined layers with superhydrophobic and oleophobic nanopatterned surfaces. MD simulations follow the dynamics of all atoms in controlled computational experiments where the bulk properties, interface geometries, mechanical loading and inter-particle interactions can be varied to explore their effect on friction. MD simulations are a standard methodology, largely exploited in the literature [15–17] to address wetting mechanisms and validate empirical and theoretical models. If the considered system contains enough liquid molecules, MD is indeed in principle capable to provide information about macroscopic parameters such as the density, surface tension, viscosity, flow, static and dynamic contact angle which eventually can be compared with experimental results.

14.3 Problem Statement and Proposed Approach

In a common hydraulic piston pump there is a set of parts that are in relative motion (see Fig. 14.1): the pistons move alternately inside the cylinder block, the slippers slide on the swashplate, the cylinder block rotate around his axis and slides on the distribution plate. Herein, the attention is focused on the slippers to reduce the

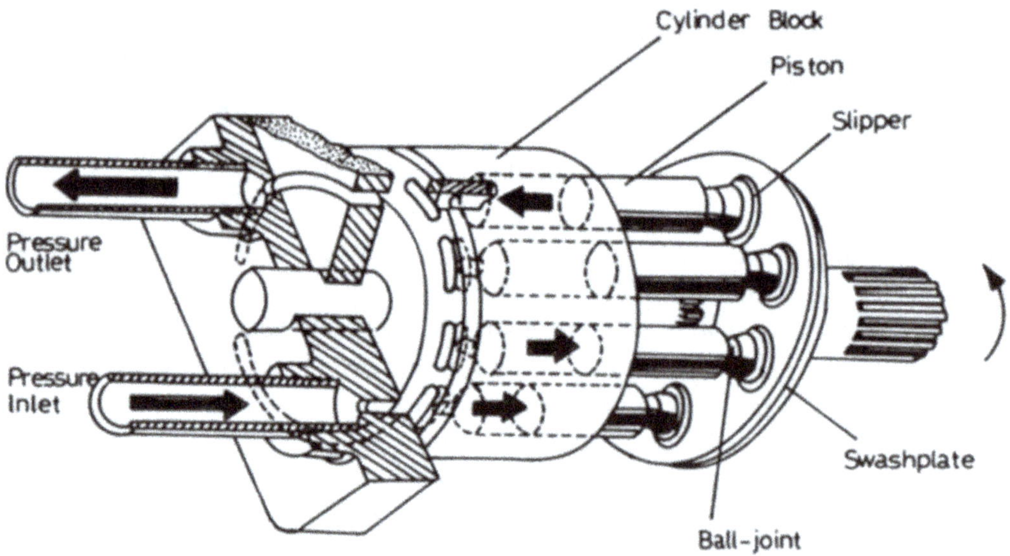

Fig. 14.1 Hydraulic piston pump

possible coating problem (geometry dimension, possible edge phenomena), and also considering the lubrication regime. The lubrication regime that is established between the slipper and the swash plate is hydrostatic, since the high pressure oil, which flows into the oil chamber under the slipper through the shock absorber orifice, generates a hydraulic force able to separate the two elements and allow their movement with low energy expenditure. The design of the slipper is focused on the balance between the hydraulic separation force and the pressure force acting on the slipper. The pressing force acting on the slipper changes continuously during the normal working condition of the pump, thus modifying the equilibrium point between the acting force and the reaction created by the lubricating film. Nevertheless, it is possible to maintain the stability of the whole system thanks to the damping effect created by the orifice. The traditional lubrication theory [18–21] shows that, between the slipper and the swashplate there is a lubricant layer. The velocity profile of this layer shows a velocity zero near the slipper (boundary layer) and a velocity equal to the swash plate linear velocity near it. In this work it was investigated the possibility to reduce (eliminate) the part of the layer where the flow velocity reduces to zero, by coating the slipper with a superhydrophobic/oleophobic layer, which enables the boundary layer to have a velocity different from zero although it moves on a fixed surface.

The problem was faced with a multi-objective approach. The first step (Sect. 14.4.1) was dedicated to the design and synthesis of functional layer, i.e. the design and synthesis of inorganic and hybrid organic-inorganic coatings able to reduce or modulate the wettability of slippers against fluids. In this context, the work was firstly focused on the synthesis of ceramic oxides nanoparticles (Al_2O_3, ZrO_2, etc.) in the form of stable colloids or nanosuspensions in different media (i.e. alcohols or water), getting a high control degree on phases, particle size and composition. Also specific synthetic approaches were developed (i.e. sol-gel), involving particle nucleation directly from precursors present into the fluid, in the meantime avoiding particle precipitation and ensuring a better control over physical variables of suspensions.

The possibility to apply the developed oxide on the slipper was addressed in the second step of the approach, by assessing the adaptation of deposition techniques to the to specific piece design (Sects. 14.4.2, 14.5.1). In this context, coating deposition by simple immersion techniques, such as dip-coating or spin-coating in controlled conditions, was investigated to produce homogeneous functional surfaces, to easily control the layer thickness in the order of nanometers, and to promote the formation of micro- and nanostructures.

The assessment of tribological performance of the obtained samples was the third step of the approach (Sects. 14.4.3, 14.5.2). The new surface treatment technology should assure the reduction of friction but, at the same time, the pump ability to work in the expected operative conditions, so it is necessary to have a complete tribological and mechanical characterization of functional layers. A non-contact profilometry, a detector of acoustic emission and a nanoindentator were used for this activity. The first instrument was used to obtain the topography of the metals and alloys coated. The acoustic detector gave the possibility to know the strength between the substrate and nanostructured coating. The second instrument allowed to evaluate the coating hardness of the surface layers.

The hydrostatic lubrication uses the fluid pressure to separate two surfaces that have a sliding motion. If the lubricating film thickness does not suffice to separate the sliding surfaces completely then wear takes place. In order to know how, the new superhydrophobic and oleophobic surface modified the hydrodynamic lubrication in comparison to the current slipper, a torque meter was used to evaluate the absorbed torque considering the different lubricating condition (low and high pressure, low and high velocity). Pressure sensor and flow meter were used to know how much flow is necessary to ensure a good lubrication conditions at different pressure. These pieces of information are relevant from a fluid power point of view, due to the possibility to evaluate the oil flow used to maintain the hydrostatic lubrication. At the same time the oil pressure and the flow necessary to guarantee the meatus using the nanocoated surface were measured and compared to that measured using the traditional surface. In addition, the absorbed torque was acquired to have information about the mechanical efficiency.

The final part of the work was dedicated to the MD simulations (Sects. 14.4.4, 14.5.3) to study, also from the computational point of view, the lubrication conditions obtained using the nanosctructured coating. The problem was faced by employing different available simulation packages and tools to build a powerful and efficient computational framework specifically aimed at performing extended MD simulations and hybrid multiscale modelling.

14.4 Developed Technologies, Methodologies and Tools

14.4.1 Design and Synthesis of Functional Layer

The first part of the work was focused on the sol-gel synthesis of ceramic oxide nanoparticles suspended in different solvents (water and alcohols) which can be used to create coatings able to modify the wettability of materials. Superhydrophobic (water repellent) and superoleophobic (repellent towards fluids with a lower surface tension than water) surfaces on different materials, particularly on ceramics, metals and alloys, were developed starting from two patents [2, 3]. Moving from the ever-growing literature [15] the activity focused on the ability to produce precursors of functional coatings through water-phase synthesis (instead of an alcohol-phase synthesis) in order to reduce environmental impact and move towards a more favourable transfer to the industrial scale in terms of costs and safety.

The precursor of the ceramic oxide is aluminium tri-sec-butoxide; in water solvent, first it undergoes hydrolysis, then it forms Al-O-Al bonds during a peptization process. To avoid a too fast hydrolysis of the butoxide with $Al(OH)_3$ precipitation, a chelating agent was dissolved in water (ethyl acetoacetate) before adding the butoxide; the chelating agent substitutes the butoxide, slowing down the hydrolysis. In order to obtain a good granulometric distribution of the suspension, which has to stay below 100 nm to lead to the formation of nanoparticles and avoid aggregates, a HNO_3 solution is gradually added to the mixture; protons bond to and charge the surface of forming particles, stabilizing them and avoiding aggregation. After 24 h at 70 °C, a transparent sol is obtained (pH = 3.64). A scheme of the reaction is reported in Fig. 14.2a.

14.4.2 Slippers Coating

After the sol preparation, the activities were focused on the slippers coating. Before the functionalization, the samples were sandblasted to create a micro-scale roughness which promotes adhesion by the coating and provides better mechanical resistance [16]. Usually, sandblasted samples show roughness of about 4–5 μ. Figure 14.2b shows the dip coated used to coat samples with different morphology and size. The methodology is easily transferrable to industrial scale. The slipper samples were dipped in the sol in two different configurations: with horizontal axis and with vertical axis.

The dipping conditions of the skid in the sol were the following:

- Dipping speed: 2 mm/s.
- Time standing still in the sol: 5 s.
- Withdrawing speed: 2 mm/s.

After emersion of the samples, the coatings required the following treatments:

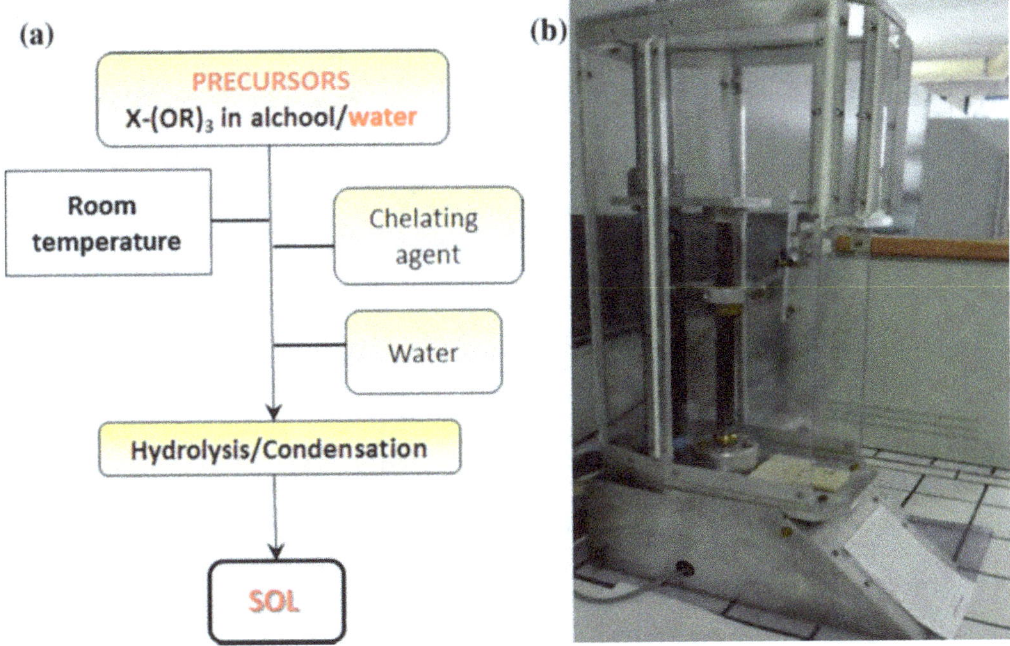

Fig. 14.2 **a** Sol-gel synthesis methodology; **b** dip coater

- First thermal treatment: 400 °C for 60 min in muffle furnace.
- Boiling in water for 30 min to create the so-called flower-like boehmite AlO(OH) nanostructure.
- Second thermal treatment: 400 °C for 10 min in muffle furnace to reinforce boehmite film.
- Dipping in a fluoroalkylsilane solution (FAS—Dynasilan SIVO Clear EC di Evonik) (withdrawing speed 2 mm/s, time standing 2 min) to lower surface energy.
- Final thermal treatment at 150 °C for 30 min to allow crosslinking by the fluorinated layer and reinforce the film.

All experimental settings were optimized to obtain the best wettability performance for the coatings. Furthermore, it was verified that the samples were dimensionally stable at the temperatures used during thermal treatments. Functionalized samples were then characterized in terms of wettability with water and lubricant Arnica 46, representative of the oils used in axial pumps.

14.4.3 Slipper Test Bench

A slipper test bench was designed and realized to test the coated slippers without needing to use a pump. Figure 14.3 shows the test bench designed to measure the friction reduction obtained using the superhydrophobic/oleophobic layer. It uses an inverse working principle compared to the real hydraulic piston pump. In fact, in the pump, the piston block rotates, moved by the transmission axle, dragging the piston,

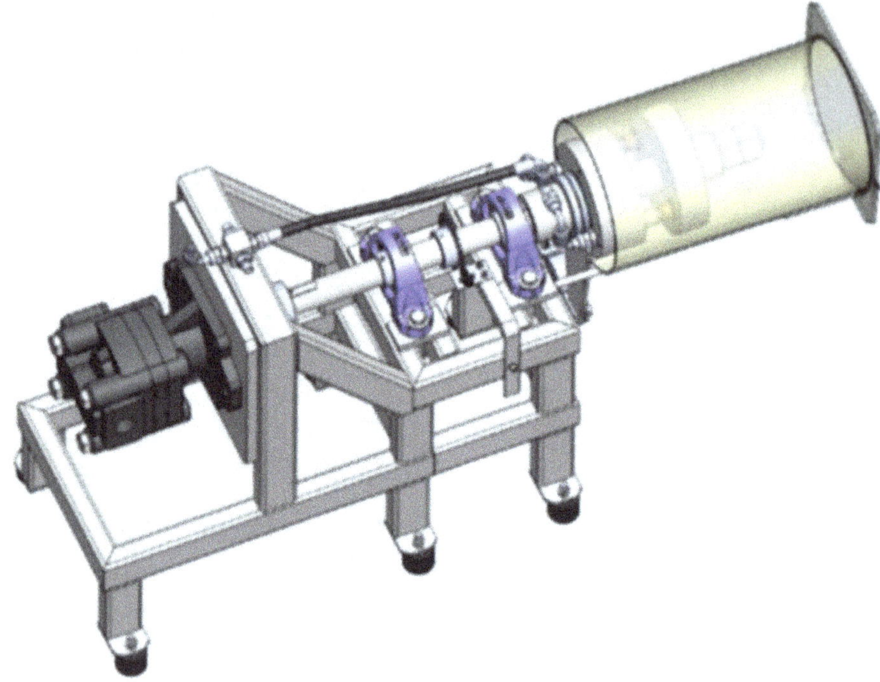

Fig. 14.3   Slipper test bench

and consequently the slippers, in his rotation. The slippers slip on the plate, which is fixed. In the test bench, due to realization problems, the principle was inverted and the plate rotates while the slippers have only the possibility to move axially on the plate. From a kinematic point of view, the movement is the same: a relative sliding between the slippers and the plate.

The frame is made by tubular profile structural steel (FE 360B). The elastic coupling, constituted by a double wheel hub and a star in polymeric material, allows the transfer of mechanical power to load. The dimensions and materials have been chosen according to the geometric dimensions of the shaft and taking in consideration that the kinematic system is subjected to a mechanical locking. The transmission shaft was made of 39NiCrMo3 tempered steel, thus guaranteeing good tenacity (typical in the tuning phase and in the case of accidental failure). The ball bearings mounted on rigid supports have been chosen in order to guarantee a maximum rotating speed of 3000 rpm. The advantages of this configuration are many: the possibility to see the slipper behaviour, the possibility to measure the friction torque and then the friction coefficient, the possibility to easily change the tested prototype without disassemble a complete pump. The results of the tests performed using the designed test bench are showed in the next section.

14.4.4 MD Simulation

Another important task is the study, at a microscopic level, of the influence of surface roughness on the static/dynamic wetting behaviour of fluids by means of numerical simulations. Our aim was to understand how and to what extent the superhydrophobicity/oleophobicity of the coatings depends on their specific structure at atomic-scale level. The MD (molecular dynamic) study was approached within the framework of a basic standard solid-fluid model, which allows us to outline and understand the main features of the fluid interaction with a nano-patterned surface in a simple and effective way. We modelled solid-solid and solid-fluid interaction with the Lennard-Jones pair potential reported in (14.1) which is suitable for neutral atoms or molecules and is composed of a steep short-range repulsive term and a smoother long-range attractive one (van der Waals type).

$$
E_{LJ} = \begin{cases} 4\varepsilon\left[\left(\frac{\sigma}{r}\right)^{12} - \left(\frac{\sigma}{r}\right)^{6}\right] & r < r_c \\ 0 & r > r_c \end{cases} \tag{14.1}
$$

The relevant quantities are the depth of the potential well (ε), the finite distance at which the interparticle potential is zero (σ), the distance between the particles (r), and the cut-off distance at which the potential vanishes (r_c). We have two sets of parameters, one for the solid-fluid interactions and one for the fluid-fluid interactions, but in the following pictures we used dimensionless units and we just had to deal with two parameters $\varepsilon_* = \varepsilon_{FS}/\varepsilon_{FF}$ and $\varepsilon\sigma_* = \sigma_{FS}/\sigma_{FF}$. In this model the ε_* parameter represents an effective interaction and accounts for all the aspects of the solid-fluid interactions due to material properties and chemical treatments. This is the parameter to be varied to outline in general the role of surface chemical condition in frictional dissipation. The role of nanopatterning from a geometrical point of view is instead studied by varying the roughness of the solid surfaces, i.e. their geometrical texture in terms of vertical deviations from the flat condition (Fig. 14.4).

We implemented also MD simulations of a fluid confined between walls with rough surfaces in order to study how roughness affects the friction at the boundaries. A shear (Couette) flow in the channel was induced by moving the upper wall with velocity v_0 and the lower wall with velocity $-v_0$. We vary the L_J parameters in a range such that the velocity profile is always $\approx$ linear. In the presence of nanostructured coatings, the interaction between fluid and wall could be described as *sliding*, i.e. a situation in which the value of the tangential component of the speed seems to be different from that of the solid surface. This behaviour is described, in the simplest way, assuming that the tangential force per unit area exerted on the solid surface is proportional to the sliding speed, i.e. $\sigma_{xz} = k v_{slip}$ where x is the direction of the flow and z the orthogonal one. Combining this with the constitutive equation for the bulk Newtonian fluid $\sigma_{xz} = \eta\vartheta_z v_x$ one gets the so-called (scalar) Navier boundary condition $v_{slip} = \eta/k\vartheta_z v_x \equiv \delta\eta\vartheta_z v_x$. The last equality defines the slip length $\delta = \eta/k$ which represents the distance inside the solid to which the velocity

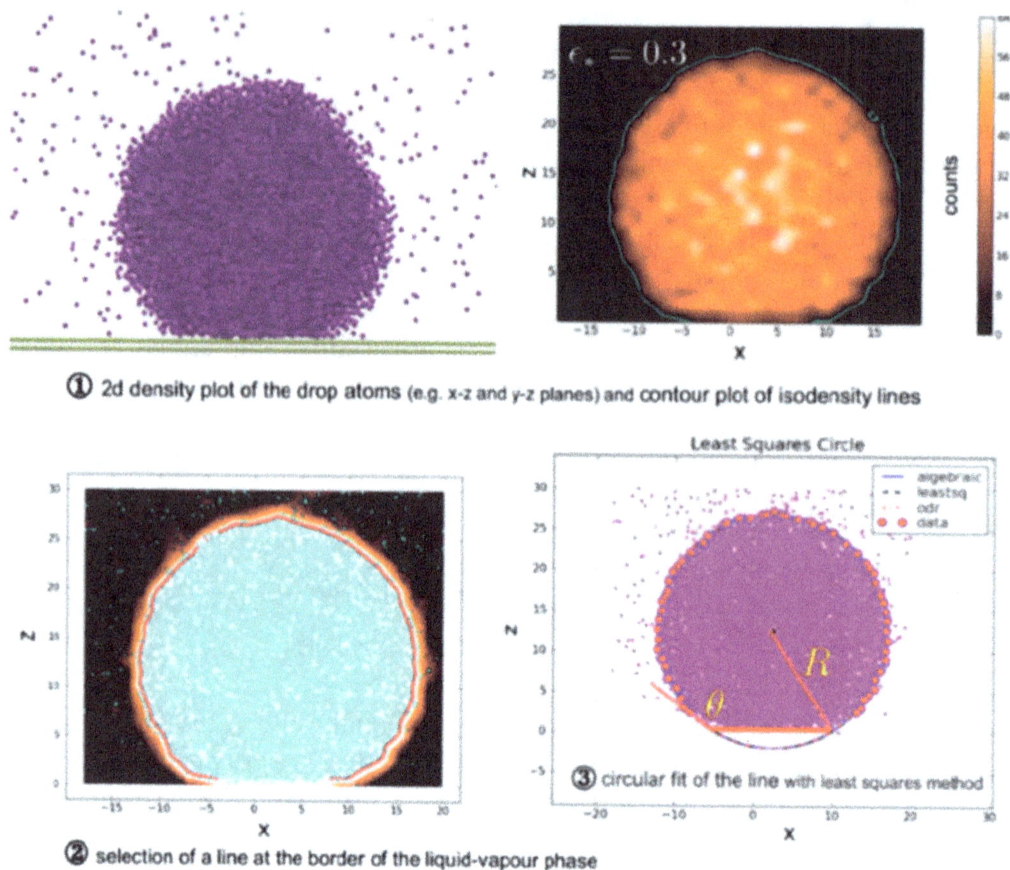

Fig. 14.4 Summary of the procedure adopted to extract the contact angle from MD results

has to be extrapolated to reach zero. This slip length represents a measure of the friction of the fluid at the solid walls and is thus influenced by the surface roughness, which is expected to reduce friction in the same fashion as it reduces the contact angle. These results basically confirm what we already found in the case of the static contact angle and conform to the existing literature: the rougher the surface (at fixed interaction potential) the better the slippage, i.e. less friction.

14.5 Experiments

Several specimens were developed during the work. The state of the art shows the possibility to coat laboratory samples having flat (or approximately flat) surface, without edges, undercut, or abrupt change of section. However, in this work the attention was focused on a real industrial component, i.e. the slipper of a hydraulic pump. Therefore, the greater part of specimens are functionalized (coated) slippers which were developed to test and investigate the superhydrophobic/oleophobic layer performance, the different coating (with and without ceramic substrate), and the

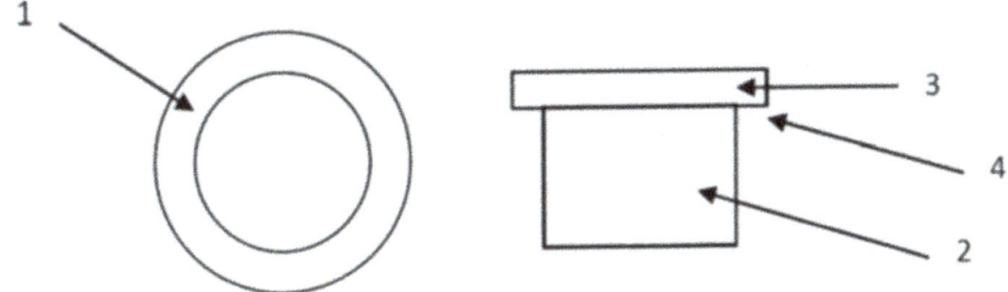

Fig. 14.5 Mapping of different *zones* were the CA characterization were performed

different slippers geometries. Furthermore, a complete pump was realized and tested in a dedicated test bench.

The slipper shows a complex surface from the coating perspective (see Fig. 14.5).

14.5.1 Fluid Surface Interaction

The first prototypes were developed starting from the slipper geometry actually used in the industrial hydraulic piston pump. No modification was introduced except the application of nanostructured coating on the sliding slipper surface. Eight couples of functionalized slippers were made by coating the sliding surface.

After the functionalization of the surfaces, the performance was characterized by measuring the WCA (water contact angle), the CAH (contact angle hysteresis), the SE (surface energy), and the OCA (oil contact angle). The results obtained can be summarized as follow:

- The mean WCA of standard sample, measured in *zone 1* and *2* were respectively $74.8° \pm 1.0°$ and $88.9° \pm 1.1°$.
- The mean WCA of functionalized samples, coated in horizontal position, was $142.6° \pm 2.6°$ degree in zone 1, $122.0° \pm 5.1°$ in zone 2, $120.8° \pm 2.7°$ in *zone 3* and $124.4° \pm 3.0°$ in *zone 4*.
- The mean WCA of functionalized samples, coated in vertical position, was $158.8° \pm 2.8°$ degree in zone 1, $123.5° \pm 3.0°$ in zone 2, $134.7° \pm 1.6°$ in *zone 3* and $112.2° \pm 1.0°$ in *zone 4*.

The results demonstrate that the nanostructured coating, developed as explained in the previous section, is able to change the surface wettability reaching contact angle typical of superhydrophobic layer ($CA > 150°$). The vertical position is more efficient in term of coating performances. The developed methodology is able to drastically modify the slipper working surface (*zone 1* in Fig. 14.5) contact angle from $74.8°$ to $158.8°$. The contact angle hysteresis (CAH), using water as the measurement fluid, was measured by needle technique (droplet volume 2 μL). The smaller the hysteresis value, the better the ability of a surface to make a drop slip (better dewetting properties [1]). CAH values measured on treated samples were:

- $4.54° \pm 2.04°$ for horizontal coating configuration
- $1.90° \pm 1.50°$ for vertical coating configuration.

The sample dipped with vertical axis proved to be more performing also in terms of hysteresis. The static contact angle was measured by diiodiomethane. Evaluations were carried out on *zones 1* and *2* both on the treated samples and on the sanded sample. Using the Owens-Wendt-Rabel-Kaelble algorithm it is possible to calculate the surface energy in the different points of the analysed samples. The measurements are made by comparing the results obtained with diiodiomethane (surface tension 50.8 mN/m) with those obtained with water. The results obtained showed the efficiency of nanostructured coating: the SE of coated sample (vertical coating configuration) results 0.47 ± 0.05 (mN/m) against a surface energy of 38.87 ± 1.00 (mN/m) measured for the standard sample. The comparison of the results obtained on the coated slipper with those obtained from the standard sample shows that the surface energy is reduced to values close to zero. Of particular relevance is the fact that the lowest SE value (0.47 mN/m) was measured on *zone 1* of the sample, the working face normally used by the slipper. Low surface energy is a fundamental parameter, together with surface nanostructure, to obtain oleophobicity [9, 10, 22], which is the repellence towards low surface tension liquids (at least lower than water). The CA measurement performed considering ARNICA 46 hydraulic oil as fluid showed again the high repellence obtained thanks to the nanostructured coating:

- CA with ARNICA 46, standard sample: $16.7° \pm 2.3°$
- CA with ARNICA 46, functionalized sample: $123.8° \pm 11.8°$.

14.5.2   Tribological Performance

The following tests were performed to assess the tribological performance:

- The scratch test in order to determine the mechanical strength adhesion, intrinsic cohesion of coating. The tests were performed with a CSM Revtest machine, applying a progressive normal load from 1 to 20 N. The total length of scratch was 5 mm, the indenter speed 10 mm/min.
- The pin on disc test in order to evaluate the wear and friction characteristics of slippers both in dry and lubricated environments. This test was performed with a CSM tribometer. The test conditions were: load 10 N, speed 20 cm/s, room temperature, pin 18MnCr5, pin diameter 2.4 mm, laps 5000.
- The surface analysis by means of a non-contact profilometer Taylor Hobson Talysurf CCI 3000Å.

A total of eight pairs of samples were tested: 2 standard (sand blasted and not sand blasted) and 6 coated samples. The coated slipper shows quite the same average roughness (1.5–2.5 µm) with only one that shows a lowest surface roughness ($\approx$0.8 µm).

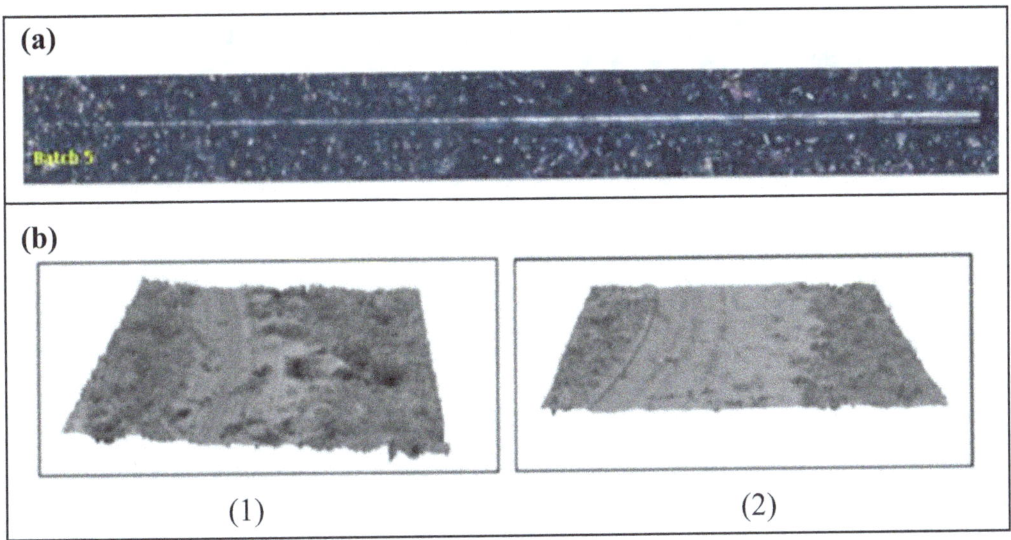

Fig. 14.6 **a** Scratch track of slipper coated with batch 5; **b** Wear test tridimensional profiles (1) batch 4; (2) batch 5

As for the adhesion results, no substantial differences, in terms of cracks typologies were found between the functionalized slippers. Nevertheless, comparing the acoustic emission, a lower value was found for the batch 1 and 6. This phenomenon suggests that lower microcraks appeared along the scratch channel.

The lubricant reduces the friction coefficient in all cases:

- WET TEST: average friction coefficient 0.145 ± 0.035
- DRY TEST: average friction coefficient 0.233 ± 0.072.

As for the surface functionalization, it does not reduce the friction coefficient. However, it must be stressed that the test cannot simulate a hydrostatic lubrication condition, as in hydraulic piston pump. An increment of the wear resistance is observed for the slippers coated with batch 2, 4 and 5. In fact, at the end of the tribological test, it was difficult to measure the track depth. SEM-EDS analysis supports tribological results. In Fig. 14.6 a summary of results is showed.

Different pairs of standard and functionalized samples were tested in a dedicated test bench, considering a constant rotation velocity (700 rpm) and different pressures from 10 to 50 bar. Figure 14.7 shows the mean friction coefficient both for standard and functionalized samples. The friction coefficient reduction is between the 40% (at lower pressure) and 25% (at higher pressure),

Other six pair prototypes were made to test the ceramic substrate influence on the slipper behaviour. In fact, it is possible to coat the slipper with fluoroalkylsilane solution (FAS—Dynasilan SIVO Clear EC di Evonik), to reduce the surface energy, without performing a complete surface functionalization (ceramic substrate + FAS coating). The coating with only FAS without the ceramic substrate, although giving comparable results in term of friction reduction in a normal (not endurance) test, do not give as good performance in term of duration, being abraded after 20 h.

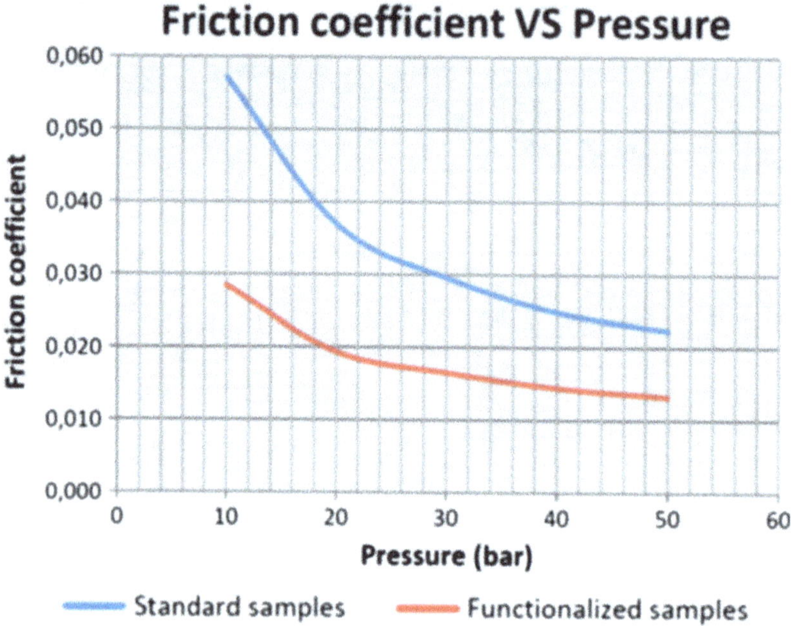

Fig. 14.7 Mean friction coefficient for standard (blue curve) and functionalized (red curve) samples

14.5.3 Testing the Computational Model

We started by analysing the static wetting mechanism and simulating liquid drops at equilibrium on different surfaces. To extract the contact angle θ from the MD result we adopted a simple fitting procedure of the density profile of the drop at equilibrium (Fig. 14.4). We then considered also a dynamical framework, with simulation of a liquid between moving flat/rough walls. We considered the influence on the contact angle due to geometrical nano-scale effects and implemented MD simulation with rough surfaces. We generated randomly rough surfaces with controlled root-mean-square roughness and correlation length and look at the variation of the contact angle as a function of roughness at fixed interaction strength ϵ. Figure 14.8 shows snapshots of drops at equilibrium for different values of the surface roughness compared to the flat case: the key feature is that the contact angle increases for increasing roughness. This corresponds to the well-known fact that surface roughness benefits the fluid-repellent properties when they are present [16, 17, 23].

Figure 14.9 shows the curves representing the contact angle for different surface roughness. It can be noticed that the contact angle increases for increasing roughness.

Regarding the MD simulations of a fluid confined between walls with rough surfaces, we obtained some velocity profiles (not showed) resulting from our simulation for varying roughness. These results confirm that the rougher the surface (at fixed interaction potential), the better is the slippage, i.e. less friction. To study the role of surface conditions in frictional dissipation from a more chemical point of view, we implemented simulations of the static contact angle on a flat surface and we looked its variation as a function of the strength of the solid-fluid interaction. The

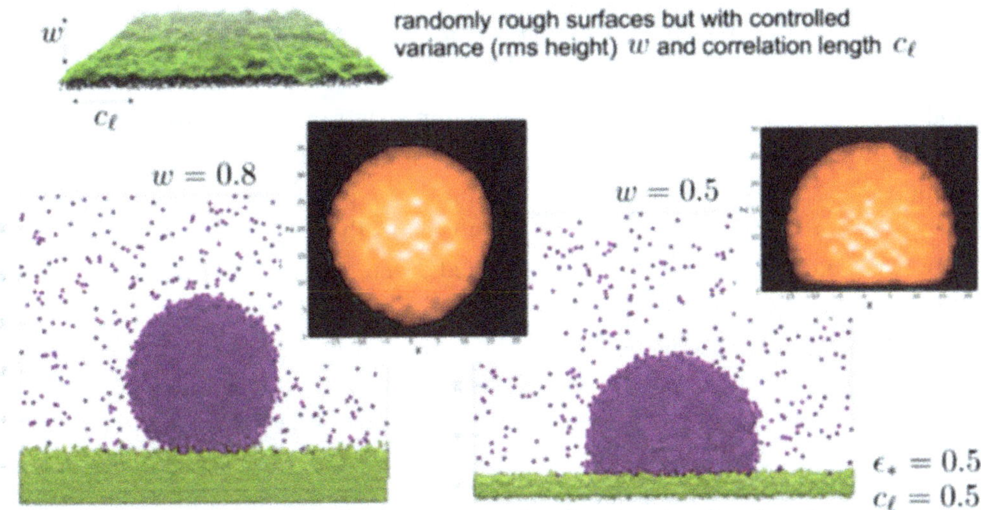

Fig. 14.8 Variation of the contact angle in the presence of a rough surface. A randomly rough surface is generated with controlled root-mean-square roughness w and correlation length cℓ (top). For fixed interaction strength ϵ, we then look at the difference in the contact angle for different values of w, in this example w$=$0.8 (left) and w$=$0.5 (right)

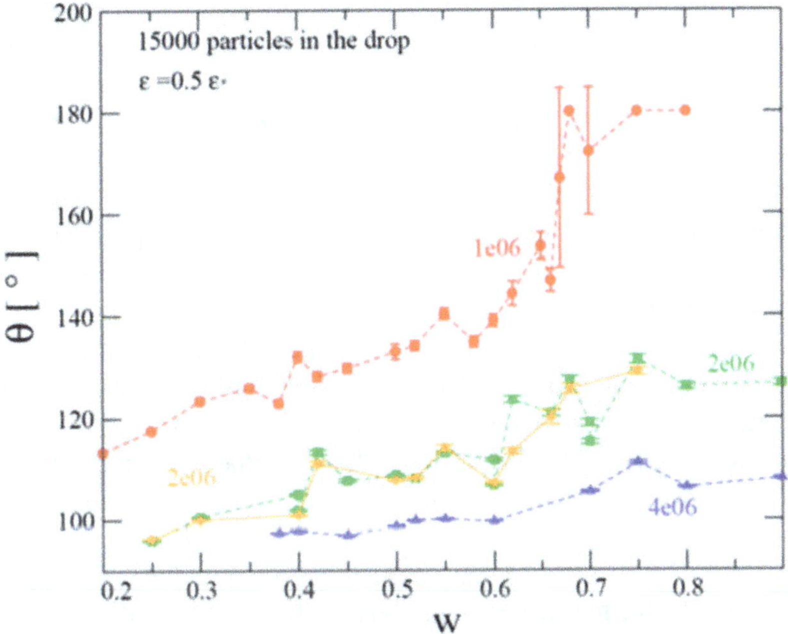

Fig. 14.9 Contact angle on a rough surface as a function of the roughness w for $\epsilon^* = 0.5$. Different curves correspond to different equilibration times

results obtained show that contact angle is inversely proportional to the strength of the potential. Superhydrophobic coatings would thus be efficiently simulated using the proposed methodology.

14.5.4 Testing the Hydraulic Pump

Finally, a prototype of hydraulic piston pump was realized, starting from a commercial pump (HP P8 made by Hp-Hydraulic, Bondioli & Pavesi Group) installing functionalized slippers instead of normal one. The pump was tested in a dedicated test bench working in closed circuit. Contrary to the expectations, the results are in contrast to that obtained in the slipper dedicated test bench because the standard slippers work better than functionalized ones. This is probably due to the fact that the slippers, mounted on the pump, are subjected to working conditions that we were not able to reproduce during the test performed in the dedicated test bench. In order to investigate this last phenomenon, other slipper prototypes were made considering geometrical modification.

The results obtained have opened a broad series of questions in which the pump manufacturer has been involved.

14.6 Conclusions and Future Research

This work investigated the potential of surface nano-coating of the piston slippers surface, with the aim of achieving better energy efficiency in a real industrial component by means of a drastic reduction of the friction coefficient. We have designed and synthesized a functional layer, obtaining a good granulometric distribution of the suspension [24–28]. We have demonstrated the feasibility of the coating processes and the adaptability of deposition techniques to the hydraulic piston pump slippers (Sect. 14.5.1). Thanks to the nanostructured coating it was possible to obtain, with water, a contact angle of $158.8 \pm 2.8°$ (superhydrophobic behaviour), whereas with oil the contact angle is $123.8 \pm 11.8°$ (oleophobic behaviour), where the standard slipper show a contact angle of only $16.7 \pm 2.3°$ (oleophilic behaviour). The samples where characterized from a tribological point of view (Sect. 14.5.2). Furthermore, wear analysis of the coatings, with particular focus on the relationship between coatings features and wear mechanisms (via SEM-EDS), have been done. Considering the adhesion results, no substantial differences, in terms of cracks typologies were found between the functionalized slippers. The lubricant reduces the friction coefficient in all cases. As for the surface functionalization, it does not reduce the friction coefficient. However, it has to be stressed that the pin-on-disc test cannot simulate a hydrostatic lubrication condition, as we have in the real pump.

The influence of the new superhydrophobic and oleophobic surface on the hydrostatic lubrication was investigated thanks to a dedicated test rig. The results [29–31] showed that, using the functionalized oleophobic coating, is possible to have a friction coefficient reduction from 20 to 30%, in function of the different working condition.

The last part of the work was dedicated to the influence on the contact angle due to geometrical nano-scale effects implementing a MD simulation with rough surfaces (Sect. 14.5.3). The key results obtained were:

- the contact angle increases for increasing roughness;
- the rougher the surface (at fixed interaction potential) the better the slippage, i.e. less friction;
- stronger fluid-solid interaction (at fixed surface roughness) means less slippage, i.e. more friction.

All these results confirm the literature, demonstrating the performed analysis validity. Future developments will be focused on the investigation of the possibility to apply the same coating on other pump component and to overcome the problems faced during this project, especially those related to the coating wearing.

Finally, the exploitation of the scientific results led to the submission and approval of a research project titled "ERCOLE on APP—Energy Recovery Coatings Originating Lubrication Effects on Axial Piston Pumps" funded under the call of Regione Emilia Romagna dedicated to the factories involved in the seismic events of 2012 (art. 12 DL 74/2012). The project was focused on antifriction solutions (not only oleophobic coating) to improve the efficiency of hydraulic piston pump.

Acknowledgements This work has been funded by the Italian Ministry of Education, Universities and Research (MIUR) under the Flagship Project "Factories of the Future - Italy" (Progetto Bandiera "La Fabbrica del Futuro") [32], Sottoprogetto 1, research projects "Surface Nano-structured Coating for Improved Performance of Axial Piston Pumps" (SNAPP) and "Axial Piston Pump Prototype Assembled with Oleophobic Surfaces Components" (APPOS).

The authors express their gratitude to the following researcher, for the precious and effective collaboration, without which all this work would not be possible: Eng. Giuseppe Rizzo, Eng. Giorgio Paolo Massarotti, Eng. Luca Pastorello, Dr. Maria Giulia Faga, Dr. Giovanna Gautier, Eng Massimo Martelli, Eng. Pietro Marani, Dr. Federico Veronesi, Dr. Magda Blosi, Dr Aurora Caldarelli, Dr. Giulio Boveri, Mrs. Guia Guarini, Dr. Carlotta Negri.

The authors would like to thank HP Hydraulic S.p.A. for its indispensable technical and methodological contribution, gaving the necessary support to perform all the tests and to develop the different prototypes.

This work would not have been possible without the foresight of Prof. Roberto Paoluzzi, in memory of whom this paper was written.

References

1. Rothstein J (2010) Slip on superhydrophobic surfaces. Annu Rev Fluid Mech (42):89–109. https://doi.org/10.1146/annurev-fluid-121108-145558
2. Rimondo M, Blosi M, Bezzi F, Mingazzini C (2011) Metodo per il Trattamento di Superfici Ceramiche per Conferire alle Stesse una Elevata Idrofobicità e Oleofobicità. Patent RM2011A000104
3. Rimondo M, Blosi M, Bezzi F, Mingazzini C (2012) Metodo di Trattamento di Superfici Metalliche per Conferire alle Stesse una Elevata Idrofobicità ed Oleofobicità. Patent RM2012A000291
4. Manringh N, Wray C, Dong Z (2004) Experimental studies on the performance of slipper bearings within axial-piston pumps. J Tribol 126(3):511–518
5. Kumar S, Bergada JM, Watton J (2009) Axial piston pump grooved slipper analysis by CFD simulation of three-dimensional NVS equation in cylindrical coordinates. Comput Fluids 38(3):648–663. https://doi.org/10.1016/j.compfluid.2008.06.007

6. Tominaga T, Yamamoto K, Matsumoto O, Hashiba M (1998) Swash plate type axial piston pump. US Patent US005809863A

7. Iwatsu R, Hyun JM, Kuwahara K (1993) Numerical simulation of three dimensional flow in a cubic cavity with an oscillating lid. ASME J Fluid Eng 115:680–686

8. Yao H, Cooper RK, Raghunathan S (2004) Numerical simulation of incompressible laminar flow over three-dimensional rectangular cavities. J Fluid Eng 126(6):919–927

9. Liu K, Yao X, Jiang L (2010) Recent developments in bio-inspired special wettability. Chem Soc Rev 39:3240–3255. https://doi.org/10.1039/b917112f

10. Tuteja A, Choi W, Ma M, Mabry J, Mazzella S, Rutledge G, McKinley G, Cohen R (2007) Designing superoleophobic surfaces. Science 318:1618–1622. https://doi.org/10.1126/science.1148326

11. Bhushan B (2011) Biomimetics inspired surfaces for drag reduction and oleophobicity/philicity. J Nanotech 2:66–68. https://doi.org/10.3762/bjnano.2.9

12. Raimondo M, (2012) Making super-hydrophobic building materials: static and dynamic behavior of nanostructured surface. In: IV ICC4-International Ceramic Conference, Chicago (USA), July 14–19

13. Savoy E, Escobedo F (2012) Molecular simulations of wetting of a rough surface by an oily fluid: effect of topology, chemistry, and droplet size on wetting transition rates. Langmuir 28(7):3412–3419. https://doi.org/10.1021/la203921h

14. Niavarani A, Priezjev N (2010) Modeling the combined effect of surface roughness and shear rate on slip flow of simple fluids. Phys Rev E 81:011606. https://doi.org/10.1103/PhysRevE.81.011606

15. Jabbarzadeh A (2013) Effect of nano-patterning on oleophobic properties of a surface. Soft Matter 9:11598–11608. https://doi.org/10.1039/C3SM52207E

16. Wang X, Zhao X, Jing C, Tao H, Han J (2005) Effects of nitric acid concentration on the stability of alumina sols. J Wuhan Univ Technol Mater Sci Ed 21(1):102–125

17. Verho T, Bower C, Andrew P, Franssila S, Ikkala O, Ras R (2011) Mechanically durable superhydrophobic surfaces. Adv Mater 23(5):673–678

18. Bassani R, Piccigallo B (1992) Hydrostatic lubrication. Elsevier, Amsterdam. ISBN 044488498x

19. Szeri A (1998) Fluid film lubrication. Cambridge University Press, Cambridge U.K. ISBN 0521481007

20. Hamrock B, Schmid S, Jacobson B (2004) Fundamentals of fluid film lubrication, 2nd edn. Marcel Dekker, Inc., New York. ISBN 0-8247-5371-2

21. Rowe W (2012) Hydrostatic, aerostatic and hybrid bearing design. Elsevier, Oxford. ISBN: 978-0-12-396994-1

22. Martell M, Rothstein J, Perot J (2010) Phys Fluids 22. https://doi.org/10.1063/1.3432514

23. Daub C, Wang J, Kudesia S, Bratko D, Luzar A (2010) The influence of molecular-scale roughness on the surface spreading of an aqueous nanodrop. Faraday Discuss 146:67–77. https://doi.org/10.1039/B927061M

24. Motta A, Cannelli O, Boccia A, Zanoni R, Raimondo M, Caldarelli A, Veronesi F (2015) A mechanistic explanation of the peculiar amphiphobic properties of hybrid organic−inorganic coatings by combining XPS characterization and DFT modeling. ACS Appl Mater Interfaces 7:19941–19947. https://doi.org/10.1021/acsami.5b04376

25. Malavasi F, Veronesi A, Caldarelli M, Zani M, Raimondo M Marengo (2016) Is a knowledge of surface topology and contact angles enough to define the drop impact outcome? Langmuir 32(25):6255–6262. https://doi.org/10.1021/acs.langmuir.6b01117

26. Raimondo M, Veronesi F, Boveri G, Guarini G, Motta A, Zanoni R (2017) Superhydrophobic properties induced by sol-gel routes on copper surfaces. Appl Surf Sci 422:1022–1029. https://doi.org/10.1016/j.apsusc.2017.05.257

27. Raimondo M, Blosi M, Caldarelli A, Guarini G, Veronesi F (2014) Wetting behavior and remarkable durability of amphiphobic aluminum alloys surfaces in a wide range of environmental conditions. Chem Eng J 258:101–109. https://doi.org/10.1016/j.cej.2014.07.076

28. Caldarelli A, Raimondo M, Veronesi F, Boveri G, Guarini G (2015) Sol–gel route for the building up of superhydrophobic nanostructured hybrid-coatings on copper surfaces. Surf Coat Technol 276:408–415. https://doi.org/10.1016/j.surfcoat.2015.06.037
29. Rizzo G, Bonanno A, Massarotti G, Paoluzzi R, Raimondo M, Blosi M, Veronesi F, Caldarelli A, Guarini G (2015) Axial piston pumps slippers with nanocoated surfaces to reduce friction. Int J Fluid Power 16(1):1–10. https://doi.org/10.1080/14399776.2015.1006979
30. Bonanno A, Masarotti G, Rizzo G, Paoluzzi R, Raimondo M, Blosi M, Veronesi F, Caldarelli A (2014) Application of nanostructured layer to improve the Energy efficiency in hydraulic piston pump. In: Proceedings of the 9th JFPS international symposium on fluid power, Matsue, October 28–31. ISBN 4-931070-10-8
31. Rizzo G, Bonanno A, Massarotti G, Pastorello L, Raimondo M, Veronesi F, Blosi M (2016) Energy efficiency improvement by the application of nanostructured coatings on axial piston pump slippers. In: 10th international fluid power conference | Dresden, Group 5-Pumps | Paper 5–6, pp 313–328
32. Terkaj W, Tolio T (2019) The Italian flagship project: factories of the future. In: Tolio T, Copani G, Terkaj W (eds) Factories of the future. Springer

Chapter 15
Monitoring Systems of an Electrospinning Plant for the Production of Composite Nanofibers

Luca Bonura, Giacomo Bianchi, Diego Omar Sanchez Ramirez,
Riccardo Andrea Carletto, Alessio Varesano, Claudia Vineis,
Cinzia Tonetti, Giorgio Mazzuchetti, Ettore Lanzarone,
Simona Ortelli, Anna Luisa Costa and Magda Blosi

Abstract Electrospinning is a versatile and promising technology for the production of polymer-based nanofibres. Composite nanofibres suitable for filtration of air and water have been developed by merging biopolymer processing and sol-gel techniques using electrospinning technology. A fine control of large-scale nanofibre formation is required to achieve reliable transfer of electrospinning-based processes into relevant industrial environments. The main goals of this work were the production of innovative multifunctional filter media (high-efficiency filtration, biocidal, self-cleaning, photo-catalytic, reactive, adsorbent properties) by integrating nanofibre membranes with inorganic nanoparticles and developing an electrospinning plant integrating sensors able to detect electrospinning fault and electrostatic alteration during the process.

L. Bonura · G. Bianchi
CNR-STIIMA, Istituto di Sistemi e Tecnologie Industriali Intelligenti per il Manifatturiero Avanzato, Milan, Italy

D. O. Sanchez Ramirez · R. A. Carletto · A. Varesano (✉) · C. Vineis · C. Tonetti · G. Mazzuchetti
CNR-ISMAC, Istituto per lo Studio delle Macromolecole, Biella, Italy
e-mail: a.varesano@bi.ismac.cnr.it

E. Lanzarone
CNR-IMATI, Istituto di Matematica Applicata e Tecnologie Informatiche, Milan, Italy

S. Ortelli · A. L. Costa · M. Blosi
CNR-ISTEC, Istituto di scienza e tecnologia dei materiali ceramici, Faenza, Ravenna, Italy

© The Author(s) 2019
T. Tolio et al. (eds.), *Factories of the Future*,
https://doi.org/10.1007/978-3-319-94358-9_15

15.1 Scientific and Industrial Motivations

Non-wovens are textile materials consisting of randomly oriented fibres linked together by bonds or physical entanglements between fibres, without any knitting or stitching. These textile materials are usually produced by putting staple fibres together to create a web and then binding them thermally (wet-laid, air-laid, carding) or mechanically or by spinning a molten polymer into continuous filaments directly distributed into a web by air streams and deflectors (melt-blown, spun-bond). Non-wovens are the primary alternative to traditional textiles as filtration media, thermal and sound insulating materials, hygienic and health/personal care textiles, automotive textiles, building materials and geo-textiles [1].

Emerging applications require fibres with decreased sizes. Since the specific surface area is proportional to the diameter of the fibre and the solid volume is proportional to the square of the diameter, the specific surface area is inversely proportional to the diameter of the fibre, thus leading to great surface area for small fibres. Moreover, apparent size of pores also depends on the diameter of the fibre, thus thin fibres produce non-wovens with small pores able to entrap and remove fine particles (penetrating particles) [1].

There are several methods for producing thin fibres with high-volume production methods like island-in-sea, fibrillation and innovative melt-blowing system or highly accurate methods such as nanolithography and self-assembly. However, combinations of a limited number of materials, elevated costs and low production rates restrict their usefulness. Instead, electrospinning is a simple and low cost process characterised an intermediate throughput. Electrospinning is an electro-hydrodynamic process that is generally considered the most promising and adaptable platform technology for the manufacturing of nanofibres. Briefly, the process takes place between a metal capillary opening (where a high voltage is applied), and a static (usually grounded) collector acting as a nanofibre gathering counter-electrode. The metal capillary is connected to a reservoir feeding the polymer solution under pressure (Fig. 15.1). The high voltage between the capillary and the grounded collector results in a nano- to micron-sized electrically-driven stream of polymer solution which is drawn out from the apex of a cone-shaped droplet (so-called Taylor cone) formed at the electrically-charged capillary opening. The solvent evaporates from the jet during the flight and, under ideal conditions, a continuous nano-sized filament is gathered on the collector in a random mode forming a 2D isotropic structure. Electrospun nanofibres are unique compared to other conventional materials due to their high surface to volume ratio, high fibre interconnectivity and nano-scale interstitial spaces. Nanofibres are therefore of considerable interest in applications where structures with a high surface area are desirable (e.g., filtration, biomedicine, composite materials). Emerging applications of electrospun nanofibres are in the field of textiles, as high-efficiency filters for air and liquids.

The first objective of the work was to produce innovative multifunctional filter media by integrating keratin nanofibre membranes with inorganic nanoparticles (as nanometals or nanometaloxides) by electrospinning. The advantage of keratin-based

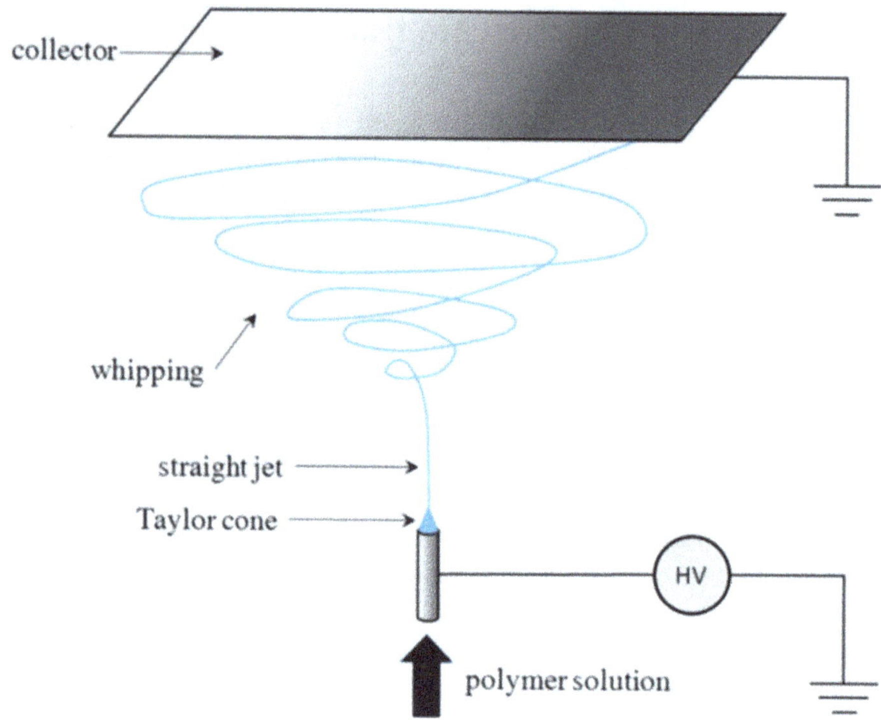

Fig. 15.1 Scheme of the electrospinning process

nanofibre membranes is that they offer both removal of suspended small particles filtration (due to the controlled porosity and small pore size) and adsorption of heavy-metals, dyes and VOCs, such as formaldehyde (due to the presence of side-chain functional groups able to chemically bind harmful chemicals) [2–6]. Nano-metals and/or nano-metal oxides improve membrane functionalities related to kill of bacteria (antimicrobial, low bio-fouling) and photocatalytic properties (removal of organic chemical compounds), respectively [7–12]. The results demonstrated that embedding inorganic nanoparticles onto keratin nanofibres makes possible to produce nano-reactive filters active for selective removal of pollutants in air and water treatments [13]. For the realization of nano-reactive filters it is necessary to prepare stable and electrospinnable keratin solutions doped with inorganic nanoparticles. In particular, keratin/Ag and TiO_2 ceramic nanoparticles dispersions were prepared by mixing keratin solutions and colloid dispersions.

Electrospinning is a technology currently at the border between basic research and industrial implementation. Recently, large-scale electrospinning systems have been developed by different research groups [14–21] in order to increase the productivity of the process at industrial level. Currently, the electrospinning process is mostly running with the operator setting the voltage and feeding flow by an open loop approach. An advanced control of nanofibre formation is required to achieve reliable transfer of electrospinning-based processes into a relevant industrial environment. The control of electrically-driven jets on large-scale electrospinning plants is one of the most important problems to be solved for real industrial application of electrospinning technology. Therefore, the second objective of the work was to introduce a monitoring and control system able to detect electrospinning fault and define the actual state of the process.

Finally, the third objective was to develop an image analysis tool for quality control of the electrospun products (measurements of fibre diameters and pore sizes, identification and quantification of defects), which is still lacking in most of the studies related to electrospinning.

15.2 State of the Art

Progresses in the comprehension of the electrospinning process have contributed to the development of functional nanofibres. The variety of polymer-based materials used in electrospinning is steadily growing and now comprises biopolymers and nano-composites. Electrospun nanofibrous materials are physically versatile and multiple chemical functionalities may be introduced using functional polymers [2–6] and/or functional nano-fillers [7–12]. The inclusion of inorganic nanoparticles in electrospinning technique matches with one of the most ambitious challenge in materials science: the creation of multifunctional material by design. Nanoparticles with antibacterial and/or with photo-catalytic/self-cleaning properties can give such new functionalities to the final polymeric product towards the creation of hybrid materials. The extremely high surface area of the electrospun nanofibres can further improve applicability and functionality of the processed materials [5, 6]. For this reason, the huge industrial potential of electrospun nanofibre-based materials claims increased activity. In particular, the proposed development of multifunctional nanofibres by electrospinning, exploitable as filtering media, represents a breakthrough approach for the development of new materials with new functions beyond the state-of-the-art.

The above-mentioned filtering media, which can offer both adsorption and high-efficiency filtration properties, can be obtained by electrospinning of functional polymers able to react with pollutant and harmful substances, such as heavy metals, dyes and/or volatile organic compounds (VOCs) like formaldehydes or phenols.

Among the functional polymers, bio-based polymers such as proteins (e.g. keratin) or carbohydrates (e.g. chitosan), offer an interesting solution for the production of nanofibrous filtering media able to adsorb and remove heavy-metal ions, dyes and VOCs. In particular, keratin [13] is characterized by chelating and ionisable groups that can form strong and stable interactions with metal ions, dyes and formaldehyde. Studies on keratin as metal ion adsorbent started in the 1950s, but in most cases they referred to wool fibres, ground wool or feathers. The benefits that could be achieved by the transformation of keratin into nanofibres are not only for filtration (membrane with small pore sizes) but also for adsorption [2, 5, 6]; in fact, due to their high specific surface area, keratin-based nanofibres exhibit enhanced reactivity with respect to wool or feathers. Furthermore, the incorporation of nanoparticles into nanofibres is an interesting strategy to confer antimicrobial, self-cleaning and photo-catalytic properties to keratin electrospun nanofibres.

On the other hand, for reaching applications of electrospinning-based processes exploitable at industrial level, some issues have to be solved [14–21]. The most important is the increase in nanofibre output related to the jet stability control. Advances in the knowledge about the complex interconnections between processing conditions/parameters and final products will lead to technical solutions for boosting a cost reliable transfer of electrospinning plants into relevant industrial environment.

The adjustable electrospinning process parameters are mostly flow rate and applied voltage [17]. The flow rate for a single needle setup is typically 0.001–1 ml/min, depending on the electrospun solution. The low production rate of a single needle electrospinning process may limit the industrial use [18]. Thus to achieve a satisfactory fibre production a high number of needles is required. Although a monitoring strategy by means of the measurement of the voltage drop across a sensing resistor for the single needle apparatus has been already applied, no solution is published for the multi-jet system. In particular, Samatham and Kim [22] measured the current on the grounded side of the system but this solution is not applicable in a multi-jet configuration where monitoring of the current flowing through every single needle, connected to the high voltage source, is needed. A relation between the type of jet, morphology of the deposed fibres and electric current was found. They observed four different types of relations: fluctuating jet associated with beaded nanofibres production, stable jet associated with the production of fibres with uniform diameter, stable jet with spikes corresponding to polymer drops associated with the production of more spherical beads than those formed with fluctuating jet and multiple jet regime associated with the production fibres similar to those obtained from fluctuating jet. The process can be monitored to check the quality of the output by elaborating the current signal in real time. A control loop for a single needle system has been designed using a CCD camera to evaluate the Taylor cone height and comparing it to a target height. When the observed height exceeded the target height, voltage was increased, otherwise it was decreased. A scalable closed loop control system that can maintain a constant source pressure at the capillary was designed and tested by Druesedow et al. [23] to improve the control over the variables of the process that affect the fibre creation. Two sensing technologies based on ultrasound and infrared ray were compared in terms of ability to measure the height of the polymer solution in the container. The air pressure above the solution was detected using a pressure transducer and adjusted by a controllable syringe pump [23].

A constant current electrospinning system has been developed by Munir et al. [24]. The electric current flowing from the collector to ground is measured and kept at a certain value by adjusting the voltage applied to the electrodes.

15.3 Problem Statement and Proposed Approach

Protein processing and sol-gel techniques, and finally production of novel composite nanofibre-based filter media suitable for the treatment of air and water require the development of specific knowledge on material science. Herein the following procedure was adopted for the production of suitable nanofibre-based filter media:

1. Nano-powders and water dispersed nanoparticles (nanosols) were produced separately.
2. The mixing of keratin and inorganic nanoparticles was carried out producing stable solutions.
3. The mixture parameters (colloidal stability, particle size distribution, surface charge) were checked and optimized to achieve electrospinnable formulations.
4. The solutions were characterized and their electrospinnability (i.e. possibility to transform keratin solutions with nanoparticles dispersion into nanofibres by means of electrospinning) was assessed.
5. Novel multifunctional composite nanofibre-based filter media were produced by electrospinning and the resulting nanofibres were tested against bacteria (antibacterial properties) and for the degradation of organic compounds (photocatalytic efficiency).

The details regarding the preparation of the electrospinnable solutions are presented in Sect. 15.4.1.

The production of electrospun nanofibers was realized with a conventional electrospinning apparatus equipped with a novel monitoring system able to measure and control several process parameters. In particular, voltage to the needle and solution flow rate are controlled, and current at the needle, current at the collector, pressure at the needle, environmental temperature and relative humidity are measured. The present monitoring system has been successfully used to detect fault during electrospinning. Moreover, it can speed up the seeking of electrospinning conditions for new polymers, polymer blends or new solvent systems.

Thanks to the measurement of process currents also at the needle, the monitoring system can better characterize process dynamics associated to jet formation and, additionally, separately evaluate each single current in a multi-needle (multi-jet) configuration. Finally, the system is the base of a closed-loop control system for electrospinning.

In order to achieve a reliable control of the electrospinning process, the electric current has been chosen as the main monitoring signal, being correlated, even if not exclusively, to the charges flowing together with the emitted polymeric solution. In order to measure the typical micro-Amperes currents while protecting the electronic circuits from the process high voltages (10–30 kV), an appropriate hardware configuration has to be developed. The main idea is to realize a measuring instrument isolated from ground that will float at the needle high voltage. It is then required to isolate both power supply and data transfer: literature from other technological sectors suggests the adoption of data link based on optical fibres and batteries or solar panels (with artificial lighting) or high voltage isolated transformers power supplies. Pearce and James developed a high voltage isolation amplifier with serial data transmission by means of optical fibre [25]. Relevant monitoring systems have been presented by Argirò et al. who developed a current monitoring system for a photomultiplier tube [26] and Sauer et al. [27]. In particular, Sauer designed a nano-ammeter floating at 25 kV, which can be supplied by a 24 V battery or by a solar panel and is protected from current spikes and short circuits.

The monitoring and control hardware system developed in this work (Sect. 15.4.2) is composed of two main electronic boxes: a floating voltage instrument box, directly connected to the system at high voltage, which collects process parameters at the needles side and control the flow pump; a grounded instrument box (low voltage) that measures the process current at the collector side and interacts with the high voltage generator, reading the monitoring signals proportional to the supplied voltage and current, and suppling the analogue signal that modulates the generated voltage. The software tool for monitoring and control is presented in Sect. 15.4.3.

In addition, a method to monitor the quality of the produced nanofibrous material has been developed (Sect. 15.4.4).

15.4 Developed Technologies, Methodologies and Tools

15.4.1 Preparation of Electrospinnable Solutions

For the production of nanocomposite functional nanofibre-based filters, keratin was extracted from wool by sulphitolysis with sodium m-bisulphite. Briefly, wool was cleaned by Soxhlet to remove fatty matter using petroleum ether, washed with distilled water and conditioned at 20 °C and 65% relative humidity for 24 h. Then, cleaned and conditioned fibres were cut into snippets of some millimetres and treated with a solution containing urea and sodium m-bisulphite at pH 6.5, for 2 h at 65 °C. The liquid obtained was filtered with a 5-μm pore-sized filter. Then, dialysis was carried out for 3 days against distilled water using cellulose tubes with 12–14 kDa of molecular cut-off. After dialysis, the resulting liquid was freeze-dried obtaining pure keratin powder. Titanium dioxide (TiO_2) sol (NAMA41, 6 wt% at pH 1.5) and silver (Ag) sol (NAMA39, 4 wt% at pH 4) supplied by Colorobbia Italia (Italy), were used to add colloids in the electrospinning feed. In order to evaluate the stability of nanosol in poly(ethylene oxide) (PEO, Mwa 400 kDa by Sigma Aldrich, Italy), some dispersion tests without keratin were performed. Aiming to assess nanoparticles stability on the PEO solutions, Dynamic Light Scattering (DLS) was used by measuring the hydrodynamic diameter of particles and assessing the particle size distribution. Measurements were carried out by Zetasizer Nanoseries apparatus (Malvern, UK) while working at the fixed angle of 173°. Samples were properly diluted in water and poured in a polystyrene cuvette. Hydrodynamic diameter measures the coordination sphere and species adsorbed on the particle surface.

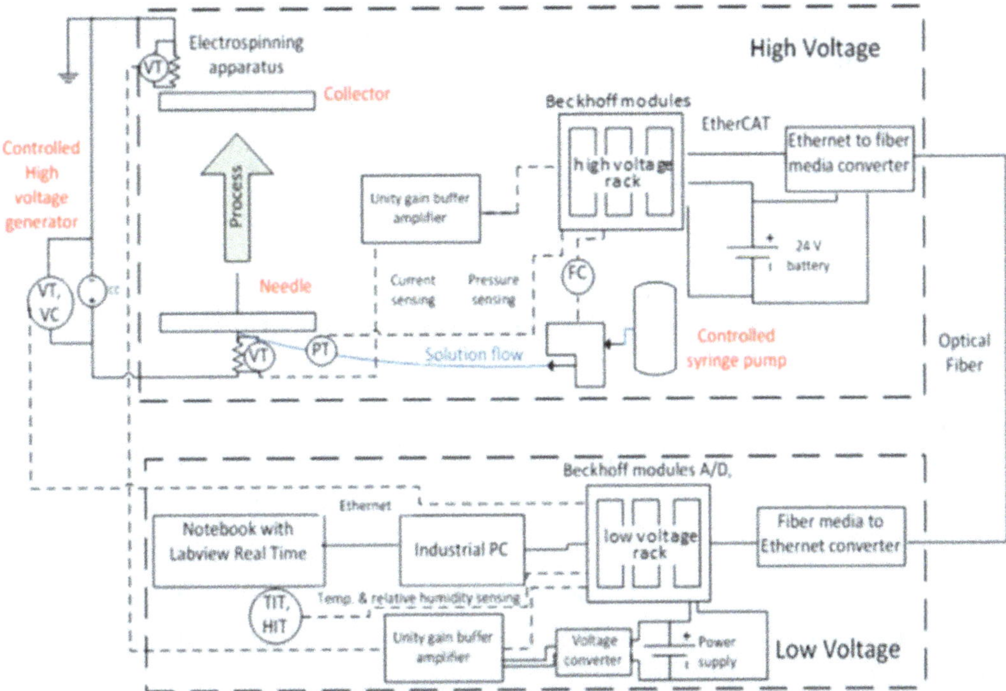

Fig. 15.2 Layout of the monitoring and control hardware system

Suspensions containing inorganic nanoparticles, PEO and keratin were prepared as follows for electrospinning. Freeze-dried keratin powder was added to de-ionized water until a concentration of 7 wt% and stirred at room temperature for 18 h. Similarly, PEO was dissolved in distilled water until a concentration of 7 wt%. After 6 h under stirring, nanosols were introduced into the PEO solution and kept under stirring for 18 h. The resulting PEO solutions with nanosols and keratin were mixed at 30:70 weight ratio, and stirred for at least 2 h before electrospinning. The nanosols were added to the PEO solutions reaching a nanoparticle concentration of 3 wt% with respect to the final keratin content. When silver nanosol was used, an anionic surfactant, i.e. sodium dodecyl sulfate, was added at a concentration of 1 wt% on the keratin content in order to avoid protein precipitation.

15.4.2 Monitoring and Control Hardware for Electrospinning

The monitoring and control hardware system for electrospinning is illustrated in Fig. 15.2 and includes:

1. High voltage instrument box.
2. EtherCAT connection via optical fibre.
3. Low voltage instrument box.

4. Given the low input impedance of the industrial A/D modules and the high value of the current metering resistors (i.e. 12 MΩ), appropriate high input impedance Instrumentation Amplifiers have been installed (4 channels).
5. An industrial PC with Real Time operating system by National Instruments. The unit acts as master on the EtherCAT bus and via a standard Ethernet line is connected to user-interface PC.
6. Standard Windows PC, with non-real time monitoring SW and User Interface, developed in Labview, National Instruments.
7. Canon EOS 60D camera, to record HD video of the needle zone.

The *High voltage instrument box* is referred to the needle high voltage (10–25 kV) and isolated from ground. It measures, for maximum three needles, their process currents and syringe supply pressures. Additionally, it controls the pump that operates on three syringes, imposing the same feeding speed. The High voltage instrument box contains:

- Beckhoff modules; i.e. EK1541 coupler with optical fibre connection; two EL3104: 4-channels analogue input (±10 V, 16 bits); EL7031 stepper motor controller, connected to a Beckhoff AS1030 stepper motor.
- 24 V battery.

The *Low voltage instrument box* is referred to ground and it measures the process current at the collector side and interacts with the high voltage generator (Spellman SL 50), reading the monitoring signals proportional to the generated voltage and current, and supplying the analogue signal that modulates the generated voltage. The Low voltage instrument box contains:

- Beckhoff modules, i.e. two EL3602 2-channel analogue input ($\pm10, \pm5, \pm2.5$ V, 24 bits), EL4132 2-channel analogue output terminals (±10 V, 16 bits).
- Ambient temperature and relative humidity sensor.

15.4.3 Software Tool for Monitoring and Control

The software tool for monitoring and control has been developed in LabVIEW and consists of two parts, one running on the non real-time PC, with Microsoft Windows, and one running on the real-time industrial PC. The former is organized in three parts:

- Graphic user interface
- Messages handling
- Target-Host communication.

It has a case-event structure (State Machine), which is handled by a standard Labview SW architecture, called *Queued Message Handler* (QMH). QMH is the combination of a producer event handler that pushes user messages onto a queue, and a consumer with an embedded state machine that can push its own messages onto the queue.

The pump can be operated manually by the operator through the pump control panel, setting the flow rate and selecting the syringe diameter, or automatically by the control SW, currently under development.

The software program running on the industrial PC consists of seven parts:

- Acquisition Module
- Elaboration Module
- Logging Module
- Pump Module
- Message Handling Loop
- Target-Host Connection.

Commands are sent from the Host to the target via the Target-Host connection and handled by the Message Handling Loop. All parts are constituted by a case structure running in a while loop. Communication between parts is handled by the QMH.

Acquisition is handled by the Acquisition Module (12 channels, 8 from the high voltage section and 4 from the low voltage one) which also sends data to the Elaboration Module, to the Logging Module and to the Pump Module. In the Elaboration Module, several scalar indexes are calculated to extract information about process. The average value, variance, number of peaks, energy, skewness, kurtosis, mean peak frequency and actual peak frequency are computed on a moving window and are shown to the operator.

The Logging Module receives data from the Acquisition and Elaboration Modules and saves it in two separate text files.

The Pump Module handles the pump speed in order to obtain the desired flow rate set by the operator. In fact the actual flow rate at the needle is equal to the pump flow rate only at steady state, because, during transients, part of the pump flow compensates for piping compliance, which is significant, compared to the small flow rate required by the process. To reduce the transient time, a pump controller has been developed based on the pressure signal measured near the needle by the installed pressure sensor. The relationship between the actual needle flow rate and the measured pressure, which depends on solution viscosity and needle geometry, was experimentally identified, performing tests with different constant pump flow rates, waiting until the steady state was reached.

15.4.4   *Image Analysis Tool for Quality Control of the Electrospun Products*

Several methods have been proposed in the literature to monitor the quality of produced nanofibrous materials. These methods can be classified according to the control strategy, i.e. (1) continuous control of the production parameters or (2) spot checks of the produced materials. Monitoring production parameters is simpler, but the quality of the produced materials can be affected by several other factors; thus, an inspection of the produced material is more suited. Two options are possible: to directly analyse the structure of the nanofibrous material by SEM (scanning electron microscope) imaging, or to assess some functional properties. As the latter option does not allow generating alerts in a short time, the most effective approach entails to attain SEM images of few material samples and observe the morphology. Outcomes from the spot analysis are then used to create a statistical inference on the entire production process [28]. However, the existing solutions to analyse SEM images of nanofibrous materials are only meant to measure fibres diameter and orientation [29–31], but it is not possible to detect the typical defects that are present in nanofibrous materials. Thus, a defect/anomaly detection solution based on a dictionary yielding sparse representations has been developed. In particular, the approach proposed in [32] to represent normal data in terms of a dictionary learned during an early training stage has been used. Then, test images are analysed in a patch-wise way, computing features to assess the conformation of each patch with the morphologies characterising the normal ones. Anomalies are recognised as outliers in the feature distribution, and the outcome of the analysis is the percentage of pixels associated with a defect [28].

The procedure previously described has been validated considering 45 SEM images (dimension 1024×696 pixels) acquired with the Field Emission Scanning Electron Microscope (Carl Zeiss Sigma NTS, GmbH Öberkochen, Germany). Indeed, specimens of 4×4 cm were collected on the produced material and placed on a metallic support. A thin gold coating of about 5 nm was applied on the sample surface to guarantee satisfactory electrical conduction by sputter-coating. All images were acquired with magnification 8000 x, extra high tension 5 kV, working distance 7 mm, brightness 45%, and contrast 52%. Our dataset contains 45 SEM images: 5 images are anomaly free, whereas 40 images contain anomalies of different size [28].

To assess the performance of the proposed method, we have considered the ratio between the False Positive Rate (FPR), i.e., the percentage of pixels erroneously identified as anomalous, and the True Positive Rate (TPR), i.e., the percentage of the number of pixels properly identified as anomalous. Performances are globally evaluated through the Receiver Operating Characteristic (ROC) curve acquired by plotting the TPR against the FPR for different values of a threshold that can be tuned by the operator. The Area Under the ROC Curve (AUC) is the common performance indicator usually adopted in the literature, which is at most equal to 1 in the case of the perfect detector (i.e., when TPR $= 100\%$ and FPR $= 0\%$) [28].

We have compared our algorithm against 5 state-of-the-art anomaly detection systems, tested under the same setting. Our approach outperforms the others, with an AUC of 0.926 against AUCs ranging between 0.619 and 0.926 for the other solutions. Details about the proposed solutions and the validation outcomes can be found in [28], while the code and some test images are available for download.[1]

---

[1] web.mi.imati.cnr.it/ettore/NanoTWICE.

15.5 Testing and Validation of Results

The system has been tested in process conditions using the materials and setup presented in Sect. 15.5.1. A set of experiment was designed to test the monitoring system (Sect. 15.5.2). Finally, a new set of conditions has been defined to validate the electrospinning prototype plant equipped with the monitoring and control system (Sect. 15.5.3).

15.5.1 Materials and Basic Electrospinning Setup Description

Poly(ethylene oxide) (PEO, average Mv 400 kDa, from Sigma Aldrich) was dissolved in distilled water at room temperature for about 12 h under magnetic stirring. All experiments were carried out at 7% wt. of concentration. PEO has been selected because of its well-known electrospinnability and huge literature available.

Suspensions containing inorganic nanoparticles, PEO and keratin were prepared as described in Sect. 15.4.1.

The solutions were electrospun using an electrospinning apparatus that comprises a plastic syringe containing about 5 mL of solution, and a high-precision syringe pump developed ad hoc by the authors. The pump is feed but the battery of the instrumentation box floating at high voltage to avoid current leakage toward ground (which can damage pump isolation and generate noise in the measured currents). Additionally, the pump speed is controlled by the supervising computer. A stainless steel needle (gauge 27) with an internal diameter of 0.2 mm was connected to the syringe and to a SL50 high-voltage generator (Spellman, UK). A stainless steel plate (20 cm by 20 cm) was positioned in front of the needle as collector, which was electrically grounded at 10–20 cm distance.

15.5.2 Experiments and Analysis of the Process

The list of experiments performed on the monitoring system is reported in Table 15.1. Values of the parameters were defined by Design of Experiments using the experiments of a previous work [33], to explore process conditions near the optimal values.

During the experiments, seven different states of the process were identified. States are described in the following as observed during the experimental campaign.

State 1 *No solution* occurs when the voltage power supply is turned on and the process voltage is applied to the tip, but there is no polymer solution available to support the formation of a jet. Even if there is no fibre formation, the current measured is comparable to the normal electrospinning process, because of significant air ionization near the needle. In that case, the pressure at the tip, as there is no polymer

Table 15.1 List of the parameters defined by design of experiments

Experiment no.	Voltage (kV)	Flow rate (ml/min)	Distance (cm)
1	23	0.02	22
2	20	0.01	22
3	20	0.02	22
5	28	0.01	22
6	28	0.02	22
8	23	0.01	22
10	20	0.01	22
11	23	0.02	22
13	18	0.005	22
14	28	0.02	22
16	20	0.02	22
18	23	0.01	22
20	18	0.005	22
21	23	0.02	22
22	28	0.01	22
24	26	0.03	28
25	26	0.03	28
26	24	0.025	28
27	28	0.025	28
28	24	0.025	28
29	28	0.025	28
30	24	0.025	25
31	28	0.025	25
32	24	0.025	25
33	28	0.025	25

solution in the system, is equal or near to the ambient pressure. A light indicator on the front panel of the monitoring software turns on, suggesting the operator to check the solution supply.

State 2 *Drop formation* occurs when the solution flow rate to the needle is too high, compared to the flow rate of the emitted jet, the drop at the needle tip increases in size. If no countermeasures are implemented, after a variable time depending on process parameters, the spinning stops. Therefore, State 2 can lead to State 6 *No electrospinning*, see below. Moreover, during State 2, it is also possible that part of the solution fed to the needle is projected from the tip as small droplets that reach the collector, forming defects, similar to a film on the nanofibre mat, as show in Fig. 15.3a. At the needle, instable jets and droplets are produced in a short sequence, as reported in Fig. 15.3b. When that happens, peaks in the current signal occur: the monitoring system identifies those peaks, as shown in Fig. 15.3c, and calculates their total number and repetition frequency.

Fig. 15.3 **a** SEM picture of nanofibers with film defects. **b** Pictures of an electrospinning tip at State 2 from the video camera. **c** Current signals measured during State 2: drop ejection

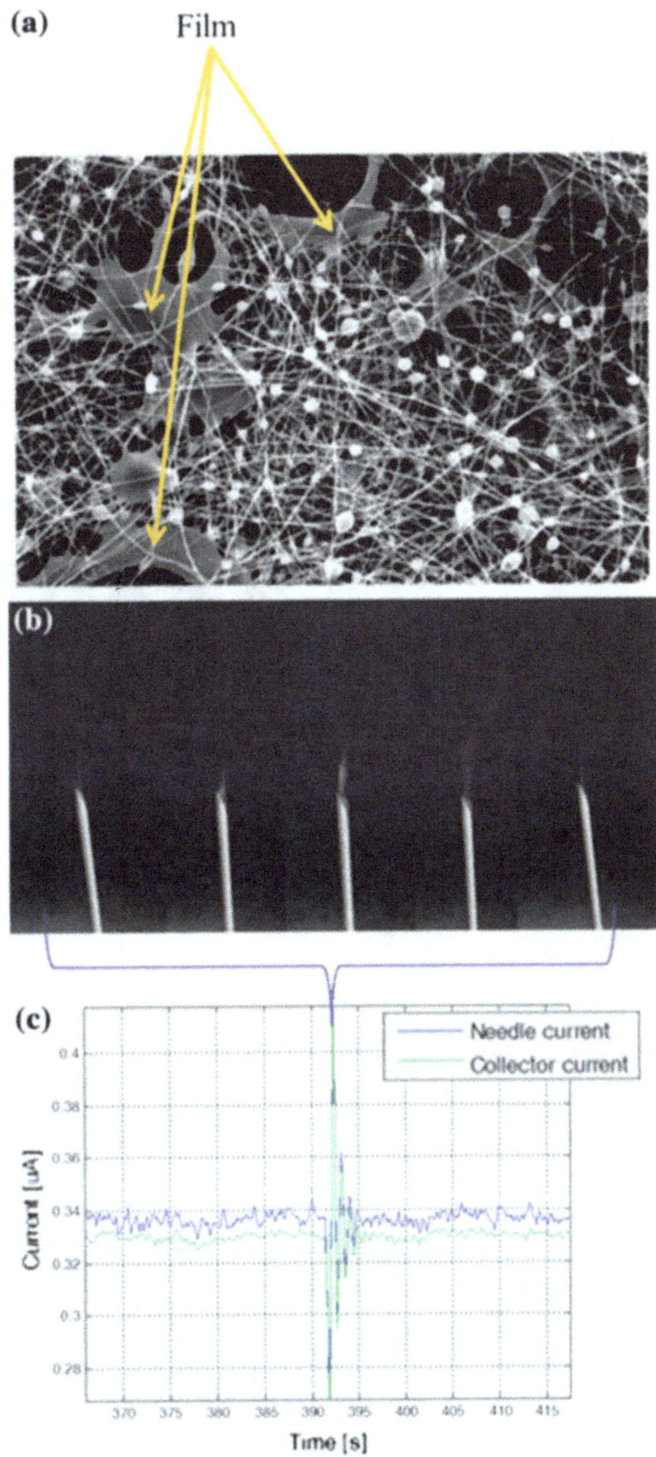

The control strategy proposed in the case of *High flow rate* is to decrease the flow rate or to increase the voltage, or a combination thereof. Particular attention must be given when the voltage is increased because the jet can reach the collector not completely dry, because of the reduced time of flight.

Fig. 15.4 SEM picture of nanofibers with beads (**a**). Pictures of an electrospinning tip at State 3 from the video camera (**b**). Current signals measured during State 3 (**c**)

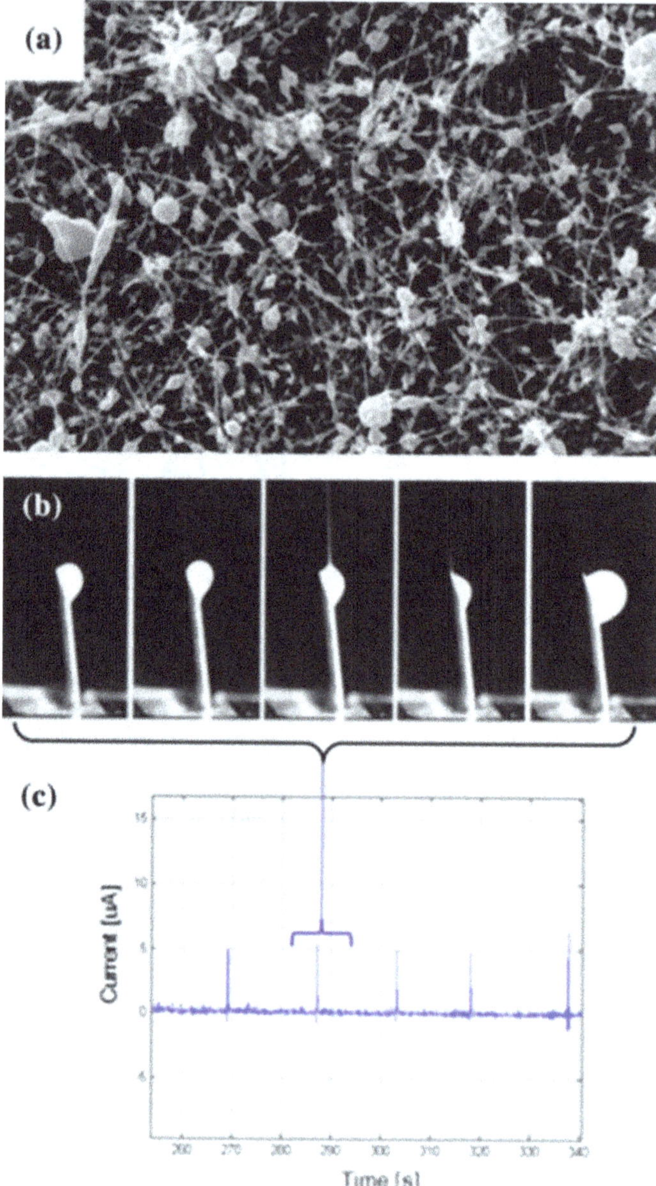

State 3 *Intermittent flow* occurs when the needle is fed with a flow rate lower than the one requested by a stable jet, the throughput of the electrospinning plant is limited and an intermittent jet can be observed. Intermittent jet is associated with a large production of beads, as shown in Fig. 15.4a. When an intermittent jet occurs (Fig. 15.4b), an intermittent current signal is measured, as reported in Fig. 15.4c. When the needle is not producing a jet, the current measured is negligible because the drop on the needle tip acts like an electric shield (compared to State 1).

State 4 is *Stable jet* and it is related to a steady electrospinning and good nanofibre production, as shown in Fig. 15.5a. A stable jet (Fig. 15.5b) is associated with a stable electric current, as Fig. 15.5c shows. In this case, nanofibres have uniform diameter with no beads and defects, according to Samatham et al. [22].

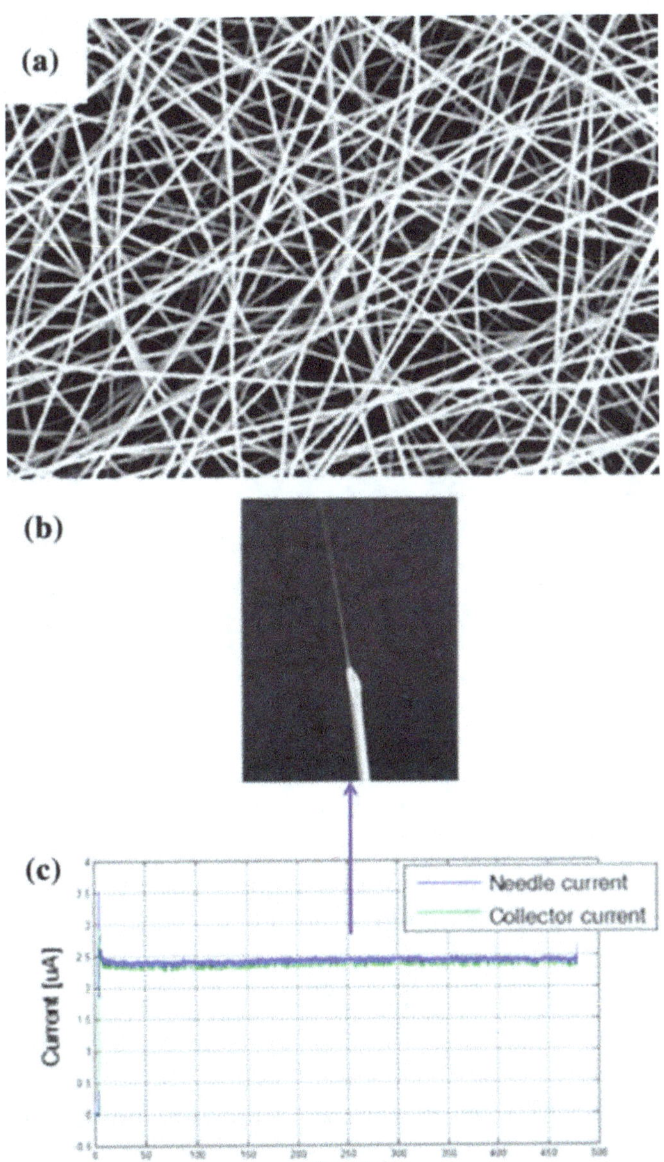

Fig. 15.5 SEM picture of nanofibers without defects (**a**). Picture of an electrospinning tip at State 4 from the video camera (**b**). Current signals measured during State 4 (**c**)

In some experimental conditions (high voltage and high flow rate), a multitude of jets were observed emerging from the needle tip (Fig. 15.6a). This is the State 5 *Multiple jets*. The electric current recorded during this phenomenon decreased. The finding disagrees with previous works [22]. The major cause of such difference is likely due to the geometry of the electrospinning equipment because the size of the collector could be not enough to intercept all the emitted jets.

State 6 is *No electrospinning* and it happens when the solution is fed to the needle at regular pressure, but no jet is ejected from the tip. Thus, the solution accumulates at the tip forming a large drop, as show in Fig. 15.6b. In this state the current signal is lower than the normal electrospinning current.

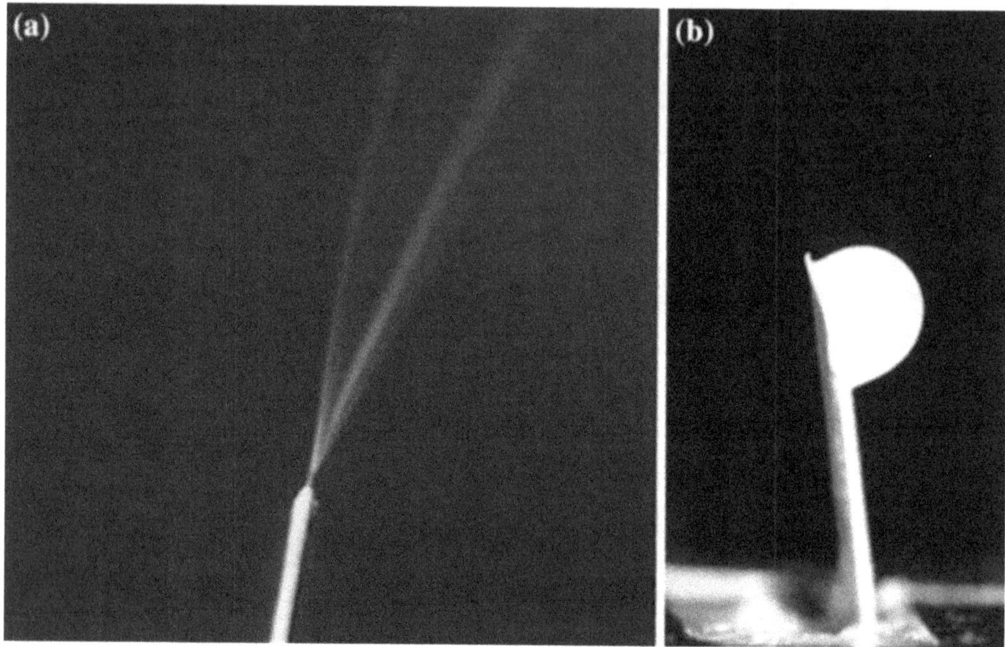

Fig. 15.6   Pictures of an electrospinning tip at State 5 (**a**) and State 6 (**b**) from the video camera

State 6 can be a worsening of the State 2. Finally, the State 7 is related to the clogging of the needle due to solidification of the solution. No electrospinning occurs, while pressure at the needle supply increases.

A new set of four experiments was carried out in order to validate the previous observations of the monitoring system employing operative conditions that lead to stable jet. Distance and flow rate were fixed at 22 cm and 0.03 ml/min respectively. Voltage was set at 22, 25, 28 and 30 kV in experiments labelled *A*, *B*, *C* and *D*, respectively. At *A–D* conditions, the process is stable and good nanofibres were produced. Moreover, the size of the drop at the needle tip decreases as the voltage increases.

15.5.3   Nanofibre-Based Filter Production and Testing

For the production of nanocomposite functional nanofibre-based filters, processing conditions were investigated (flow-rate from 0.001 to 0.05 ml/min, voltage from 20 to 35 kV and tip-to-collector distance from 15 to 30 cm). In electrospinning, temperature and relative humidity can affect process stability and nanofibre morphology; therefore, environmental conditions of the electrospinning room were recorded using an Escort RH iLog datalogger and adjusted accordingly. The temperature was maintained between 25 and 30 °C, and the relative humidity was maintained in the range 40–50%. Based on the previous test results, optimal experimental electrospinning

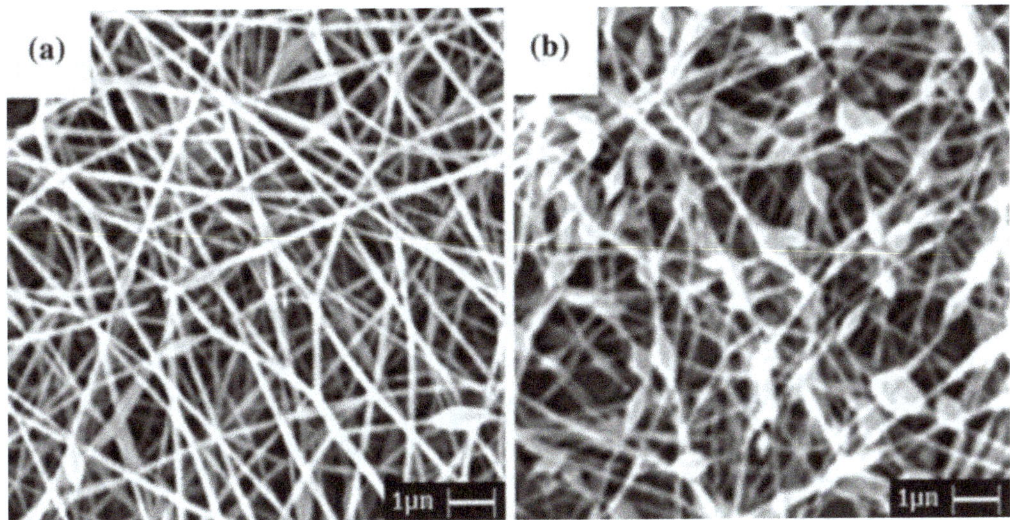

Fig. 15.7 SEM pictures of keratin nanofibers with silver nanoparticles (**a**) and titania nanoparticles (**b**)

parameters were: applied voltage 25 kV, flow rate 0.01 ml/min and tip-to-collector distance 22 cm, in good agreement with the model [33].

Nanofibres were characterized by infrared spectroscopy and SEM before and after the heat treatments and after water filtration. Infrared spectra were carried out by means of a Thermo Nicolet Nexus spectrometer by Attenuated Total Reflection (ATR) technique using a Smart Endurance accessory (diamond crystal ZnSe focusing element). The spectra were collected in the range from 4000 to 550 cm^{-1} with 100 scansions at 4 cm^{-1} of band resolution. Omnic 6.2 software tool was employed to record the spectra and to perform ATR baseline correction. The presence of keratin was assessed and morphology of the nanofibres does not change due to treatments, as Fig. 15.7 shows.

Antibacterial tests against Escherichia coli (ATCC 11229) on electrospun nanofibres with silver and titanium dioxide were based on the AATCC 100 Test Method. The procedure was the following:

(a) The electrospun keratin/PEO samples with titanium dioxide and the corresponding reference samples without nanoparticles were contacted with 1 ml of bacteria inoculum after activation by UV irradiation. The bacteria inoculum contains $1.5–3 \times 10^5$ CFU/ml. The samples were exposed to UV light at 20 °C and 65% RH for 1 h. After this time, 100 ml of sterile buffer were added on each sample. The buffer was then diluted ten times and pleated on Petri dishes with agar. The Petri dishes were incubated at 37 °C for at least 24 h, and then the bacteria colonies were counted.

(b) The samples with silver nanoparticles and on the corresponding reference sample without nanoparticles were contacted with 1 ml of bacteria inoculum containing $1.5–3 \times 10^5$ CFU/ml. Then the samples were incubated for 1 h at 37 °C. After this time, 100 ml of sterile buffer were added to each sample. The buffer was then dilute ten times and pleated on Petri dishes with agar. The Petri dishes were incubated at 37 °C for at least 24 h, and then the bacteria colonies were counted.

Bacterial reduction is calculated according to Eq. (15.1), where A represents the number of colonies of the samples with nanoparticles, whereas B represents the number of colonies of the samples without nanoparticles (references).

$$\% \text{ bacterial reduction } = \frac{B - A}{B} \cdot 100 \tag{15.1}$$

Photocatalytic tests were performed measuring the degradation of the Rhodamine B dye (RhB) in water as a model organic compound. The photocatalytic degradation of RhB by UV-irradiation was carried out at room temperature using samples of electrospun nanofibres containing 3% wt. of TiO_2. Before irradiation, the system was kept in the dark for 1 h to stabilize an adsorption-desorption equilibrium between nanofibres and RhB dye. The solution containing the electrospun nanofibre sample was irradiated with intensity of 9 W/cm^2 (Osram ULTRA-Vitalux 300 W lamp) keeping the solution stirred. In order to ensure a steady spectral emission the UV lamp was switched on 30 min before the photocatalytic test. The degradation reaction was regularly checked by taking 3 mL of solution for the measure of the absorbance at the wavelength of maximum absorbance of RhB (at 554 nm) with a single beam spectrophotometer (UV/Vis Spectrophotometer S-22, Boeco, Germany). According to the Lambert-Beer law, the absorbance is proportional to the RhB concentration, this way the photocatalytic efficiency was calculated as conversion (%) over the time of irradiation.

The so-obtained conversion refers to the ratio between the amount of reagent consumed and the starting amount of reagent following Eq. (15.2), where A_0 is the initial absorbance and A_t the absorbance after a certain irradiation time t.

$$\text{Conversion}(\%) = \frac{A_0 - A_t}{A_0} \cdot 100 \tag{15.2}$$

A sample without the addition of TiO_2 was considered as reference. Antibacterial tests were carried out using AATCC 100 Test Method on nanofibres with silver and titanium dioxide and compared to the nanofibres without nanoparticles. The results were 97% of bacteria reduction for nanofibres with titania nanoparticles and 95% of bacteria reduction for nanofibres with silver nanoparticles. Nanofibres showed excellent antibacterial properties as a sign that the nanoparticles maintain their functions even if inserted in the keratin matrix of the electrospun nanofibres.

Photocatalytic tests based on model reaction of RhB dye degradation showed promising photocatalytic properties for the hybrid electrospun nanofibres. The results in terms of conversion values were 53% for keratin nanofibres with titania nanoparticles, 14% for keratin nanofibres without nanoparticles and 12% for control solution (without electrospun materials).

An approach to optimize the process parameters in order to fit some requirements on the produced material while reducing its defectiveness has been developed. It is based on a statistical Design of Experiments (DOE) framework with three response variables related to both the defects and the structural characteristics of the fibres i.e., the defect ratio, the fibre diameter, and the size of pores. DOE can help to study the correlation between fibre diameter and process factors [34], but the analyses were limited to factors that exhibit an influence only on one process output and built up regression models to inspect only these specific relationships.

A more flexible tool has been developed. In particular, exploiting the Response Surface Methodology (RSM) idea, a 23 central composite design based on 20 runs is adopted [35], thus obtaining a regression model for each output variable in function of both controlled process parameters (polymer concentration, polymer feed rate, and applied voltage) and uncontrolled environmental factors, (room temperature and relative humidity), which are known to affect the fibre mat structure. Output data are obtained from SEM images of electrospun nanofibrous materials, as described above, while input data are either imposed or measured during the process.

The regression models are combined into a constrained optimization problem: response surfaces of pore and fibre diameter are used to detect an acceptable region, where the minimum of the defect surface is searched. The use of process parameters obtained in this manner produce a fibre mat with the desired morphology (porosity and diameter are in the desired range of values) and guarantee that presence of defects is as less as possible.

This approach has been applied to polyethylene oxide (PEO) on the prototype electrospinning apparatus. In particular, three examples of constrained optimization were analysed in that work to show outcomes of the proposed optimization approach. Large admissible intervals for fibre diameter and the size of pores were defined in the first test to explore the optimal surface on a wide domain, while narrower intervals were considered in the following tests, to represent realistic and specific production requests. Optimization outcomes are in line with the experience and allow to properly define the process parameters to set based on the desired material properties. Detailed results of the tests and some insights on the estimated surfaces can be found in [33].

15.6 Conclusions and Future Research

Functional inorganic nanoparticles of silver and titania were embedded into a protein matrix of keratin in order to produce hybrid organic/inorganic nanofibres by electrospinning. The use of colloids in solutions suitable for electrospinning allows the production of multi-functional nanofibres. Keratin (a protein extracted from wool) has been used as polymer matrix because of its properties in adsorption of dye, VOCs and metals, titania was selected for its excellent photocatalytic and antibacterial capacity, and silver for antibacterial and antifouling properties.

This work demonstrates the possibility to use in electrospinning water, instead of harmful chemicals, as solvent for the production of keratin-based nanocomposite nanofibres with multifunctional properties suitable for both air and water treatments and depuration. On the other hand, there are clear practical advantages in using aqueous colloids instead of nano-powders (e.g. attain good dispersion of nanoparticles in the polymer nanofibres, overcome safety issues). The main practical benefit of embedding colloids in electrospun nanofibres is that both unprocessed and final materials can be handled more easily than nanoparticle powders.

Finally, functionalities of the nanoparticles embedded in the nanofibres in degrading organic compounds and/or in killing microorganisms were assessed. The results demonstrate that electrospun keratin-based nanofibres with a high surface area guarantee preserving nanoparticle properties.

As for the analysis of the images and the evaluation of the optimal working conditions, the developed methodology for analysing the defects proved to be effective and to accurately map pixels of a SEM image associated to a defect in few minutes. From that, it is possible to obtain indicators mediated over the acquired images to be used as final quality control of the production.

The developed monitoring system can monitor the needle current by an innovative instrument floating at high voltage. Tests have shown that current at the needle side carries more information about the jet producing project, compared to the current at the collector side, usually used for process monitoring in the literature. This allows an early diagnosis of solution accumulation at the needle tip, useful to avoid drop emission and intermittent and/or low quality electrospinning. The developed monitoring system integrates needle and collector current measurements with supply pressure measurement at the needle. The system has been designed to monitor in parallel up to three emitting needles, to support multi-jets configurations.

Acknowledgements This work has been funded by the Italian Ministry of Education, Universities and Research (MIUR) under the Flagship Project "Factories of the Future—Italy" (Progetto Bandiera "La Fabbrica del Futuro") [36], Sottoprogetto 1, research projects "composite Nanofibres for Treatment of air and Water by an Industrial Conception of Electrospinning" (NanoTWICE) and "Automated electrospinning plant for industrial manufacturing of functional composite nanofibres" (AUTOSPIN).

References

1. Varesano A, Carletto RA, Mazzuchetti G (2009) Experimental investigations on the multi-jet electrospinning process. J Mater Proc Technol 209(11):5178–5185
2. Aluigi A, Vineis C, Tonin C, Tonetti C, Varesano A, Mazzuchetti G (2009) Wool keratin-based nanofibres for active filtration of air and water. J Biobased Mater Bioen 3:311–319
3. Baek DH, Ki CS, Um IC, Park YH (2007) Metal ion adsorbability of electrospun wool keratose/silk fibroin blend nanofiber mats. Fibers Polym 8:271–277
4. Aluigi A, Tonetti C, Vineis C, Varesano A, Tonin C, Casasola R (2012) Study on the adsorption of chromium (VI) by hydrolyzed keratin/polyamide 6 blend nanofibres. J Nanosci Nanotechnol 12:7250–7259
5. Aluigi A, Tonetti C, Vineis C, Tonin C, Mazzuchetti G (2011) Adsorption of copper(II) ions by keratin/PA6 blend nanofibres. Eur Polym J 47:1756–1764

6. Aluigi A, Rombaldoni F, Tonetti C, Jannoke L (2014) Study of Methylene Blue adsorption on keratin nanofibrous membranes. J Hazard Mater 268:156–165

7. Linsebigler AL, Lu G, Yates JT (1995) Review-photocatalysis on TiO_2 surfaces: principles, mechanisms, and selected results. Chem Rev 95:735–758

8. Hoffmann MR, Martin ST, Choi W, Bahnemann DW (1995) Environmental applications of semiconductor photocatalysis. Chem Rev 95:69–96

9. Costa A, Ortelli S, Blosi M, Albonetti S, Vaccari A, Dondi M (2013) TiO_2 based photocatalytic coatings: from nanostructure to functional properties. Chem Eng J 225:880–886

10. Fujishima A, Honda K (1972) Electrochemical photolysis of water at a semiconductor electrode. Nature 238:37–38

11. Ortelli S, Blosi M, Albonetti S, Vaccari A, Dondi M, Costa AL (2013) TiO_2 based nano-photocatalysis immobilized on cellulose substrates. J Photochem Photobiol A Chem 276:58–64

12. Rai M, Yadav A, Gade A (2009) Silver nanoparticles as a new generation of antimicrobials. Biotechnol Adv 27:76–83

13. Varesano A, Vineis C, Tonetti C, Sánchez Ramírez DO, Mazzuchetti G (2014) Chemical and physical modifications of electrospun keratin nanofibers induced by heating treatments. J Appl Polym Sci 131:40532

14. Tomaszewski W, Szadkowski M (2005) Investigation of electrospinning with the use of a multi-jet electrospinning head. Fibres Text East Eur 13(4):22–26

15. Kim GH, Cho YS, Kim WD (2006) Stability analysis for multi jets electrospinning process modified with a cylindrical electrode. Eur Polym J 42(9):2031–2038

16. Kumar A, Wei M, Barry C, Chen J, Mead J (2010) Controlling fiber repulsion in multijet electrospinning for higher throughput. Macromol Mater Eng 295(8):701–708

17. Varesano A, Rombaldoni F, Mazzuchetti G, Tonin C, Comotto R (2010) Multi-jet nozzle electrospinning on textile substrates: observations on process and nanofibre mat deposition. Polym Int 59(12):1606–1615

18. Yang Y, Jia Z, Li Q, Hou L, Liu J, Wang, L et al (2010) A shield ring enhanced equilateral hexagon distributed multi-needle electrospinning spinneret. IEEE Trans Dielectr Electr Insul 17(5):1592–1601

19. Angammana CJ, Jayaram SH (2011) The effects of electric field on the multijet electrospinning process and fiber morphology. IEEE Trans Ind Appl 47(2):1028–1035

20. Zheng Y, Zhuang C, Gong RH, Zeng Y (2014) Electric field design for multijet electropsinning with uniform electric field. Ind Eng Chem Res 53(38):14876–14884

21. Zheng Y, Gong RH, Zeng Y (2015) Multijet motion and deviation in electrospinning. RSC Adv 5(60):48533–48540

22. Samatham R, Kim KJ (2006) Electric current as a control variable in the electrospinning process. Polym Eng Sci 46(7):954–959

23. Druesedow CJ, Batur C, Cakmak M, Yalcin B (2010) Pressure control system for electrospinning process. Polym Eng Sci 50(4):800–810

24. Munir MM, Iskandar F, Khairurrijal OK (2008) A constant-current electrospinning system for production of high quality nanofibers. Rev Sci Instrum 79(9):093904

25. Pearce JW (1978) Optically coupled high voltage isolation amplifier. Rev Sci Instrum 49(11):1562

26. Argirò S, Camin DV, Destro M, Guérard CK (1999) Monitoring DC anode current of a grounded-cathode photomultiplier tube. Nucl Instrum Methods Phys Res Sect A 435(3):484–489

27. Sauer BE, Kara DM, Hudson JJ, Tarbutt MR, Hinds EA (2008) A robust floating nanoammeter. Rev Sci Instrum 79(12):126102

28. Carrera D, Manganini F, Boracchi G, Lanzarone E (2017) Defect detection in SEM images of nanofibrous materials. IEEE Trans Ind Inf 13(2):555–561

29. Ziabari M, Mottaghitalab V, McGovern S, Haghi A (2008) Measuring electrospun nanofibre diameter: a novel approach. Chin Phys Lett 25(8):3071

30. Milasius R, Malasauskiene J (2014) Evaluation of structure quality of web from electrospun nanofibres. Autex Res J 14(4):233–238

31. Electrospinz SEM analyser, Electrospinz Ltd, Blenheim, New Zealand. www.electrospinz.co.nz
32. Boracchi G, Carrera D, Wohlberg B (2014) Novelty detection in images by sparse representations. In: Proceedings of IEEE symposium on intelligent embedded systems, pp 47–54
33. Borrotti M, Lanzarone E, Manganini F, Ortelli S, Pievatolo A, Tonetti C (2017) Defect minimization and feature control in electrospinning through design of experiments. J Appl Polym Sci 134:44740
34. Coles SR, Jacobs DK, Meredith JO, Barker G, Clark AJ, Kirwan K, Stanger J, Tucker NJ (2010) A design of experiments (DoE) approach to material properties optimization of electrospun nanofibers. Appl Polym Sci 117(4):2251–2257
35. Box GEP, Draper NR (1987) Empirical model-building and response surfaces. In: Wiley Series in Probability and Statistics. Wiley, USA
36. Terkaj W, Tolio T (2019) The Italian flagship project: factories of the future. In: Tolio T, Copani G, Terkaj W (eds) Factories of the future. Springer

Chapter 16
Plastic Lab-on-Chip for the Optical Manipulation of Single Cells

Rebeca Martínez Vázquez, Gianluca Trotta, Annalisa Volpe,
Melania Paturzo, Francesco Modica, Vittorio Bianco, Sara Coppola,
Antonio Ancona, Pietro Ferraro, Irene Fassi and Roberto Osellame

Abstract Lab-on-chips (LoCs) are microsystems capable of manipulating small amounts of fluids in microfluidic channels. They have a huge application potential, from basic science to chemical synthesis and point-of-care medical analysis. Polymers are rapidly emerging as the substrate of choice for LoC production, thanks to a low material cost and ease of processing. Two breakthroughs that could promote LoC diffusion are a microfabrication technology with cost-effective and rapid prototyping capabilities and also an integrated on-chip optical detection system. This chapter proposes the use of femtosecond laser micromachining combined with microinjection moulding as a novel highly-flexible microfabrication platform for polymeric LoCs with integrated optical detection, for the realization of low-cost and truly portable biophotonic microsystems. We demonstrate a LoC for the relevant application of non-invasive and contactless mechanical phenotyping of single cancer cells.

R. M. Vázquez · R. Osellame (✉)
CNR-IFN, Istituto di Fotonica e Nanotecnologie, Milan, Italy
e-mail: roberto.osellame@polimi.it

G. Trotta · F. Modica
CNR-STIIMA, Istituto di Sistemi e Tecnologie Industriali Intelligenti per il Manifatturiero
Avanzato, Bari, Italy

A. Volpe · A. Ancona
CNR-IFN, Istituto di Fotonica e Nanotecnologie, Bari, Italy

M. Paturzo · V. Bianco · S. Coppola · P. Ferraro
CNR-ISASI, Istituto di Scienze Applicate e Sistemi Intelligenti "Eduardo Caianiello",
Pozzuoli, NA, Italy

I. Fassi
CNR-STIIMA, Istituto di Sistemi e Tecnologie Industriali Intelligenti per il Manifatturiero
Avanzato, Milan, Italy

© The Author(s) 2019
T. Tolio et al. (eds.), *Factories of the Future*,
https://doi.org/10.1007/978-3-319-94358-9_16

16.1 Scientific and Industrial Motivations

Point-of-care analysis is one of the great challenges in many application fields, from healthcare to environmental monitoring [1]. The widespread diffusion of electronic portable devices enables data analysis in-field; however, the bottleneck is often represented by the miniaturization of the sensor. While electric sensors, e.g. accelerometers, are effectively integrated already on cell-phones and watches, microfluidic devices still suffer from a limited diffusion, albeit they could represent the most interesting devices for personalized healthcare, since they give access to biological fluid monitoring. Such miniaturized microfluidic devices are often called Lab-on-Chip (LoC) because they aim at integrating in a small chip a whole set of analysis on a given fluid that would normally require an entire laboratory [2, 3]. Among the advantages of a LoC approach, it is worth highlighting the point-of-care use, the limited amount of liquid sample required, the full automation of the analysis process that can be performed by non-specialized personnel. Notwithstanding their huge application potential, LoC penetration in the market is limited by the development cost and by the difficulty in integrating the analysis apparatus on the same chip where the fluid manipulation is performed. Two improvements that could promote LoC diffusion are:

 (i) A microfabrication technology with low-cost and rapid prototyping capabilities.
(ii) On-chip optical components for detection and/or manipulation of the fluidic sample.

These two features will significantly decrease the cost for LoC development and greatly enhance the compactness and portability of the devices, respectively. In the following we will introduce and outline the main ingredients that one may want to combine into a new generation of LoC devices.

Most of the commercial LoCs are made in polymeric materials due to their low cost and easy processing [4]. Amorphous polymers are completely transparent to wavelengths in the UV-visible range, and their optical properties can be easily tuned by changing the composition. Many of them are already commercially available products, but an almost unlimited number of modifications in the composition are feasible for a further optimization of the chosen structure.

The most widely employed manufacturing technology is microinjection moulding since it enables large volume and high precision production of plastic devices. This technology requires the fabrication of moulds and tools that are quite expensive and limit the reconfigurability of the LoC layout.

A completely different fabrication approach is based on laser micromachining as a flexible and rapid prototyping process for the creation of a variety of microstructures in a broad range of materials. In particular, femtosecond lasers, due to their peculiar light-matter interaction regime, offer unique microfabrication capabilities both with transparent (e.g. polymeric) and non-transparent (e.g. metallic) materials [5]. Recent technological advances have dramatically improved the performance and reliability of femtosecond laser systems, making them available as a practical micromachining tool [6, 7]. However, this technology is mainly used only in academic laboratories because it is too expensive for large batch production.

The analysis of biological fluids often requires optical means, from microscope imaging to fluorescence detection. These operations are currently performed with external and bulky equipment, thus frustrating the miniaturization of the fluidic handling of the sample [8]. Digital Holographic Microscopy (DHM) is a powerful method to visualize biological cells [9]. Not only it provides an enhanced view of transparent objects but it also allows their three-dimensional reconstruction. In addition, it enables software-based autofocus and aberration correction thus greatly simplifying the optical imaging system; this makes DHM a very suitable approach for an integrated, highly-informative and low-cost microscopy set-up.

Herein, we propose the combination of the above elements to produce a highly flexible microfabrication platform for polymeric LoCs with integrated optical imaging, for the realization of low-cost and truly portable biophotonic microsystems.

The unique integration of photonics and microfluidics, offered by the assembly of optical components in the biochips, will enable to implement a wealth of novel functionalities. In this work, we will focus on a specific polymeric device that has relevant application potentials and that demonstrates the main features of our new approach. We will present a LoC for mechanical phenotyping of single cells exploiting optical forces on the chip.

The chapter is organized as follows. In Sect. 16.2 we will discuss the state of the art of all the techniques that we will exploit in this work. In Sect. 16.3 we will detail the basic concepts behind the innovative approach we are proposing, and we will discuss how it will allow to overcome the current limitations in the LoC manufacturing industry. In Sect. 16.4 we will introduce in details the technologies, methods and tools of our new approach. In Sect. 16.5 we will present the results that we achieved and, in particular, we will present the LoC device for single-cell mechanical phenotyping as an example of a complex analysis that can be performed in a simple device efficiently manufactured. Finally, in Sect. 16.6 we will draw some conclusions and discuss the perspectives of our work.

16.2   State of the Art

The increasing interest in microfluidic devices for healthcare and diagnostic applications, such as Lab on Chips (LoCs), pushed the research towards the evaluation of alternative manufacturing technologies capable of reducing the production costs and guaranteeing a mass production of ready-to-use micro structured devices. Among different micro-moulding techniques, micro-injection moulding (μIM) is a promising process for manufacturing these polymeric disposable microfluidic devices [10].

μIM is a process capable of reproducing the micron or submicron features of metallic moulds to a polymeric product. The granular material is transferred from a hopper into a plasticizing unit to be made molten and soft. Then the material is forced inside a mould cavity where a holding pressure is applied for a specific time to compensate the shrinking of the material. After enough time, the material freezes into the mould shape, then it is ejected to repeat the cycle.

Polymers are a wide family of materials characterized by a large spectrum of properties, thus it is relatively easy to find the proper material that has the required properties for processing and application. A typical polymer can completely fill and accurately replicate small features down to tens of nanometres, when the optimum processing conditions are applied [11]. Among all polymers, Poly(methyl methacrylate) (PMMA) is appropriate for lab-on-chip applications because it is mechanically strong, optically transparent, biocompatible and it has good chemical stability. Micro-injection moulding of PMMA is a well-established process and it is possible to obtain LoCs with extremely small features [12–14], if the process is combined with UV lithography coupled with electrodeposition (LIGA process) for the manufacture of nickel mould inserts.

Femtosecond laser micromilling is an ablation procedure causing the vaporization of the material in a layer-by-layer fashion because of the interaction between the laser beam and the work piece that is machined. This technology is becoming important for applications in rapid prototyping and manufacturing of miniaturized metallic, semi-conductor as well as ceramic and glass components or devices. Femtosecond pulses provide high precision and reduced thermal damages and/or structural modifications with respect to longer laser pulses [15, 38].

Besides pulse duration, several parameters have a direct impact on the accuracy of the laser milling process and the quality of the microstructured surfaces, i.e. the pulse repetition rate and laser fluence. Even in the ultrashort pulse regime, it is always advisable to work with near threshold fluence because for higher average powers the ablation mechanism is once again dominated by melting [15]. Furthermore, several studies have clearly shown that the process strategy significantly affects the resulting surface quality and edge sharpness [16, 17]. In particular, the surface roughness and its homogeneity are highly influenced by the beam-scanning pattern.

Femtosecond laser ablation offers several advantages for the fabrication of microfluidic chip on polymer substrates. Besides its geometrical flexibility and 3D capabilities, the ultrashort deposition of energy in the material causes vaporization before thermal diffusion can occur, thus minimizing the heat-affected zone and increasing the accuracy of the ablated structures (if compared with longer pulses or continuous wave laser micromachining) [18]. This technology is particularly useful for the direct and rapid prototyping of devices, since it does not need the use of moulds or masks. Thus, the LoC is created by laser ablation along the pattern of the desired microfluidic scheme.

Besides material removal, femtosecond lasers are also used for welding different transparent materials at their interface by exploiting localized melting and resolidi-fication below the ablation threshold. Several studies have demonstrated the ability of focused femtosecond laser pulses to weld fused silica [19], borosilicate and/or dissimilar glasses [20]. Tamaki et al. [21] achieved a first demonstration of fs-laser welding of silica glass in the low repetition rate regime (1–200 kHz). In their study, the laser focus was accurately placed at the interface of two glass specimens. The focal region was elongated along the optical axis and across the two material interface due to filamentation generated by the balance between self-focusing and diffraction [22]. The intensity in the filamentary region was high enough to induce localized melting and subsequent resolidification and joining.

Another reported approach to weld fused silica exploits ultrashort laser pulses at repetition rates in the MHz regime. In this case no filamentation occurs and the energy accumulation mechanism between subsequent pulses is at the basis of the material modification. High temperature hotspots close to the interface to be welded are produced, causing once again material melting and resolidification over volumes larger than the focal one. Even though the laser is focused on one material, heat diffusion causes melting and resolidification of both materials across the interface. Femtosecond laser welding of glass substrates in this regime has been reported by several works [19, 23]. Welding with high repetition rate lasers has several advantages compared to the low repetition rate regime because a much higher processing speed is accessible and the size of the structural modification can be easily tuned by adjusting the writing speed.

The capability to observe the samples inside LoCs is crucial for monitoring and extracting information from the processes occurring inside the microfluidic channels. To this aim, the standard imaging techniques have been adjusted to be applied for the investigation of biological specimens in LoCs, novel diagnostic tools specifically tailored for LoCs have been proposed and new strategies to support the integration of the diagnostic and imaging functionalities on chip have been investigated. The novel imaging strategies to study biological samples in microfluidic channels need to satisfy the following requirements:

- quantitative information is needed to better understand the cell structure and behaviour;
- compact portable devices are needed to use LoCs for point-of-care diagnostics;
- a high-throughput imaging system needs to be developed to obtain statistically relevant information by analysing multiple specimens in a reduced time period;
- label-free methods are favourite to elude sample pre-treatment and undesired alterations of the specimens.

Up to now, the two main advanced techniques for on-chip microscopy are the Digital in Line Holography (DILH) and the Holographic Opto-fluidic Microscopy (HOM), both based on a lensless approach. The use of DILH technique as imaging tool for LoC devices is quite limited because of the cost and size of coherent sources [24].

In this work we aim at demonstrating a final device for mechanical phenotyping of single cells. This requires the capability of characterizing the viscoelastic response of single cells to an external mechanical stimulus. In fact, cell deformability is an important aspect in biological analysis of cell functionalities and health state, resulting in a sensitive and label-free marker to monitor physiological or pathological changes in the cell. The study of cell mechanics can be a powerful tool to predict complex cell changes in a fast and more comprehensive way than the single analysis of biochemical or morphological parameters. This approach has already been validated for the detection of malignant change in cells [25]. Mechanical analysis of single cells is performed by exploiting optical forces in a double counterpropagating optical trap [26, 27]. Two optical waveguides or fibres, perfectly aligned on the two sides of a microfluidic channel, can trap a transparent particle through the optical forces exerted by the light emitted from the two sides. Increasing the optical power in the trap it is possible to also stretch the cell. By continuously imaging the cell contour one can retrieve the cell deformation and thus characterize the deformability cell by cell. Optical cell trapping and stretching has been predominantly a scientific tool, with no commercial or biomedical impact yet. However, cancer diagnosis with the optical stretcher has the potential for a broad impact on public health, since it can analyse very small number of cells with high sensitivity in a fast and reliable way, thus enabling early detection of a disease.

16.3   Problem Statement and New Approach

The new approach that we present in this work aims at satisfying the following requirements:

- use of polymers as materials for LoCs;
- development of a rapid prototyping technique;
- need for a low-cost fabrication technique for mass production;
- need for integrated imaging system for on-chip analysis.

Although μIM has demonstrated to be an excellent technique for mass production of polymer devices, it shows its weaknesses in a still non-mature market, where prototyping of new devices and small batch productions is the most common requirement. The limit of μIM mainly resides in the low reconfigurability of the mould itself and in the high precision required to reproduce the micron-sized features. The common effects of polymer creep or shrinkage in the final products imposes an accurate design optimization of both the device and the mould. In addition, improved mould manufacturing strategies and control of the moulding conditions are required to allow μIM to become a reliable mass production technology that helps the spreading of microfluidic devices.

In this work, we present a variation of the traditional μIM process that makes it a versatile micro-manufacturing platform for the production of LoCs. In fact, we will cope with the weak points of μIM thanks to a synergic combination with femtosecond-laser-micromachining, thus creating a manufacturing process that is both effective for the prototyping and for the mass production.

Femtosecond laser micromachining will be employed in different steps of the LoC production cycle in order to provide the following advantages: (i) rapid prototyping of LoCs by direct laser ablation of a polymeric substrate; (ii) tool fabrication by laser milling of steel for mass production of the LoC by microinjection moulding; (iii) direct welding of polymeric slides for sealing the microfluidic network of the LoC without the addition of any glue.

The possibility to use femtosecond lasers for rapid and precise structuring of steel is particularly effective when associated to a novel approach to microinjection moulding, based on the use of application-tailored removable inserts in a standardized mould. This approach will greatly enhance the versatility of μIM. In fact, this approach will require the substitution of only small parts in the mould (the inserts) when changing the geometrical features of the device. The small steel inserts can be effectively micromachined by high-resolution femtosecond laser ablation without the drawback of long processing time that would be required to machine the whole mould.

A further critical point in the fabrication of plastic microfluidic devices is the sealing of the channels. Currently, the most widely used method for joining polymer parts is adhesive or thermal pressure bonding. The critical aspects of these approaches are the blocking of the channel by adhesive overflow and the damage of the microstructures due to the high temperatures and pressures that are applied. For these reasons, we have developed a novel fs-laser-based welding technique that can be exploited to join the fabricated PMMA slabs to obtain the final, closed, LoC device [28].

As previously discussed, holographic methods have been demonstrated to be quite successful for the imaging of specimens flowing inside microfluidic chips. However, the traditional lensless approach represents an inner limitation to the available resolution. Proper algorithms, mainly based on subpixel shifting, are needed to overcome the limit imposed by the pixel size [29]. As an alternative, we propose a lens-based approach that integrates the imaging functions on the chip by fabricating the lenses directly on top of it. We will use a method based on a pyro-electrohydrodynamic ink-jet technique that allows to deposit polymer drops on a polymer substrate [30].

16.4 Developed Technologies, Methodologies and Tools

In this section we will provide a technical discussion of the technologies and methodologies that we exploit in our new approach for LoC manufacturing. We will discuss the removable inserts concept (Sect. 16.4.1) and their micromachining with femtosecond lasers (Sect. 16.4.2.2). We will also discuss the use of femtosecond lasers to seal the LoCs (Sect. 16.4.3) and the use of forward pyro-printing to integrate microlenses on the chip (Sect. 16.4.4).

16.4.1 Removable Inserts Concept

As previously discussed, the main limitation of μIM technology is the low reconfigurability of the mould once its geometry is defined. In fact, any modification to the mould cavity is very expensive. Our approach is based on a fixed master mould and several exchangeable inserts, which can be easily replaced to produce devices with different functionalities. In this way, the complexity of the mould increases, but it allows a significant improvement in its versatility. The combination of different micro-manufacturing technologies has been evaluated in order to produce inserts that can be easily assembled in the mould and that can provide the microfeatures required by the LoC design.

Femtosecond laser ablation of steel has been used to produce the negative of the microfeatures required on the plastic devices. This technology allows to process reasonably large surfaces with small depth features and at high removal rate. In addition, it provides geometrical precision and surface roughness values that meet the requirements of the microfluidic application. To implement easy connection of the LoC with external tubing, complex 3D micro features with high aspect ratio have been produced with micro Electro Discharge machining (μEDM) technology.

In the following subsections the design of the LoC, the mould design and the tool insert manufacturing are described in detail.

16.4.2 Device Design

The targeted device is an optical stretcher, consisting of two optical fibres, facing each another on the opposite sides of a microfluidic channel. The counter-propagating light beams delivered by the two fibres exert optical forces on the passing cells, trapping and deforming them. A microscopic imaging of the cell deformation for calibrated optical forces allows retrieving the mechanical compliance of each cell. The conceptual design of the device (Fig. 16.1) is made by two shells (19.5 mm length, 7.5 mm large and 0.7 mm thick) assembled together by femtosecond laser welding [28]. On the lower shell, a microchannel of 100 μm width and 90 μm depth with final reservoirs of 1 mm diameter is produced. V–grooves have been realized for the optical fibre housing with depth and width of 180 μm and 235 μm, respectively. Connectors on the upper shell have been conceived to reduce the number of external components involved in the assembly, thus enabling a direct connection of external capillaries to the chip.

16.4.2.1 Mould Design

The mould consists of two cavities, corresponding to the two shells depicted in Fig. 16.1, implemented on two removable inserts (Fig. 16.2b, c).

A filling and packing phase simulation with Autodesk Moldflow Insight 2016® has been preliminary performed to verify the device geometry and to define the gate design. It is important to avoid incomplete filling weld lines and voids (Fig. 16.3a) and to validate the set of the identified parameters through the analysis of the shrinkage distribution (Fig. 16.3b), which is expected to be uniform along the whole component.

The versatility of the mould design is ensued by the possibility of modifying each micro feature independently from the others simply replacing the related insert with a new one. Furthermore, the cavities are independent from the feeding system and can

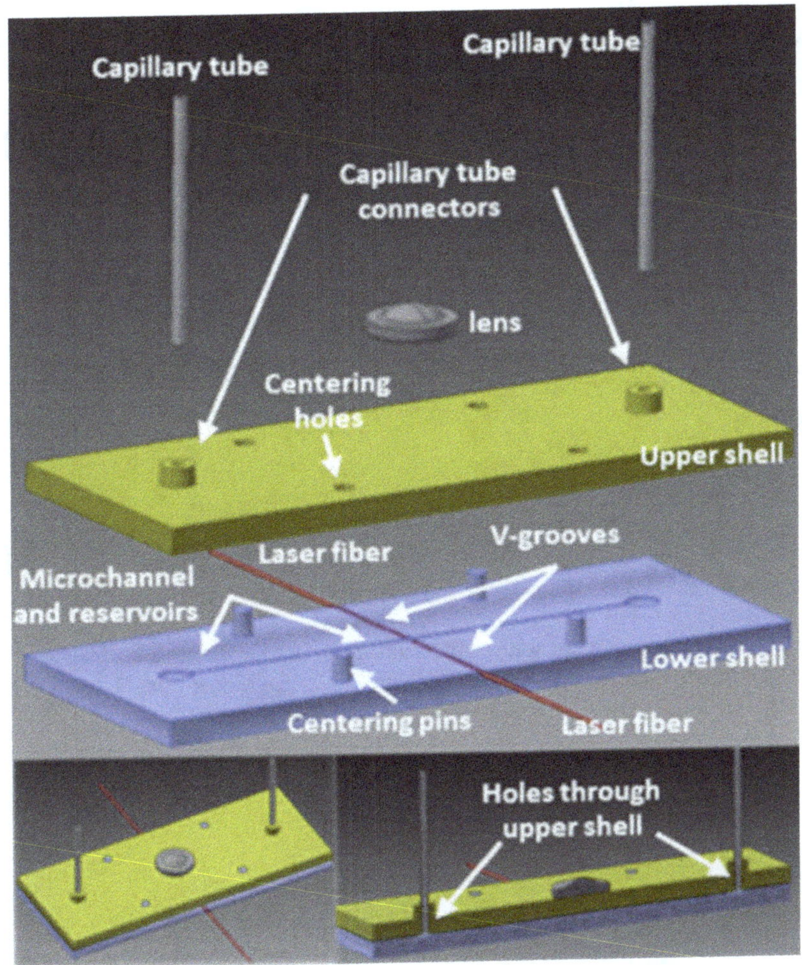

Fig. 16.1 Final design of the optical stretcher LoC developed in this work

be easily replaced thanks to the single sprue that is divided into a double symmetric gate facing the inserts with cavities (Fig. 16.2a).

Since the volume of polymer required for each cavity (about 125 mm^3) is close to the limit capacity of the micro injection moulding machine (150 mm^3), a couple of interchangeable pins having different height are used alternatively to close a runner while the other one is open (Fig. 16.2a—inset) so that one cavity at a time is filled.

16.4.2.2   Removable Inserts Fabrication

A synergic combination of different technological approaches has been adopted for the fabrication of the removable inserts. Traditional milling has been used to machine the tool with its plates and ejection systems, while the bulk cavity has been realized by die-sinking EDM. The inserts most critical features like microchannel, reservoirs and V-grooves have been machined by femtosecond laser micromilling.

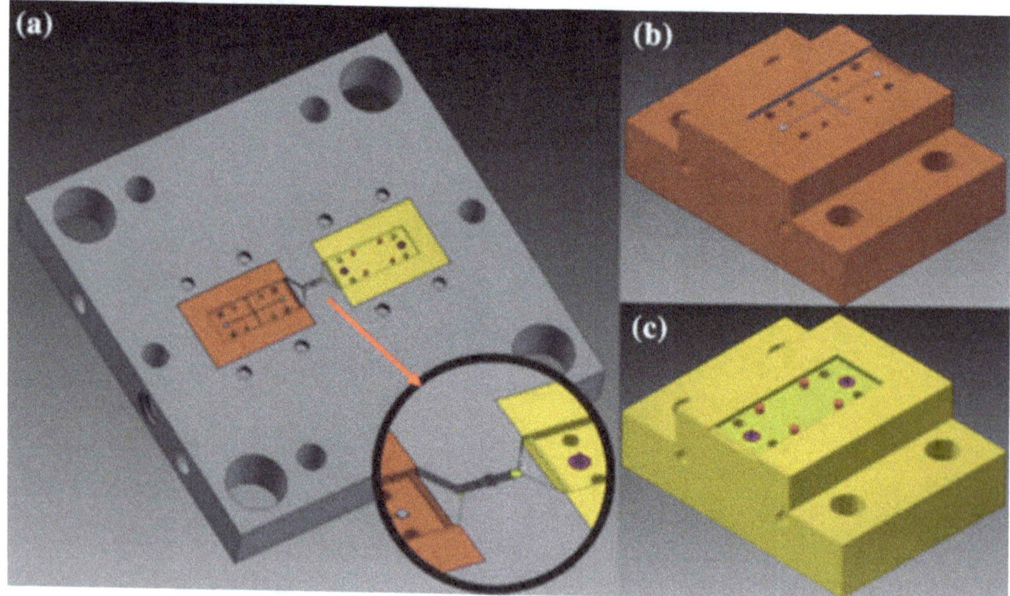

Fig. 16.2 Design of the mould plate with runners and gates (**a**). Detailed view of the interchangeable pins used to switch between cavities ((**a**)—inset) and of the cavities inserts (**b, c**)

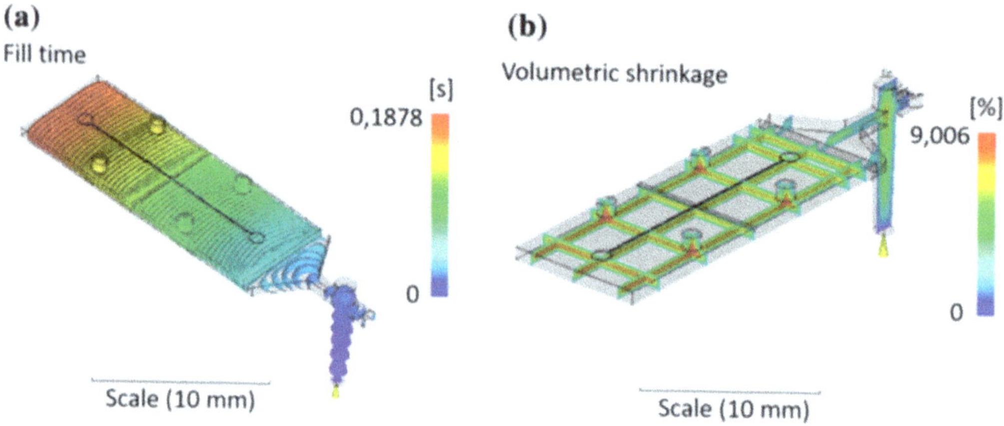

Fig. 16.3 **a** Polymer filling distribution (time contour plot) and **b** volumetric shrinkage (section probe plot)

In order to simplify the connection of capillary tubes to the LoC, moulds for access holes were created on the insert corresponding to the top PMMA slide by micro Electro Discharge Machining (μEDM) technology. In Fig. 16.4a the design geometry and dimensions of the tube housing insert are reported. Starting from a cylindrical workpiece, assembled and referred to the main mould, the central conical pin is obtained by eroding the cavities following a layer-by-layer strategy of 0.004 mm thickness to obtain a smooth surface and avoid step-like effects. Then, a smaller electrode tool, having a diameter of 0.15 mm, has been used to machine the narrow cavity floor at the pin root.

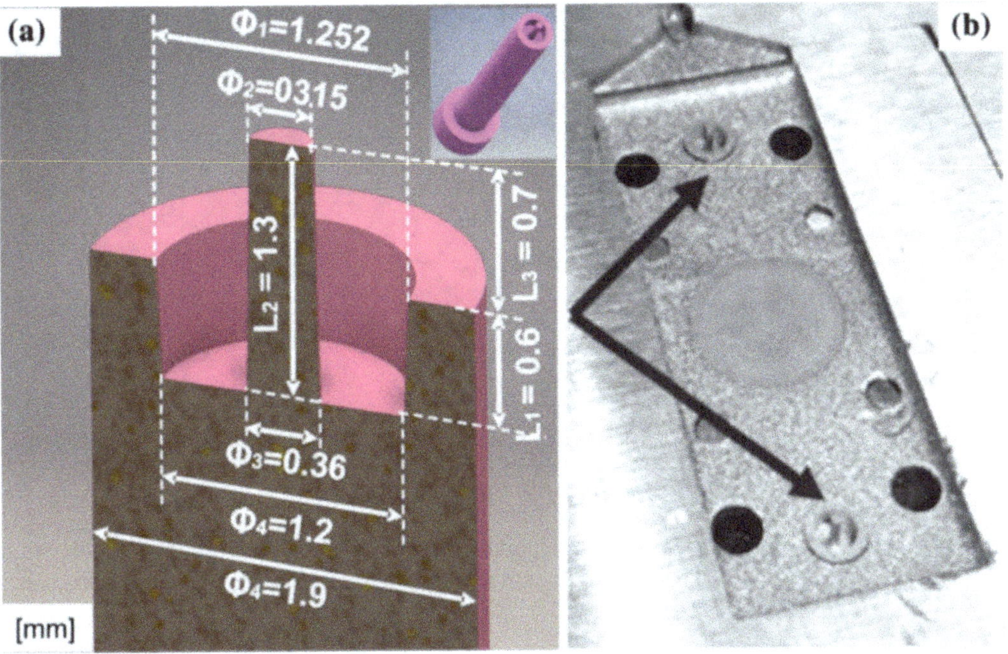

Fig. 16.4 Design of the capillary tube housing insert (**a**) and picture of its implementation in the μEDM-fabricated insert (**b**)

The same process has been adopted for realizing a smooth circle area of 3 mm diameter in the centre of the insert (Fig. 16.4b) in order to reduce roughness where the lens, required for the imaging of the samples flowing in the microfluidic channel, will be located.

Femtosecond laser micromachining was used to mill the metal insert containing the complex 3D microfeatures of the bottom layer of the LoC.

Preliminary fs-laser milling tests on steel have been carried out to find the best combination of process parameters, namely the average power, the translation speed and the spacing of the hatch d, to get the targeted surface roughness. Each test has been executed on a 3 mm^2 area, moving the beam along vertical and horizontal parallel lines. After laser milling, the specimens were ultrasonically cleaned with isopropanol for 5 min to remove the recast debris. Then, depth and roughness measurements were performed with a confocal microscope. The calculation of the roughness is based on the DIN EN ISO 4287 standard.

The pulse repetition rate has been set at the lowest value delivered by the ultrafast laser source, i.e. 50 kHz. In spite of the reduced material removal rate, working at such a low repetition rate produces smoother surfaces because of the reduced thermal load delivered onto the irradiated area [31]. The laser beam was focused and scanned on the insert surface by a galvo-scan head, equipped with an F-theta lens of 100 mm focal length, so that the beam spot size on the metal surface is about 25 μm.

Firstly, the beam translation speed was fixed and the influence of the laser average power on the surface quality was investigated. The surface roughness dramatically increases with the laser fluence. For laser power values below 600 mW a gentle

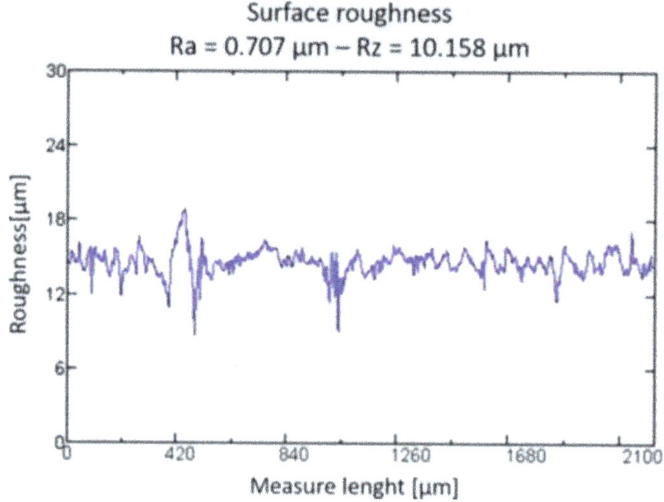

Fig. 16.5 Surface roughness measurement obtained through confocal microscopy (magnification 100x) of the milled surface (200 scanning loops; rep. rate: 50 kHz; average power: 0.4 W; hatch: 2 µm)

ablation regime is established producing smooth surfaces. As soon as the laser fluence is increased, ablation occurs through a process referred to as phase explosion, characterized by homogeneous nucleation of bubbles in vapour phase leaving significant pores on the irradiated area and a surface with orange peel morphology and a higher roughness [31].

The influence of the scan speed from 50 mm/s to 1 m/s on the surface quality has been also investigated. It was noticed that the slower the speed, the less homogeneous is the ablated surface. Furthermore, for the slowest investigated speed values (50 and 100 mm/s) some grooves and pores appear on the surfaces. Therefore, the best results in terms of surface quality are obtained for scan speeds higher than 300 mm/s, as already confirmed by previous works [31, 32]. Finally, the inter-line spacing d of the hatch has been optimized. Values of d comprised between 2 and 20 µm have been investigated, finding that the most homogeneous and smooth surfaces are obtained for $d = 2$ µm.

It has been furthermore verified that the ablation depth of the micromilled area can be finely controlled with micrometre precision by adjusting the number of scanning loops while keeping the surface quality almost unchanged.

Based on the results of the preliminary fs-laser micro-milling tests, the following set of laser parameters has been selected for the fabrication of the mould insert: 50 kHz for the pulse repetition rate, 400 mW of laser average power, 600 mm/s of scanning speed and 2 µm of hatch distance. Results of the measurement of surface roughness Ra of a fs-laser milled area of about 600 µm length are shown in Fig. 16.5. Ra values below 1 µm can be obtained with this set of laser parameters.

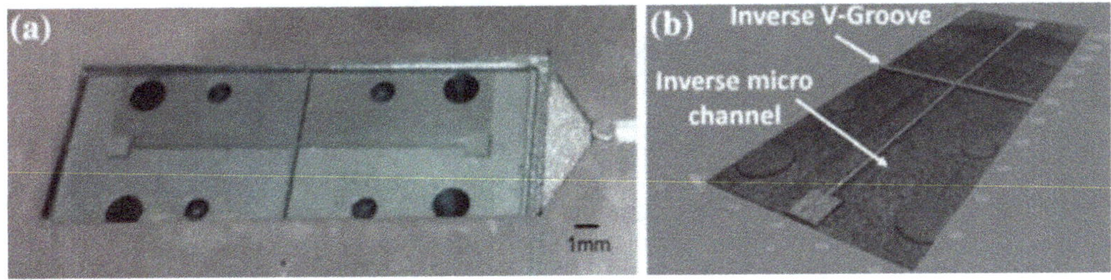

Fig. 16.6 **a** Picture of the insert machined by fs-laser. **b** Characterization of the 'negative' microchannel and V-grooves produced on the insert (Zeiss Axio CSM 700)

A picture of the final fs-laser-fabricated mould insert is shown in Fig. 16.6a, whereas Fig. 16.6b shows a confocal image of the same surface. The targeted micro-features sizes have been successfully reproduced with a tolerance of about 2 μm in the worst case.

16.4.3 Sealing of the Microfluidic Chip

The two polymeric shells were finally assembled by direct fs-laser welding. Although already successfully demonstrated for glasses, this kind of approach was never applied to transparent polymeric substrates.

Firstly, the combinations of process parameters producing melting of the laser focal volume at low pulse energies and high repetition rates were estimated by solving the heat conduction equations describing the pulsed-laser irradiation of a PMMA substrate [28]. Then, the modifications induced by focusing fs-laser pulses in the volume of the bulk material were experimentally investigated. The goal of this study was to identify an appropriate window of process parameters producing homogeneous and continuous melting, spatially localized in the focal region. This time the laser beam focus was achieved through a 0.3 numerical aperture lens and positioned 1.5 mm below the surface of a 3 mm-thick PMMA specimen. The beam was then moved along single and multiple parallel lines. Several tests were conducted by operating the laser source at 200 kHz, 500 kHz, 1 MHz, 5 MHz with average power ranging from 0.1 to 1 W. The translation speed was varied from 0.01 to 10 mm/s. Each test was repeated with 650 fs and 18 ps pulse durations in order to understand the influence of this parameter on the material modification.

The tests demonstrated that the peak power has a significant role in triggering changes of the material. Therefore, we scaled the pulse energies at the two pulse durations to have the same peak power. At both pulse durations, it was evident that heat accumulation is enhanced at higher repetition rates (>1 MHz) and low translation speeds (<1 mm/s), in agreement with the simulation results. We observed that better quality results are achieved with 650-fs pulses rather than with 18-ps pulses, therefore we chose the shorter pulses for performing this process.

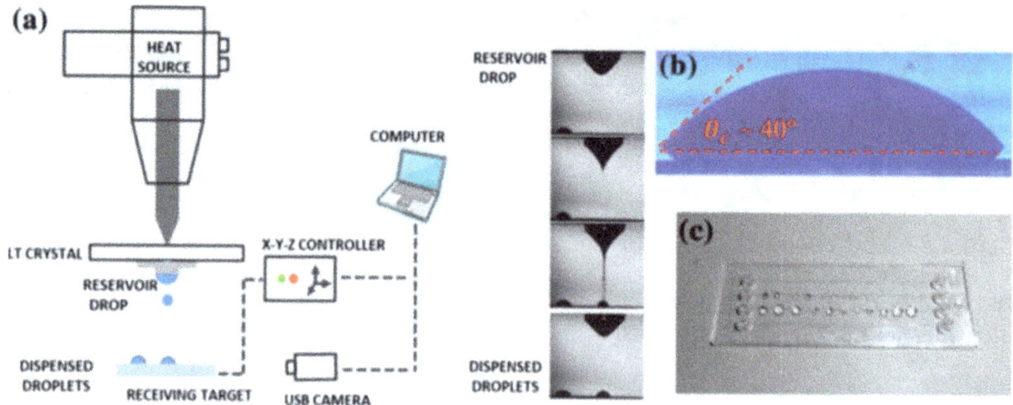

Fig. 16.7 **a** A schematic set-up of the pyro-EHD printing of polymer microlenses onto microfluidic chip and side view images of a typical sequence of dispensed microlenses. **b** Geometrical properties of the deposited lens. **c** A pattern of microlenses with different sizes produced on top of a microfluidic device

Based on the results of these preliminary tests, a suitable set of laser parameters was selected producing homogeneous melting in the focal region. The laser source was operated at 5 MHz with pulse energy of 0.4 µJ. The two injection-moulded PMMA shells (Fig. 16.7) composing the optical stretcher were mechanically clamped to achieve an air gap of a few micrometres in the welding area. The laser beam was focused at the interface between the two plates and was moved with a translation speed of 0.1 mm/s. The laser-welding path surrounded the microfluidic network composed of the two reservoirs, the channel and the two V-grooves.

16.4.4 On-chip Imaging

As a preliminary step, the micro-lenses were integrated onto a commercial PMMA microfluidic device (ChipShop GmbH) through forward pyro-printing [33]. Pyro-EHD printing is a novel technique to develop micro-lenses integrated on LoCs while guaranteeing high chemical compatibility, good optical characteristics and flexibility in the design of the realized devices [34].

Micro-lenses fabrication was performed using the printing configuration showed in Fig. 16.7a. A polydimethylsiloxane drop (PDMS Dow Corning Sylgard 184, 10:1 mixing ratio, base to curing agent Midland; $\eta = 3900$ cP, $\varepsilon p = 2.65$ from datasheet), was deposited as a reservoir drop on the pyroelectric crystal and placed above the PMMA device. Once an external heating source is applied onto the crystal, the PDMS drop starts to deform assuming a conical shape, known as Taylor's cone, and dispensing micro-nano and pico-drops [35]. Figure 16.7c shows pictures of a microfluidic device with PDMS micro-lenses pyro-ink-jet-printed on it. Micro-lenses could present different shapes and dimensions, as a function of the pyro-inkjet printing parameters. Once the lenses were deposited on the channel, the device was heated at 80 °C for 30 min, in order to cure the polymer.

A fluorosilane agent (Fluorolink S10, Solvay Solexis) was used to modify the substrate wettability to avoid excessive spreading of the PDMS lens onto the PMMA substrate. Moreover, the antireflection properties of the coating guarantee good optical properties of the final device. Furthermore, fluorosilane has good chemical resistance and endures abrasion. An O_2 plasma treatment (Femto System, Diener Electronic GmbH & Co. KG, Ebhausen, Germany) was performed on the microfluidic device to improve the adhesion between the fluorosilane agent and the PMMA substrate.

16.5 Testing and Validation of Results

In this section we will present the characterization of the structures fabricated with the techniques and methodologies described in the previous section. We will discuss morphological properties, shape and roughness (Sect. 16.5.1), as well as functional properties, both fluidic (Sect. 16.5.2) and optical (Sect. 16.5.3), and finally we show the operation of the optical stretcher LoC with red blood cells (Sect. 16.5.4).

16.5.1 PMMA Device Characterization and Strategies to Reduce the Roughness

The µIM production of the two PMMA shells is based on the approach proposed in [36]. The process parameters defined with the simulation and used for the production of the components are reported in Table 16.1 where T_{melt} is the melt temperature, T_{mould} is the mould temperature, V_{inj} is the injection speed, P_{hold} is the packing pressure, t_{hold} is the packing time, t_{cool} is the cooling time and Run is the plunger injection run.

The µIM machine used for the device production is a DESMA Tec Formica Plast 1 K and the polymer Acyrex CM211 PMMA was selected as material for the device fabrication.

The adopted strategy to reduce mould roughness was to use a very low spark energy level during µEDM milling and to adjust the number of scanning loops during femtosecond laser ablation. With the selected parameters (reported in previous sections) we obtained microfluidic channels with a roughness (Ra of 0.59 µm) and a

Table 16.1 Set of process parameters

Material	T_{melt} (°C)	T_{mould} (°C)	V_{inj} (mm/s)	P_{hold} (bar)	t_{hold} (s)	t_{cool} (s)	Run (mm)
PMMA	250	80	100	1000	5	5	10

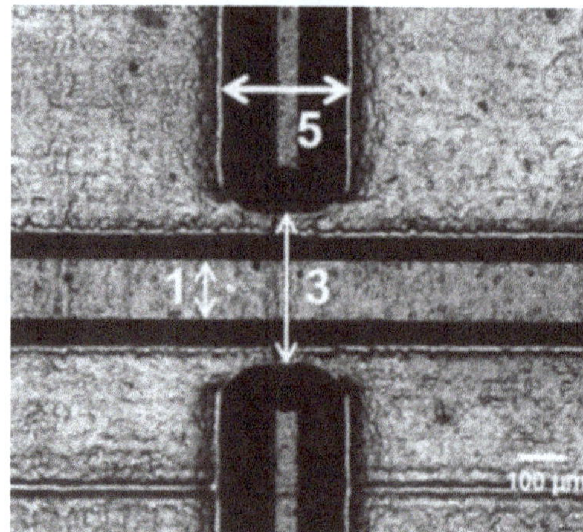

Fig. 16.8 Optical microscope image (left) and confocal microscope image (right) of the microchannels and V-grooves. The critical parameters are: 1: channel width, 2: channel depth, 3: V-groove fibre distance, 4: V-groove depth and 5: V-groove width. The obtained dimensions are within 6% of the expected ones (measurements done with confocal microscope ZEISS AXIO CSM 700)

shape that match the designed values (see Fig. 16.8). The surface of the access holes on the upper shell presented a relatively low roughness (Ra of 0.37 µm) if compared with raw material. The capillary tube was easily inserted in the access hole providing a good housing adhesion when the fluid flew.

Finally, the V-grooves were realized orthogonal and coaxial to each other (Fig. 16.8) to allocate two standard optical fibres perfectly aligned and emitting two counter-propagating beams at a height of 20 µm with respect to the bottom of the microchannel. This height was chosen to maximize the number of cells intercepted by the optical trap. The obtained feature dimensions (see Fig. 16.8) are within 6% of the expected ones.

To perform optical imaging of the specimens inside the microfluidic channels, it is mandatory to have a channel bottom surface with optical quality. We developed a chemical post-treatment to smoothen the channel surface: the chip was exposed to Chloroform vapour (at 40 °C) for a few minutes and then annealed at low temperature (80 °C) for 20 min [37]. In this way, the roughness of the surface was reduced (see Fig. 16.9, showing the microchannel bottom before and after the treatment). The improved quality of the surface enabled accurate imaging of the elements flowing in the channel. In addition, this smoothening process is very simple and can be applied in parallel to several chips. Therefore, it relaxes the specifications on the quality of the micromachined sample allowing faster and cost-effective fabrication of the devices.

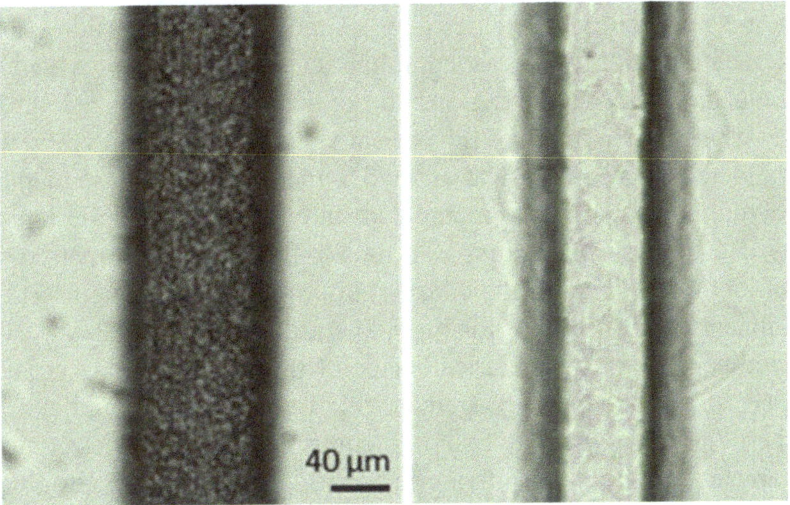

Fig. 16.9 Microscope image (in transmission) of a fs-laser-micromachined channel before (left) and after (right) the chloroform vapour treatment (the roughness is reduced from ~1–2 µm to ~300–400 nm)

16.5.2   Laser Bonding Characterization

Firstly, static leakage tests were carried out. A dead-end reservoir, composed of two laser-bonded PMMA slides, one of them preliminarily laser-drilled, was connected to a microfluidic pump (MFCS-EZ, Fluigent) [28]. A blue-coloured liquid was injected at a pressure between 100 mbar and 1 bar with steps of 100 mbar, retaining each value for 3 min. The possible occurrence of fluid leakages was monitored through an optical microscope.

Further leakage tests were executed on a laser milled microchannel, which is the main core of any LoC. The microchannel was fabricated on the surface of a 1-mm-thick PMMA layer by direct femtosecond laser ablation, resulting in a square cross section of 80 µm side and a length of 1 mm. The channel connected two laser micro-machined reservoirs, each of them provided with a fs-laser drilled hole. This micro-machined layer was laser bonded to a standard PMMA slide all along the perimeter of the channel and of the reservoirs. A blue liquid was pumped in one reservoir, while the other end was hermetically closed. The leakage test was repeated several times on the same device to demonstrate that the laser bonded contour was able to withstand pressure over time [28].

16.5.3   Integrated Lenses Characterization

The lenses were characterized interferometrically. In fact, an interferometric characterization based on Digital Holography (DH) microscopy was carried out to evaluate the optical aberrations of the microlenses and their focusing properties. The optical set-up was a classical Mach-Zehnder interferometer in transmission configuration [33]. Once the hologram was recorded, numerical propagation provided the object complex field in whatever plane along the optical axis z, from which the intensity and the phase distributions of the optical wave field transmitted by the sample could be extracted. A 2D fitting was applied to the recovered phase map by using a Zernike polynomial expansion to model and study the optical aberrations produced by the lens. The analysis of the first 10 orders of the linear combination of Zernike polynomials showed that, after the constant offset (P = 1), the main contribution to the development of the Zernike function was a defocus term (P = 5). This result was expected because of the spherical shape of the lens. Furthermore, a tilt along the y-axis (P = 3) was identified together with a coma aberration along the x-axis (P = 8). The other terms gave a negligible contribution [33]. After completing the optical characterization of the polymeric micro-lenses, their imaging capabilities were tested to show one of their possible applications on a LoC platform. Fibroblast cells were inserted to flow inside the microfluidic channel where the lenses were deposited and aligned. The images of the cells created by the on-chip lens were further imaged by an optical microscope [33]. Figure 16.10a, b show the images of the cells inside and outside the lens area. In particular, these are clearly in-focus only when they flow below the lens, while these are out-of-focus in all the other portions of the channel. In Fig. 16.10a it is also possible to notice the magnifying action operated by the lens, as the channel itself appears to be enlarged. Moreover, in order to proof the lens magnification, a test resolution target placed behind the chip is imaged. It is evident that the smallest target segments in the red box of Fig. 16.10c are not visible at all, while these get resolved when observed through the lens (see the area in the red box in Fig. 16.10d) [33].

16.5.4 Validation of the LoC Device: Stretching of Red Blood Cells

The integrated optical stretcher (Fig. 16.1) consists of a central microfluidic channel (width 100 μm, depth 90 μm) which is crossed perpendicularly by the light beams coming from the two fibres. After some preliminary experiments, it was clear that the fibres should arrive directly to the microfluidic channel without any plastic wall in front of them. In fact, the power required to trap and stretch cells is sufficient to deform and damage the thin polymeric material that was initially left in front of the fibre to improve the channel sealing. The removable insert was thus modified with the femtosecond laser in order to remove the small wall between the fibre ducts and the central microfluidic channel.

An image of the device is shown in Fig. 16.11. The cross-shaped trace of the laser bonding is clearly visible. It should be noted that the welding trajectory of the fs-laser has been changed several times depending on the specific chip design, further evidencing the simple reconfigurability of femtosecond laser processing. The

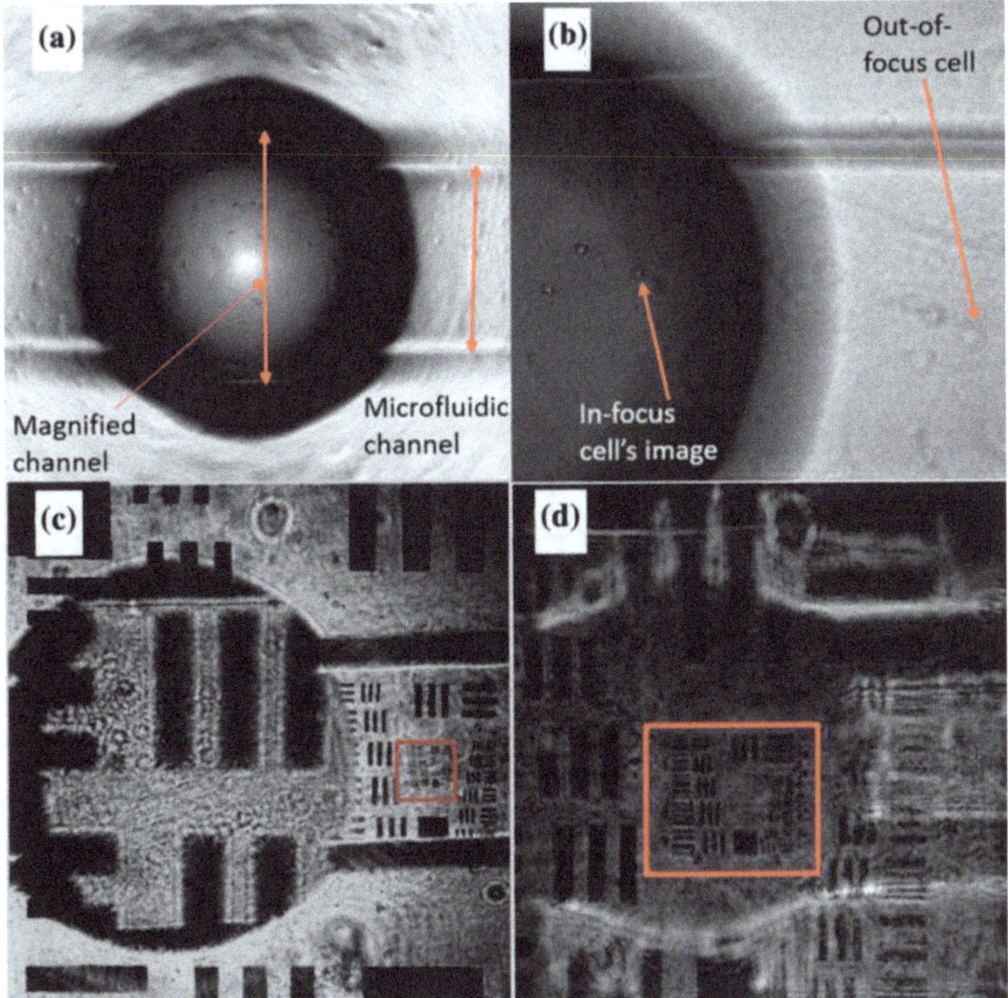

Fig. 16.10 Image of cell flowing inside a microfluidic channel with a plastic lens deposited on the chip surface. **a** and **b** details of the flowing cells, **c** and **d** imaging of a test resolution target

fibres were glued in their specific holders. It is important to point out that no further alignment of the fibres is needed as they are forced to follow the ducts that are precisely moulded.

Figure 16.12a shows the experimental set-up that was designed to demonstrate the effectiveness of the integrated optical stretcher. A continuous ytterbium fibre laser (YLD-5 k-1070, IPG Fibertech), characterised by an emitting power up to 5 W at 1070 nm, was used as light source. The beam coming from the laser was split into two branches by means of a 50–50% fibre coupler, all the components were single mode at the working wavelength as well as the spliced bare end-fibres (HI-1060-Corning) [27]. The chip was mounted on an inverted microscope, equipped for phase contrast microscopy (Leica), to image the cells in the integrated microfluidic channel. Phase

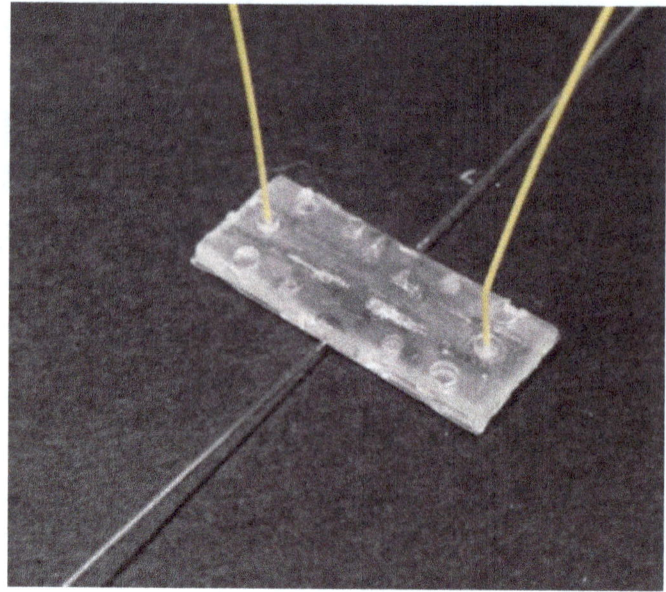

Fig. 16.11 Picture of the plastic optical stretcher after the final assembly

contrast images of optical trapping and stretching of cells were acquired by a CCD camera (DFC310 FX, Leica) through a bandpass optical filter rejecting the infrared light (FGS550, Thorlabs).

We tested the trapping and stretching capabilities of the prototype by using red blood cells. We prepared a solution of 10 µL of blood in 8 mL of hypotonic solution, in which the Red Blood Cells (RBC) acquired a quasi-spherical shape with a radius of ~4 µm. The cell suspension was transported in the microfluidic channel by an external microfluidic pump (MFCS-EZ, Fluigent). The value of the cell speed was set in the 10–50 µm/s range for an easy imaging of the flowing cells. RBCs optical trapping was achieved with an estimated optical power at each fibre output of about 30 mW. When a single cell was stably trapped in the microchannel it could be stretched along the trap axis by simultaneously increasing the optical forces applied to the cell by the two counter-propagating beams [27]. Figure 16.12 shows the image of a trapped RBC inside our chip (b), an increase of the optical power in the two fibres clearly produced the stretching of the cell (c). The current device suffered from the limitation of the surface roughness (measured to be ~600 nm), which did not allow to achieve optimal quality images. This limit could be overcome by making the chloroform vapour treatment to the channel before welding it to the top slab, as explained in Sect. 16.5.3. Figure 16.12d shows the image of a stretched RBC inside a microfluidic channel after the chloroform treatment, indeed the imaging performance was extremely improved. With this image quality, it will be possible to digitally analyse the deformability of the trapped cells.

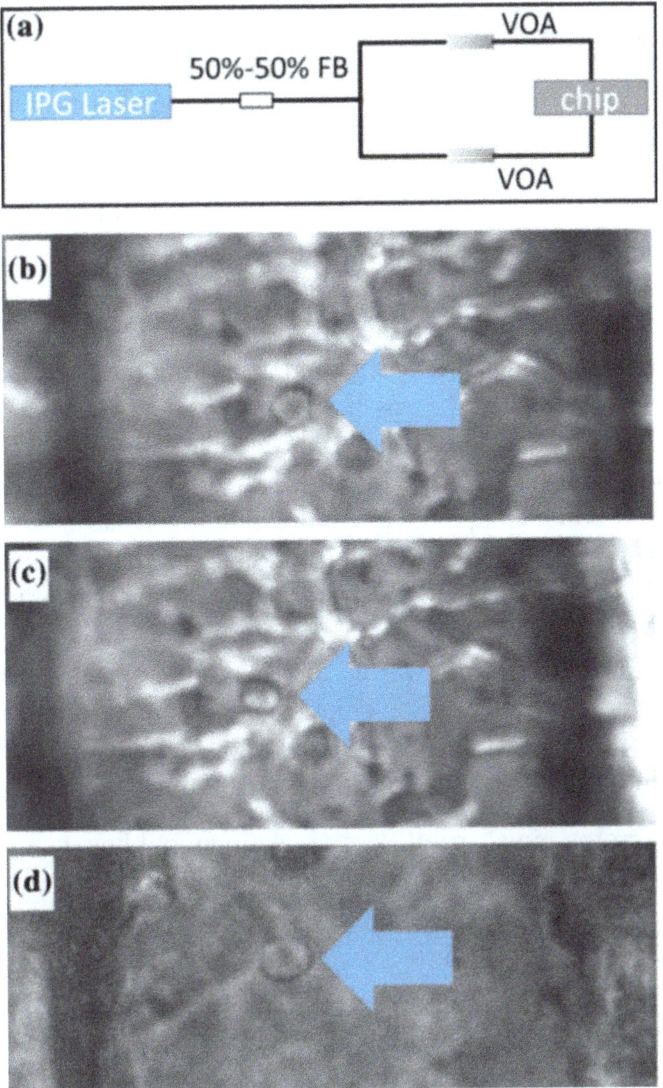

Fig. 16.12 a Scheme of the experimental setup used to demonstrate the LOC working principle.
b Microscope image of a trapped red blood cell **c** microscope image of a stretched red blood cell
d image of the stretched red blood cell inside a chloroform treated channel (radius of a RBC ~4 μm)

16.6   Conclusions and Future Perspectives

The main goals of work have been successfully achieved. The concept of a mould
tool with removable inserts has been fully demonstrated. The steel inserts have been
ablated by a femtosecond laser according to the design of the integrated optical
stretcher and to accurate numerical simulations of the micro injection moulding
process.

The processing parameters for the laser ablation of the inserts have been carefully
optimized in order to achieve the best surface roughness and the highest fidelity to
the design.

We have designed and demonstrated a new access-hole concept for easy plugging external capillaries to the chip. These access-holes were directly moulded in the top shell of the LoC and make the device extremely compact and easy to connect to the external microfluidic circuit.

In addition to the design flexibility allowed by femtosecond laser micromachining, we have also taken advantage of micro-injection moulding by producing hundreds of devices in a short time. In particular, we have created two different generations of inserts for the top and bottom halves of the integrated optical stretcher during the research activity. This would not have been possible, in terms of both time and cost, if it were not for the new manufacturing approach introduced here.

We have also demonstrated the possibility to weld the two slides together with a femtosecond laser, without the need of any intermediate adhesive layer. This confirms the femtosecond laser as a very versatile tool, capable of solving several important tasks that perfectly complement the injection moulding of lab-on-chip devices.

For the integrated imaging of the microchannel content, we have developed a method to fabricate polymer microlenses directly on top of microfluidic chips. This method, based on a pyro- electrohydrodynamic ink-jet technique, allows one to deposit droplets of PDMS on top of the microfluidic chip that will act as optical lenses. Through the translation of the chip, using a computer-controlled XY stage, an array pattern of polymer microlenses with various shapes was formed.

We can resume the main technological outputs of this work in three points:

- Adding rapid-prototyping capabilities to injection moulding, using the removable insert approach together with femtosecond laser machining of the inserts.
- Demonstrating that optical integration on-chip is possible, fabricating plastic lenses directly on the device surface.
- Increase the portability and reduce the cost of the biosensing systems, by the demonstration of an optical cell manipulator with different functionalities integrated on it.

In terms of perspectives, the developed approach was further exploited in the demonstration of a new prototype of a fully portable plastic LoC for flow cytometry coupled to a cell-phone [38]. This second application required a completely different LoC layout, which was developed and optimized in a very short time exploiting femtosecond laser micromachining of the removable inserts. The versatility of μIM enhanced by fs-laser micromachining has the potential of greatly reducing the development time and cost of new LoC devices. These advantages will be particularly beneficial for small-medium enterprises that are very dynamic in developing new products, but do not have the economic capability to produce many different moulds for the optimization of the LoC layout. In addition, this manufacturing versatility will favour custom-oriented solutions that are particularly relevant in the LoC market where there is not yet a single application requiring large volume productions, but more diversified applications requiring different small productions.

Acknowledgements This work has been funded by the Italian Ministry of Education, Universities and Research (MIUR) under the Flagship Project "Factories of the Future - Italy" (Progetto Bandiera "La Fabbrica del Futuro") [39], Sottoprogetto 1, research projects "Plastic Lab-on-a-chip fabricated by Ultrafast laser Sources" (PLUS) and "Imaging Citometry in Plastic Ultra-mobile Systems" (IC+).

References

1. Whitesides GM (2006) The origins and the future of microfluidics. Nature 442:368–373
2. Abgrall P, Gue AM (2007) Lab-on-chip technologies: making a microfluidic network and coupling it into a complete microsystem—a review. J Micromech Microeng 17:R15–R49
3. Figeys D, Pinto D (2000) Lab-on-a-chip: a revolution in biological and medical sciences. Anal Chem 72:330–335
4. Kangning Ren K, Zhou J, Wu H (2013) Materials for microfluidic chip fabrication. Acc Chem Res 46:2396–2406
5. Malinauskas M, Žukauskas A, Hasegawa S et al (2016) Ultrafast laser processing of materials: from science to industry. Light Sci Appl 5:1–14
6. Osellame R, Hoekstra HJ, Cerullo G et al (2011) Femtosecond laser microstructuring: an enabling tool for optofluidic lab-on-chips. Laser Photon Rev 5:442–463
7. Sugioka K, Xu J, Wu D et al (2014) Femtosecond laser 3D micromachining: a powerful tool for the fabrication of microfluidic, optofluidic, and electrofluidic devices based on glass. Lab Chip 14:3447–3458
8. Mogensen KB, Klank H, Kutter JP (2004) Recent developments in detection for microfluidic systems. Electrophoresis 25:3498–3512
9. Charrière F, Marian A, Montfort F et al (2006) Cell refractive index tomography by digital holographic microscopy. Opt Lett 31:178–180
10. Giboz J, Copponnex T, Méle P (2007) Microinjection molding of thermoplastic polymers: a review. J Micromech Microeng 17:R96–R109
11. Attia UM, Marson S, Alcock JR (2009) Micro-injection moulding of polymer microfluidic devices. Microfluid Nanofluid 7:1–28
12. Guber AE, Heckele M, Hermann et al (2004) Microfluidic lab-on-a-chip systems based on polymers-fabrication and application. Chem Eng J 101:447–453
13. Becker H, Locascio LE (2002) Polymer microfluidic devices. Talanta 56:267–287
14. Sun Y, Kwok YC (2006) Polymeric microfluidic system for DNA analysis. Anal Chim Acta 556:80–96
15. Chichkov BN, Momma C, Nolte S et al (1996) Femtosecond, picosecond and nanosecond laser ablation of solids. Appl Phys A 3:109–115
16. Schille J, Schneider L, Loeschner et al (2011) Micro processing of metals using a high repetition rate femtosecond laser: from laser process parameter study to machining examples. In: Proceedings of ICALEO: 30th international congress on applications of lasers & electro-optics, Orlando, FL
17. Neuenschwander B, Bucher GF, Hennig G et al (2010) Processing of dielectric materials and metals with ps laser pulses. In: Proceedings of ICALEO: the 29th international congress on applications of lasers & electro-optics, Anaheim, California
18. Eaton SM, De Marco C, Martinez Vazquez R et al (2012) Femtosecond laser microstructuring for polymeric lab-on-chips. J Biophotonics 5:687–702
19. Richter S, Döring S, Tünnermann A et al (2011) Bonding of glass with femtosecond laser pulses at high repetition rates. Appl Phys 103:257–261
20. Watanabe W, Onda S, Tamaki T et al (2007) Joining of transparent materials by femtosecond laser pulses. In: Proceedings of SPIE 6460, commercial and biomedical applications of ultrafast lasers VII, 646017, 13 Mar 2007

21. Tamaki T, Watanabe W, Nishii J et al (2005) Welding of transparent materials using femtosecond laser pulses. Jpn J Appl Phys 44:687–689
22. Watanabe W, Tamaki T, Ozeki Y et al (2010) Filamentation in ultrafast laser material processing. In: Yamanouchi K, Gerber G, Bandrauk, AD (eds) Progress in ultrafast intense laser science VI. Springer Series in Chemical Physics, vol 99. Springer, Heidelberg, pp 161–181
23. Tamaki T, Watanabe W, Itoh K (2006) Laser micro-welding of transparent materials by a localized heat accumulation effect using a femtosecond fiber laser at 1558 nm. Opt Express 14:10460–10468
24. Bishara W, Zhu H, Ozcan A (2010) Holographic opto-fluidic microscopy. Opt Express 18:27499–27510
25. Guck J, Schinkinger S, Lincoln B et al (2005) Optical deformability as an inherent cell marker for testing malignant transformation and metastatic competence. Biophys J 88:3689–3698
26. Lincoln B, Schinkinger S, Travis K et al (2007) Reconfigurable microfluidic integration of a dual-beam laser trap with biomedical applications. Biomed Microdevices 9:703–710
27. Bellini N, Vishnubhatla KC, Bragheri F et al (2010) Femtosecond laser fabricated monolithic chip for optical trapping and stretching of single cells. Opt Express 18:4679–4688
28. Volpe A, Di Niso F, Gaudiuso C et al (2015) Welding of PMMA by a femtosecond fiber laser. Opt Express 23:4114–4124
29. Sobieranski AC, Inci F, Tekin HC et al (2015) Portable lensless wide-field microscopy imaging platform based on digital inline holography and multi-frame pixel super-resolution. Light Sci Appl 4:e346
30. Ferraro P, Coppola S, Paturzo M et al (2010) Dispensing nano–pico droplets and liquid patterning by pyroelectrodynamic shooting. Nat Nanotechnol 5:429–435
31. Bartolo P, Vasco J, Silva B et al (2006) Laser micromachining for mould manufacturing: I. The influence of operating parameters. Assem Autom 26:227–234
32. Lee S, Yang D, Nikumb S (2008) Femtosecond laser micromilling of Si wafers. Appl Surf Sci 254:2996–3005
33. Vespini V, Coppola S, Todino M et al (2016) Forward electrohydrodynamic inkjet printing of optical microlenses on microfluidic devices. Lab Chip 16:326–333
34. Coppola S, Vespini V, Nasti G et al (2014) Tethered pyro-electrohydrodynamic spinning for patterning well-ordered structures at micro- and nanoscale. Chem Mater 26:3357–3360
35. Vespini V, Coppola S, Grilli S et al (2011) Pyroelectric adaptive nanodispenser (PYRANA) microrobot for liquid delivery on a target. Lab Chip 11:3148–3152
36. Trotta G, Volpe A, Martínez Vázquez R et al (2016) Design and fabrication of a mould with multiple inserts for a polymeric microfluidic device. In: International conference on multi material micro manufacturing—4M2016. https://doi.org/10.3850/978-981-11-0749-8 733
37. De Marco C, Eaton SM, Martínez Vázquez R et al (2013) Solvent vapor treatment controls surface wettability in PMMA femtosecond-laser-ablated microchannels. Microfluid Nanofluid 14:171–176
38. Martínez Vázquez R, Trotta G, Volpe A et al (2017) Rapid prototyping of plastic lab-on-a-chip by femtosecond laser micromachining and removable insert microinjection molding. Micromachines 8:328–336
39. Terkaj W, Tolio T (2019) The Italian flagship project: factories of the future. In: Tolio T, Copani G, Terkaj W (eds) Factories of the future. Springer

Chapter 17
CIGS-Based Flexible Solar Cells

Edmondo Gilioli, Cristiano Albonetti, Francesco Bissoli, Matteo Bronzoni, Pasquale Ciccarelli, Stefano Rampino and Roberto Verucchi

Abstract This chapter reports the progress on the fabrication of thin film CIGS-based solar cells by means of the low temperature pulsed electron deposition technique. The innovative and multidisciplinary approach aims to solve the main issues preventing a possible industrial scale up of the process, i.e. the need of a fast, reliable and automated process, suitable for both static and dynamic deposition of CIGS solar cells on flexible substrates. The final goal is to open new opportunities, particularly in the emerging field of the building-integrated photovoltaic.

17.1 Scientific and Industrial Motivations

The global economic situation of the renewable energies and the photovoltaic (PV) market is in continuous evolution. The economic analysis goes beyond the scope of this work, but the analysts agree that renewable sources will gradually but unavoidably fill the gap between energy demand and supply, due to the decreasing availability of fossil fuels. A complete transition between fossil fuels-based technology and the mix of different renewable energies will be accomplished by 2050 [1].

During the Convention on Climate Change held in Paris (COP21, 2015), the world's policy leaders decided to limit the global warming below 2 °C and to strive eventually to 1.5 °C. These targets are very ambitious but this optimistic view is justified by the awareness that there are cost-effective technical solutions already

E. Gilioli (✉) · F. Bissoli · M. Bronzoni · S. Rampino
CNR-IMEM, Istituto dei Materiali per l'Elettronica ed il Magnetismo, Parma, Italy
e-mail: edmondo.gilioli@cnr.it

C. Albonetti
CNR-ISMN, Istituto per lo Studio dei Materiali Nanostrutturati, Bologna, Italy

P. Ciccarelli
Consorzio Hypatia, Rome, Italy

R. Verucchi
CNR-IMEM, Istituto dei Materiali per l'Elettronica ed il Magnetismo, Povo, TN, Italy

© The Author(s) 2019
T. Tolio et al. (eds.), *Factories of the Future*,
https://doi.org/10.1007/978-3-319-94358-9_17

"

available today and further efforts must be done in this direction to match the Paris goals. Solar power plays a major role in this effort.

From the industrial point of view, the penetration of PV in the electricity market has rapidly increased thanks to the introduction of incentives in several countries. In Italy the PV fraction at peak time has already passed the 20% and it is still growing despite of the cut of the feed-in-tariff scheme. The cost of own-produced PV electricity for the Italian domestic user is already well lower than the average price of grid electricity (grid parity) even in absence of incentives [2].

The availability of new PV products specifically designed for the building integration will cause a further significant reduction of the costs associated with the design and the installation of domestic PV systems. Given that domestic, commercial and public buildings together account for more than 50% of the overall electricity consumption in Italy, the emerging field of the building/product integrated photovoltaic (BIPV & PIPV) in combination with the diffusion of smart grids, surely will have a major impact on the overall PV diffusion.

Thin film solar cell (TFSC) is the most suitable technology for BIPV applications and $Cu(In,Ga)Se_2$ (CIGS) seems to be the most promising candidate for its good compromise between conversion efficiency, aesthetics, stability and possibility to be grown on different substrates, including flexible ones.

Being a complex material, one of the main limitations is represented by its cost, due to the expensive and complicated processes to grow it. Several industries and research centres worldwide are trying to develop production processes alternative to the commonly used multi-stage co-evaporation/co-sputtering followed by the selenization. In the search for simple and cost effective methods for the deposition of TFSC, a new process for the deposition of TFSC has been recently developed at IMEM-CNR in Parma, based on the Pulsed Electron Deposition (PED). This technique, already demonstrated the possibility to realize, on laboratory scale, CIGS-based TFSC with high PV conversion efficiency using a very cheap and simple method [3]. PED has important advantages, such as low installation and running costs and the capability to transfer of complex materials from a bulky target to the growing film in a single-stage process, without any post growth treatment. Even more important, PED allows the growth of high-quality CIGS at low temperature (the technique is also known as LT-PED), enabling the use of a large number of substrates, including polymeric and thermo-labile ones. Therefore, PED is an effective method to grow complex materials and to realize unconventional devices, such as flexible solar cells.

This chapter reports the achievements to solve specific issues that hinder the industrial scale-up of this novel process, ambitiously aiming to transfer the PED technology to a (pre)industrial level and to re-create an Italian industry in the strategic PV sector. To do this, the following issues must be addressed:

- Adapt the conventional PED for the deposition on flexible substrates.
- Provide a better control of the deposition rate via automation of the film growth
- Increase the deposition rate of the PED process.
- Realize a prototype system for the continuous (roll-to-roll) deposition of flexible solar cells.

The ultimate goal is the fabrication of a PED system prototype to deposit flexible CIGS-based solar cells, both in a *static* and in a *dynamic* process.

17.2 State of the Art

Commercially available TFSC suitable for BIPV applications are made of: cadmium telluride (CdTe), amorphous or microcrystalline silicon (a-Si) or CIGS.

CdTe has been the first thin film technology to emerge as a cost competitive alternative to crystalline silicon. The PV efficiency reached recently 22.1% in early 2016, 16.4% on commercial modules [4].

The American company First Solar is currently among the top 5 PV manufacturers although their products based on large glass substrates still do not differ significantly from the standard Si flat panels. Furthermore, CdTe is a controversial material for the environmental risks associated with both production process and end-of-life disposal of the PV modules. As a matter of facts, First Solar has been primarily responsible of the TFSC price drop since 2006.

As far as a-Si concerns, Swiss EPFL's Institute of Microengineering reached a remarkable 10.7% efficiency single-junction microcrystalline silicon solar cell [5], beyond the previous record of 10.1% held by the Japanese company Kaneka Corporation since 1998. However, the efficiency seems to be still too low for a significant market share of a-Si based TFSCs.

CIGS-based solar panels are commercial products. However, despite CIGS solar cells have reached 22.6% efficiency on the lab scale [6], the industrial scale-up of this technology is very challenging. Basically, only the Japanese company Solar Frontier has a significant CIGS volume production, while several companies in the last years (e.g. Solibro, MiaSolé, Solyndra, Avancis, Bosch, Global Solar Energy, Soltecture, Nanosolar, AQT, HelioVolt, Ascent Solar, SoloPower, TSMC, Manz, NuvoSun, Siva Power, etc.) have suffered serious financial problems. Flisom, a spin-off company of EMPA (the Swiss Material Science and Technology research centre), is specifically working on flexible CIGS, but it is not on the market, yet.

The main reasons are the complexity of the production processes and the cost of the production lines. Different deposition processes have been developed for the production of CIGS solar cells [7]. Among them, thermal co-evaporation [8], sputtering/selenization [9] and electrodeposition [10] are considered as the best options for industrial development, even though they are still characterized by severe limitations. Multi-stage co-evaporation, the so-called three stage process, holding the lab-scale PV record efficiencies [6, 11], is not suitable for mass production because of its complexity and limited composition reproducibility in large manufacturing systems [12]. Sputtering of metal precursors alloy followed by a post-growth selenization step seems to be more promising for industrial scale up. However, the selenization stage [13] has a strong environmental impact because of the use of highly toxic H_2Se gas. Moreover, the need of high temperature processing [14] makes both methods extremely costly, complex and time consuming.

In the last years, several single-stage processes based on sputtering [15] and co-evaporation have been proposed. The efficiencies reported for lab-scale cells are usually lower than 12.5% and the only exception is represented by an in-line co-evaporation process at a substrate temperature $T_{sub} = 550$ °C, able to achieve 18% efficiency [16].

The PED technique developed at IMEM-Parma is very efficient in reducing the deposition costs [3, 17]. CIGS-based thin film solar cells with 17% efficiencies by using a low-temperature single-stage PED process, have been recently reported [3] on a small area, resulting in a dramatic simplification of the deposition process compared to the conventional methods. It is remarkable to note that the active layers were deposited by PED at 270 °C, thus suggesting the possibility to use polymeric flexible substrates that usually do not stand high temperatures; this is extremely appealing for the BIPV/PIPV growing market.

PED is based on high-power electron-beam ablation of a bulk material (target) having the desired composition and stoichiometry. Electron beam pulses of about 100 ns dissipate a power density in the order of 10^8 W/cm^2 within a depth of $\cong 1$ μm of the target surface. This leads to the rapid, non-equilibrium evaporation of all the elemental components of the target, regardless the single element melting point or vapour pressure and the formation of a could of ions (plasma plume) with a large average kinetic energy, as schematically shown in Fig. 17.1. Therefore, the deposition of good quality CIGS films occurs at a much lower substrate temperature compared to sputtering or single-stage co-evaporation.

Although the reported result is extremely promising on a laboratory level, the technique still suffers some limitations that prevent a real industrial application of the CIGS-based solar cells grown by means of the PED technique; the goal is to fill this gap.

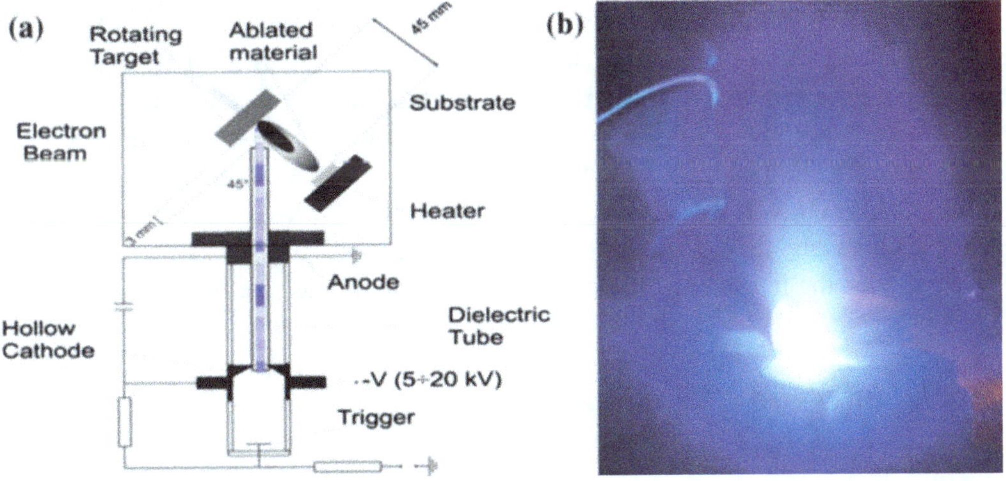

Fig. 17.1 Representation of the pulsed electron deposition (PED) technique: **a** schema, **b** in operation

17.3 Problem Identification and Proposed Approach

This section presents the different issues that must be addressed to enable the fabrication of CIGS-based TFSC by PED beyond the lab-scale area and on flexible substrates.

17.3.1 Improvement of CIGS Solar Cell Efficiency by Controlling the Substrate/Mo/CIGS Interfaces

A CIGS solar cell is a multilayer device where the interfaces significantly affect the performance. The typical structure is: substrate (glass, 1–3 mm)/back contact (Mo 1 μm by sputtering)/absorber (CIGS, 1.8 μm by PED)/buffer layer (CdS, 70 nm, by chemical solution)/window layer (transparent conductive oxides, TCO, ZnO-ZnO:Al, by sputtering).

The Mo/CIGS interface is identified as the critical interface to improve the electrical performance of the solar cell, in particular of flexible devices during the bending.

17.3.2 Selection of Flexible Substrates and Optimization of Substrate Holder

The selection of suitable flexible (polymeric) substrates is based on three parameters: (a) temperature resistance (up to 300 °C); (b) thermal expansion coefficient (TCE) and (c) stress stability. Their surface morphology has to be comparable to the glass (roughness ≈6 nm).

Kapton is chosen as benchmark for testing the Mo growth, being cheap and easily available. Other polymeric substrates, such as polyimide (PI) or polyethylene naphthalate (Teonex) from Dupont, perfluoroalkoxy alkanes (a copolymer of tetrafluoroethylene, PFA), fluorinated ethylene propylene (FEP) and ethylene tetrafluoroethylene (ETFE), provided by Saint-Gobain, were analysed in the temperature range 165–260 °C. Besides, metallic foil (stainless steel, SS 430B) and carbon fibres were also tested as possible substrates.

Obviously, the flexible substrate requires a tailored substrate holder.

17.3.3 Film Growth Automation

The film thickness and growth rate are the fundamental parameters to be monitored and controlled. The oscillating quartz microbalance, which is commonly used in physical vapour deposition (PVD) techniques, cannot be utilised with highly energetic pulsed beams. An alternative method, using an infrared (IR) pyrometer, can be used to measure the heat irradiated by the sample during the deposition. The total radiated power per unit area, called the photon radiant emittance, is constant at constant temperatures (Stefan-Boltzmann law), but it is subjected to interference phenomena when irradiated from a thin film, leading to changes of the radiant emittance not depending on temperature. By analysing the emittance profile, the interference fringes due to the CIGS thickness variation appear, enabling the measurement of growing film thickness [18]. Once the film reaches the desired value, a dedicated software moves the substrate.

Optical Emission Spectroscopy (OES) has been employed to monitor in real time the composition of the PED plasma plume. OES is based on the detection of the optical emission (excitations and ionizations) occurring in the plasma. Since the plasma consists of the elements composing the targets, the absolute intensity of the emission peaks gives information on the ablation rate, while the ratio between emission peaks from different elements reveals the plasma stoichiometry. Both the ablation rate and the stoichiometry transfer depend on the PED acceleration voltage and the gas flow.

17.3.4 PED Source Current Stability

To achieve the physical *ablation* of the CIGS target, it is necessary to heat it suddenly, preventing the thermal *evaporation*. Being CIGS an incongruent melting material, the *slow* heating leads to the target melting, resulting in the deposition of unwanted phases. In short, this is the reason why PED is so efficient to deposit complex and incongruent melting materials such as CIGS.

The energy power (i.e. e-beam acceleration voltage multiplied by electron current) is the main parameter for controlling the mechanisms of PED ablation (congruent/incongruent evaporation, plume distribution and expansion range, stoichiometry transfer, adatom energy, particulate density, etc.) [19]. The e-beam acceleration is nominally the voltage needed to fully charge the capacitance connected with the hollow cathode and is set by the operator in commercial PED sources. The electron current depends on the combination of charge voltage, source geometry (tube length, cathode aperture), trigger type, gas type and pressure. While the former parameters are constant during the deposition, the gas pressure inside the source changes due to the warming of the cathode and the dielectric tube. The evaluation of the current drop induced by the change of the local pressure and the way to keep constant the energy power transferred to the target is essential for long time depositions. This measurement, not provided by the commercial PED system, can be made by using a contactless toroidal coils, called *Rogowski coils*, placed around the dielectric tube.

The variation in time of the discharge current of the pulsed current (pulse duration ~100 ns), generates a variation of magnetic flux inside the Rogowski coil, proportional to the variation of the discharge current in time. The automatic feedback of the discharge current measurement stabilizes the signal, by remotely controlling the flux-meter injecting the gas into the chamber.

17.3.5 Increase the PED Deposition Rate

The deposition rate in the PED technique depends on many factors, among which the electrical parameters of the source, the material to ablate and, above all, the pulse repetition rate. For CIGS, a typical deposition rate value is 0.25 nm/pulse over a 3 square-inch substrate (implying about 10′ to deposit 1.6 μm thick CIGS layer). This process must be much faster for a pre-industrial development.

The easiest approach would be to increase the pulse frequency; since a single pulse has a duration of only 100–200 ns, the maximum frequency could be much higher than 20 Hz, that is the maximum repetition rate provided by the only 2 companies commercializing the PED sources (Organic Spintronic, Italy and Neocera, USA). While a faster electronic could be in principle easily implemented, the main problem is the presence of the dielectric tube guiding the electron beam from the PED body that cannot stand higher frequency for the rapid heating, leading to the melting of the tube.

An innovative solution is to drive the electron beam using a supersonic gas beam, instead of a tube [20]. Further advantages of the tubeless solution, are: (i) easier process (the tubes are the only parts that require periodic cleaning and substitution), (ii) removal of geometrical constraints, and (iii) reduction of contaminants, typically silicon and carbon.

17.4 Developed Technologies, Methodologies and Tools

The main achievements to solve the problems identified in Sect. 17.3 are hereafter reported. The adopted solutions are selected on the basis of their feasibility rather than the scientific innovation, in view of a possible integration in a production line.

17.4.1 Optimization of Molybdenum Films on Flexible
Substrates

Molybdenum (Mo) is generally used as back contact in TFSC to collect the photo-generated charges. While the deposition of Mo on rigid glass does not represent a major issue, the use of flexible substrates poses serious limitations to the solar cell performance. Mo films sputtered on flexible substrates are investigated by atomic force microscopy (AFM) and compared to the film grown on glass at room temperature (RT), as reference. Typically, films have a surface roughness of approx. 2 nm and composed of grains with an average diameter of 60 nm. The roughness parameter H [21], which describes how the film morphology changes in the plane, is approx. 0.3, indicating that films are quite smooth over large area.

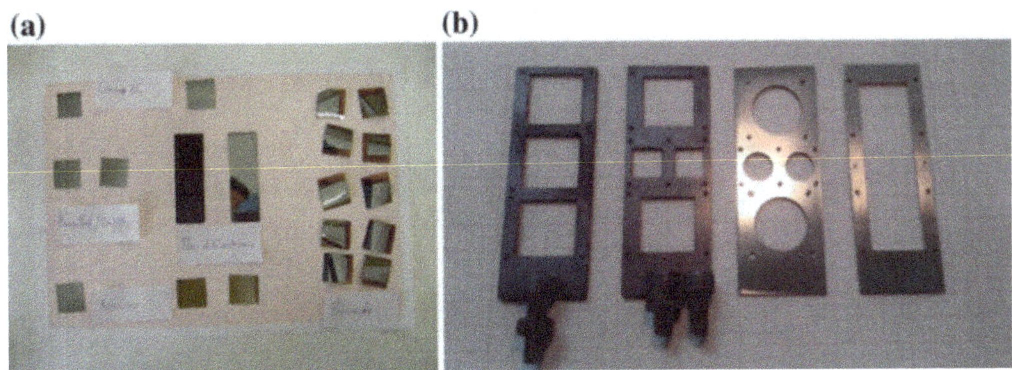

Fig. 17.2 **a** Different Mo coated flexible substrates and **b** sample holders suitable for CIGS deposition by PED

Mo films were firstly sputtered on kapton, by using the same experimental conditions used for glass (film thickness 500 nm, substrate temperature RT). Such films are rougher ($\approx$6 nm) and composed of bigger grains with an average diameter of 160 nm. H isapprox. 0.04, indicating a *jagged* surface, in agreement to the roughness increases. In order to decrease it, kapton was heated at 120 °C during the deposition. Doing so, the surface roughness significantly decreases to approx 3 nm, but severe cracks appear in the film due to the permanent bending of the polymeric substrate generated by the heating. In order to remove such cracks, suitable sample holders for flexible substrates are designed, fabricated and successfully tested (Fig. 17.2). Other polymeric substrates (PFA, FEP, ETFE and PI) exhibiting smooth surfaces (roughness ranging from 0.5 to 4 nm) and temperatures resistance from 165 to 260 °C, were tested. Among them, preliminary results on adhesion and conductive properties of Mo films sputtered on FEP in the same experimental conditions (film thickness 500 nm, substrate kept at RT) indicate that it is a promising substrate for fabricating flexible CIGS-based TFSCs.

The exploitation of the Mo layer required the development of a radically new method for the substrate heating, based on the application of a DC electrical power directly through the Mo back contact of the cell, thus converting electrical energy into heat by Joule effect (Fig. 17.3). The very efficient heat transfer to the thin (<1 μm) Mo layer requires a low electrical power density (few W/cm^2) to achieve the required deposition temperature, that is much lower compared to the traditional resistor- or lamp-based external heaters. Joule-heated CIGS based TFSCs shows comparable or slightly better PV properties than conventional heaters [22].

17.4.2 Film Growth Automation

The real-time control of the CIGS film thickness is carried out by analysing the heat radiation emitted "in situ". A pyrometer (model Marathon MM LT), located outside the chamber measures the IR radiative power emitted from the substrate

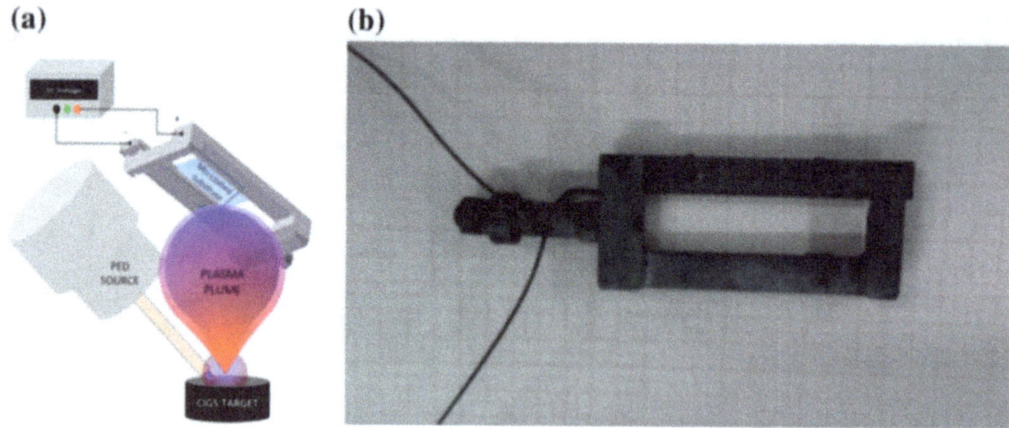

Fig. 17.3 Joule heated flexible substrates: **a** scheme and **b** sample holder

surface through an IR-transparent ZnSe viewport. A LabView software calculates the $T = f(t)$ function, stopping automatically the deposition when the CIGS thickness achieves the desired pre-set value.

The temperature is acquired and filtered every 200 ms and the average temperature value is calculated every 5 points (1 s), limiting the signal noise. Both 1st and 2nd derivative of the T versus time function are instantaneously calculated. When the first inflection point (2nd derivative = 0) is detected, the average deposition rate and the expected time for maximum peak and deposition end are calculated. These calculations are repeated when the maximum (corresponding to a CIGS thickness = 1018 nm) and the second inflection point are reached. Finally, when the curve reaches the minimum (roughly corresponding to the desired final thickness = 2036 nm), the software shuts down the PED source, automatically stopping the deposition process.

The system is very precise and it has been installed in all the deposition chambers currently in use at IMEM-CNR equipped with PED.

The deposition rate is recorded by analysing the temperature versus time, as shown in Fig. 17.4. To further increase the sensitivity in the calculation of the deposition rate, a real-time evaluation of the PED parameters by OES is employed, using a Hamamatsu mini-spectrometer (250–800 nm range). Due to the pulsed nature of PED process, 20 subsequent spectra taken every 300 ms are averaged (every 6 s). The dependence of the OES spectrum on the acceleration voltage is shown in Fig. 17.5. By increasing the voltage, the intensity of the In peak at 325 nm, hence the deposition rate, reaches a maximum at 16 kV, then slowly decreases for higher voltages. The ratio between Ga (417 nm) and In peak becomes constant (0.5) at voltages greater than 12 kV, while the ratio between the Cu (521 nm) and In peaks saturates at 0.30 for values exceeding 14 kV. These trends confirm that an optimal stoichiometry transfer is achieved at voltages higher than 14 kV, while the maximum ablation rate occurs at 16 kV. During the deposition process, when a change in the OES spectrum and/or a variation in the deposition rate are detected, the PED parameters (voltage and gas flow) can be varied in order to recover the optimal conditions. It worth noting that the gas flow is an essential parameter for controlling both the ablation rate and the

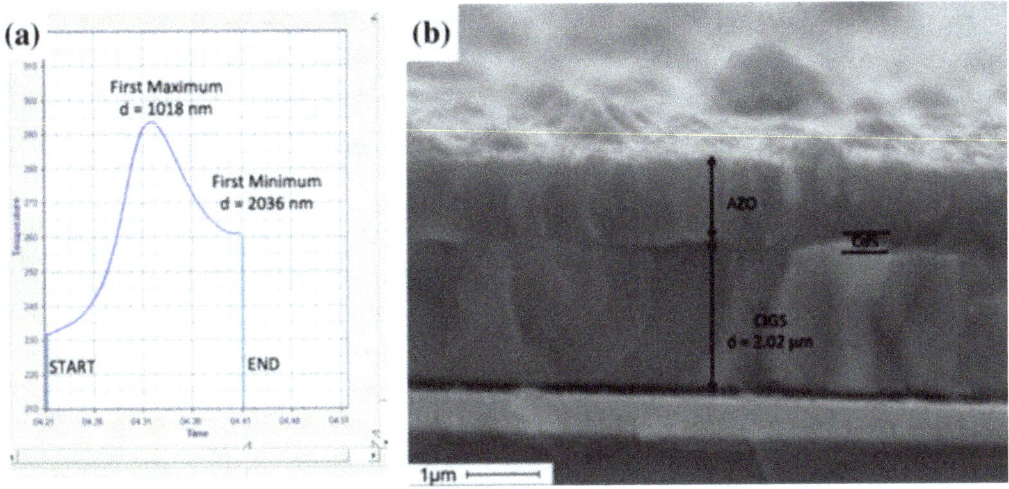

Fig. 17.4 **a** Substrate temperature as measured by the optical pyrometer during the deposition of 2 micron-thick CIGS by PED and **b** corresponding SEM cross-section of the final cell

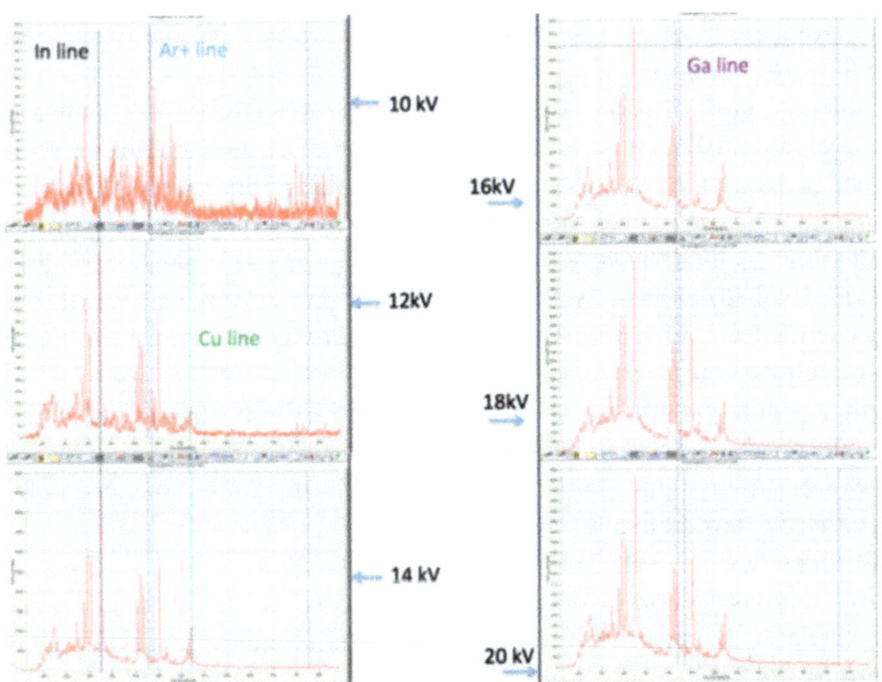

Fig. 17.5 Optical emission spectra of CIGS plasma at different acceleration voltages

stoichiometry transfer. The latter can be always obtained at low Ar flows (from 5 to 8 sccm) and in this range both Cu/In and Ga/In peak ratio are not affected by the gas flows.

17.4.3 PED Current

By measuring the PED e-beam current and voltage with the Rogowsky coil, the beam energetic distribution (*fast electrons* > ablation, *slow electrons* > evaporation) can be unambiguously identified.

This information is important to design and realize a new PED source, with a more efficient and highly energetic electron beam, optimized for the ablation of the CIGS target.

17.4.4 Faster Deposition

Aiming to increase the PED deposition/growth rate (currently limited to 15–20 Hz pulses frequency), the innovative idea was first to use a tubeless supersonic beam instead of a ceramic tube to guide the pulsed electron beam. The main problem is that a micrometric nozzle is required for the generation of the supersonic beam that rapidly tents to be blocked.

Commercial, custom made molybdenum aperture, selected for its chemical inertia, i.e. a disc of about 10 mm diameter, thickness around 0.1 mm and nozzle aperture of 20, 40 and 50 μm, have been tested in different working conditions. As expected, the use of the 20 μm nozzle gives the best results in terms of electron beam propagation and minimum gas load; however, the argon pressure in the supersonic beam source as high as 1000–2000 mbar is necessary, with a pressure increase in the vacuum chamber from 10^{-6} mbar to 10^{-4}–10^{-3} mbar. Clearly, the larger the nozzle, the more difficult the control of the beam.

During the ablation process at high voltage (>14 kV), in particular in presence of oxygen, the nozzle aperture only lasts a few minutes before obstructing, while using low power electron beams, no damaging of the Mo diaphragm has been observed, also in case of exposure to air.

Unfortunately, the threshold for the CIGS target ablation is 16 kV, (Fig. 17.6) and in these conditions the nozzle lifetime is limited to 30 min; after that, unavoidably the occlusion forms, due to massive erosion and material melting, irreversibly closing the aperture.

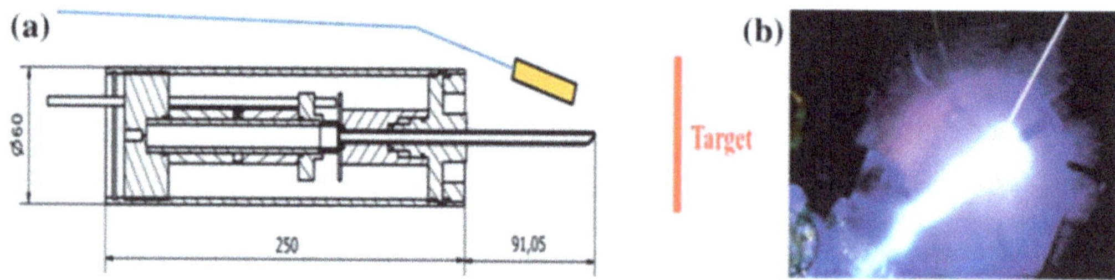

Fig. 17.6 **a** Scheme of the supersonic beam assisted PED and **b** test in vacuum

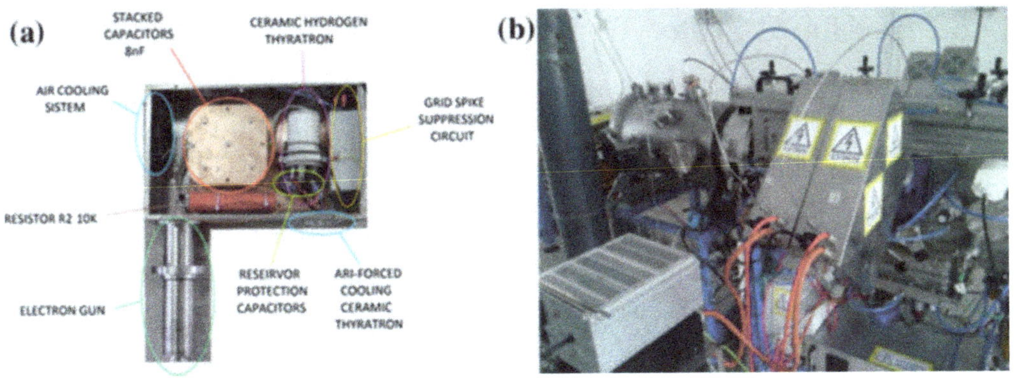

Fig. 17.7 **a** The new PED source and **b** the multi-PED system (4 new PED sources) installed at IMEM for large area CIGS-based TFSC fabrication

The same problem occurs by using a source made of quartz with a 50 μm nozzle. Again, its maximum lifetime is around 30–40 min, then the gas pressure sudden increases due to the formation of cracks around the nozzle.

Although the principle of the supersonic beam has been successfully demonstrated, this solution is considered not suitable for the production scale-up.

An alternative way to increasing PED deposition rate is to provide a more efficient control of the pulse generation to increase the pulsed frequency. The difficulty is mainly related to the necessity to manage a quick (transient of few ns) and fast (width of 100–200 ns) impulse of current at very high voltages (10–20 kV) and peak current (1000–2000 A). High voltage (HV) power supply can be only used up to 40 Hz, because of its high internal intrinsic resistance. A gas HV switch Thyratron is finally adopted, by far more complex to control, requiring different low/middle voltage supplies for proper working, but enabling a reliable and efficient control unit.

The radically new PED sources, fabricated in collaboration with Noivion s.r.l.,[1] show a good performance in the working conditions at frequencies up to 150 Hz (Fig. 17.7).

17.4.5 Deposition Chamber

The installation of the new PED sources could not fit the existing chamber, therefore a new deposition system has been designed and built. Aiming at the pre-industrial scale-up, it hosts 4 sources, able to deposit solar cells as large as 160×160 mm^2, on both rigid (glass) and flexible substrates (Fig. 17.8).

The preliminary results confirms that, by adjusting the PED sources geometry, it is possible to homogeneously deposit on 160×160 mm^2, corresponding to the size of a conventional Si-based solar cells.

---

[1] www.noivion.com

17.5   Developed Prototypes

The final upgrade provides a solution enabling a *dynamic* deposition from the *static* deposition of the multi-PED chamber in Fig. 17.8, by installing a flexible substrate holder composed of rotating wheels (motorised reel-to-reel apparatus, Fig. 17.9). The rotating speed is driven by the "in situ" film thickness measuring system, allowing the tape movement when the film reaches the desired thickness. The reel-to-reel motor is also connected to the PED source, switching on/off when the tape starts/stops.

The installation of the reel-to-reel substrate holder requires major changes in the vacuum chamber that has to be completely redesigned and built.

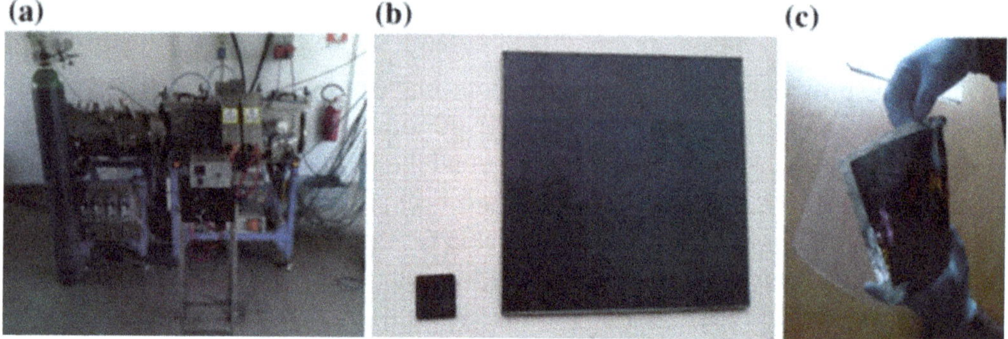

Fig. 17.8   a The new multi-PED system for large area $(16 \times 16 \text{ cm}^2)$ CIGS-based TFSC. **b** The larger cells are deposited on aluminium foil and **c** transparent polymers (TEONEX)

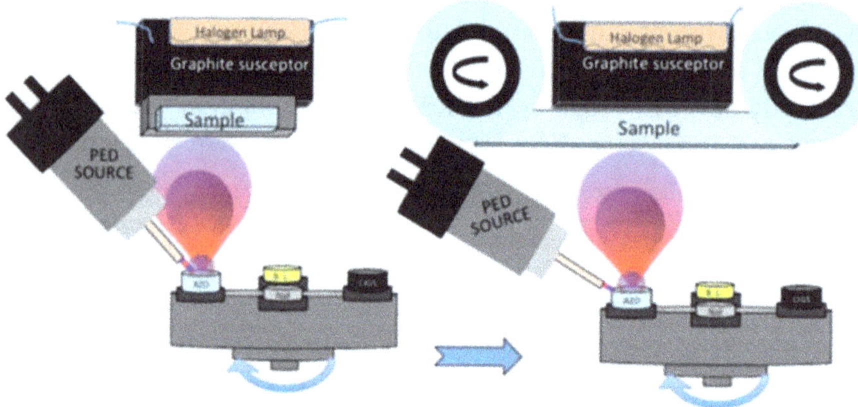

Fig. 17.9   Scheme of the evolution from static to dynamic CIGS deposition

17.5.1 Wheel Size Specifications

To design the radius of the wheels, an AFM combined with a mechanical buckling system (Fig. 17.10) is used to measure, in situ and in real time, the maximum bending curvature of CIGS-based TFSCs deposited on a 1 cm wide flexible stripe [23]. To determine the maximum curvature to which the device is irreversibly damaged, both topographical changes and electrical performances are tested during the deformation induced by the bending.

FEP bare stripes were first investigated to understand how the surface roughness changes with the angle subtended by the circular arc formed by the bended film (α). The roughness for $\alpha = 0°$ (flat stripe) is 10 nm, which increases linearly to 20 nm for $\alpha = 180°$ (maximum bending angle) due to the relatively large deformation of the polymeric stripe. In order to test different polymeric substrates in different conditions, PI (roughness 4 nm) was coated with 500 nm of Mo to test the back electrode behaviour versus bending, while ETFE was coated with ITO (indium tin oxide, 40 μm) and CIGS (4 μm) to test the behaviour of the CIGS solar cell. Mo film is invariant with respect to bending up to $\alpha = 70°$, then the surface roughness increases suddenly from 4 to 6 nm and the electrical performances irreversibly change. CIGS films are morphologically invariant versus bending up to $\alpha = 180°$, while the electrical resistance changes drastically after $\alpha = 70°$.

On the base of these experimental results, a critical angle α_C of 70° is set as limit to preserve both mechanical and electrical properties of Mo and CIGS films. This angle corresponds to a wheel radius of 1.5 cm, but a safer roll-to-roll system with 5 cm wheel radius has been realized to make it suitable for any polymeric substrates.

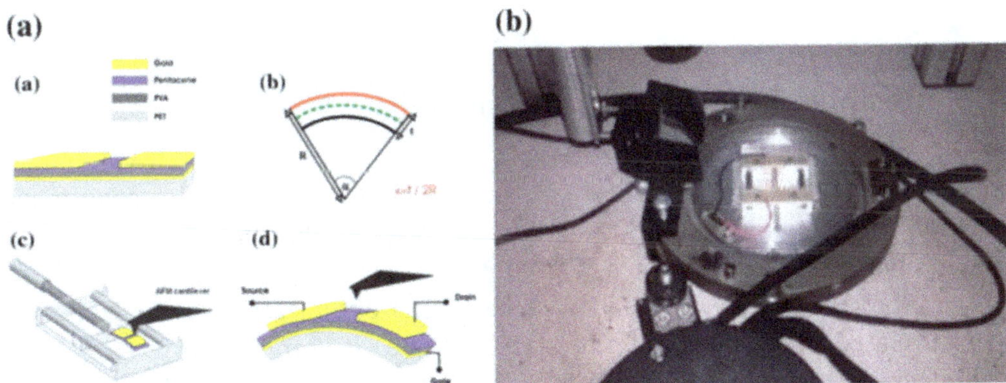

Fig. 17.10 **a** Scheme of the system to study the performance of flexible substrates versus bending and **b** mounted in the AFM basal plane

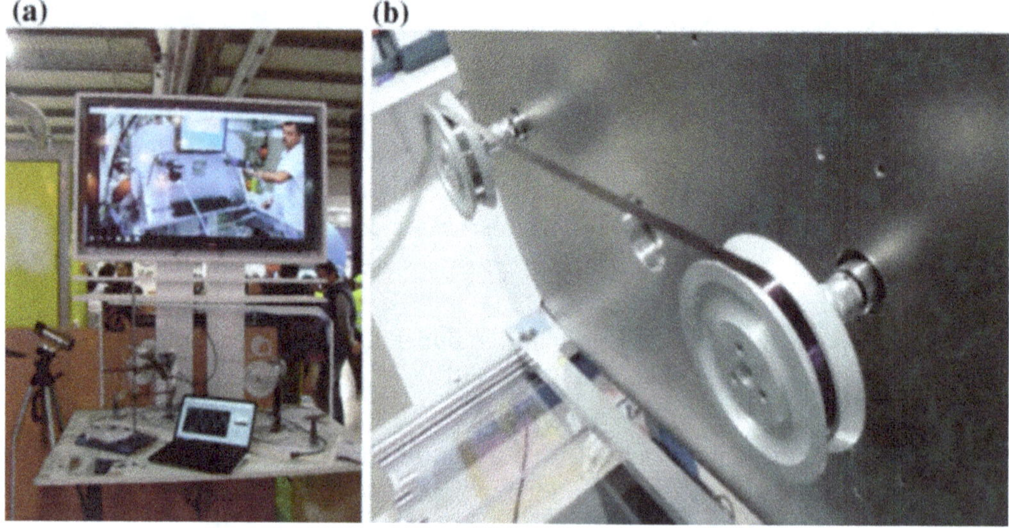

Fig. 17.11 Rolling CIGS prototype. **a** Simplified version presented at BIMU and **b** in operation

17.5.2 Automated Roll-to-Roll Movement

An enabling signal on the HV unit, switches on/off the pulsed electron ablation. The growth of thin films determines the movement of the flexible substrate, by interconnecting the PED source, the reel-to-reel motors and the thickness monitoring system. All the operations are controlled by a LabView program and a friendly user screenshot visualizes the cell final length and the time required to perform the whole process, allowing to change the parameters during the operation.

17.6 Testing of the Prototype System

Once defined the reel-to-reel wheel size and the hardware/software to control and synchronise the motor movement, the final prototype of the deposition system has been designed, realized and tested (Fig. 17.11).

The prototype is fully operating and the further optimization of the deposition of CIGS-based TFSC on various flexible substrates is currently in progress. The preliminary results, although still significantly lower than the 17% PV efficiency reported for small area, glass substrates, are promising. The main issue is represented by the bending of the substrate, since it induces some kind of defects. Indeed, despite the cells seem to be intact, the PV efficiency drops when the polymeric substrates bend. Further characterizations are in progress, both on the material and the device, such as DLTS (deep level transient spectroscopy) and EBIC (electron induced charge current). The first results also indicate that the adhesion of the interfaces (in particular Mo/CIGS) is critical, because the films frequently detach.

A simplified version of the prototype has been presented during the 30° BIMU Exhibition (Milano, October 6–8, 2016).

17.7 Conclusions and Future Research

The development of a new method to fabricate low cost, high efficient thin film solar cells is extremely ambitious, aiming to assign again an important role to the Italian companies in the strategic PV market, eventually using proprietary technologies. In this sense, the potential advantages of the PED compared to the currently available techniques can be summarized in a drastic simplification of the deposition process leading to a low-cost fabrication of CIGS-based TFSC. PED is particularly suited for the use of flexible substrates, thanks to the very low working temperature, allowing the deposition on a large range of thermo-labile polymeric substrates.

External and somehow unpredictable factors may determine the real impact of the presented results. In the PV market, often non-scientific factors strongly affects the diffusion of the solar plants, such as the rapid price changes of the PV cells/modules, the national economic policies (feed-in-tariffs) on the PV installations, the geo-political situations, the economic decision of the oil companies, etc. In any case, the direction clearly points to the use of the renewable energy.

Besides the thin film, flexible solar cells, it is worth recalling that there are other innovative devices with promising applications in the fields of BIPV and PIPV, such as the thin film semi-transparent bi-facial solar cells, where the PV conversion efficiency can be significantly increased by collecting light from both front and rear side of the solar cell. Again, by combining the low temperature PED and the use of unconventional transparent substrates, innovative PV products can be developed [24, 25].

Moreover, the space sector, where the Power/Weight ratio (Wp/gr) is the key factor may benefit from the development of solar cells on flexible, light substrates. It is important to stress, however, that in terms of efficiency, the CIGS-based flexible TFSC are not yet performing like the conventional rigid panels and the scientific community is strongly committed to improve the PV properties and the uniformity over large area.

Acknowledgements This work has been funded by the Italian Ministry of Education, Universities and Research (MIUR) under the Flagship Project "Factories of the Future—Italy" (Progetto Bandiera "La Fabbrica del Futuro") [26], Sottoprogetto 1, research projects "Mature CIGS technology for solar cells" (MaCISte) and "Roll-to-roll deposition of flexible CIGS-based solar cells" (Rolling CIGS).

All the reported activities have been endorsed by the *Consorzio Hypatia* (F. Lucibello and M. Zarcone), that also gave financial support for the construction of the new deposition chamber and to P. Ciccarelli in the framework of the grant: *Torno Subito*.

A special mention goes to F. Annoni, M. Calicchio, A. Kingma and F. Pattini (IMEM-CNR, Parma) for their precious contribution in the deposition and characterization of CIGS-based solar cells and to M. Murgia (ISMN-CNR, Bologna) for manufacturing the mechanical buckling system. AFM images were collected in the SPM@ISMN facility.

Some activities have been partially funded by the "PED4PV-Pulsed Electron Deposition for PhotoVoltaic" project (N° EE0100062), of the Italian Ministry of the Economical Development, *Industria-2015* program.

References

1. Breyer C, Gerlach A (2013) Global overview on grid-parity. Prog Photovolt Res Appl 21:121–136
2. Barnham K, Knorr K, Mazzer M (2012) Progress towards an all-renewable electricity supply. Nat Mater 11:908–909
3. Mazzer M, Rampino S, Gombia E et al (2016) Progress on low-temperature pulsed electron deposition of $CuInGaSe_2$ solar cells. Energies 9:207
4. First Solar Press Release (2014) First solar builds the highest efficiency thin film PV cell on record. http://www.firstsolar.com/
5. Tamang A, Hongsingthong A, Sichanugrist P et al (2014) Light-trapping and interface morphologies of amorphous silicon solar cells on multiscale surface textured substrates. IEEE J Photovolt 4:16–21
6. Jackson P, Wuerz R, Hariskos D et al (2016) Effects of heavy alkali elements in Cu(In, Ga)Se$_2$ solar cells with efficiencies up to 22.6%. Phys Status Solidi RRL 10:583–586
7. Rampino S, Armani N, Bissoli F, Bronzoni M, Calestani D, Calicchio M, Delmonte N, Gilioli E, Gombia E, Mosca R, Nasi L, Pattini F, Zappettini A, Mazzer M (2012) 15% efficient Cu(In, Ga)Se$_2$ solar cells obtained by low-temperature pulsed electron deposition. Appl Phys Lett 101:132107
8. Repins I, Contreras MA, Egaas B et al (2008) 19.9%-efficient ZnO/CdS/CuInGaSe$_2$ solar cell with 81.2% fill factor. Prog Photovolt Res Appl 16:235
9. Caballero R, Guillén C, Gutierrez MT et al (2006) Cu(In, Ga)Se$_2$-based thin-film solar cells by the selenization of sequentially evaporated metallic layers. Prog Photovolt Res Appl 14:145
10. Ganchev M, Kois J, Kaelin M et al (2006) Preparation of Cu(In, Ga)Se$_2$ layers by selenization of electrodeposited Cu–In–Ga precursors. Thin Solid Films 325:511–512
11. Jackson P, Hariskos D, Lotter E et al (2011) New world record efficiency for Cu(In, Ga)Se$_2$ thin-film solar cells beyond 20%. Prog Photovolt Res Appl 19:894
12. Dhere N (2011) Scale-up issues of CIGS thin film PV modules. Sol Energy Mater Sol Cells 95:277–280
13. Bosio A, Romeo N. Menossi D et al (2011) The second-generation of CdTe and CuInGaSe$_2$ thin film PV modules. Cryst Res Technol 46(8):857–864
14. Dhage SR, Kim H-S, Hahn HT (2011) Cu(In,Ga)Se$_2$ thin film preparation from a Cu(In,Ga) metallic alloy and Se nanoparticles by an intense pulsed light technique. J Electron Mater 40:122–126
15. Chen CH, Shih WC, Chien CY et al (2012) A promising sputtering route for one-step fabrication of chalcopyrite phase Cu(In, Ga)Se$_2$ absorbers without extra Se supply. Sol Energy Mater Sol Cells 103:25–29
16. Lindahl J, Zimmermann U, Szaniawski P et al (2013) Inline Cu(In, Ga)Se$_2$ Co-evaporation for high-efficiency solar cells and modules. IEEE J Photovolt 3(3):1100–1105
17. Rampino S, Bissoli F, Gilioli E, Pattini F (2013) Growth of Cu(In, Ga)Se$_2$ thin films by a novel single-stage route based on pulsed electron deposition. Prog Photovolt Res Appl 21:588–594
18. Bronzoni M, Stefancich M, Rampino S (2012) Role of substrate temperature on the structural, morphological and optical properties of $CuGaSe_2$ thin films grown by pulsed electron deposition technique. Thin Solid Films 520(24):7054–7061
19. Strikovski M, Kim J, Kolagani S (2010) Springer handbook of crystal growth. Springer, Berlin, Heidelberg, pp 1193–1211

20. Rampino S, Dilecce G, Verucchi R (2010) Metodo e apparecchiatura per il trasporto di fasci elettronici. Italian Patent 1399182, 28 Jan 2010
21. Valle F, Brucale M, Chiodini S et al (2017) Nanoscale morphological analysis of soft matter aggregates with fractal dimension ranging from 1 to 3. Micron 100:60–72
22. Rampino S, Annoni F, Bronzoni M et al (2015) Joule heating-assisted growth of Cu(In, Ga)Se$_2$ solar cells. J Renew Sustain Energy 133:013112
23. Scenev V, Cosseddu P, Bonfiglio A et al (2013) Origin of mechanical strain sensitivity of pentacene thin-film transistors. Org Electron 14:1323–1329
24. Mazzer M, Rampino S, Spaggiari G et al (2017) Bifacial, CIGS solar cells grown by low temperature PED. Solar Energy Mater Solar Cells 166:247–253
25. Cavallari N, Pattini F, Rampino S et al (2017) Low temperature deposition of bifacial CIGS solar cells on Al-doped zinc oxide back contacts. Appl Surf Sci 412:52–57
26. Terkaj W, Tolio T (2019) The Italian flagship project: factories of the future. In: Tolio T, Copani G, Terkaj W (eds) Factories of the future. Springer

Chapter 18
Mechano-Chemistry of Rock Materials for the Industrial Production of New Geopolymeric Cements

Piero Ciccioli, Donatella Capitani, Sabrina Gualtieri, Elena Soragni, Girolamo Belardi, Paolo Plescia and Giorgio Contini

Abstract The reduction of CO_2 emission from cement industry represents a priority task in the roadmap defined for the year 2020 by the European Union (EU) Commission for a resource efficient Europe. Several research projects have been undertaken aimed at developing non-hazardous materials as partial substitute of clinker in cement formulations, but also new, low-carbon, cements fully replacing clinker. Among the new cementing materials, Si–Al geopolymers seem the most promising, in terms of CO_2 emission and mechanical and thermal properties. In this chapter, mechano-chemical processing of kaolin clays to produce metakaolin (MKA) for the synthesis of Si–Al geopolymers is proposed as an alternative process to replace thermal treatments performed at 650–850 °C. Results obtained show that the mechano-chemical process is also suitable to make low cost blended Si–Al geopolymers where 40% of MKA is replaced by mechano-chemically activated volcanic tuffs. The compatibility of mechano-chemistry with industrial production was investigated by building a prototype milling system that was tested in a small industrial facility producing zeolites from industrial wastes. The degree of automation allowed the prototype to work unattended for 10 months. Based on the results obtained from these tests, a milling system for a full scale production of mechano-chemically activated rock materials was designed, and its performances analysed.

P. Ciccioli (✉) · D. Capitani
CNR-IMC, Istituto di Metodologie Chimiche, Rome, Italy
e-mail: piero.ciccioli@cnr.it

S. Gualtieri · E. Soragni
CNR-ISTEC, Istituto di Scienza e Tecnologia dei Materiali Ceramici, Faenza, Italy

G. Belardi · P. Plescia
CNR-IGAG, Istituto di Geologia Ambientale e Geoingegneria, Monterotondo Scalo, RM, Italy

G. Contini
CNR-ISM, Istituto di Struttura della Materia, Rome, Italy

© The Author(s) 2019

T. Tolio et al. (eds.), *Factories of the Future*,
https://doi.org/10.1007/978-3-319-94358-9_18

18.1　The Importance of Manufacturing Industry, Global Perspective

The reduction of CO_2 from industrial processes is a priority task in the roadmap defined for the year 2020 by the European Union (EU) Commission for a sustainable Europe [1]. In the EU countries most of the CO_2 emission comes from sectors belonging to the Emission Trading System (ETS), where cement industry is one of the most important. The case of Italy is paradigmatic, because with 0.35 Mt y^{-1} of CO_2 [2], Italy is one of the major CO_2 emitting countries in Europe, and is ranked as 19th in the world. While the targets fixed for the year 2020 in the reduction of CO_2 emission from non-ETS sectors have been already met by Italy, those fixed from industrial activities of the ETS sector are still far to reach, since Italy is the second cement producer in Europe [3]. High emission of CO_2 occurs in cement production, because the primary component of cement used in building industry is clinker, obtained by heating at 1,450 °C stoichiometric mixtures of carbonate and aluminosilicate rocks in rotary kilns. Although CO_2 emission from fuel combustion is relevant, that produced by thermal decomposition of $Ca(CO_3)$ in carbonate rocks is the most important.

To reduce CO_2 emission, the European Cement Association [4] suggests to replace part of the clinker with supplementary cementitious materials (SCM), with a high pozzolanic activity. The most suitable ones are some industrial wastes, such as carbon fly-ashes, blast furnace slag, silica fume and rice husk, and natural materials, such as clay-rich or zeolite-rich rocks [4]. Larger emphasis has been given to the use of industrial wastes in blended cements [4] because this allows their safe recycling. Since the use of SCM depends upon the availability of wastes from other industrial sectors, not all countries can profitably adopt this practice. This is the main reason why in the last 5 years the Italian production of blended cements using 40–60% of industrial wastes was still limited to 3.6% [5]. By considering the large volcanic deposits of tuffs lithified by zeolites in Central-Southern Italy [6], it is not clear why the production of cements using natural pozzolanic materials is also limited to 12.6% in Italy. Indeed, 84% of the cement production in Italy is still made of cements with low SCM contents [5].

In the last 50 years, enormous efforts have been made to find new wastes acting as SCM, but also to find new cements for building industry. Among them, Si–Al geopolymers [7] are the most promising for the years to come, as they could fully replace the clinker-based ones. While clinker uses the hydraulic reaction to form a nano-material with randomly oriented hydrated calcium silicate crystals, different aluminosilicate materials can make cements by reacting in alkaline solutions, through the same pozzolanic reaction used by the ancient Romans to make concrete [7]. As shown by Davidovits [7], this approach is more flexible than the hydraulic one, because different 3D polymers can be obtained depending on the Si/Al ratio of the reagents, and the type of compensatory cations introduced in the randomly oriented nano-crystal structure of Si–Al geopolymers. Today, amorphous metakaolin (MKA) obtained by dehydration of kaolinite crystals (KA) in kaolin clays is, by far, the most used reagent for the synthesis of Si–Al geopolymeric cements, because it reacts faster than any other aluminosilicate rock material in alkaline solutions [7]. With respect to clinker-based cements, the use of MKA as primary reagent offers noticeable advantages in terms of CO_2 emission, because it is obtained by heating kaolin clays at temperatures (600–850 °C) much lower than those used in clinker production. Since in the thermal treatment of kaolin clays CO_2 emission almost exclusively comes from fuel combustion, Si–Al geopolymers are often termed as *green cements*. The most interesting aspect of MKA is that it can also be obtained at near ambient temperature and in a very short time by mechano-chemical processing of kaolin clays [8, 9]. Since milling systems are usually powered by electricity, CO_2 emission can be further reduced in the production of Si–Al geopolymeric cements, by using renewable resources for electricity generation.

Until now the use of mechano-chemistry has been confined to a laboratory scale, because no adequate investigations have been made to assess the feasibility of this process at an industrial scale. In particular, data collected so far do not allow to develop industrial milling system for the production of this new type of cements. To fill this gap, a detailed study was undertaken to get the information required to make the industrial production of Si–Al geopolymers possible, through the mechano-chemical processing of rock materials necessary to produce them.

In this work, the grinding conditions necessary to obtain Si–Al geopolymers with good mechanical and thermal properties from kaolin clays and, for the first time, from volcanic tuffs, were investigated. The possibility to produce them at an industrial scale was studied by building a prototype system for the mechano-chemical treatment of rock materials, using the results obtained at a laboratory scale. Tests performed in a small industrial facility led to the design of a milling geometry for the mechano-chemistry of rock materials at an industrial scale, and it was possible to evaluate energy consumption required to activate them.

The chapter is organised as follows. Section 18.2 gives and overview of the state of the art and Sect. 18.3 presents the proposed approach. The development of Si–Al geopolymers from mechano-chemically treated rock materials is addressed in Sect. 18.4, the industrial aspects of this new technology are reported and discussed in Sects. 18.5 and 18.6. The conclusions are drawn in Sect. 18.7.

18.2 State of the Art

The amorphization of kaolinite (KA) crystals to produce MKA by mechano-chemical treatment of kaolin clays is known since long time [8], but is still confined to a laboratory scale, because the high energy requested by this phase transition can only be reached by mills, such as planetary ball mills, that are impossible to use for an industrial scale production. Several types of industrial mills for mineral treatment exist on the market, but their rotation speed is so small that they are unable to transmit a sufficiently high specific energy ($\geq 6\,\mathrm{J\,g^{-1}\,s^{-1}}$) to kaolin particles, to fully amorphize KA crystals in a time shorter than 1 h. Because of this, they are used most for the comminution of rock materials with a large size range. The selection of the most suitable milling geometry for industrial application was thus fundamental in our research. Only having defined this, the milling parameters necessary to get the most reactive materials for the synthesis of Si–Al geopolymers could have been assessed through laboratory investigations performed on mills able to reproduce, at a smaller scale, the same grinding conditions existing in industrial mills. With this approach, data collected at a laboratory scale could have been used to assess the performance afforded by a mill for the industrial production of MKA, and other rock materials.

The amorphization of KA crystals illustrates well the number and type of parameters necessary to design a milling system for the synthesis of Si–Al geopolymers at an industrial scale. The energy required for this phase transition is, in fact, well known from the thermal treatment of kaolin clays, from which MKA is produced at an industrial scale. With the thermal treatment, the formation of MKA from kaolin clays strongly depends on temperature and time [7, 10–12]. The same is true with mechano-chemical processing, where MKA formation depends upon the grinding time [8, 9], if other parameters, such as the grinding ratio and the rotation speed of the moving bodies (ω) are constant. The grinding ratio plays an important role in the mechano-chemical processing of minerals, as it determines the time needed to induce phase transitions in the grinded material. Defined as the ratio M/m, where M is the mass of the moving bodies in the mill and m that of the material to grind, the grinding ratio inversely relates with the grinding time. The less material is grinded in a mill with a given mass M rotating at a given speed ω, shorter is the time to fully convert KA crystals into amorphous MKA. This inverse relation is easy to understand, because the less is the mass to grind, higher is the specific energy the crystals receive by collision. It must be noted that to convert KA into MKA at an industrial scale, grinding ratios not higher than approx. 250 can be used at a rotation speed of 1,500 r.p.m. in order to optimize the production. This means that a moving mass of 250 kg is needed to covert 1 kg of KA into MKA in the order of minutes. The huge mass of the moving bodies, combined with the high speed to which they must rotate, highlight the technological difficulties encountered in the construction of industrial mills for mechano-chemical treatment of minerals in rocks.

Another critical point is the wide interval of energy in which amorphous MKA exists. In this range, the glassy network of MKA evolves as a function of the energy, either thermal or mechanical, absorbed by the material [7, 10–12]. In particular, local structures with a mullite-like composition progressively form at high energy until γ-alumina and mullite crystallize from them [10, 11]. In the thermal treatment of kaolin clays, MKA starts to form at approx. 650–850 °C, whereas γ-alumina crystallizes above 1,000 °C [10, 11]. Since different amorphous structures are formed in the interval of existence of MKA, the most reactive and cost-effective must be identified. According to Davidovits [7], the highest reactivity is reached in MKA when the intensity of the band of 5 coordinated aluminium atoms (Al), indicated as Al(V), determined by [27]Al Magic Angle Spinning Nuclear Magnetic Resonance (^{27}Al MAS NMR), is higher than those of 4 and 6 coordinated Al, indicated as Al(IV) and Al(VI), respectively. This rule, which is still strictly followed by scientists working on the synthesis of geopolymers, has been questioned by some authors [12], who have raised doubts about its general validity. These doubts arise from recent observations indicating that also low reactive structures with mullite-like composition can generate an intense Al(V) band in MKA [13]. Because of this, results provided by ^{27}Al MAS NMR MKA are not sufficient to identify the most reactive MKA structure, and also those provided by other techniques must be used. The combination of different techniques is also required because the energy to form MKA strongly depends upon the degree of disorder of KA crystals, which, in turn, depends on the impurities present in them. Differences between 100 and 200 °C have been actually observed in the formation of MKA by thermal treatment as a function of the degree of disorder of KA crystals [14]. Emphasis is given here to the mechano-chemistry of MKA, because this material is of great technological interest, since is also used for the synthesis of zeolites in alkaline environments [15]. While in the synthesis of zeolites no silicon (Si) sources are added to MKA, proper amounts of liquid alkali silicates must be added to the reaction mixture to make geopolymers with Si/Al ratio of 1, 2 or 3 [7]. Formation of zeolites from MKA further emphasizes the need of an industrial mill for the mechano-chemical treatment of KA in kaolin clays [16].

Since different types of geopolymers can be synthesized from MKA [7], a specific one needs to be selected to test the reactivity of thermally and mechano-chemically treated products. Poly-siloxo sialate (PSS) geopolymers with a Si/Al ratio equal to 2 are definitely the best candidates, because they exhibit the highest cementing properties for building applications [7]. In particular, those having K$^+$ as compensatory cation (K-PSS from now on), combine a good mechanical resistance to unconfined uniaxial compression (UCS) with a very high thermal stability [7]. Melting temperatures higher than 1,400 °C have been reported for these geopolymers [7], against approx. 850 °C reached by clinker cements, although the UCS values usually fall in the same range (42–55 MPa). This means that also the thermal resistance of K-PSS geopolymers must be assessed to corroborate the high reactivity of MKA. This further emphasizes the need to certify Si–Al geopolymers obtained by mechano-chemical processing of kaolin clays with different techniques. Only through this approach it is, in fact, possible to define adequate protocols for quality assurance (QA) and quality control (QC) of geopolymers produced at an industrial scale with this process.

Mechano-chemical studies exist on the phase changes occurring in other rock materials than kaolin clays, but none of them refer to volcanic tuffs, although it has been shown that those cemented by zeolites exhibit a sufficient pozzolanic activity to act as SCM in clinker cements [17]. The possibilities afforded by the mechano-chemical treatment of these materials for the synthesis of K-PSS geopolymers were also investigated, by considering their low price with respect to that of kaolin clays.

18.3   Problem Statement and Proposed Approach

The selection of the milling geometry is fundamental to assess the feasibility of mechano-chemistry for the industrial production of Si–Al geopolymers. Definitely, the one with the lowest number of moving bodies rotating in a horizontal plane around a vertical axis, was the best suited for an industrial production, as it does not pose problems that cannot be solved with currently available technologies. Since this geometry does not substantially differ from that of ring mills used for mechano-chemical processing of materials at a laboratory scale, data collected with this type of mills could have been used to develop an automated grinding system for a small scale industrial production of MKA, and other activated materials. Based on the results obtained with the prototype, it was thus possible to design a milling system for a full scale production, using proper models.

Ring mills simulate well the kinematic motions of the type of mill better suited for industrial production, because they consist of a metal jar containing one or more concentric rings and a central cylinder as moving bodies, together with the material to grind. The jar is closed by a cover capable to maintain the motion of the colliding bodies in a horizontal plane, and keep the material inside the mill during grinding. A tight fixing of the jar with the metal housing obtained with O-rings, allows transferring an eccentric rotating motion to the mill from the shaft of an electric engine, whose speed can be controlled. As a function of the energy provided by the engine, the rings and cylinder, that are free to move inside the jar, can reach such a high centrifugal forces to induce phase transitions on the grinded material by collision. Pressures proportional to M and ω are generated between the moving bodies and the internal walls of the jar. The fact that ring mills with jars of 200 mL capable to reach a ω value of 1,500 r.p.m. were available on the market, provided a simple way to identify the optimal grinding conditions necessary to get a sufficient amount of reactive materials to synthesize Si–Al geopolymers from a commercial kaolin clay, and two volcanic tuffs. For modelling purposes, M/m ratios one order of magnitude lower than those commonly used in laboratory investigations were used in the ring mill. This allowed getting indications on the typical grinding time required in an industrial mill, and estimating the energy consumption necessary to produce a given mass of product from it.

A commercial kaolin clay, known to produce MKA with a high pozzolanic activity by thermal treatment, was selected as test material. Phase transitions occurring at increasing values of the grinding times were followed by looking at the structural changes in the grinded material. This was done by using X-ray powder diffraction spectroscopy (XRD), Fourier-transform infrared spectroscopy (FT-IR), X-ray fluorescence (XRF), X-ray photoelectron spectroscopy (XPS), ^{1}H, ^{29}Si and ^{27}Al MAS NMR spectroscopy, and various types of thermo-gravimetric (TG) determinations, such as thermo-gravimetric analysis (TGA), differential-thermo-gravimetry (DTG), and differential thermal analysis (DTA). This, combined with the fact that the kinetic of MKA formation by thermal processing was also followed on the same test material, represented a distinctive feature of our study. Structural information provided in previous studies were, in fact, not detailed enough to unambiguously identify the conditions in which the most reactive and cost-effective MKA was formed. In our study, thermal and mechano-chemical transformations of kaolin clay were followed with a sufficient detail to know which MKA was the most suitable for the synthesis of K-PSS geopolymers. Proper amounts of liquid potassium silicate solutions with known composition (K-Sil), acting as supplementary source of K^{+} and Si, needed to be added to MKA and KOH [7] to get K-PSS geopolymers with a general formula reported in (18.1), where n is the degree of polymerization and y the number of water molecules coordinated by the structure.

$$K^+ - \left[-(SiO_2)_2 - AlO_2-\right]n \cdot yH_2O \qquad (18.1)$$

Various formulations were thus tested with MKAs obtained with different processes to find those better approaching a Si/Al = 2.

Since geopolymers obtained from mechano-chemically and thermally treated MKAs exhibit different mechanical properties [9], only those with UCS falling in the range of commercial clinker cements (42–55 MPa) were tested for thermal resistance. Among them, only those melting at temperatures $\geq 1,400$ °C were considered sufficiently competitive with clinker cements, to justify a production at an industrial scale. The structure and composition of K-PSS geopolymers was confirmed with the same techniques used for the characterization of MKA, to get products to be used as reference materials for the QA and QC of industrial products.

The same approach was followed to test the ability of pure and mechano-chemically treated volcanic tuffs cemented by zeolites (lithified tuffs) or glass (welded tuffs), to form K-PSS geopolymers. No thermal treatments were made on them, because the phase transitions of these rocks are too complex to follow, due the number and type of minerals present in them, and the presence of glass that melts at rather low temperatures.

Similarly to what is commonly done with clinker cements, the possibility to use mechano-chemically treated volcanic tuffs as SCM in the synthesis of Si–Al geopolymers was also investigated. This was done because zeolites contained in lithified tuffs have a sufficient pozzolanic activity to act as SCM in clinker cements [17]. The partial replacement of MKA with activated tuffs in Si–Al geopolymer production was considered an important aspect to investigate, because it could have dramatically reduced the cost of the final product, if the grinding time required to activate tuffs was much lower than that of MKA.

Data obtained with ring mills were used to build a prototype system with a sufficient degree of automation to be used in a small scale industrial facility. Since it was impossible to build a demonstration plant for geopolymer production, the prototype was tested in an industrial facility for zeolite production where mechano-chemical processing of minerals was requested for at least 3 months. In this way, enough data became available to design a milling system for industrial applications.

18.4 Si–Al Geopolymers from Mechano-Chemically Treated Rock Materials

The initial part of our study was devoted to define the optimal grinding parameters and formulations necessary to synthesize K-PSS geopolymers from commercial kaolin clay, and from two volcanic tuffs. This was done through laboratory experiments performed on ring mills, by posing an upper limit to the grinding ratio used. Although a value of 120 was considered the best compromise to maximize the production, also a value of 250 was tested. The results obtained with mechano-chemically activated kaolin clay are described in Sect. 18.4.1, whereas those obtained with mechano-chemically activated tuffs or by suitable combinations of activated materials are reported in Sect. 18.4.2.

18.4.1 Geopolymers from Activated Kaolin Clays

After a detailed chemical, mineralogical and structural characterization of the kaolin clay used as test material, aliquots were grinded at increasing times, at rotation speeds of 900 and 1,500 r.p.m. Products obtained were then analysed with the same techniques, to follow the structural changes induced by the mechano-chemical process. The grinding time was increased until very little structural changes were detected in the grinded material. The same protocol was followed in the thermal treatment. In both cases, materials collected until a steady state was reached were checked for their ability to remove $Ca(OH)_2$ (portlandite) from alkaline solutions over a time higher than 28 days, according to the pozzolanic test described in [7, 17]. This allowed to get indications on their reactivity in alkaline solutions. For each of the materials showing a good pozzolanic activity, 3 different formulations were used to obtain K-PSS geopolymers with a Si/Al$=2$. The paste obtained from the alkaline reaction was poured in a plastic mould, to get products with a cylindrical shape, whose composition, structure and mechanical and thermal properties were checked after 28 days, when more than 90% of the material was polymerized [7].

A kaolin clay Argical-BS4 from AGS-Mineraux, with a Si/Al$=0.98$, and a SSA$=$ 18 m^2g^{-1} was used in these investigations. Information on the oxide composition and physical features can be found in [12]. Grinding was performed on ring mills from Sepor Inc. (Bleuler rotary mill) and Herzog (HSM vibration disk mill), both equipped with steel jars of 200 mL, and a mass M of 3 kg. Solid state NMR spectra of the raw and grinded materials were recorded on a Bruker ASX 200 spectrometer at a spinning rate of 12 kHz, and data processed to assess the relative contents of Al(IV), Al(V) and Al(VI) in the sample. XRF data were collected on a Philips PW 1480 WDS, while XRD diffraction spectra were collected on a Bruker D8 X-ray equipped with a Cu source. FTIR-AR spectra were collected on a Thermo Scientific Nicolet iS10 FTIR-AR equipped with a diamond window. TG determinations were all performed on a SDT Q600 apparatus from TA Instruments.

Figure 18.1 shows the XRD, FTIR-AR, NMR and TGA profiles of the kaolin clay used as test material in our investigations, together with indication of the bands of KA and the other minerals present in it. The thermal conversion of KA into MKA was followed from 300 °C up to 850 °C, while the mechano-chemical one was followed from 10 min to 1 h of grinding time, using a grinding ratio of 120 and a rotation speed of 1,500 r.p.m. With the thermal treatment, MKA started to form at 550 °C, but only

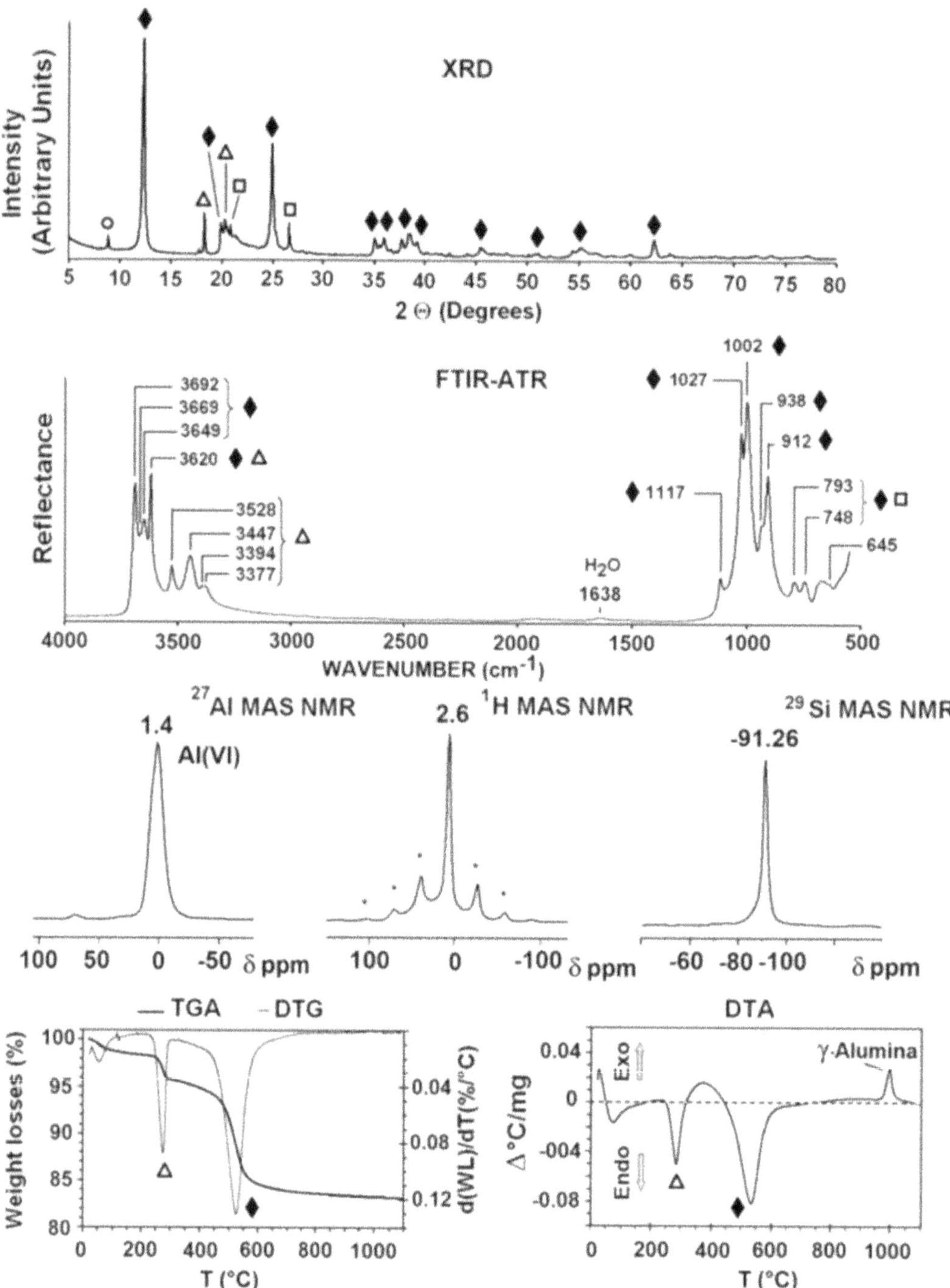

Fig. 18.1 XRD, FTIR-AR, solid state NMR and TG profiles of the clay Argical-BS4 used in this study to produce MKA by thermal and mechano-chemical treatment. Symbols are used to indicate the bands generated by different minerals in kaolin clay. Filled diamond = Kaolinite, empty circle = Illite, empty triangle = Gibbsite, empty square = Quartz

at 650 °C a steady state was reached above which the populations of Al(V), Al(IV) and Al(VI) did not change appreciably [7]. With the mechano-chemical treatment, a steady state was achieved in 30 min.

Figures 18.2 and 18.3 show the spectral profiles recorded when treated BS-4 reached a steady state, in which a complete amorphization of KA was observed. In both cases, the amorphization was indicated by the disappearance of the KA reflections and bands generated in the XRD and FT-IR spectra, and by the broadening of the Si band in the NMR spectra. The unambiguous formation of MKA was indicated by the presence of the Al(IV) and Al(V) bands in the ^{27}Al MAS NMR spectra. Since in both cases, only 73–75% of the original Al(VI) was converted in the other bands, the Al(VI) band still remained the most intense in MKA obtained with the independent processes, in evident contrast with the rule theorized by Davidovits [7]. The intensity of the other Al bands were, instead, quite different in the MKAs obtained from the two processes. While the Al(IV) band accounted for 12.6% of total Al in thermally treated BS-4, a value of 24.5% was measured in mechano-chemically treated BS-4. This effect, never detected before, indicated that some structural differences existed between the MKAs obtained with the two processes.

Consistently with data reported in the literature [9], substantial differences in the water content were observed in the MKAs obtained with the two processes. While few percent losses were measured by TG determinations on thermally treated BS-4, more than 25% weight losses were recorded on the mechano-chemically treated material, indicating that dehydrolixylated water released from KA remained inside the MKA structure. Differences in the water content are clearly visible in the FTIR-AR spectrum of Fig. 18.3, where a series of intense bands between 3,000 and 4,000 cm^{-1}, similar to those produced by zeolitic water in aluminosilicate rocks, were detected. These data are consistent with the ^{1}H MAS NMR spectrum of Fig. 18.3, where intense sidebands, very small in thermally treated BS-4, were observed around the main one.

Profiles of Figs. 18.2 and 18.3 were shown here because they refer to the most reactive MKAs as measured by the pozzolanic test, and by the alkaline reaction leading to the formation of K-PSS geopolymers. Some differences in reactivity existed, however, between the material reported in Fig. 18.2 (from now on indicated as BS-4 TT) and Fig. 18.3 (from now on indicated as BS-4 MT), because aluminium in BS-4 MT was 15% more reactive than that present in BS-4 TT. While 10 g of BS-4 MT needed to be mixed with 14.48 g of K-Sil and 2.39 g of KOH to get the K-PSS geopolymer, only 8.6 g of K-Sil and 2.57 g of KOH were required for the same amount of BS-4 TT. The fluid paste obtained by mixing the reagents for 10 min was cast in a plastic mould, and products analysed after 28 days.

XRD, FTIR-AR and ^{27}Al MAS NMR spectral profiles reported in Fig. 18.4 confirmed that all materials obtained consisted of randomly oriented nano-crystals, characterized by the typical Al(IV) coordination of PSS geopolymers [7].

XRF and XPS determinations indicated also that the Si/Al ratio was between 1.9 and 2, and K^{+} was definitely the compensatory cation. As indicated by the ^{1}H NMR spectra, differences in the water content existed between the two K-PSS geopolymers, and they were confirmed by TG determinations. Results obtained with mercury intrusion porosimetry, performed on a Thermo Finnigan 240 instrument, indicated that the geopolymer formed from BS-4 TT was characterized by an accessible porosity of 8–9% and a total pore volume of 50–52 mm^{3}g^{-1}, whereas the one obtained from

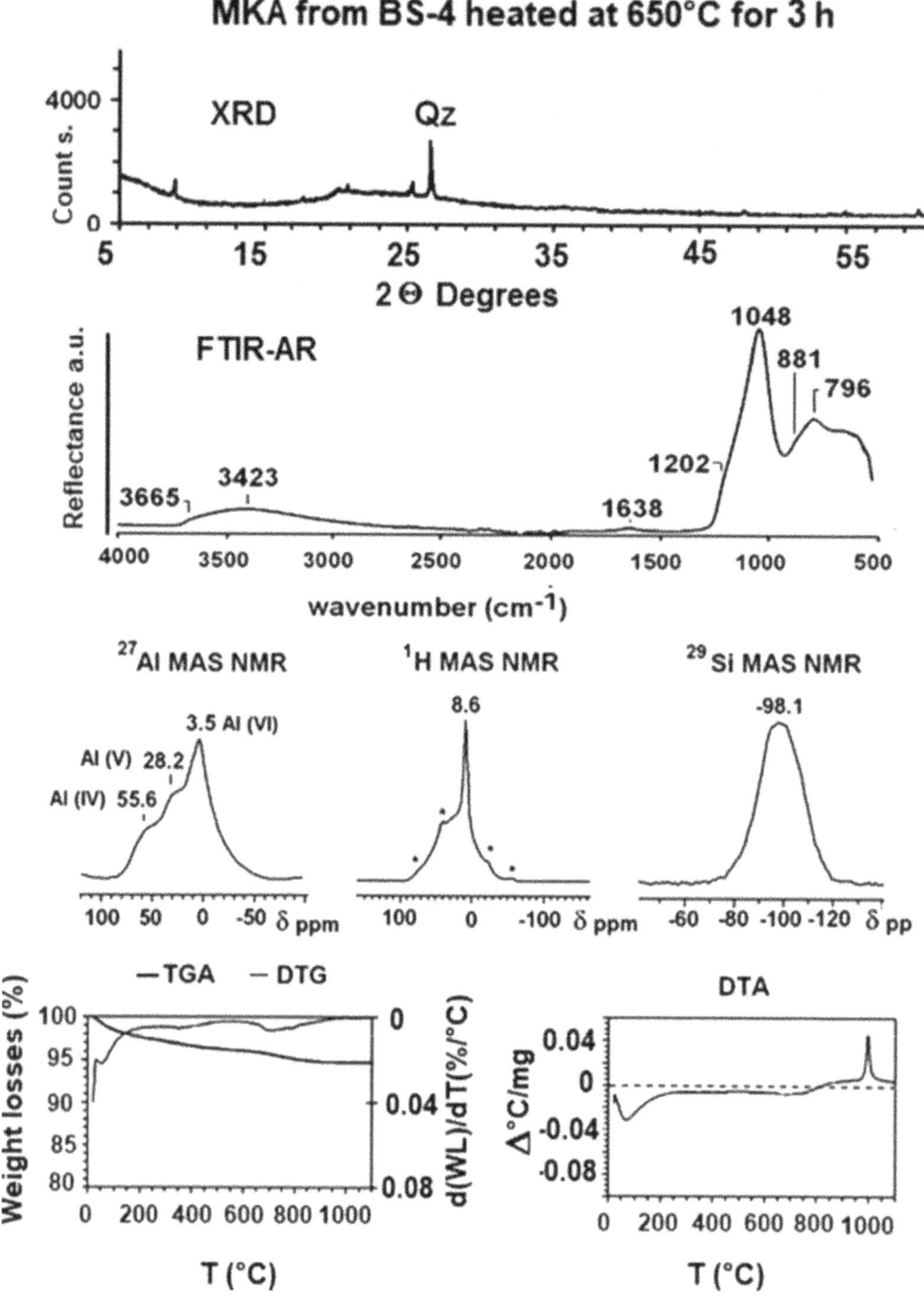

Fig. 18.2 XRD, FTIR-AR, ^{27}Al, ^{1}H and ^{29}Si MAS NMR and TG profiles of MKA produced by thermal treatment of kaolin clay BS-4. Qz = quartz

BS-4 MT showed an accessible porosity of 25% and a total pore volume of 180–190 mm^{3}g^{-1}. This was explained by the faster reactivity of BS-4 MT, producing clusters of reacting material enclosing water that formed pores when water was released as

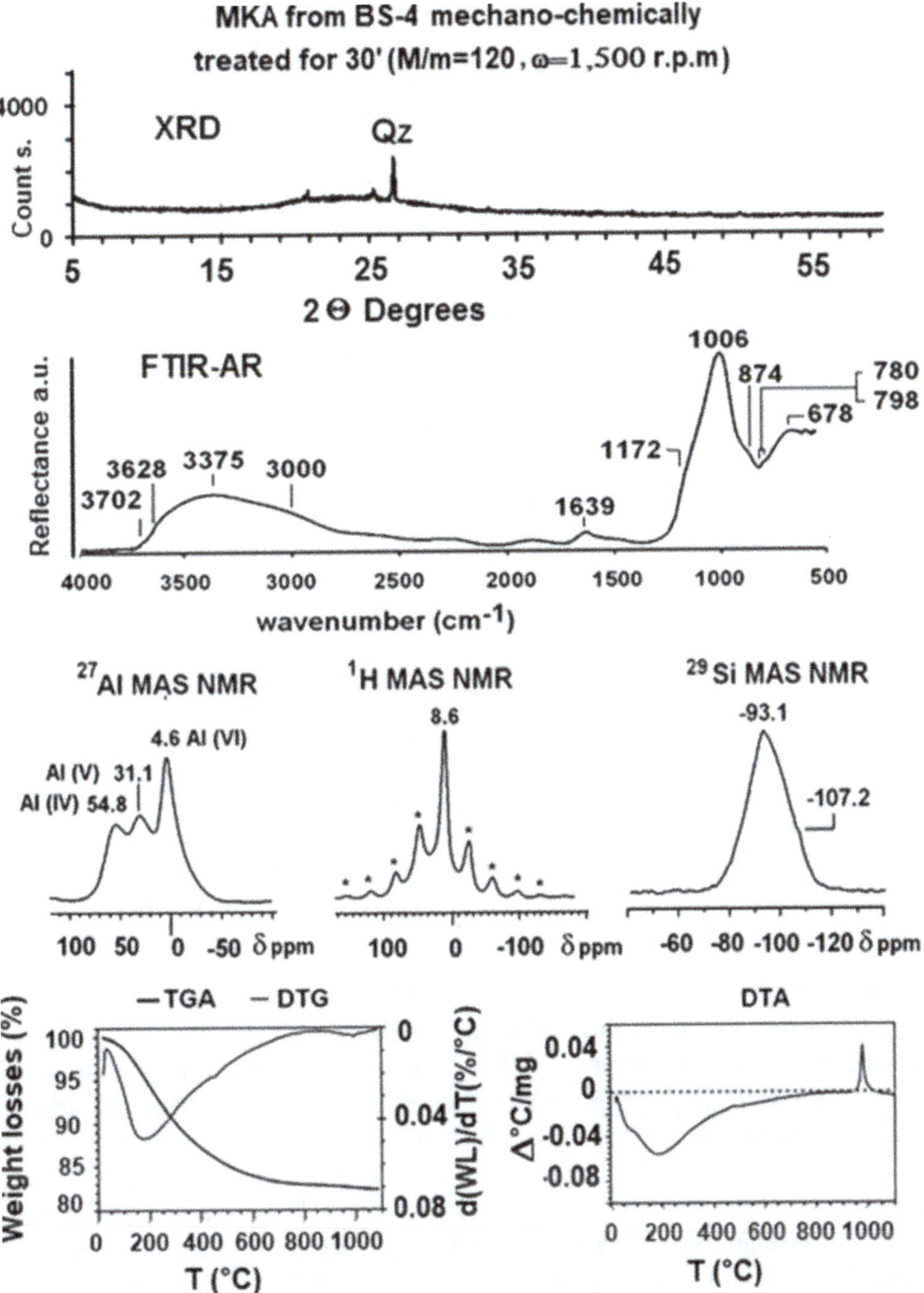

Fig. 18.3 XRD, FTIR-AR, ^{27}Al, ^{1}H and ^{29}Si MAS NMR, and TG profiles of MKA produced by mechano-chemical treatment of the same kaolin clay of Fig. 18.2. Qz = quartz

vapour during the exothermal polymerization step. Differences in porosity explained why geopolymers obtained from BS-4 MT showed a lower UCS value after 28 days (46.5 ± 4 MPa), than those obtained from BS-4 TT (UCS = 55.7 ± 9 MPa). Also the

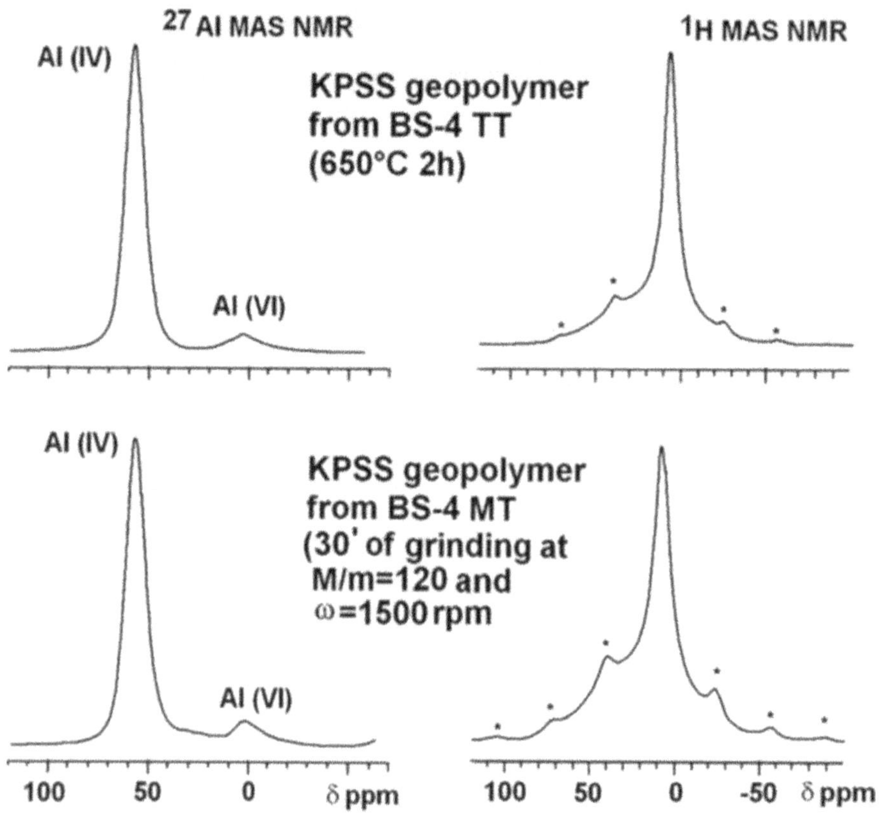

Fig. 18.4 ^{27}Al and ^{1}H MAS NMR profiles of K-PSS geopolymers synthesized from BS-4 MT and BS-4 TT. The formation of a K-PSS geopolymer made of randomly oriented nano-crystals with Al in coordination 4 is evident in both cases. Also evident is the different content of H present in the material as free or adsorbed water

values of the flexural strength of the geopolymers obtained from BS-4 TT (9.5 MPa +1.2) were better than those obtained from BS-4 MT (5.5 ± 0.4).

Since the UCS of all geopolymers fell well within the ranges of commercial clinker cements, their thermal properties were determined using a hot stage microscopy MISURA M3M1600/80/2 from Andersen Lt. It was found that the melting temperatures of K-PSS from BS-4 MT and BS-4 TT were comparable, as all ranged between 1,575 and 1,595 °C, although the sintering process of K-PSS from BS-4 MT started 150 °C earlier than that of BS-4 TT (1,205 °C), indicating a larger reduction in volume due to the different porosity and bulk density of the two products.

The fact that the thermal stability of K-PSS geopolymers obtained from BS-4 MT was close to the highest ever obtained with these geopolymers [7], confirmed the suitability of mechano-chemically formed MKAs to make hard geopolymeric cements with excellent thermal properties.

18.4.2 Geopolymers from Activated Tuffs and Activated Rock Mixtures

A protocol analogous to that described in Sect. 18.4.1 was followed with tuffs, with the only difference was that no thermal treatments were made on them, and all materials grinded were used to synthesize K-PSS geopolymers. A lithified tuff cemented by zeolites (TRSN) with a Si/Al of 2.45, and a SSA $= 5$ m^{2}g^{-1}, and a vitric tuff cemented by glass (PTV), with a Si/Al $= 3.3$ and a SSA of 1.75 m^{2}g^{-1} were selected as test materials. Details on their origin, petrographic features, and practical applications can be found in [18], together with the average oxide composition and other features, including the spectroscopic ones.

Figure 18.5 highlights well the differences between the two tuffs. The lithified nature of TRSN is well visible from the intense XRD reflections generated by the cementing zeolites chabazite, and phillipsite, accounting for 56–58% and 12–14%, respectively, of the whole sample. Their presence explains the high thermal losses of zeolitic water measured by TG determinations, and the signals of the Si(nAl) pentads generated in the ^{29}Si NMR spectrum. TRSN is a porous and soft material (UCS $=$ 25–30 MPa) with an average pore radius of 2.74 μm, an open porosity of ca. 40%, and an apparent density of 1.97 g cm^{-3}. Thanks to the high zeolite content, it displays a rather good pozzolanic activity [17].

The vitric nature of PTV was indicated by the fact that cementing glass accounted for 60% of the whole material, whereas minerals for no more than approx. 35%. The glassy nature of the cementing matrix is well visible from the broad bands generated in the ^{27}Al and ^{29}Si MAS NMR spectra. PTV is a quite hard material (UCS $=$ 40–60 MPa) with an average pore radius of 0.89 μm, an open porosity of 14.4%, and an apparent density is 2.7 g cm^{-3}.

Figure 18.6 shows the loss of crystallinity of the main minerals present in TRSN and PTV as a function of the grinding time. To better follow the kinetic of crystal transformations they were grinded in the ring mill at a lower rotation speed and at a higher grinding ratio than those used in the experiments reported in Sect. 18.4.1.

The results in Fig. 18.6 show that the amorphization of minerals proceeded to a different extent as a function of the grinding time, highlighting the role played by the mineral hardness in determining their amorphization rates. The two zeolites present in the TRSN were softer than sanidine, as they were amorphized faster than this mineral, which was still 80% in the crystal phase after 18 min of grinding. Also in PTV, sanidine resulted the hardest mineral, but it was amorphized by more than 45% in 16 min due to the similar hardness of plagioclases anorthite and bitwonite, quite abundant in PTV. In both tuffs, micaceous biotite was the fast mineral to be amorphized, due to its phyllosilicate structure. The partial amorphization of minerals was confirmed by solid state NMR and FTIR-AR.

Tuffs processed at a grinding ratio of 120 and 900 r.p.m. were used for making K-PSS geopolymers, because it was found that no substantial increase in the pozzolanic activity occurred on TRSN grinded above 6 min, and on PTV grinded above 12 min.

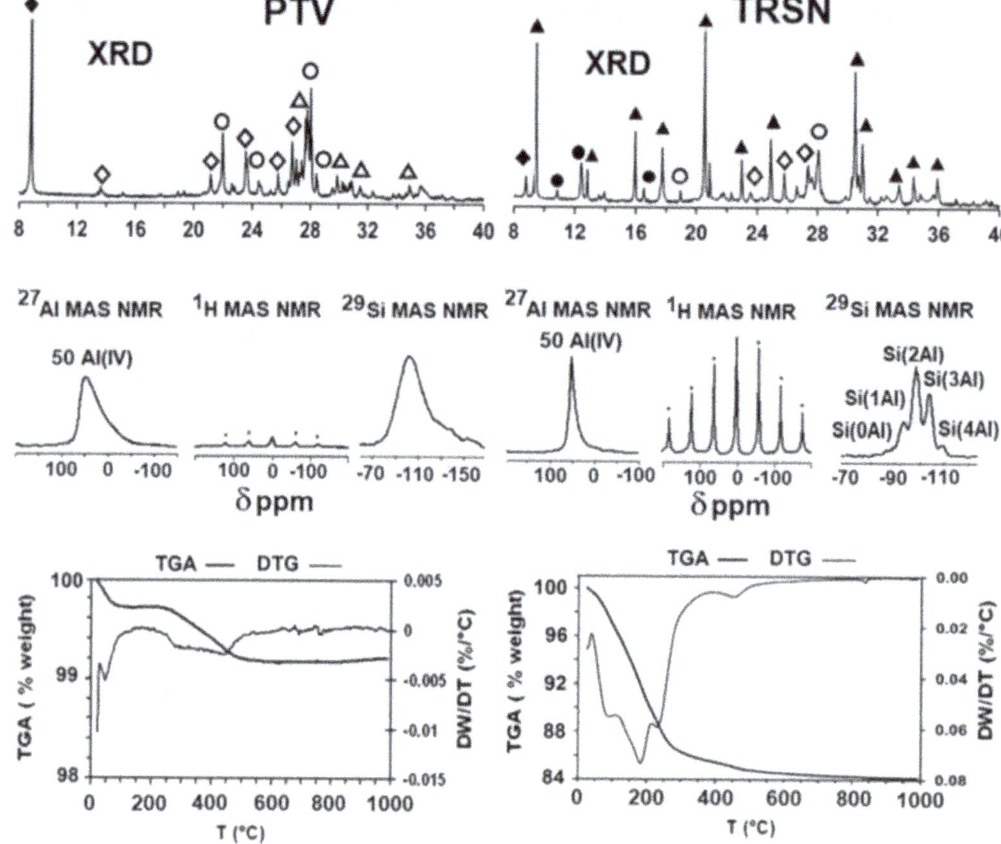

Fig. 18.5 Distinctive features of the welded tuff (PTV) and lithified (TRSN) as recorded by XRD, solid state NMR and TG determinations. Different symbols are used to indicate the minerals present in them. Filled diamond = Biotite, filled triangle = Chabazite, filled circle = Phillipsite, empty diamond = K-felspar Sanidine, empty circle = Plagioclase Bitownite and Anorthtite, empty triangle = Clynopyroxene Augite

Various formulations were tested with untreated, and mechano-chemically treated tuffs.

The only geopolymeric product worth to mention was a K-disiloxo-sialate (K-DSS) geopolymer with a Si/Al ratio of 3, obtained from TRSN grinded for 6 min. It showed an UCS of 26.8 ±2.6 MPa, a flexural strength 6.80 ± 0.10 MPa, a density of 1.51 ± 0.05 g cm^{-3}, and a melting temperature of 1,520 °C. The good thermal stability of this product, but also of the weaker obtained with PTV, suggested that both activated materials could be added to BS-4 MT to make blended K-PSS geopolymeric cements. Using different combinations of the grinded materials with BS-4 MT, it was found that up to 40% of activated tuffs could have been used to make K-PSS geopolymers.

Table 18.1 compares the thermal and mechanical properties of K-PSS geopolymers obtained by using BS-4 TT, BS-4 MT and those of the blended ones obtained by mixing 60% of BS-4 MT with 40% of mechano-chemically activated TSN (TRSN-MT) and PTV (PT-MT), both grinded for 6 min. By considering the lower price

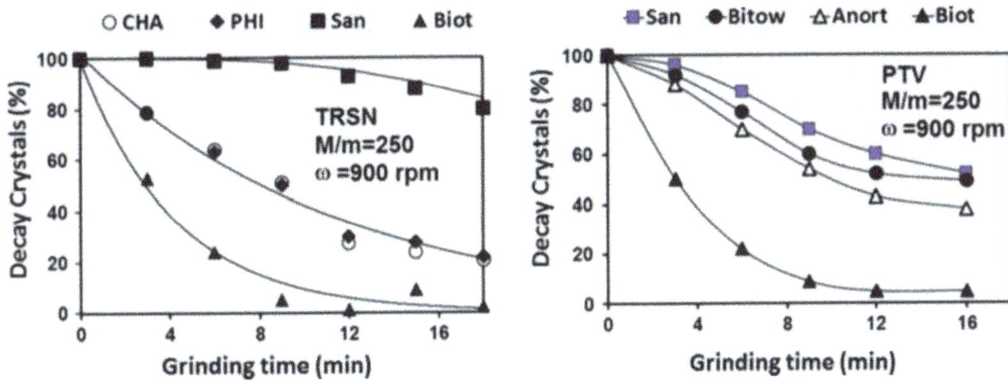

Fig. 18.6 Amorphization rate of minerals in TRSN and PTV as function of the grinding time. Different symbols are used to indicate the trends followed by the various minerals. Empty circle = zeolite Chabasite (CHA), solid diamond = zeolite Phillipsite (PHI), solid square = K-feldspar Sanidine (San), solid circle, plagioclase Bitownite (Bitow), empty triangle = plagioclase Anorthite, solid triangle = mica Biotite (Biot)

Table 18.1 Mechanical and thermal properties of blended K-PSS geopolymers obtained by mixing mechanically treated tuffs with BS-4 MT, with respect to those obtained using BS-4 MT and BS-4 TT

Type of KPSS geopolymer	UCS (MPa)	Flexural strength (MPa)	Sintering T ($°C \pm 0.5\ °C$)	Softening T ($°C \pm 0.5\ °C$)	Melting T ($°C \pm 0.5\ °C$)
BS-4 MT	46.55 ± 4.05	5.95 ± 0.55	1,052	1,530	1,599
BS-4 TT	55.75 ± 4.87	5.95 ± 1.25	1,205	1,545	1,590
60% BS-4 MT+40% TRSN MT	46.05 ± 5.6	6.72 ± 0.99	990	1,390	1,595
60% BS-4 MT+40% PTV MT	31.43 ± 5.10	7.81 ± 2.10	1,050	1,420	1,585

of tuffs with respect to BS-4, and the short grinding time required for the activation of TRSN (that can be reduced to 3–4 min at a rotation speed of 1,500 r.p.m.) blended geopolymers made with this activated tuff were particularly cost effective with respect to those obtained from BS-4 MT and BS-4 TT.

18.5 Prototype Milling System for the Mechano-Chemical Processing of Rocks at a Small Industrial Scale

Results presented in Sect. 18.4 suggested that it was possible to develop a prototype system for a small scale production of mechano-chemically activated rocks for geopolymer production, but also for other applications, such as zeolite production.

To work in an industrial environment, the prototype was designed in such a way that some operations, such as the charge and discharge of material from the mill or the reduction of the size of the raw material, could have been performed in an automated way. Particular attention was paid to the comminution of the raw material down to the size (200–100 μm) that was the most suitable for mechano-chemical activation. The possibility to directly treat quarried materials in the mill would have reduced the time and costs of the whole process, by eliminating the use of a dedicated mill for the comminution of the sample, that require a quite long time to reduce the size of rocks down to a sandy fraction, due the low speed at which it is operated.

The prototype was built using a continuous ring mill Model ESSA by Labtech that was specifically modified for semi-continuous operations. Since the mass of the moving bodies was 15 kg, and could rotate at same maximum speed (1,500 r.p.m.) reached in laboratory experiments, the maximum energy transferred to particles was 4 times higher. This allowed treating more material in a shorter time. Although limited to 30 kg h^{-1}, the maximum amount of material produced by this prototype was sufficient to test the system in a small industrial facility in which mechano-chemical treatment was needed to make products of technological interest from wastes and rocks.

As indicated in Fig. 18.7, the prototype consisted of two jars set in series. The top jar, which is fed with the raw material by a hopper, contains a rotating disk of 20 kg, named *Flying Saucer*, that enables a fine grinding the raw material to get only a sandy fraction ($D_{50} \approx 100$ μm) out of it. From the top jar, the sandy fraction falls in a second jar equipped with 2 rings and 1 cylinder as moving bodies, where the material is subjected to mechano-chemical processing. The bottom of the second jar is equipped with a sieve with a defined mesh size range, so that the grinded material can be discharged outside the mill by gravity.

The system was modified to maintain the powder inside the second jar for the time necessary to activate the material. This was done by keeping the top jar under depression to prevent the discharge of material from the second jar. Only when the grinding time necessary to activate the material was reached, the depression was removed and the activated material discharged and utilized. An electric valve was used to maintain the depression inside the system. With these modifications, the ring mill basically worked as a fluidized mechano-chemical reactor, able to optimize the whole mechano-chemical processing of the material.

18.6 Prototype Testing and Design of a Milling System for a Full Scale Industrial Production

This section reports the results obtained by testing the prototype system in a representative industrial environment (Sect. 18.6.1) and how these results were exploited to design a milling system for large industrial applications (Sect. 18.6.2).

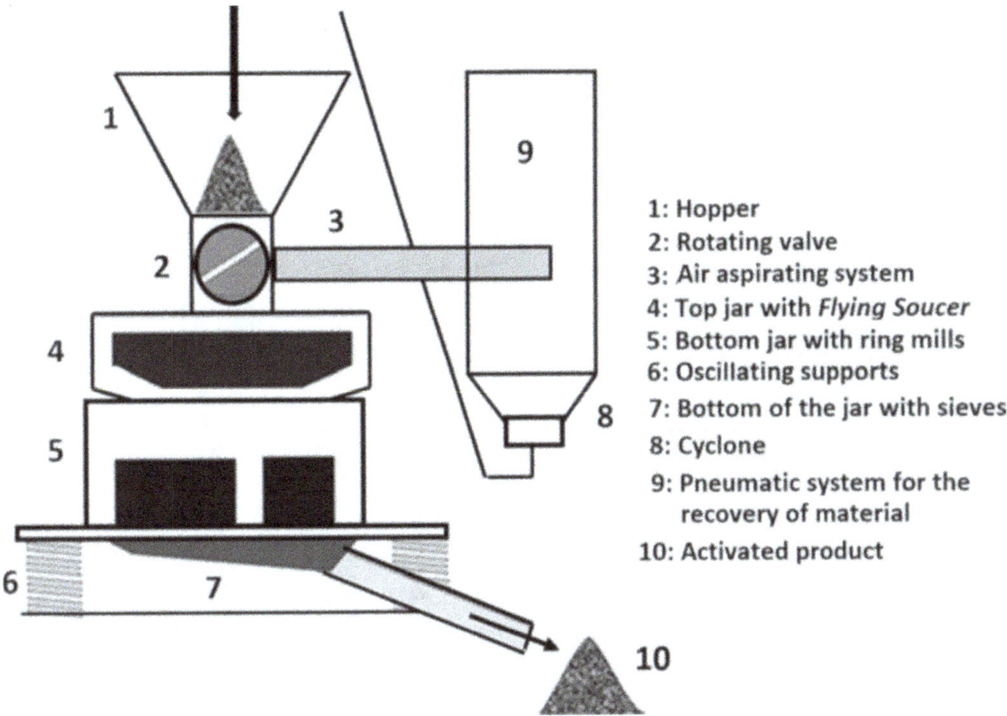

Fig. 18.7 Schematic diagram of the prototype milling system built for mechano-chemical processing of minerals and rocks for a small-scale production of K-PSS geopolymers and other technological products

18.6.1 Prototype Testing

The prototype system described in Sect. 18.5 was tested in a demonstration plant for the production of zeolites from industrial wastes built by the Sicily region in Melilli. Figure 18.8 shows a schematic diagram of the various processes used by the plant to produce zeolites, and the position where the prototype for the mechano-chemical activation of wastes was located. When exhausted catalysts from the cracking of crude oil were treated, the mechano-chemical unit was placed after the unit separating the iron-containing fraction of wastes.

When the proper degree of activation was reached, the material was transferred in a closed reactor kept at 90 °C that was filled with NaOH 1 M, from which zeolites were formed by hydrothermal reaction. Using an M/m ratio of 400, the mill was able to activate these wastes in 2 min. The degree of activation was such that grinded products reacted so quickly with alkaline solutions to form zeolites in less than 10 h. This was a relevant result, because more than one month was necessary to produce the same amounts of zeolite from the untreated wastes. Since the zeolite formed from these wastes was the same present in volcanic tuffs, the product showed a good pozzolanic activity, and could have been used for the production of pozzolanic cements and blended Si–Al geopolymers.

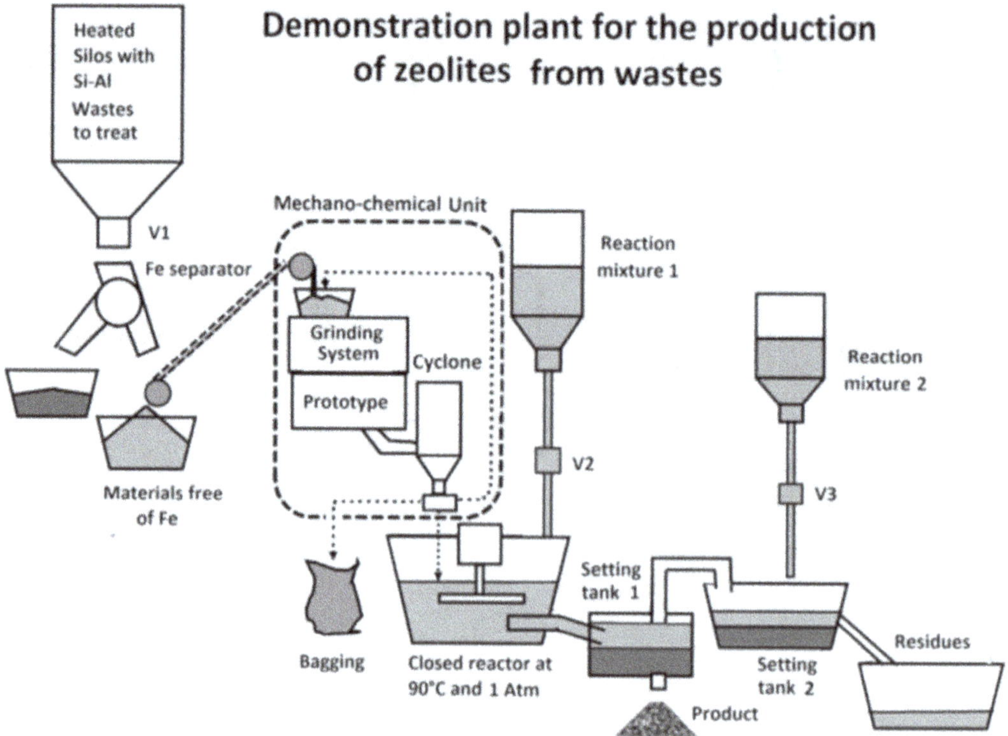

Fig. 18.8 Schematic diagram of the demonstration plant for zeolite production from wastes where the prototype milling system was tested. A thick dotted line is used to identify the sector of the plant where the mechano-chemical unit was placed

In the Melilli plant, the prototype was able to treat 144 kg of industrial wastes per day, with no substantial problems during 10 months of testing. The high versatility of the milling system was demonstrated by the fact that for a short period it was fed with volcanic ashes produced by the Etna volcano, that were also converted into zeolites different from those obtained from industrial wastes. In the treatment of volcanic ashes, the presence of the *Flying Saucer* shown in Fig. 18.7 resulted particularly useful to reduce the time and cost of the final products, because of the wide and different size range of the raw material used.

Preliminary results showed that 20 min of grinding were needed to make the volcanic ashes sufficiently reactive for the synthesis of zeolites. Due to the high content of hard minerals (plagioclases) in the ashes, a long activation time was needed to reach the degree of amorphization necessary to produce zeolites out of them.

18.6.2 Design of an Industrial Mill for Mechano-Chemical Treatment of Rocks

Although the results reported in Sect. 18.6.1 showed the ability of the mechano-chemical system to work in an industrial plant, the amount of treated material was still too low for a full scale industrial production, where 5 tons of material per day, corresponding to 420 kg h^{-1} for 12 h, can be regarded as an acceptable target.

To achieve this goal, a further research was made to improve the geometry of the milling system. It was aimed at drastically reducing the heavy vibration problems posed by ring mills, where moving bodies are forced to rotate in an eccentric mode. The geometry shown in Fig. 18.9 was selected, because it ensures the same collision and friction forces than a ring mill, but it does not induce heavy vibrations in the horizontal and vertical plane. It is composed of a jar with 4 grinding masses connected to the shaft of an electrical motor through mechanical arms, letting the mass free to move toward the cylindrical walls by centrifugal forces. In this way the same shearing and friction forces acting in a ring mill can be simultaneously applied to the crystal grains.

To calculate the dimensions and operational parameters required by the mill, a model was used to estimate the specific energy transferred by the system to the grinded material. The time needed to produce 5 tons per day of MKA from BS-4 was thus derived using the data obtained from laboratory experiments. This was necessary, because present models are unable to predict with sufficient accuracy the impact that co-grinding effects, arising the simultaneous presence of different minerals in the material, have in affecting the enthalpy of the phase transitions occurring on each of them.

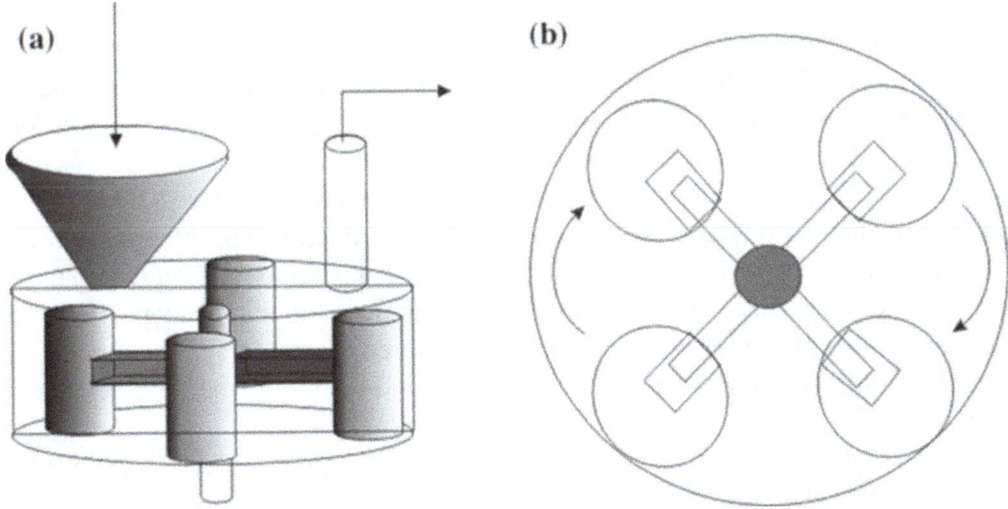

Fig. 18.9 Different views of the geometry proposed for the mechano-chemical treatment of rock materials at an industrial scale

The power (P in Watt) required to move the grinding bodies in this type of mill was calculated with the model developed by Founti et al. [19] reported in Eq. (18.2), where d is the jar diameter in m, ω is the rotation speed in Hz, μ is the friction coefficient and Q is the centrifugal force in N.

$$P = \mu Q \pi / (15\, d\omega) \tag{18.2}$$

In the configuration of Fig. 18.8, Q is equal to $\pi^3/3600 \cdot M\omega^2 a$, where a is the distance in m between the centre of the mill and the wall of the jar, and M the mass of the moving bodies in kg. Calculations were made by considering a mill with 4 jars, one on top of the other, having a diameter of 1.2 m and a volume of 340 L each, where only 2 grinding bodies of 1,161 kg were present. From the data listed in Table 18.2, the system can produce 5 tons of MKA per day from BS-4, because is able to fully convert 14 kg of KA into MKA in 2 min. According to the engineers of a mechanical company, this geometry, if properly built, can safely work up to a rotating speed of 2,000 r.p.m., thus further reducing the grinding time.

Based on the data reported in Table 18.2, 7,000 kW h^{-1} t^{-1} were estimated to produce 5 tons of MKA per day, by considering the time required for the discharge and charge of the mill. These figures strongly differ from those provided by some authors, according to which only 1,000 kW h^{-1} t^{-1} are required to convert KA in kaolin clays into MKA. Since no industrial mills for the production of MKA are presently in operation, it is not clear how these estimates were made. The estimate of the energy necessary to produce MKA by thermal processing appear more realistic, because a value of 2,000 kW h^{-1} t^{-1}, including that required for the comminution of the raw material, was given. Although these figures can be affected by more than 40% uncertainty depending upon the type of kiln and fuel used, still the mechano-chemical processing appears more energy demanding.

Only if the total costs are analysed, the mechano-chemical treatment of rocks becomes competitive with the thermal one. The investment and operating costs of a plant for the grinding and thermal processing of kaolin clays are in fact much higher than those of mechanical processing, if the expenses for the control of combustion, and those for the monitoring and abatement of atmospheric pollutants in the emission, primarily NOx, VOC, CO and particulate matter, are also considered. These costs can be quite high also because rotary kilns for thermal treatment of minerals in rocks must be maintained into a continuous operation, even when the product demand is low, and the price of fuel is high. This does not happen with the mechano-chemical treatment, where the production can be better tailored to the market demand, as the mill can be turned off when the demand is low, and operated when the demand is high.

If the composition of the raw material changes, a large amount of material is consumed with the thermal treatment before the optimal conditions to get a new MKA with comparable reactivity is obtained. These limitations do not apply to the mechano-chemical treatment, because the system can be more easily fine-tuned as a function of the material to be treated, as it was shown by the tests performed in

Table 18.2 Working parameters of an industrial mill using the geometry of Fig. 18.9 designed for the mechano-chemical production of 5 tons per day of MKA from kaolin clays

Dimensions and technical properties of the industrial mill	Value	Units
Jar diameter	1.20	m
Jar height	0.30	m
Jar volume	0.339	m^3
Diameter of the cylindrical moving body no. 1	0.60	m
Height of the cylindrical moving body no. 1	0.29	m
Volume of the cylindrical moving body no. 1	0.082	m^3
Area of the cylindrical moving body no. 1	0.546	m^3
Perimeter of the cylindrical moving body no. 1	1.994	m
O.D. of the grinding body no. 2	1.000	m
I.D. of the grinding body no. 2 (ring)	0.850	m
Height of the grinding body no. 2 (ring)	0.290	m
Volume of the grinding body no. 2 (ring)	0.0632	m^3
Total volume of the grinding bodies	0.1451	m^3
Total mass of the grinding bodies	1,161	kg
Rotation frequency	25.00	Hz
Revolution per minute	1,500	r.p.m.
Circumference	3.770	m
Maximum theoretical speed of the moving bodies	94.20	$m\ s^{-1}$
Density of the material (stainless steel)	8,000	$kg\ m^{-3}$
Weight of the material to treat	14.0	kg
M/m ratio	82.9	n
Number of grinding bodies	2	n
Kinetic energy produced	$3.61\ 10^6$	$J\ s^{-1}$
Energy supplied for grinding	$8.99\ 10^5$	$J\ s^{-1}$
Kinetic energy lost	$2.71\ 10^6$	$J\ s^{-1}$
Kinetic energy transferred to the material to grind by collision	$3.61\ 10^3$	$kJ\ s^{-1}$
Total energy transferred in 2 min grinding	$8.0\ 10^5$	kJ

Melilli plant, where in 1–2 h it was possible to switch from industrial wastes to Etna ashes.

The fact that a mechano-chemical mill with the same size of that described in Table 18.2 can be transported near the extraction plant, and powered with electricity produced through renewable resources, can further reduce the costs of transport, which can be also high in the case of thermal treatment, and contribute to the total CO_2 emission.

18.7 Conclusions and Future Research

The results of this research highlighted some of the capabilities afforded by the use of the mechano-chemical treatment of different materials for making pure and blended K-PSS geopolymeric cements with good mechanical properties and an extraordinary thermal resistance. In particular, the possibility to replace 40% of kaolin clay with mechano-chemically activated tuffs can reduce the cost of the cementing product by more than 50%, due to the lower price and much smaller energy required for activating lithified tuffs with respect to any kaolin clay. The testing of the small prototype has emphasized the capability and reliability of the mechano-chemical treatment to work unattended in an industrial facility, where the possibility to rapidly switch from one raw material to another to produce different zeolites was also demonstrated.

In spite of the advantages offered by geopolymeric cements, the idea to produce them at an industrial scale has not raised so much enthusiasm in the Italian cement industry, probably due to the fact that the persistent crisis in the building industry has so decreased the profits, that funds available for research and development are not sufficient to build a demonstration plant for new types of cements. The mechano-chemical processing of materials has raised, instead, great interest, although the lack of grinding systems operating at a large industrial scale has posed some limits to it.

From a scientific point of view, this is the first study in which mechano-chemically treated kaolin clays and volcanic tuffs were mixed together to make low price geopolymeric cements with high mechanical and thermal properties. From a theoretical point of view, the use of solid state NMR in combination with other techniques has allowed to identify structural differences between MKA obtained by thermal and mechano-chemical treatment that can explain the different reactivity of the two materials in alkaline media. Our result show also that not necessarily the most reactive MKA has a structure in which the Al(V) population is always dominant, and, consequently, the rule proposed by Davidovits [7] cannot be considered a universal one. Since the chemistry of geopolymers strongly relies on this rule, our results open a new perspective on the possible relations between the reactivity and structure of MKA, suggesting that, in spite of the outstanding progress made in recent years with different modelling approaches, the structure of MKA is still poorly understood, and certainly deserves further investigations to be clarified.

From the results obtained in our study, the mechano-chemical activation of industrial wastes from which various products can be obtained is certainly a promising field for future investigations. In particular, the co-grinding of different raw materials to get new and more reactive phases is a highly promising. While the co-grinding of clinker and different SCM can lead to new clinker-based cements, the co-grinding of kaolin clays and tuffs can lead to MKA structures better suited for the synthesis of Si–Al geopolymers.

Acknowledgements This work has been funded by the Italian Ministry of Education, Universities and Research (MIUR) under the Flagship Project "Factories of the Future—Italy" (Progetto Bandiera "La Fabbrica del Futuro") [20], Sottoprogetto 1, research projects "Mechano-chemistry: an innovative process in the industrial production of poly-sialate and poly-silanoxosialate geopolymeric binders used in building construction" (MECAGEOPOLY). The technical support of Laura Lilla and Emanuela Tempesta is also acknowledged. We would like to thank the chemists of Colacem S.p.A. (Gubbio, Italy) and the engineers of Officina Meccanica Molinari S.r.l. (Bergamo, Italy) for the helpful discussions, and the Sicily Region which allowed us to test the prototype in the demonstration plant of Melilli (Siracusa, Italy).

References

1. Roadmap to a Resource Efficient Europe Documents (2011) CSST/2011/1463. COM/2011/571/FINAL. http://www.europarl.europa.eu/meetdocs/2009_2014/documents/com/com_com(2011)0571_/com_com(2011)0571_en.pdf
2. Olivier JG, Janssens-Maenhout G, Muntean M, Peters JHAW (2016) Trends in global CO_2 emissions—2016. JRC report, PBL publication number 2315. http://edgar.jrc.ec.europa.eu/news_docs/jrc-2016-trends-in-global-co2-emissions-2016-report-103425.pdf
3. European Environment Agency (2015) Trends and projections in the EU ETS in 2015. EEA technical report no 14/2015. ISSN 1725–2237. https://www.eea.europa.eu/publications/trends-and-projections-eu-ets-2015
4. Cembureau, Cements for a low carbon Europe (2012) Report D/2012/5457/ November 2012. https://cembureau.eu/media/1501/cembureau_cementslowcarboneurope.pdf
5. Associazione Italiana Tecnico Economica del Cemento (2015) Relazione annuale 2015. https://www.aitecweb.com/Portals/0/pub/Repository/Area%20Economica/Pubblicazioni%20AITEC/Relazione_Annuale_2015.pdf
6. De' Gennaro M, Langella A (1996) Italian zeolitized rocks of technological interest. Miner Deposita 31(6):452–472
7. Davidovits J (2011) Geopolymer chemistry & applications. Institute Géopolymere, S. Quentin, France
8. Takahashi H (1959) Effects of dry grinding on kaoilin minerals. Bull Chem Soc Jpn 32:235–263
9. Sugiyama K, Filio JM, Sajto F, Waseda Y (1994) Structural change of kaolinite and pyrophyllite induced by dry grinding. Mineral J 17:28–41
10. Gualtieri A, Bellotto M (1998) Modelling the structure of the metastable phases in the reaction sequence kaolinite-mullite by X-ray scattering experiments. Phys Chem Miner 25(6):442–452
11. He HP, Guo JG, Zhu JX, Hu C (2003) ^{29}Si and ^{27}Al MAS NMR study of the thermal transformations of kaolinite from North China. Clay Miner 38(4):551–559
12. Fabbri B, Gualtieri S, Leonardi C (2013) Modifications induced by the thermal treatment of kaolin and determination of reactivity of metakaolin. Appl Clay Sci 73:2–10
13. Temuujin J, MacKenzie KJD, Schmucker M, Schneider H, McManus J, Wimperis S (2000) Phase evolution in mechanically treated mixtures of kaolinite and alumina hydrates (gibbsite and boehmite). J Eur Ceram Soc 20:413–421
14. Bellotto M, Gualtieri A, Artioli G, Clark SM (1995) Kinetic study of the kaolinite-mullite reaction sequence. Part I: kaolinite dehydroxylation. Phys Chem Miner 22(4):207–214
15. Cundy CS, Cox PA (2005) The hydrothermal synthesis of zeolites: precursors, intermediates and reaction mechanism. Microp Mesop Mater 82(1–2):1–78
16. Prokof'ev VY, Gordina NE, Efremov AM (2013) Synthesis of type A zeolite from mechano activated metakaolin mixtures. J Mater Sci 48(18):6276–6285
17. Mertens G, Snellings R, Van Balen K, Bicer-Simsir B, Verlooy P, Elsen J (2009) Pozzolanic reactions of common natural zeolites with lime and parameters affecting their reactivity. Cem Concr Res 39(3):233–240
18. Ciccioli P, Plescia P, Capitani D (2010) ^{1}H, ^{29}Si, and ^{27}Al MAS NMR as a tool to characterize volcanic tuffs and assess their suitability for industrial applications. J Phys Chem C 114(20):9328–9343

19. Founti M, Zannis G, Makris P (2008) Power aspects of a horizontal ring mill pulverizer under continuous comminution of olivine. Int J Min Process 85(4):85–92
20. Terkaj W, Tolio T (2019) The Italian flagship project: factories of the future. In: Tolio T, Copani G, Terkaj W (eds) Factories of the future. Springer

Chapter 19
Silk Fibroin Based Technology for Industrial Biomanufacturing

Valentina Benfenati, Stefano Toffanin, Camilla Chieco, Anna Sagnella,
Nicola Di Virgilio, Tamara Posati, Greta Varchi, Marco Natali,
Giampiero Ruani, Michele Muccini, Federica Rossi and Roberto Zamboni

Abstract Natural biomaterials are more and more used for the development of
high technology solutions, setting the scene for a bio-based material economy that
responds to the increasing demand of environmentally friendly products. Among natural biomaterials, silk fibre protein called silk fibroin (SF) produced by the Bombyx
mori L. insect, recently found a broad range of applications in biomedical field. SF
substrates display remarkable properties like controlled biodegradability, flexibility,
mechanical resistance and optical transparency, solution processability. These properties combined with the water-based extraction and purification process make SF a
promising material for sustainable manufacturing enabling to partially replace synthetic, plastic-based and non-biodegradable material use. The use of SF interfaces in
biocompatible electronic or photonic devices for advanced biomedical applications
has been recently highlighted. However, the use of a natural biomaterial is challenging due to the complex nature of the biological molecule, and it requires to tightly
control biomaterial properties during all the manufacturing steps. In this work, we
show the results obtained by in loco production of raw-material, defining the best
condition for *silkworm* selection and growth. The assessment and standardization of
extraction/purification methodology are reported with reference to the high purity and
remarkable performance in terms of chemo-physical property and biocompatibility
of the obtained SF products. Finally, we demonstrate the fabrication, characterization
and validation of microfluidic and photonic components of a lab-on-a-chip device
for biodiagnostic based on biomanufactured SF.

V. Benfenati (✉) · T. Posati · G. Varchi · R. Zamboni
CNR-ISOF, Istituto per la Sintesi Organica e la Fotoreattività, Bologna, Italy
e-mail: valentina.benfenati@isof.cnr.it

S. Toffanin · M. Natali · G. Ruani · M. Muccini
CNR-ISMN, Istituto per lo Studio dei Materiali Nanostrutturati, Bologna, Italy

C. Chieco · N. Di Virgilio · F. Rossi
CNR-IBIMET, Istituto di Biometeorologia, Bologna, Italy

A. Sagnella
Laboratory MIST E-R, Bologna, Italy

© The Author(s) 2019
T. Tolio et al. (eds.), *Factories of the Future*,
https://doi.org/10.1007/978-3-319-94358-9_19

19.1 Scientific and Industrial Motivations

A major challenge toward bio-based economy consists in the replacement of fossil fuels on a broad scale, not only for energy applications, but also for material, clothing and plastic application. In this view, the worldwide trend to a low-carbon economy and sustainable primary production stimulates the use of naturally derived raw materials as alternative resource to oil-based plastics for manufacturing. However, natural materials must be renewable, available, recyclable, and biodegradable to be competitive and they need to be processed by green and sustainable approaches. Moreover, natural biomaterial products should provide a higher technical performance for selected application in comparison to synthetic and plastic based counterpart. In this context, the manufacturing of biomedical devices for diagnostic or therapeutics offers a potential market opportunity for the use of natural biomaterials. Indeed, natural biomaterials due to their well-known intrinsic characteristics such as biodegradability, and high biocompatibility are ideally suited to develop innovative biomedical products such as those for tissue engineering and also medical sensing. Indeed, during the last three decades, the use of natural biomaterials in tissue engineering is rapidly evolving and innovative devices are able to support and recover the structure and the functionalities of injured hosting tissues [1, 2]. Several evidences have consolidated the use of biomacromolecules for development of scaffold devices that enable the regeneration of different tissues (i.e. nerve, cartilage, bone) [1, 2]. Natural biopolymers mainly includes proteins (i.e. collagens, gelatine, zein, silk fibroin, elastin), polysaccharides (i.e. chitin, alginates, chitosan, cellulose derivate), and amides (i.e. starch). Silks are natural proteins polymers produced by different species of insects, such as spiders, scorpions, silkworms [3]. Among silks, Silk fibroin (SF) produced by Bombyx mori cocoon have been used clinically as sutures for centuries [3]. Nonetheless, in recent years, SF-based materials have been extensively studied in tissue engineering and drug delivery due to their biocompatibility, slow degradability and remarkable mechanical and optical properties [1–3]. Notably, through a process of reverse engineering Rockwood et al. [4] have defined a water-based and sustainable process that enable to obtain an aqueous-based SF solution, called regenerated silk fibroin (RSF), from the cocoon fibre. The use of the RSF solution is particularly interesting in the context of biomedical application [1–4] because it can be processed in various formats (films, fibres, nets, meshes, membranes, gels, sponges) retaining exceptional chemo-physical and biological properties. In this context, SF displays the potential to be exploited as a raw material to become a technological material platform [3, 5] for eco-sustainable manufacturing. However, some chemical and physical post-processing treatments of SF could damage/denature the protein and completely modifying its primary structure properties, thus altering the properties of the obtained substrates. Nonetheless, when dealing with naturally derived products and biomedical application it is highly desirable to establish and control the whole product lifecycle: from the raw material production to substrate preparation to technology validation. In this view, the goal of our work is to define and control the whole silk chain by in loco production of the raw-material, the assessment and standardization of extraction/purification methodology the characterization of the chemo-physical and biocompatibility properties of the obtained SF products. Herein, we report on state of the art methods and protocols to use SF as new material for advanced bio-technological application and sustainable manufacturing (Sect. 19.2). The proposed approach is presented in Sect. 19.3 and further detailed in Sect. 19.4, whereas the experiments are reported in Sect. 19.5. In particular, we demonstrate the importance of the definition and control of the whole chain for biomanufacturing underpinning the silk fibroin-based technology. Furthermore, we demonstrate and report the results obtained on the fabrication, characterization and validation of microfluidic and photonic components of a lab-on-a-chip device for biodiagnostic based on biomanufactured SF.

19.2   State of the Art

The progress in material science, biotechnology, photonics and electronics is promising for the development of novel technologies that will dominate the future of healthcare and improve the quality of life. Lab-on-a-Chip (LOC) devices have demonstrated to be highly performing for analytical purposes in genetic sequencing, proteomics, and drug discovery applications [6]. However, they are not yet compatible with point-of-care diagnostics applications, where cost and portability are of primary concern. Recent advance in plastic electronics revealed the promising future for oil-based organic (plastic) material for on chip integration of optic, photonic and microfluidic structure to fabricate portable, low-weight and low cost devices. However, global trends towards sustainable development stimulate the use of renewable and biodegradable raw materials and innovative environmental friendly manufacturing of the technology of the future. Naturally occurring (bio-derived) materials provide a compelling template to reinterpret modern manufacturing while rendering it sustainable and green. The challenge to make bio-derived materials a credible alternative to current plastic and synthetic materials and inorganic substrates is to identify materials that, while being environmentally sustainable, display properties that combine chemo-physical, technological end economic need for sustainable biomanufacturing. As an example, the material has to be widely available and cost-competitive within the global supply chain and, at the same time it should maintain adequate chemo-physical and biocompatibility properties to successfully interface with technology intended for biomedical application [1–3, 5]. In this context, natural fibres can provide several benefits such as low cost, *green* origin, abundant and renewable, carbon dioxide neutral, lower densities, recyclability, biodegradability, moderate mechanical properties. Moreover, their mechanical and technological performance shows high potentials in their applications [1–3, 5].

Among natural fibres, SF produced by the silkworm *Bombyx mori* L. has high potentials for a wide range of innovative, green and high-tech applications [3]. Silk fibroin solution, extracted from the cocoon by a water based eco-friendly method [4], could be solution-processed (by spin coating, ink-jet printing, spray draying) in different forms and substrates with exceptional properties in terms of mechanical flexibility, dielectric properties, biocompatibility, optical transparency and possibility of functionalization with optically/biologically active molecule (drugs, dye, nanoparticles). Thus, silk fibroin offers a unique possibility to enable multiple and complementary approaches for technological and biomedical application.

Among different forms of SF-based biomaterials, SF films can be obtained from regenerated solution RSF with defined thickness (with a range spanning from hundreds nanometers to tens micrometres) by means of liquid processing such as spin coating, inject printing, doctor blade that are suitable for industrial manufacturing. SF films can be also nanostructured by soft lithography, contact printing or nanoimprinting. Nonetheless, silk fibroin can be doped, blended and functionalized to obtain a numerous amount of substrates for advanced technological applications in the biomedical field, such as organic electronic sensors based on field effect transistors [5, 7] for electrophysiological recording of neural cells or optical device such as photonic sensors and optoelectronic devices [8]. In particular, optical and biological properties of silk fibroin films can be enhanced thanks to several processes that are able to functionalize and modify SF with optically active dyes, doping molecules, growth factors and/or chemicals or nanoparticles [9–11]. Among the others, a green and sustainable method to produce innovative modified silk substrates is feeding cocoon with a diet characterised by the addition of doping molecules that enable the biological incorporation of dopant into SF [12–14]. The approach indeed avoids the need of an additional chemical process, and post-treatments associated with it [12–14]. A series of fluorescent dyes have been successfully incorporated in silk fibre by addition of colorant compounds to the mulberry diet [12–14].

The feasibility of using SF materials has been demonstrated in vitro and in vivo for opto-electronic biomedical applications and for the use in peripheral nerve tissues repair [7, 8, 15–21]. It has been shown that optically transparent, ultrathin and flat silk films can be integrated as dielectric in organic field effect transistor (OFET) and organic light emitting transistor (OLET) architectures [7], also used to stimulate and record cells functional properties [20, 21]. Also, it was demonstrated that silk films can be nano- and micro-patterned to obtain mechanically robust structures with desired optical wavelength size [8]. Organic lasing from a stilbene-doped silk film spin-coated onto a one-dimensional distributed feedback grating (DFB) was demonstrated, with the lasing threshold lower than that of others organic DFB lasers based on the same active dye.

19.3 Problem Statement and Approach

The industrial scale development of silk fibroin based technologies requires the optimization and control of all the silk-fibroin production steps, from the raw material to the technological application. The lack of such control hampers the use of SF at industrial scale and limited the demonstration of silk-based devices at a lab-scale. The aim of our work to establish and control the whole-chain underpinning the fabri-

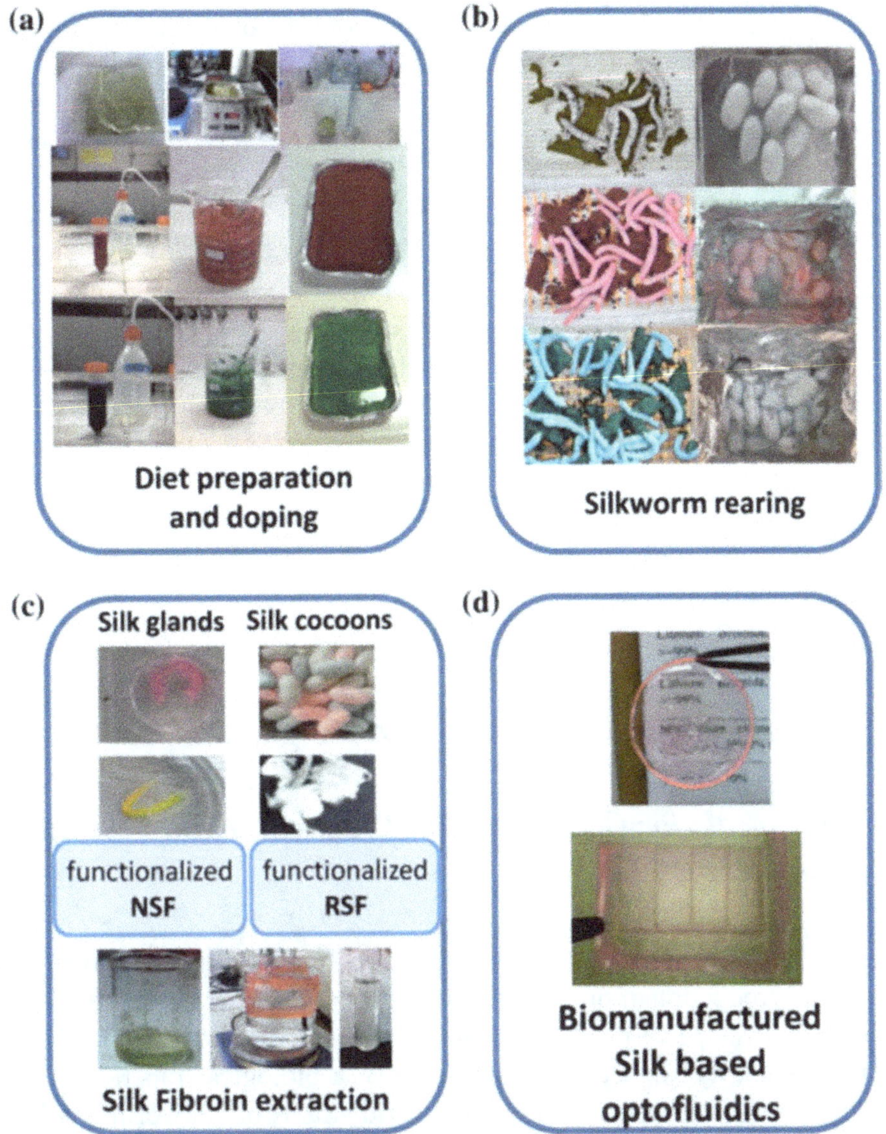

Fig. 19.1 Schematic picture of the biomanufacturing approach to obtain SF based technological substrates

cation of the silk fibroin based technology (Fig. 19.1).The technology of interest is a SF-based innovative lab-on-a-chip (LOC) device for high throughput bio-diagnostic. In particular, the optical based LOC is intended for fluorescence optofluidic detection and sorting of tumour cells.

The design of the proposed biomanufacturing approach consists of the following steps:

(1) Diet preparation (Fig. 19.1a) and optimization of silkworm breeding (see Fig. 19.1b) to obtain a reproducible, controlled and high quality functionalized silk fibroin. Notably, the production of the raw-material described in this work includes the standardization of a novel process that enables direct func-

tionalization of SF fibres by feeding the cocoon with a diet doped with optically active molecule of interest (Sect. 19.4.1).

(2) Standardization of the extraction, purification, and characterization of functionalized silk fibroin (native and regenerated) from different races of *B. mori* (see Fig. 19.1c; Sect. 19.4.2).

(3) Protocols to obtain regenerated silk fibroin (RSF) solutions and then produce RSF films (top of Fig. 19.1d; Sect. 19.4.3).

(4) Fabrication, characterization and validation of a silk-based photonic and microfluidic component of LOC device (see bottom of Fig. 19.1d; Sect. 19.4.4). Optical and photonics components of the silk based device have been fabricated by using in loco produced and extracted silk fibroin, characterized and validated for the use in optofluidic lab-on-a-chip. Soft lithography and nano-imprinting approaches have been developed for the fabrication. Also, results on AFM topography, optical transparency of the obtained substrate are reported. Biological properties in terms of cell adhesion growth and alignment have been evaluated [22].

19.4 Developed Technologies, Methodologies and Tools

19.4.1 *Protocols for Optimized Silk Production and Fibroin Biodoping*

In order to obtain a reproducible, controlled and high quality functionalized silk fibroin for the production of opto-fluidic devices, optimization of conventional silk-worm breeding was performed. By feeding the silkworm larvae with a diet added with a fluorescent dye molecule, Rhodamine B, a biological functionalization of the silk fibroin was promoted. White polyhybrid strains of silkworm coming from germplasm collection of the CRA-API (CRA, Honey bee and Silkworm Research Unit, Padua seat) were reared in plastic boxes (different in size according to the larvae age) placed in a room under controlled conditions (relative humidity >85%; 12:12 L:D photoperiod). During the whole larval stages the insects were fed *ad libitum* with artificial diet provided by CRA-API and prepared avoiding as much as possible the alteration of nutrients contained in it [23].

For protocol definition, the silkworm survival and length of the larval cycle (time between the incubation of the eggs and the cocoons spinning) were monitored at three different breeding temperature conditions (24 °C ± 1; 26 °C ± 1; 28 °C ± 1). A total number of 185, 408 and 295 larvae were tested at 24 °C ± 1, 26 °C ± 1 and 28 °C ± 1, respectively.

Starting from the 3rd day of the 5th instar, the Rhodamine B has been added to the artificial diet at two different concentrations: 0.05, 0.03 g in 100 g of powder diet. Three groups of ten larvae were used for each concentration against 30 larvae fed with artificial diet without any dyes representing *the blank* control treatment [23].

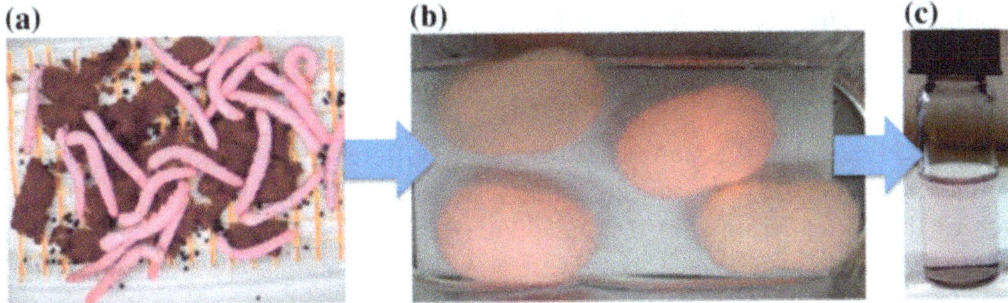

Fig. 19.2 Biomanufacturing of Rhodamine biodoped silk fibroin solution

19.4.2   Protocol for Native Fibroin Extraction

Among different methodologies for fibroin extraction of silk fibroin from *B. mori* cocoon, the reverse engineering methods, developed by Rockwood et al., enable to produce an aqueous regenerated silk fibroin solution, avoiding the use of strong and toxic organic solvent [4]. However, the process is optimized for a lab-scale production and not for extraction of silk containing xenobiotics. The current protocol was modified and optimized to promote up-scaling and to gain the highest concentration of xenobiotic in the silk extracted protein. Methods for extraction of native fibroin from silk glands of a mature 5th instar larvae of silkworm have been reported by Tansil et al. [12, 13]. This method of extraction applied to functionalized silk fibroin cocoons promises to maintain the naturally existing properties of protein and, moreover, allows to investigate the distribution of the dye into silk gland. Suitable revision of the methodology of extraction are reported, that have been carried out in order to optimize and adapt the procedure of protein functionalization and to obtain a high quality fibroin for the production of advanced optical devices (Fig. 19.2).

The whole silk glands were extracted from 5th instar larvae one day before they started to spin. The silkworm was anesthetized by using chloroform. A dorsal incision was carried from the head to sacral side of the larva. After gut removal whole silk glands were pulled out from the abdominal side of the worm. The gland was separated in two parts, the middle part and the posterior part that were treated separately [23].

Two pairs of posterior silk glands (coming from two larvae) was placed into a baker containing 3 ml of distilled water, cut in small pieces, gently shaken for 1 h and then kept in refrigerator overnight. The day after, the protein released into the water by the glandular tissue was collected and stored in refrigerator.

The middle gland was washed in deionized water and individually placed into a glass Petri dish. After the epithelium removal, the proteins were left for 2 h in 3 ml of distilled to dissolve the most soluble sericin. After sericin removal, the remaining protein, was added with 3 ml of distilled water and the solution was kept at ambient temperature for one day until the total dissolution of fibroin. The latter was defined as native silk fibroin solution (NSF).

19.4.3 Protocols to Obtain Regenerated Silk Fibroin (RSF) Solutions and Films

The protein extraction was performed testing two different methodologies: the extraction directly from the silkworm silk gland (Native Silk Fibroin, NSF) and the extraction of the protein from silkworm cocoons (Regenerated Silk Fibroin, RSF). The cocoons obtained as described above were processed to obtain RSF, optimizing Rockwood et al., protocol [4]. Cocoon were degummed in boiling 0.02 M Na_2CO_3 (Sigma-Aldrich, St Louis, Mo) solution for 45 min. The SF fibres were then rinsed three times in Milli-Q water and dissolved in a 9.3 M LiBr solution at 60 °C for 6 h. Subsequently, the SF water-solutions were dialyzed (dialysis membranes, MWCO3500) against distilled water for 48 h and centrifuged to obtain pure regenerated SF solutions (ca. 6–7 w/v %). The SF water-solutions were stored at 4 °C [23].

The films were fabricated as follows to characterize chemophysical and biological properties of SF films: a 160 μL aliquot of RSF or Native Silk Fibroin (NSF) dropped on 19 mm diameter glass coverslips and successively dried for 4 h in a sterile hood generating films with a thickness of around 20 μm. A support/mould of polydimethylsiloxane (PDMS) was used as a substrate to obtain free-standing films. The resulting RSF films were used for transmittance test [23].

According to the adopted approach for the realization of a silk-based optofluidic Lab-on-a-Chip device, the microfluidic structures have to accommodate photonic components such as diffraction gratings and optical filters. Thus, the dye-doped regenerated silk fibroin thin-films that are implemented in the micromoulding process have to show the proper characteristics in terms of optical features. In particular, a process protocol was optimized to obtain free-standing fibroin films with the suitable thickness and superficial smoothness in order to guarantee high optical gain and efficient waveguide effect.

A detailed study was performed to fabricate micrometre-thick films of regenerated silk fibroin doped with lasing dye Rhodamine B (RhB) by varying the drop-cast process conditions, substrates (size and material) and post-process thermal treatment. Indeed, it was observed that the drying process of the films has to be performed at 50 °C in a vacuum oven overnight at a pressure ranging from 600 to 800 mbar when using silk fibroin water solution at 5–10% w/v concentration regardless the dye concentration.

Figure 19.3a, b show the images of the film that were obtained from drop-casting RhB-doped regenerated silk fibroin (SF) water-solution at 5% w/v concentration onto high-quality quartz. By preserving the dimensions of the templating substrates (~4 cm^2), it was possible to control the film thickness by varying the volume of RhB-doped SF solution. Indeed, films of 20 μm (in Fig. 19.3a) and 50 μm (in Fig. 19.3b) thickness were obtained by implementing solution volumes of 0.6 and 1.3 mL, respectively.

Even though it was chosen to use the thicker films with lower dye concentration for the following fabrication steps because of the expected high gain in the optical amplification characterization, it was observed that the glass substrate enables the

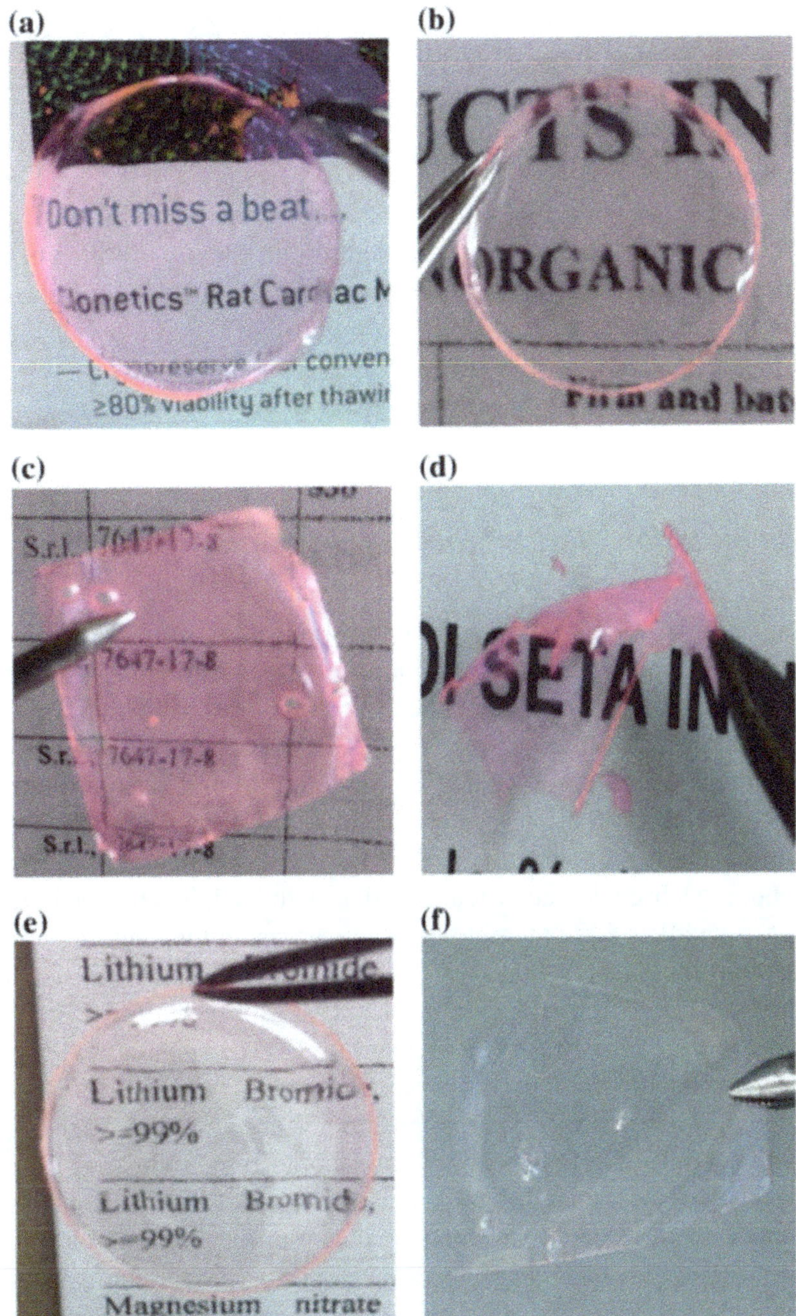

Fig. 19.3 SF films that are obtained by drop-casting a different amount of solution on various substrates and slow drying

realization of homogenous, defect-free and smooth free-standing silk-fibroin films regardless the deposition conditions.

On the other hand, it was observed that the use of different templating substrates did not guarantee the expected quality in silk filmability. Figure 19.3c, d show the images of RhB-doped free standing films obtained by drop-casting 2 mL of SF water-solution at 5% w/v onto 6 cm^2-wide patterned PDMS (Polydimethylsiloxane) (in Fig. 19.3c) and polycarbonate (in Fig. 19.3d) substrates. Indeed, the deposition onto PDMS results into heterogeneous films where millimetre-size air bubbles are clearly visible; the deposition onto polycarbonate results in a very thin film which cracks into pieces due structural rigidity.

Finally, it was demonstrated that the modification of the silkworm diet by artificial doping with dye-laser molecules did not deteriorate the structural properties and processability features of the silk fibroin thin-films. Figure 19.3e, f show that the best performing substrate for drop-casting silk fibroin water-solution is high-quality quartz (in Fig. 19.3e), while the use of PDMS substrate results in fabrication of non-homogeneous and porous thin films as expected (in Fig. 19.3f).

19.4.4 Fabrication of Silk Fibroin Microfluidic Structures

After having optimized the protocol for drop-casting silk fibroin water solutions onto technologically relevant flat substrates with different adhesion properties, a protocol was developed to fabricate microchannels into silk fibroin films for engineering complex microfluidic devices.

The implemented lamination strategy is based on drop-casting silk fibroin water solution to bond replica-moulded water-insoluble silk films. Replica-moulded silk fibroin films cast on PDMS negative moulds could be produced in rapid succession while maintaining a high degree of feature fidelity. Indeed, it is reported in literature that features as small as 400 nm could be produced using this method.

It was demonstrated that the adhesion properties and, thus, the consequent delamination efficiency are strictly dependant on the substrate features. Indeed, unsatisfactory results were obtained by implementing PDMS substrates. Moreover, the accuracy of the replica-moulding technique with patterned substrates is mainly correlated to the specific quality of the substrate surface in terms of roughness and grade of purity. Thus, in collaboration with Laboratory of Industrial Research and Technology Transfer of the High Technology Network of Emilia-Romagna (MIST-ER), high-quality PDMS patterned substrates were produced. The micromoulds for PDMS were realized using xurography, a rapid prototyping technique for creating microstructures in various films with a cutting plotter. A cutting plotter with a resolution of 100 μm was used. This technique has the advantage of being inexpensive and permits rapid prototyping of microfluidic devices. This microfabrication technique uses the cutting plotter to directly create microstructures in vinyl films, without photolithographic processes or chemicals. Then, positive and negative microstructures in vinyl films were fabricated, with channels width ranging from 150 to 250 μm. After cutting, unnecessary parts are peeled off the release liner. An application tape is used to hold structures in place while transferring into a petri dish. PDMS was moulded on positive and negative microstructures cut in vinyl. PDMS pre-polymer was mixed with a curing agent (SYLGARD 184, Dow Corning) at a 5:1 weight ratio and then poured into the petri dish. The PDMS mixture was degassed to remove

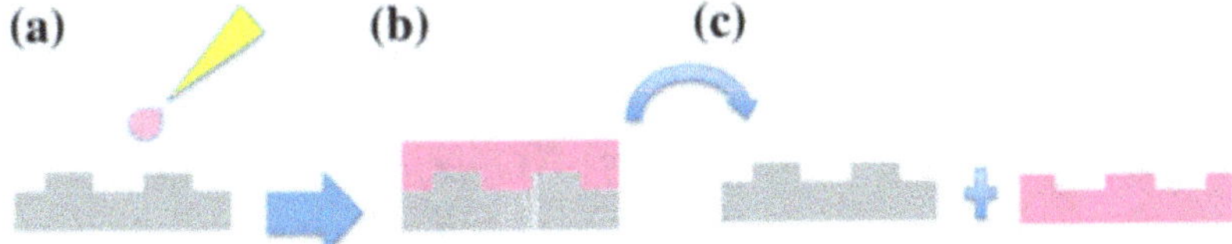

Fig. 19.4 Fabrication procedure for silk fibroin microfluidic channels: **a** drop casting of SF on micropatterned PDMS mould, **b** slow drying and film self-assembly, **c** peal-off of micropatterned silk film

air bubbles and cured at 70 °C. The cured PDMS was then peeled away from the moulds. Finally, micro-patterns were realized on the dye-doped films of silk fibroin according to the process flow that is sketches in Fig. 19.4.

Samples were fabricated from dye-doped and blank regenerated silk fibroin water solutions at 5% w/v. The previously-optimized deposition protocol was implemented in order to obtain films with approximately 100 microns in thickness by controlling the volume to surface area ratio during casting.

Silk solutions were cast on both micromoulded PDMS negative moulds and flat PDMS substrates through water evaporation that was obtained by introducing the films in a vacuum oven overnight at temperature of 50 °C and pressure which is regulated in the range between 600–800 mbar. In this way it is possible to produce water-stable films, which are easily delaminated without any treatments. Differently from what previously reported in literature, it was not necessary to use any solutions for modifying the hydrosolubility of silk-fibroin films or for increasing the delamination efficiency.

Figure 19.5 shows the images of the micromoulded RhB-doped and blank films of silk fibroin after delamination. As it is evident, the fidelity in the reproduction of the features of the negative moulds is very high onto cm^2-wide area. Though, further improvements in the procedure have to be achieved in order to avoid the trapping of water bubbles in the films during the evaporation in the oven.

The dye-doped microfluidic device has been characterized in order to verify if the photonic properties of the blend are preserved in channel regions where the optical stimulation and detection are located in the final optofluidic Lab-on-a-Chip.

The silk microfluidic devices were further characterized by performing imaging onto moulded microchannels by implementing optical microscopy in transmission mode in the case of the white silk fibroin sample so that the signal collected is the photoluminescence of the RhB dye molecules dispersed in the silk fibroin film (Fig. 19.5e–f).

The geometrical features of the negative mould are preserved in the laminated films. In fact, a channel width of 70 μm was measured as expected. The walls of the channels are sharply defined for all the samples fabricated even though some material removed from the channel is still deposited onto the wall borders (darker areas in Fig. 19.5e).

Therefore, it was possible to engineer and optimize: (i) a drop-cast deposition technique for obtaining optically high-quality, water-insoluble and free-standing dye-

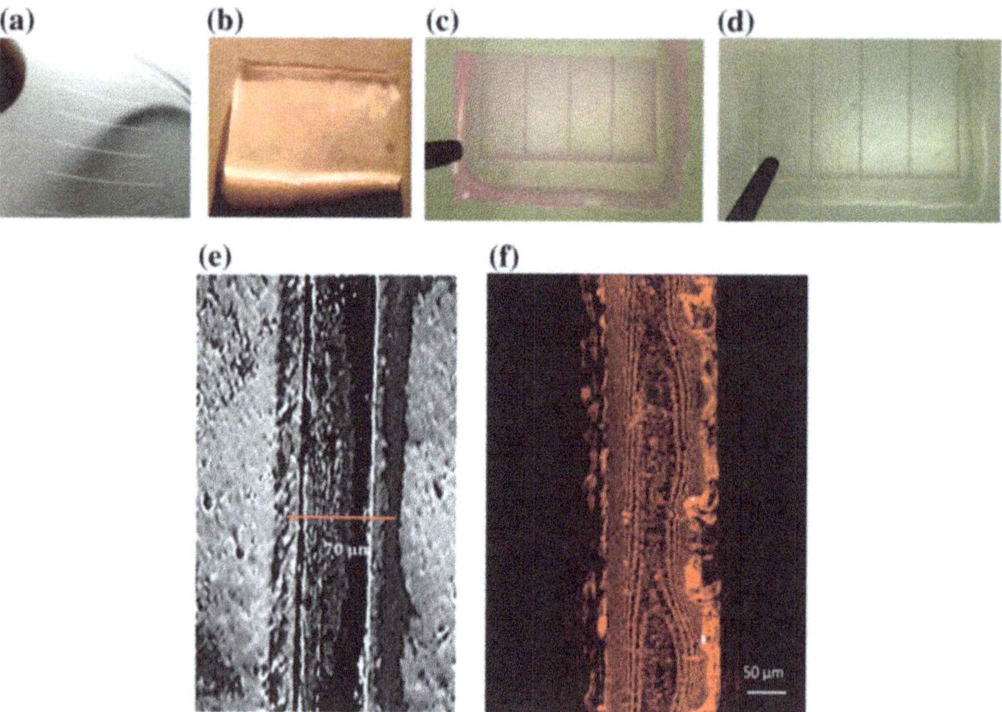

Fig. 19.5 Images of the negative moulds in PDMS. Images of the micromoulded silk fibroin films obtained from RhB-doped (**c**) and white (**d**) silk fibroin water solutions. **e** Optical microphotograph of the channel moulded into the blank (reference) sample. **f** Confocal Laser Scanning Microscope image of the channel moulded into the RhB-doped sample

doped silk fibroin films and (ii) a device component fabrication approach capable for the rapid and scalable production of silk-fibroin-based microfluidic devices for cell-seeding without the need for harsh processing conditions or cytotoxic compounds. The techniques employed in this experimental approach are scalable by designing systems with increased surface area and lamination of multiple layers.

19.5 Testing and Validation

19.5.1 Validation of Protocols for Fibroin Production and Functionalization

The first step of our study was to optimize the protocols to rear the *B. mori* worm for the production of raw material for high technological purposes (see Sect. 19.4.1). In particular, each step of breeding was monitored and the best breeding condition were selected [23]. Among different parameters monitored (humidity, temperature, amount of feed, light/dark exposure), we verified that temperature is the most important parameter to be controlled for successful and high yield breeding. In particular,

Table 19.1 Data on larval mortality (%) and on duration of the larval stage (gg)

Temperature (°C)	Mortality (%)	Mortality in 5th instar (%)	Length cycle (gg)
24	29.3	0	45
26	13.2	0	35
28	69.6	34.9	32

Table 19.2 Data on mortality in 5th instar larvae feed with diet added with different Rhodamine B concentration

Concentration (%)	Mortality in 5th instar (%)
0.03	3.33
0.05	30

our results demonstrated that the optimal temperature for *B. mori* L. breeding is 26 °C ± 1 [23]. From the experiments carried out, it resulted that the higher larval survival percentage was indeed at 26 °C. With temperature of 28 °C the percentage of dead resulted almost of 70%, with a 34% of dead occurring during the 5th instar. At 28 °C the duration of the larval cycle was minimum, while at 24 °C cycle duration was too long to be acceptable (Table 19.1). These data demonstrate that length cycle was correlated with the temperature of breeding.

Our results suggested that 26 °C ± 1, even if determinate a little longer larval cycle duration respect to 28 °C, is an optimal temperature for an efficient and scalable silkworm breeding.

The results about the association between the larvae mortality and the concentration of the Rhodamine B are shown in Table 19.2.

The dye was administrated to the larvae only during the last larval instar (5th), when the greatest synthesis of fibroin and sericin into the silk glands occurs. Administrated at 0.05% Rhodamine B caused a mortality of 30% in larvae of 5th instar. An acceptable percentage of dead (3.3%) was recorded using lower dye concentration. So, the protocol for production of biologically functionalized fibroin with Rhodamine B has been set up at the concentration of 0.03%.

19.5.1.1 Chemo-Physical Properties of Regenerated and Native Silk Fibroin Solution and Film Obtained from SILK.IT Fab

Chemo-physical properties of Regenerated and Native Silk Fibroin solution and film obtained from in loco produced raw material (*B.mori* cocoon) were analysed. Nuclear Magnetic Resonance (NMR) and FourierTransformed-InfraRed spectroscopy (FT-IR) have been performed to define the composition and the conformational structure of silk fibroin extracted from RSF and NSF.

In particular, the ^{1}HNMR spectrum of RSF in deuterated water (Fig. 19.6a), recorded at room temperature, confirmed that the sample obtained by our procedure

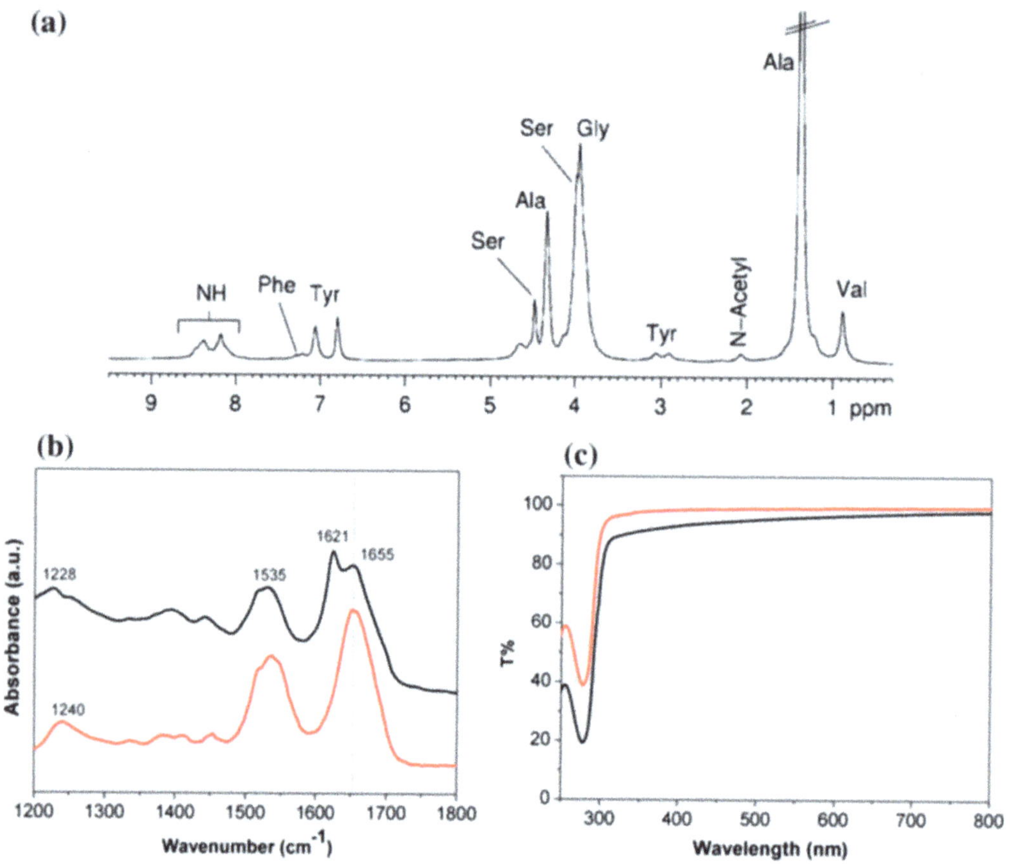

Fig. 19.6 Chemo-physical properties of in loco produced RSF and NSF solution and films [23]

is pure and that the aminoacids composition is in accordance with previous literature report [23, 24].

SF conformation influences the biodegradation, mechanical and optical properties of the SF films. In particular, it is known that a conformational structure with dominance of random coils and alpha-helices (called silk I structure) is highly water soluble and highly biodegradable with respects to the silk II structure, characterized by elevated presence of secondary structures and crystalline formats such as β-sheet.

In order to get further insight on the conformational structure of SF protein in SF films obtained from NSF and RSF, we performed comparative analyses of FT-IR spectra on films prepared from NSF and RSF (Fig. 19.6b). The results reported in Fig. 19.6, revealed that NSF, together with the bands assigned to silk I conformation, display peaks generally attributed to silk II structure. These results indicated that a β-sheet structure is present in NSF (Fig. 19.6b). In this view, it is expected that the biodegradability of NSF is slower that the one of RSF, and that optical properties can be modified. Accordingly, analyses of transmittance of regenerated and native SF films in the UV-visible range, revealed that the regenerated SF film is transparent in the visible region (300–800 nm). Silk fibroin Aromatic aminoacid Tyrosine (Tyr) and Tryptophan absorbance at 274–277 nm account for the decrease observed at the corresponding wavelength. The transparency of native SF within the same region was 70–90%, thus lower than regenerated SF film (Fig. 19.6c).

19.5.1.2 Properties of Biodoped Silk Films

We next demonstrated that the cocoons of *Bombyx mori* fed with a Rhodamine B (RhB) added diet were coloured and enabled the extraction and purification of RSF and SF films containing RhB [13]. Films were optically transparent and fluorescent with UV-Vis properties typical of RhodamineB added to white natural RSF. Comparative analyses of optical and vibrational features of RhB biodoped SF solution and films with those of white SF blended with RhB were performed, revealing significant difference, suggesting that silkworms' metabolism could be involved in binding mechanism of SF with RhB (for details see [14]).

19.5.1.3 Biological Properties of Silk Films Prepared by Different Methods

The control of the chemo-physical properties of SF films plays a key role in the preparation of tailored SF-based biomaterials and composites targeting optoelectronic devices and photonics technologies intended for biomedical applications such as cell phenotyping. We have then demonstrated that the application of different fabrication methods to prepare SF films influences the chemo-physical properties of SF films and their interaction with neural cells. In particular, we found that hydrophobic surfaces induce proliferation of astrocytes and neuron adhesion, whereas hydrophilic surfaces promote a remarkable neurite outgrowth [24].

Interestingly, surface properties such as roughness and wettability, mechanical resistance, dissolution and degradation profiles can be affected by fabrication method. These properties drive the interaction of SF films with primary neural cells, namely astrocytes and neurons [24]. Also we studied the effect of micro and nanopatterning on cell growth. Notably, micropatterned silk films promoted strong alignment of primary astrocytes with a high dependence from the grove depth and width (Fig. 19.7).

19.5.1.4 Integration of Silk-Based Photonic Structures into Microfluidic Devices and Validation

We aim at estimating quantitatively the amount of laser dye molecules present in the silk fibroin thin-films naturally functionalized with Rhodamine B (RhB), which is necessary for allowing optical amplification process. As we have already discussed in the previous section, we succeeded in demonstrating the feasibility of the process of directly feeding larvae of *B. mori* with specific dye molecules in order to achieve optically active silk fibroin substrates.

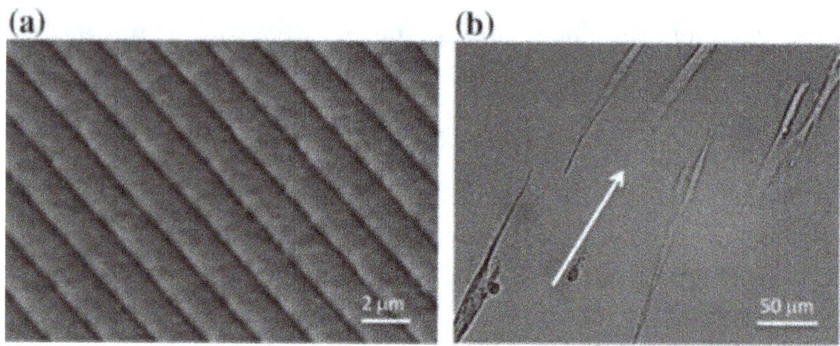

Fig. 19.7 **a** SEM image of micropatterned SF substrate. **b** Micrograph of astrocytes aligned on the substrate of silk micropatterned substrate

After having discriminated which is the most efficient protocol for doping silk fibroin substrates, we demonstrated the maintenance of the photonic features of the most performing dye-doped silk fibroin substrate once structured into a microfluidic device.

We report on the amplified stimulated emission (ASE) characterization performed on silk-fibrin films obtained from blending the regenerated silk fibroin water-solutions at 5% w/v with RhB at different concentrations (i.e. 10^{-6}, 5×10^{-6}, 10^{-5} M): we compared the extrapolated figures of merit with the ones obtained by characterizing the naturally-doped silk fibroin substrates. By means of this comparison we are able to estimate empirically the amount of RhB dye molecules in naturally-doped samples. The ASE characterization was carried out as follows. The samples were measured under vacuum to prevent photo-degradation upon irradiation. The laser beam of a Q-switched Nd:Yag laser was focused onto the sample using a cylindrical lens in order to obtain a rectangular stripe of about 100 μm width. The light excitation was the 532 nm frequency-doubled line of the laser with temporal pulse width of 4 ns and 10 Hz of repetition rate. The ASE signal is collected in fibre at 90° with respect the impinging laser by using an optical multichannel analyser. The samples were cut into strips of 4 mm width and about 10 mm length. Moreover, the DIO Board was set to generate the trigger signal to start the spectral acquisition once the excitation pulse was delivered. Thus, the ASE signal is excited by laser single-shot and collected in an automated mode, thus avoiding any possible degradation of the sample.

Figure 19.8 shows the curves displaying the functional dependence from energy per pulse of the full-width-at-the-half maximum (FWHM) values of the RhB emission peak for the three different doped silk-based films. In particular, a silk fibroin film doped at 10^{-5} M with RhB (a), a RhB naturally-doped silk fibroin film obtained from regenerating white polyhybrid strain silkworm (b) and RhB naturally-doped silk fibroin films obtained from regenerating Nistari silkworm (c). In the case of the 10^{-5} M RhB doped silk fibroin film the expected threshold-like behaviour is evident with an abrupt onset of the ASE process at around 70 μJ/pulse.

Indeed, we performed the ASE characterization protocol on different doped silk fibroin films by varying the dopant concentration (after having verified that the film

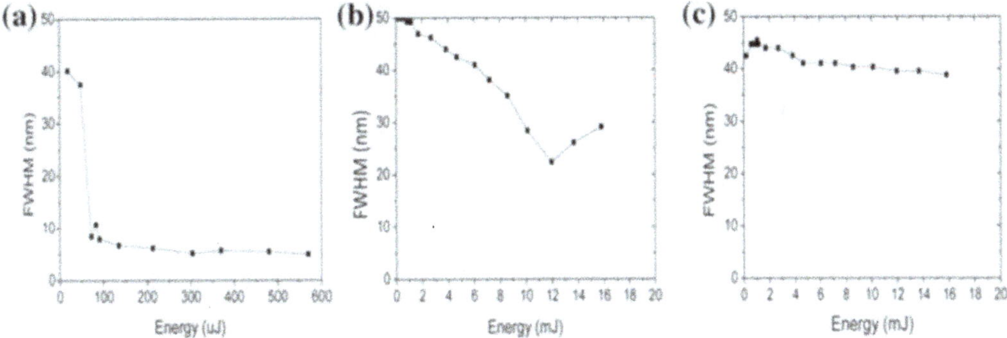

Fig. 19.8 Full-Width-at-the-Half-Maximum (right) *versus* the impinging energy per pulse graphs in the case of **a** silk fibroin sample doped at 10^{-5} M with RhB **b** RhB naturally-doped silk fibroin films obtained from regenerating white polyhybrid strain silkworm and **c** RhB naturally-doped silk fibroin films obtained from regenerating Nistari silkworm

Table 19.3 ASE threshold values in doped silk fibroin films as a function of the RhB dye concentration

Concentration M	Threshold energy (μJ)	FWHM (nm)
10^{-6}	>4000	16
5×10^{-6}	700	7.5
10^{-5}	71	5.5

thickness for the different samples is comparable). In Table 19.3 we report the ASE threshold values we estimated as a function of the RhB dye concentration.

We have to mention that in the case of the 10^{-6} M sample a clear threshold of ASE process cannot be identified, given that the FWHM decreases from an initial 45 nm value to about 15 nm almost linearly as the input energy/pulse increases. On the other hand, for samples with concentration at 5×10^{-6} M the beginning of amplification process is observed for energies above 700 μJ. Also in this case, the onset is not abrupt and the samples tend to degrade quickly upon further measurements. Then we compared the ASE threshold values of the blended films with ones of the naturally functionalized films. In particular, we implemented the regenerated silk fibroin solution extracted from the silkworm of two different races: the standard worm (white polyhybrid strain) and the Nistari race.

Figure 19.8b, c show the curves of FWHM as a function of the input energy/per pulse for the standard-worm and the Nistari naturally-doped silk fibroin thin-films, respectively. As it can be observed, no ASE threshold is achieved even at the highest possible fluences with the available laser pump; only in the case of regenerated silk fibroin obtained from the standard-race silkworm an incipient peak narrowing up to 30 nm is detectable before the sample is severely damaged. This result suggests that the amount of Rhodamine B molecules present in the naturally-functionalized silk fibroin films is at least less than 10^{-6} M, possibly as low as 10^{-7} M from the empirical estimation reported in Fig. 19.9.

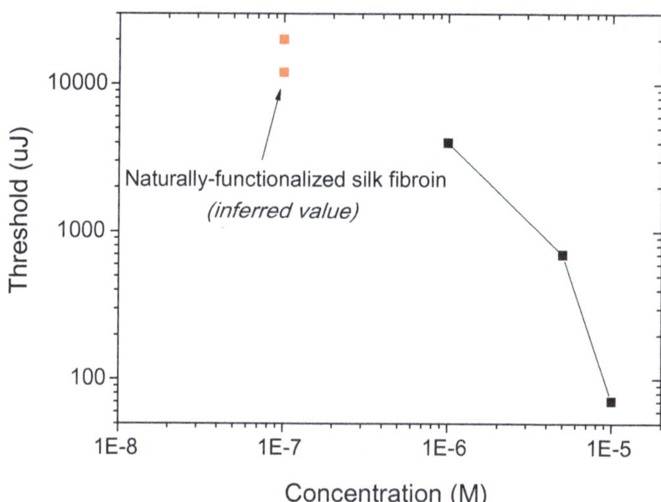

Fig. 19.9 Estimation of the RhB dye concentration in naturally-functionalized silk fibroin inferred from the measured ASE threshold trend (red squares in the graph)

From these results, considered that a minimum concentration of about 5×10^{-6} M is necessary for the blended sample to be operated in population inversion, we can infer that the naturally-functionalized silk fibroin substrates cannot be efficiently implemented ad gain material in biocompatible nanostructured organic lasers. Thus, we implemented RhB-doped silk fibroin substrate obtained from blending regenerated silk fibroin water solution with RhB water solution at 10^{-5} M in the micro-moulded fabrication protocol for achieving microfluidic devices. Indeed, in the Lab-on-a-Chip (LOC) scheme that we propose the microfluidic device component is endowed with the optimized photonic structure for allowing higher degree of miniaturization. For this purpose, we decided to implement in the fabrication of the LOC the best performing silk fibroin substrate in terms of photonic characteristics.

As a final validation of the entire fabrication protocol, we checked the preservation of the photonic characteristics of the RhB-doped silk fibrin material in the channel region of the microfluidic device. Indeed, by using a 100 μm-width laser spot it is assured to probe only the region where the micromoulding is present.

As expected, we observed that the efficiency of ASE process is preserved in the channels of the microfluidic device as it is evidenced by the maintenance of the threshold values with respect to unmoulded samples (data not reported), regardless the location of the channels in the device. Only a variation in the maximum achievable output intensity is observed from region to region but this spot-location dependence is expected within the fabricated devices variability. Thus, we can conclude that the micromoulding process that we engineered did not increase the photon losses or degrade waveguiding properties inherent to the optimized RhB-doped silk fibroin substrate.

19.6   Conclusions and Future Research

In this work we addressed the assessment of whole chain production and validation of SF-based substrates and technology for biomedical applications. Our goal was to standardize and optimize methods and protocols to use this silk fibroin as new material for advanced bio-technological application and sustainable manufacturing.

A protocol was developed for fibroin cocoon breeding and feeding for functionalization via bio-doping [23]. The biological incorporation of doping dye molecules into SF by means of feeding cocoon avoid the need for an external chemical process associated with the use of toxic solvents, thus it is an eco-friendly and innovative method to produce doped silk substrates. Tansil et al. [12, 13] demonstrated the ability to include several xenobiotics in the silk fibre by feeding the cocoon with died modified by addition of selected molecule. We reported results showing the effect of xenobiotic addition to the diet on the larval survival as well as on modification and optimization of breeding conditions [23].

Protocols were defined for the extraction, purification and chemophysical characterization of regenerated fibroin from standard and biodoped cocoon [14, 25]. In this work we saw that adequate extraction and preparation of the RSF and NSF solution coming from the diet doped cocoon results in colour solutions and films [14, 25]. The incorporation by doping and blending of organic dyes in water based silk fibroin [26] might be problematic due to lipophilic nature of the majority of these high quantum yield optically active molecules. In this view, analyses and comparison of samples obtained by blending and addition to SF solution of the selected compound with samples obtained by doping diet method revealed that the sample obtained by doping diet method displays a similar efficiency of functionalization of the substrates. Moreover the method does not affect SF structural, chemophysical and biological property [14, 24, 27, 28].

We have previously shown that DFB and multilayer photonic crystal can be fabricated by silk fibroin doping with stilbene. Herein, we have shown that the same structure can be obtained by using doped and biodoped silk fibroin obtained by in loco produced, extracted and purified raw material. Also soft lithography procedures can be applied to biodoped SF film, to obtain microfluidic chip [25, 29, 30] and his open the view for industrial scale up of biomanufacturing of SF.

An important issue when dealing with biomedical applications is achieving a true biocompatibility. In this view, we have reported here that plating neural cells, called astrocytes, on micro-nanostructured SF enables their alignment and growth. For biomedical application targeting direct interaction of biological samples with the biomaterials, it is fundamental to study and to understand the effect of chemophysical properties of the biomaterial on cell adhesion, growth, differentiation and behaviour [24]. To reach the goal of integrating SF in a LOC optofluidic device for single cell analyses, it was fundamental to observe if the analysed astroglial cell could be driven to a specific position by following the patterned structure. The feasibility of the proposed approach indicated by the observation reported in Fig. 19.7, complete the picture to validate the use of nanopatterned SF, fabricated in loco with controlled condition in advanced technological device for cell biosensing.

Studies to integrate optofluidic component in one LOC for diagnosis of brain pathologies such brain tumours are under investigation. Moreover, we are exploring the possibility of exploitation of a direct application in biomaterials and biomedical field of coloured silk fibroin extracted from coloured cocoons [31, 32]. The results reported here were the basis for application for national, private sector supported projects, for the development of silk-based anti-counterfeiting technologies and for European initiatives regarding biomedical use of silk fibroin.

Acknowledgements This work has been funded by the Italian Ministry of Education, Universities and Research (MIUR) under the Flagship Project "Factories of the Future—Italy" (Progetto Bandiera "La Fabbrica del Futuro") [33], Sottoprogetto 1, research projects "SILK Italian Technology for industrial biomanufacturing" (SILK.IT).
Assunta Pistone, Susanna Cavallini, Giovanni Donati, Simone Bonetti are acknowledged for the work provided within the project. Simone Sugliani from Laboratory MIST E-R is acknowledged for the support on microfluidics fabrication. Marco Caprini and Alessia Minardi from FABIT Dept. of the University of Bologna are acknowledged for the support in the preparation and maintenance of cell culture according to approved Ethical protocol from Italian Ministry of Health (360/2017-PR, approved 05/2017).

References

1. Guarino V, Benfenati V, Cruz Maya IM, Saracino E, Zamboni R, Ambrosio L (2018) Natural proteins for 3D scaffolds in tissue engineering. book chapter. In: Elsevier Press. https://doi.org/10.1016/b978-0-08-100979-6.00002-1

2. Guarino V, Benfenati V, Cruz Maya I, Borrachero-Conejo AI, Zamboni R, Ambrosio L (2018) Bioinspired scaffolds for bone and neural tissue engineering. book chapter. In: Elsevier Press. https://doi.org/10.1016/b978-0-08-100979-6.00003-3

3. Tao H, Kaplan DL, Omenetto FG (2012) Silk materials—a road to sustainable high technology. Adv Mater 24:2824–2837

4. Rockwood DN, Preda RC, Yücel T, Wang X, Lovett ML, Kaplan DL (2011) Materials fabrication from Bombyx mori silk fibroin. Nat Protoc 6:1612

5. Bettinger C, Bao Z (2010) Biomaterials-Based Organic Electronic. Polym Int 59(5):563–567

6. Webb AB, Chimenti M, Jacobson MP, Barber DL (2011) Dysregulated pH: a perfect storm for cancer progression. Nat Rev Cancer 11(9):671–677

7. Capelli R, Amsden JJ, Generali G, Toffanin S, Benfenati V, Muccini M, Kaplan DL, Omenetto FG, Zamboni R (2011) Integration of silk protein in organic and light-emitting transistors. Org Electron 12(7):1146–1151

8. Toffanin S, Kim S, Cavallini S, Natali M, Benfenati V, Amsden JJ, Kaplan DL, Zamboni R, Muccini M, Omenetto FG (2012) Low-threshold blue lasing from silk fibroin thin films. Appl Phys Lett 101:091110

9. Kim DH, Kim YS, Amsden J, Panilaitis B, Kaplan DL, Omenetto FG (2009) Silicon electronics on silk as a path to bioresorbable, implantable devices. Appl Phys Lett 95(13):133701

10. Kim DH, Viventi J, Amsden JJ, Xiao J, Vigeland L, Kim YS, Blanco JA, Panilaitis B, Frechette ES, Contreras D, Kaplan DL, Omenetto FG, Huang Y, Hwang KC, Zakin MR, Litt B, Rogers JA (2010) Dissolvable films of silk fibroin for ultrathin conformal bio-integrated electronics. Nat Mater 9(6):511–517

11. Melucci M, Zamboni R (2015) Organic materials—silk fibroin synergies: a chemical point of view In: Andrews DL, Grote JG (eds) New horizons in nanoscience and engineering. SPIE Press Book
12. Tansil NC, Li Y, Teng CP, Zhang S, Win KY, Chen X, Liu XY, Han MY (2011) Intrinsically colored and luminescent silk. Adv Mater 23(12):1463–1466
13. Tansil NC, Koh LD, Han MY (2012) Functional silk: colored and luminescent. Adv Mater 24(11):1388–1397
14. Sagnella A, Chieco C, Di Virgilio N, Toffanin S, Posati T, Pistone A, Bonetti S, Muccini M, Ruani G, Benfenati V, Rossi F, Zamboni R (2014) Bio-doping of regenerated silk fibroin solution and films: a green route for biomanufacturing. RSC Adv 4:33687–33694
15. Altman GH, Diaz F, Jakuba C, Calabro T, Horan RL, Chen J, Lu H, Richmond J, Kaplan DL (2003) Silk-based biomaterials. Biomaterials 24:401
16. Yang Y, Ding F, Wu J, Hu W, Liu W, Liu J, Gu X (2007) Development and evaluation of silk fibroin-based nerve grafts used for peripheral nerve regeneration. Biomaterials 28:5526
17. Madduri S, Papaloïzos M, Gander B (2010) Trophically and topographically functionalized silk fibroin nerve conduits for guided peripheral nerve regeneration. Biomaterials 31(8):2323
18. Benfenati V, Toffanin S, Capelli R, Camassa LM, Ferroni S, Kaplan DL, Omenetto FG, Muccini M, Zamboni R (2010) A silk platform that enables electrophysiology and targeted drug delivery in brain astroglial cells. Biomaterials 31:7883–7891
19. Benfenati V, Stahl K, Gomis-Perez C, Toffanin S, Sagnella A, Torp R, Kaplan DL, Ruani G, Omenetto FG, Zamboni R, Muccini M (2012) Biofunctional silk/neuron interfaces. Adv Fun Mater 22:1871
20. Benfenati V, Toffanin S, Bonetti S, Turatti G, Pistone A, Chiappalone M, Sagnella A, Stefani A, Generali G, Ruani G, Saguatti D, Zamboni R, Muccini M (2013) A transparent organic transistor for bidirectional stimulation and recording of primary neurons. Nat Mater 12:672–677
21. Toffanin S, Benfenati V, Pistone A, Bonetti S, Koopman W, Posati T, Sagnella A, Natali M, Zamboni R, Ruani G, Muccini M (2013) N-type perylene-based organic semiconductors for functional neural interfacing. J Mater Chem B 1(31):3850
22. Tung YC, Huang NT, Oh BR, Patra B, Pan CC, Qiu T, Chu PK, Zhang W, Kurabayashi K (2012) Optofluidic detection for cellular phenotyping. Lab Chip 12:3552–3565
23. Sagnella A, Chieco C, Di Virgilio N, Toffanin S, Cavallini S, Posati T, Pistone A, Varchi G, Muccini M, Ruani G, Benfenati V, Zamboni R, Rossi F (2015) Silk.it project: Silk italian technology for industrial biomanufacturing. J Composite B 281–287
24. Sagnella A, Pistone A, Bonetti S, Donnadio A, Saracino E, Nocchetti M, Dionigi C, Ruani G, Muccini M, Posati T, Benfenati V, Zamboni R (2016) Effect of different fabrication methods on the chemo-physical properties of silk fibroin films and on their interaction with neural cells. RSC Adv 11:9304–9314
25. Cavallini S, Toffanin S, Chieco C, Sagnella A, Formaggio F, Pistone A, Posati T, Natali M, Caprini M, Benfenati V, Di Virgilio N, Ruani G, Muccini M, Zamboni R, Rossi F (2015) Naturally functionalized silk as useful material for photonic applications. J Composite Part B 71:152–158
26. Posati T, Melucci M, Benfenati V, Durso V, Nocchetti M, Cavallini S, Toffanin S, Sagnella A, Pistone A, Muccini M, Ruani G, Zamboni R (2014) Selective MW-assisted surface chemical tailoring of hydrotalcites for fluorescent and biocompatible nanocomposites. RSC Adv 4:11840–11847
27. Sagnella A, Zambianchi M, Durso M, Posati T, Del Rio A, Donnadio A, Mazzanti A, Pistone A, Ruani G, Zamboni R, Benfenati V, Melucci M (2015) APTES mediated modular modification of regenerated silk fibroin in a water solution. RSC Adv 5:63401–63406
28. Posati T, Benfenati V, Sagnella A, Pistone A, Nocchetti M, Donnadio A, Ruani G, Zamboni R, Muccini M (2014) Innovative multifunctional silk fibroin and hydrotalcite nanocomposites: a synergic effect of the components. Biomacromol 15:158–168
29. Prosa M, Sagnella A, Posati T, Tessarolo M, Bolognesi M, Posati T, Toffanin S, Cavallini S, Benfenati V, Seri M, Ruani G, Muccini M, Zamboni R (2014) Integration of a silk fibroin based film as luminescent down-shifting layer in ITO-free organic solar cells. RSC Adv 84:44815–44822

30. Benfenati V, Martino N, Antognazza MR, Pistone A, Toffanin S, Ferroni S, Lanzani G, Muccini M (2014) Photostimulation of whole-cell conductance in primary rat neocortical astrocytes mediated by organic semiconducting thin films. Adv Healthc Mater 3:392–399
31. Pistone A, Sagnella A, Chieco C, Posati T, Varchi G, Formaggio F, Bertazza G, Saracino E, Caprini M, Bonetti S, Toffanin S, Divirgilio N, Muccini M, Rossi F, Ruani G, Zamboni R, Benfenati V (2016) Silk fibroin film from Golden-Yellow Bombyx mori is a biocomposite that contains lutein and promotes axonal growth of primary neurons. Biopolymers 105:287–299
32. Dionigi C, Posati T, Benfenati V, Sagnella A, Pistone A, Bonetti S, Ruani G, Dinelli F, Padeletti G, Zamboni R, Muccini M (2014) Nanostructured conductive bio-composite of silk fibroin/single walled carbon nano tube. J Mater Chem B 10:1424–1431
33. Terkaj W, Tolio T (2019) The italian flagship project: factories of the future. In: Tolio T, Copani G, Terkaj W (eds) Factories of the future. Springer

Part VII
Conclusions

Chapter 20
Key Research Priorities for Factories of the Future—Part I: Missions

Tullio Tolio, Giacomo Copani and Walter Terkaj

Abstract This chapter investigates research priorities for factories of the future by adopting an approach based on mission-oriented policies to support manufacturing innovation. Missions are challenging from a scientific and technological point of view and, at the same time, are addressing problems and providing results that are understandable by common people. Missions are based on clear targets that can help mitigating grand challenges. Based on the results of the Italian Flagship Project *Factories of the Future*, this chapter proposes seven missions while identifying the societal impact, the technological and industrial challenges, and the barriers to be overcome. These missions cover topics such as circular economy, rapid and sustainable industrialisation, robotic assistant, factories for personalised medicine, internet of actions, factories close to the people, and turning ideas into products. The accomplishment of missions asks for the support of a proper research environment in terms of infrastructures to test and demonstrate the results to a wide public. Research infrastructures together with funding mechanisms will be better addressed in the next chapter of this book.

20.1 Mission-Oriented Research and Innovation Policies

The importance of the manufacturing industry both for developed and developing countries has been assessed in several works [1–4], which highlight the need of continuous innovation to cope with societal grand challenges [5]. The results of inno-

T. Tolio
Director of the Italian Flagship Project "Factories of the Future", Direttore del Progetto Bandiera "La Fabbrica del Futuro", CNR - National Research Council of Italy, Rome, Italy

T. Tolio
Dipartimento di Meccanica, Politecnico di Milano, Milan, Italy

G. Copani · W. Terkaj (✉)
CNR-STIIMA, Istituto di Sistemi e Tecnologie Industriali Intelligenti per il Manifatturiero Avanzato, Milan, Italy
e-mail: walter.terkaj@stiima.cnr.it

vation actions strictly depend on the policies that are designed and implemented, mainly at public level. Cantner and Pyka [6] proposed a framework to classify technology policies by considering two axes: (a) *vicinity to the market* and (b) *specificity of a policy measure*. The first axis differentiates between basic and applied research, whereas the second axis specifies how the policy is precise in describing the research goal. High vicinity to the market leads to mission- or diffusion-oriented policies in case of precise or broad goals, respectively. *Diffusion-oriented policies* tackle a wide range of heterogeneous technologies to be further developed and applied in no specific economy sector. *Mission-oriented policies* are characterised by a relatively high degree of specificity in terms of both target technologies and economic application [6].

A mission-oriented policy for research and innovation shares common traits with Public Procurement for Innovation (PPI) [7], i.e. when a public organisation places an order for the fulfilment of certain functions within a defined time horizon. Therefore, the goal of PPI is not primarily to support the development of new products, but to provide solutions for human needs or societal problems. This kind of demand-side policy can be considered as a mission-oriented innovation policy that mitigates grand challenges, since a grand challenge is too broad to be tackled as a whole. Moreover, PPI facilitates the interactions among organisations (e.g. the procurer and supplying companies) and this is particularly relevant because companies almost never innovate in isolation. These interactions can be further enhanced by means of *focus groups* or *task forces* [7]. PPI however cannot be applied in all the sectors and tends to blur in the same step the creation of innovative solutions with the creation of a market for it and with its application in a real day by day operating context, thus putting some additional constraints that are not strictly required to achieve a mission.

It must be stressed that, in any case, the realisation of mission-oriented policies requires public sector organisations to set the *direction of change*, thus assuming a leading role in the economy that goes beyond fixing problems related to market failures [8]. The goal of a mission-oriented policy is to pave the way for new ambitious approaches and to demonstrate their feasibility instead of introducing new products or services to the market. Typically, only public sector organisations can afford high-risk investments and pursue goals like satisfying human needs and solving societal problems.

A mission-oriented approach is currently under discussion within the Directorate-General for Research and Innovation of the European Commission in view of the 9th EU Framework Programme for Research and Innovation [9]. Such approach is thought to be a solution to make the European economy sustainable and competitive. The involvement of civil society is needed to identify the main challenges and also to assess the impact of the mission results. The final objectives of the mission-oriented policy include the re-industrialisation of Europe, the creation of new jobs, the solution of societal problems, the attainment of goals that project Europe into the future, the enhancement of the European knowledge base, and the improvement of skills. This type of mission must be [9]:

- Bold, inspirational with wide societal relevance, e.g. improving health, environment, nutrition, welfare, etc. for a large share of the population.
- Defining a clear direction that is targeted, measurable and time-bound. Clear goals should be understood by the civil society.
- Ambitious but realistic. Possible (theoretical) solutions should be identified.
- Cross-disciplinary, cross-sectoral and cross-actor innovation.
- Requiring multiple, bottom-up solutions.

In this scope, the Directorate-General for Research and Innovation identified 14 examples of missions, ranging from industry renewal to bio-manufacturing, from energy independence to clean and safe mobility [10].

This chapter, on the basis of the results attained within the Flagship Project *Factories of the Future* [5], takes into consideration the mission-oriented approach in order to identify, among the others, seven missions for future manufacturing research (Sect. 20.2). These missions will foster the role of manufacturing as a backbone for the employment and wealth of European economies.

For the accomplishment of missions, a fundamental role is played by research and innovation infrastructures, as well as by effective and sustainable research and innovation funding mechanisms. These topics are anticipated in the conclusions (Sect. 20.3) and better discussed in the next chapter of this book [11].

20.2 Missions for Manufacturing Industry

The growth and sustainability of European economy requires a change from competition based on cost reduction towards high value added activities by adopting a competitive sustainable manufacturing (CSM) paradigm [12]. Indeed, European countries need continuous innovation to keep their position in a global market place while creating value in a sustainable society [13, 14]. The development of the sustainability paradigm has translated over the years into a series of global initiatives with broad objectives but without precise quantitative targets. For example, in 2015 the General Assembly of the United Nations (UN) adopted the 2030 Agenda for Sustainable Development that includes the definition of 17 Sustainable Development Goals (SDGs) [15]. The SDGs aim at promoting prosperity for all countries while protecting the planet; indeed, also the European Union is committed to monitor and improve the performance with respect to the SDGs [16]. While the agreement on the 17 goals is a big step forward, more defined actions and goals are needed in order to attain them. Missions are one of the possible ways to better specify the goals.

Adopting a mission-oriented approach (Sect. 20.1) and elaborating on the results emerged from the Flagship Project Factories of the Future [5], this section proposes seven missions for future manufacturing industry, while making references to the SDGs. The proposed missions are:

- Circular Economy (Sect. 20.2.1)
- Rapid and Sustainable Industrialisation (Sect. 20.2.2)

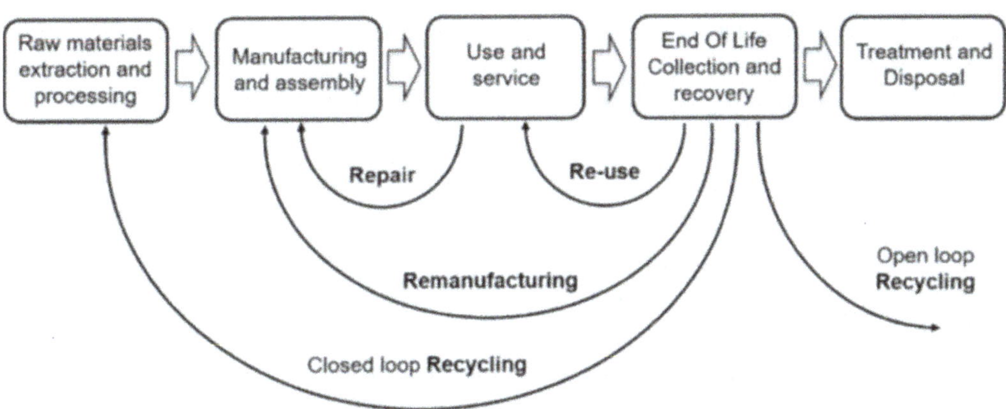

Fig. 20.1 Circular economy cycles (adapted from [19])

- Robotic Assistant (Sect. 20.2.3)
- Factories for Personalised Medicine (Sect. 20.2.4)
- Internet of Actions (Sect. 20.2.5)
- Factories close to the People (Sect. 20.2.6)
- Turning Ideas into Products (Sect. 20.2.7).

For each mission, a general description, the societal impact, the industrial and technological challenges, and the barriers to be overcome are presented. The need to face industrial and technological challenges will presumably lead to a set of projects and bottom-up experimentations, i.e. *mission projects* as defined in [9]. Barriers are meant as contextual aspects (e.g. regulations, standardization, market demand and supply, business organizations and relationships, change management) that typically are not addressed by specific mission projects.

20.2.1 Circular Economy

20.2.1.1 Definition

Circular Economy is a new paradigm fostering sustainable growth and progress while taking in due consideration limited natural resources and environment degradation [17, 18]. Circular Economy aims at overcoming the traditional linear production and consumption model that entails production, use and disposal with consequent need of large quantity of natural resources for new products, high generation of waste and negative impacts on the environment due to disposal processes. A set of *restorative* processes and practices must be put in place with reference to the traditional linear manufacturing chain (Fig. 20.1) [19].

Processes enabling Circular Economy include:

- *Re-use*, when product or components functionalities allow additional use cycles without performing any restorative operation.

- *Repair*, when ordinary or extraordinary repair operations can restore initial products functionalities.
- *Remanufacturing*, when product has to be re-built to bring it at initial specifications (using a combination of new, reused and repaired parts) [20]. Recently, the concept of *remanufacturing with upgrade* has also been proposed in addition to traditional remanufacturing [21], as well as the option to use in other products the functions provided by some components of the de-manufactured product.
- *Recycling*, when the state of the product is not anymore compatible with restore of functionalities and its materials are used to produce secondary raw materials that can be re-used in the same production process (closed loop recycling) or in other processes (open loop recycling).

Based on products characteristics and conditions of End-of-Life products, the process able to retrieve maximum value should be selected every time in a hierarchical approach [19, 22] in which re-use, remanufacturing and recycling are in a decreasing order of priority.

These processes are carried out by *Circular Factories* [17, 23] that should work in alignment (or being integrated) with the existing production factories to maximise the benefits of Circular Economy. Thus, the implementation of Circular Economy requires a systemic perspective of manufacturing, de-manufacturing, and re-manufacturing networked factories that collaborate from the product design phase up to the End-of-Life.

The proposed mission for the effective implementation of Circular Economy in manufacturing is to *Design factories and industrial networks for the management of a new class of fully circular products according to innovative circular business models*.

Circular product service systems (CPSS) are totally conceived and managed during their whole life cycle according to the Circular Economy paradigm. These CPSS will be designed and used to be efficiently re-used, remanufactured and recycled, according to new processes and technologies that will make it possible. CPSS will embed smart sensors that will collect information on the use conditions and on their state, so that the optimal circular strategies can be planned and implemented after each use phase. The de-manufacturing network will include companies specialised in the various End-of-Life processes adopting new smart de-manufacturing technologies. Companies in the network can be seen as the entities providing the *"energy"* that allows to continually circulate the materials and restore the functions carried by the CPSS. Therefore the factories of the future will be part of a Circular Factory Network (CFN).

New circular business models will be based on offering service levels rather than products. Manufacturers will turn into service providers (or will establish strategic partnerships with them) and will develop innovative product-service bundles where product ownership will not be necessarily transferred to customers. The latter will have the guarantee of product performance and will have the option of pay per use or per result according to innovative financial mechanisms. Service suppliers will also provide continuous performance upgrade that will be obtained through remanufacturing with upgrade enabled by a careful product design. Circular business models will revolutionise consumption behaviours and will improve quality of life of people, increasing manufacturing sustainability at the same time [24].

20.2.1.2 Societal Impact

Circular Economy has the potential of generating significant societal impacts. In the circular model, products and materials are kept in the use cycle as long as possible, thus maximising their possible exploitation and value that is generated through initial production processes. With this concept, the need for new resources, new production processes and the recourse to landfilling is reduced (cf. SDG 12.5 [15]). In 2015, the generation of waste was 1716 kg per capita in Europe and recycling of municipal waste was only 45% of total waste [25]. Indeed, it is estimated that in the consumers' goods sector, nearly 80% of the value of materials used in the production is lost globally every year [26]. Circular Economy will reduce the need for material input for production processes. It is estimated that a reduction between 17% and 24% of virgin material consumption can be achieved in European manufacturing industry through resources optimisation by 2030 [27] (cf. SDG 12.2 [15]). The reduced need of virgin materials, together with lower energy needs and emissions of manufacturing processes will strongly contribute to reduce CO_2 emissions and to positively affect climate change (cf. SDG 12.4 [15]).

Circular Economy is also an enormous opportunity for European economy and welfare. European companies will become more competitive in the global market because of the reduction of production costs due to lower cost of raw materials, energy and disposal. It is estimated that Circular Economy will provide a saving of production resources expenditure amounting at 600 billion euro per year in EU [27]. In addition, Circular Economy offers companies the opportunity to innovate their business model in the direction of *servitisation* [28], thus increasing the added value—and profitability—of their offering while limiting environmental impact at the same time [21] (cf. SDG 8.4 [15]).

From a strategic point of view, circular economy will make Europe less dependent from countries extracting raw materials and will offer the opportunity to develop a new leadership on advanced technologies for the implementation of circular economy processes, leveraging on the already recognised leadership on production technologies. Overall, increased company competitiveness and future potential leadership in Circular Economy will generate new jobs for people in Europe, both in advanced and converging regions [29–31] (cf. SDG 8.5 [15]).

In societal terms, circular economy will establish production and consumption models that will make modern products affordable for a wider class of world population, thus improving quality of life of people and stimulating global progress (cf. SDG 10.2 [15]). For example, according to the concept of *Frugal Innovation*, manufacturers can offer affordable high-quality products in emerging global markets [32]. Remanufacturing with upgrade, through which new advanced functionalities can be added to products during remanufacturing operations, coupled with advanced non-ownership based services, can offer customers products with frequently upgraded performance avoiding the need to buy new expensive products [21].

20.2.1.3   Industrial and Technological Challenges

The development of Circular Factories involves heterogeneous sources of information and asks for distributed data gathering and cyber-physical systems to increase availability and traceability of information. Decision support tools will be based on advanced techniques such as data analytics. These smart manufacturing and Industry 4.0 topics are already commonly applied to manufacturing systems, but in case of de- and remanufacturing they have been investigated in a less structured approach. Specific challenges to be addressed include [18]:

- *Circular Economy engineering* shall be developed combining strategies for recovering the highest residual value from post-use products [33], e.g. via disassembly [34, 35]. A new set of manufacturing system engineering methods will be needed to design circular factories.
- *Zero-defect de- and remanufacturing*. The profitability of Circular Economy businesses and product acceptance by customers strictly depends on the quality of recovered materials and remanufactured products. Therefore new systemic zero-defect solutions for remanufacturing and recycling processes should be developed.
- *Flexible automated technologies for adaptable de- and remanufacturing systems*. Automated technologies of circular factories should be flexible enough to be adapted to various applications that will change unexpectedly over time due to technology trends and product conditions [36].
- *Digital factory for circular factories*. The integrated design and management of de- and remanufacturing processes and systems will need the support of digital platforms. Simulation of process capabilities will help to set process parameters. A digital twin of the real plant will support tactical and operational decisions.
- *New circular business models and value-chains*. Systematic methodologies to identify optimal circular strategies, to design circular business models and to quantitatively assess business performances and risks for the various actors of the supply chain should be developed. Also, organisational guidelines and best practices to shift towards circular service-based business models should be made available to companies.
- *Full tracing of products components and materials during multiple lives*. Products will become the means to bring functionalities where needed. Their characteristics and interaction with the environments (manufacturing, de- and remanufacturing, transport, storage, usage, reconditioning, etc.) during their multiple lives should be continuously traced and analysed in order to guarantee the desired performance and to plan the following phases of their multiple lives.

20.2.1.4 Barriers

On top of technological challenges and opportunities, there are several non-technical systemic pre-conditions and barriers that need to be addressed in favour of a large-scale implementation of new circular business models.

Circular Economy requires a systemic perspective and high cooperation among actors at all levels of the manufacturing and de-manufacturing supply chain [37]. In case of industrial symbiosis, where waste generated by an industrial ecosystem can be used as source of materials and energy for other industrial and non-industrial ecosystems [38], there is also the need for cooperation among different supply chains. Compared to current practices, one of the most relevant barriers is thus the establishment of trustful conditions among all supply chain actors enabling such an effective cooperation. Multilateral information systems for life-cycle management will enable to manage the whole product life-cycle from supplier to end-user, while adding value and maximizing the resource utilization.

Circular Economy is becoming relevant in the worldwide political and research agendas, e.g. the *Alliance on Resource Efficiency* launched by in 2015, the initiative *Closing the loop—An EU action plan for the Circular Economy* launched by the European Commission in December 2015, initiatives promoted in the Chinese *Five Year Plan* since 2006, and similar initiatives in the US, Japan, and Australia.

Legislation constitutes an additional barrier to the implementation of Circular Economy businesses. Currently, a sector-oriented approach for recycling has been adopted by the European Union. The directive on end-of-life-vehicles (2000/53/EC) and the directive on reusability, recyclability and recoverability of waste vehicles (2005/64/EC) define standards for the quality of reused parts and materials in automotive industry. Similarly, directives were issued for electrical and electronic equipment (2002/95/EC and 2011/65/EU). However, remanufacturing is still legally undefined in most countries, even though ongoing initiatives are addressing *end-of-waste* regulation [39]. At European level, specific actions were started to identify potential legislative barriers for the development of new Circular Economy businesses [40].

The lack of understanding on how to manage the uncertainties associated with Intellectual Property is another significant barrier for companies wishing to adopt a remanufacturing approach [41].

Finally, profitable circular businesses can be developed only if also cultural aspects are properly addressed. Indeed, the market acceptance of reconditioned products and the collection of post-use products are crucial factors for the transition towards use-oriented businesses. While the focus of research is mainly on technical enablers, more attention should be paid to social mechanisms of value creation in circular economy, e.g. network externality, public goods, social dilemma, and lifestyles aspects [13].

20.2.2 Rapid and Sustainable Industrialisation

20.2.2.1 Definition

A sustainable manufacturing industry [1–5] is fundamental to support the societal and economic growth of any country [42]. Industry allows the transformation of materials and energy into products with useful functions; therefore, the absence of industry prevents the full development especially of highly populated regions and creates enormous societal problems leading to low quality of life and frequently a sense of impotence and desperation. At the same time, the industrialization should not come at the expenses of safety and environment otherwise an immediate advantage may turn into a long term problem. Therefore, guaranteeing in each region the opportunity of sustainable industrial development is one of the ultimate challenges of humanity. This is critical in regions that did not develop an industrial system or in regions that due to catastrophic events lost, at least partially, their industrial base. The rapid installation of production capacity is particularly important in these cases since it prevents hopeless situations that in turn trigger emigration. An effective industrialization should be extremely rapid in order to revert the trend and restore hope.

Therefore, the proposed mission is *Rapid and Sustainable Industrialisation.* One possible declination of the described mission could be *to establish factories and connect them to industrial networks in a six-month period and make the newly installed production capacity sustainable in a three-year horizon.*

The availability of technologies, methodologies, and tools for the novel establishment of manufacturing capacity will provide a key instrument of foreign policy that developed countries in the European Union can exploit to build strong economic relations with *low income* countries around the world. Moreover, similar considerations apply also to regions that are not yet characterized by a significant manufacturing output in some parts of the European Union periphery, thus failing to reach an effective economic convergence. Consequently, this mission will contribute to create wider inter-regional value chains for the economic and societal progress of the whole Europe.

The technologies for rapid industrialisation can be exploited also to make already existing industries more resilient in case of natural disasters (e.g. earthquakes, flooding, eruption of volcano, major fire) that can cause serious damages to factories and infrastructures (e.g. roads, railways, airports) with consequent interruption of the production. Indeed, the economic relevance of manufacturing (Sect. 20.1) [1–5] leads to high social and private costs whenever industrial capacity and industrial networks are damaged or even destroyed in a specific region.

20.2.2.2   Societal Impact

The proposed mission will contribute to reach SDG 9 [15], and in particular Goal 9.2 (*Promote inclusive and sustainable industrialization*) while creating positive international collaborations (cf. SDG 10 [15]). In case of a region or a whole country without (sufficient) manufacturing industry, the establishment of production capacity in a short time horizon represents a boost that can have a large multiplier effect. Agriculture, mining and services risk to collapse or to be significantly depreciated if a manufacturing industry is missing. Indeed, manufacturing transformations add value to raw materials and agricultural products, whereas interactions with services take place along the whole manufacturing value chain.

The impact of fast industrialization in depressed areas is potentially huge, because the societal and economic development would be enhanced; also skilled labour force would be required and the vicious circle related to the human capital flight could be mitigated [43].

Similarly, the loss of production capacity is linked with a direct and indirect loss of employment. Whenever a relevant natural disaster happens, after the fundamental *search and rescue* operations, the restoration of the production capacity (primary, secondary, and tertiary sectors) of the region should be one of the first actions to be scheduled. This is needed so that the resident population can have a future in the stricken area thanks to economic activities that are not exclusively based on government subsidy. Otherwise, the mid- and long-term effect will be a forced mass emigration and the actual economic desertification of the region, thus creating a hardly reversible void in the emigration area and potential societal problems in the immigration area. Typically, the immigration converges to large cities, thus making the urbanization process hardly sustainable (cf. SDG 11.3 [15]).

Production capacity can be established (or restored) only if factories are built (or repaired) together with temporary infrastructures that are necessary for any trade. These tasks generate high public costs, even though also the demand-side stimulus to the economy should be considered. This type of actions can be supported only by strong government policies and international organizations. A country with the ability to quickly install production capacity would have a relevant competitive advantage that can be exploited in international negotiations as a valuable alternative to traditional policies based mainly on subsidies or military intervention.

20.2.2.3   Industrial and Technological Challenges

The fast realization of manufacturing capacity poses serious and multi-faceted technological and organisational problems. Factories implementing rapid industrialisation evolve in a short time period and need to be highly reconfigurable [44], changeable [45], and scalable [46].

If new factories must be built in a short time, then inspiration can be taken from the building and construction domain where prefabricated elements have been used for a long time [47, 48], thus leading to the novel concept of *prefabricated factory*. In this case, a critical issue is related to the high number of functional elements composing a factory. Rapid prototyping techniques are already employed in manufacturing, but they could be scaled up for applications in factory construction as recently done with 3D printing of buildings and building components [49, 50]. In addition, a whole factory or a part of it could be transported to the destination site as a *motorfactory* (i.e. the factory version of a motorhome [51]). Much research and innovation is needed to realize a working *motorfactory* with very short ramp-up times.

Logistics and supply chain are fundamental to establish an effective manufacturing capacity. At the beginning, a rapid industrialisation will be necessarily missing most elements of a self-sustaining industrial environment, therefore just one or few industrial plants in a region need to be connected to a larger supply and distribution chain that can be far away. Initial high transportation costs will be gradually reduced as soon as missing or damaged transport infrastructures are built together with the addition of further industrial plants as close nodes of the supply chain. However, in the short-term period it will be necessary to find quick solutions to (re)activate transport channels, even if with above-market costs; for instance, autonomous vehicles (e.g. Unmanned Aerial Vehicles) could be employed in case of critical demand [52].

Rapid industrialisation must gradually evolve also from an organisational perspective. The use of advanced technologies in new factories requires skilled employees that could be missing in a region with limited manufacturing tradition. Therefore, at the beginning only a subset of the functional areas will be covered in the new plant (e.g. production, basic maintenance), whereas others (e.g. design, planning, advanced maintenance) will be initially provided remotely. This transition and extension of on-site functions will be supported by enabling technologies such as:

- Formalization, sharing and transfer of production knowledge. Semantic Web and ontologies are candidate technologies [53].
- Remote assistance for the configuration and maintenance of devices, machines and systems. Augmented and Virtual Reality (AR/VR) represent promising technologies that can be further enhanced for industrial applications (see also Sect. 20.2.5) [54, 55].
- Evolutionary planning and control that does not take the production facilities as a fixed asset but as an evolving tissue connecting several sites [56].

A new evolving mix of high-tech and low-tech [57] components needs to be designed to continuously guarantee a good fit with the improving conditions of the target area.

Energy supply is highly critical in a place without power plants and possibly without a working distribution network. Future research will address mobile power plants that can be scaled up and upgraded, while making use of renewable sources [58] (see also SDG 7 [15]).

In the particular case of natural disasters, methodologies and tools are needed to assess the performance of a damaged manufacturing plant and then support its repair.

20.2.2.4 Barriers

Initiatives of local and international governments are needed to develop the proposed mission, because of the large investment costs and the strategic trigger that normally involves political decisions.

The availability of natural resources depends on the characteristics of the country (e.g. geology and climate), but in case of manufacturing it is possible to take political decisions to enhance the capacity and thus the independence of a region, even if taking in due consideration the characteristics of every region.

The fast realization of manufacturing capacity has to deal with problems related to its long-term sustainability. New manufacturing plants will be initially supported by dedicated policies, but they must be gradually turned into self-sustaining sites by adopting appropriate business models.

The connection to a global supply chain is a decisive step for newly installed manufacturing capacity because an economically depressed area may cause problems to find proper suppliers and customers.

International organizations formally support cooperation among countries but an effective cooperation promoting a higher independence of low income countries is much more difficult to be implemented.

20.2.3 Robotic Assistant

20.2.3.1 Definition

Robotics technology is becoming ubiquitous in a wide range of applications including manufacturing, healthcare, agriculture, civil, commercial and consumer, transport and logistics.

The proposed long-term mission is to *Provide Robotic Assistants to everybody that can take advantage of their services in workplace and home environments.* More specifically, a declination of this mission for a mid-term horizon is to *provide a Robotic Assistant to human operators involved in manufacturing operations, logistics or in maintenance activities of machine tools and production systems.* Indeed, manufacturing systems (e.g. assembly lines [59]) offer key advantages for the development of robotic assistants because they are characterized by structured environments, clear applications, well-defined safety regulations, and continuous monitoring of performance indicators. Moreover, industrial companies can afford high investments in capital goods that will be needed to provide robotic assistants. The use of robotic assistants will be extended to other domains as soon as it is successful in manufacturing.

Independently from the application domain or technology cluster, the overall robot system performance can be characterised in terms of *abilities*: perception ability, configurability, adaptability, manipulation ability, motion ability, cognitive ability, decisional autonomy, interaction ability, dependability [60].

The interaction between humans and robots (*interaction ability*) constitutes an important aspect in current robotic applications aimed at assisting people, instead of replacing them [59]. A fruitful collaboration can be achieved only if the robot has a full understanding of the human behaviour also in unstructured contexts, including intentions, emotions, and desiderata (*cognitive ability*).

The use of a robot in real environments (e.g. advanced manufacturing, surgical room, outdoor or indoor applications in agriculture) implies the control of the operation field and of the surrounding environment (*perception ability*).

Decisional autonomy is a key feature to make robots useful in real applications to fruitfully collaborate with people also in dangerous and unstructured environments.

20.2.3.2 Societal Impact

In the future, various artificial embodied agents will populate human living and working environments. The smooth integration of these intelligent agents generates a wide range of societal and technological issues to be addressed.

The regulatory and technological evolution of robotics will bring relevant benefits for our society, because people will be supported in repetitive, unhealthy and dangerous tasks. Indeed, empowering humans is highly demanded in industry because several onerous tasks (e.g. lifting and installation of heavy components) are still manually performed, causing musculoskeletal disorders due to non-ergonomic postures (cf. SDG 8.5 [15]). Cooperative robots (e.g. wearable robotics and collaborative manipulators) will empower human operators in industrial tasks by improving ergonomics and reducing musculoskeletal stress [61]. However, it is also necessary to carefully analyse the societal impact of a co-working robot in an industrial environment [62]. The concept of *robotic assistant* implies that humans will keep a central role as orchestrators; the more skilled is the human operator, the more numerous and sophisticated will be the orchestrated robots. Empowered humans will be able to focus on tasks that can be hardly automated (e.g. reaction to unforeseen events and search of new solutions), while being free to orchestrate the robots within certain limits to better reach the production goals. Accurate and reliable robotic assistants will further improve human skills, such as in precision manufacturing, surgery, etc. (cf. SDG 4.4 [15]).

Robots acting and interacting in human contexts will have a new specific social role that will take into account different aspects such as safety, wellbeing, health, and productivity related to human companions and collaborators.

Beyond manufacturing, robotic assistants will empower humans in innovative health applications both in medical and home environments. Also the fruition and protection of public and commercial spaces (e.g. cultural heritage sites, cf. SDG 11.4 [15]) will benefit from the support of cognitive and social robots.

20.2.3.3 Industrial and Technological Challenges

The development of advanced perception systems also using alternative sensing modalities is fundamental to enhance the autonomy and safety level of robotic platforms operating in dynamic semi-structured and unstructured environments like in manufacturing (e.g. process control, surveillance, assembly and disassembly [30]) and transport (e.g. autonomous vehicles and advanced driver-assistance systems). Main challenges include the design and development of multi-sensor platforms and multi-sensor processing algorithms to be integrated on-board unmanned ground vehicles for tasks, such as multi-modal map building, situation awareness, and traversability estimation [63, 64]. Alternative sensing modalities like radar, depth-sensors, and cameras sensing outside of the visible spectrum (e.g. hyperspectral cameras or thermal cameras) and their intelligent combination and fusion, need to be further investigated for autonomous navigation under field conditions [65].

Research challenges also deal with the design and development of novel estimation and cooperative perception strategies for robotic networks to perform tasks, such as cooperative mapping, cooperative manipulation, target tracking, and environmental monitoring [66, 67].

Autonomous vehicles and cooperative robots need improved fast and safety-critical compliant communication networks and protocols, both on the intra- and inter-machine levels, for effective and safe task execution. The presence of human bystanders/co-operators also requires new conceptual frameworks and practical standardized procedures for a high-level safety validation [68]. Joint design approaches have been proposed to combine safety and security requirements in communication networks, in the typical modern scenario of ubiquitous connectivity [69]. Already existing and upcoming standards will be key enablers for the application of the Internet of Things (IoT) paradigm in a wider context [70].

Control algorithms will play a key role to enhance high-performance and high-precision human-robot cooperation, while guaranteeing safety, in particular in industrial environments (e.g. assembly tasks [71]). Learning-from-demonstration algorithms can be employed to directly teach a task to a robot. Impedance-based algorithms can improve the physical guidance of manipulators, while involving the human dynamics estimation/measurement in the control loop [72]. Machine Learning techniques allow to deal with uncertain interactions due to the robot itself or to the surrounding environment (such as in assembly tasks), thus enabling the auto-tuning of the robot control parameters [73].

The analysis and evaluation at the various levels of abstraction of the human-robot interaction through cognitive models and architectures will enable complex social interactions and effective task cooperation. The cognitive architecture of the robotic assistant will enable its *human orchestrator* to teach and activate complex tasks and behaviours by means of verbal and non-verbal (e.g. human gesture) interactions, thus creating effective and continuously evolving work teams.

Further challenges are related to the design and development of robotic systems able to mimic biological systems (e.g. bio-mimetic robotic vehicles, robotic arms capable of mimicking the soft-behaviour of human arms) in unstructured environments so that robots can cooperate in a flexible way without rigid constrains.

Automated planning and scheduling constitute a research challenge to address intertwined task planning and execution in robotics. Timeline-based planning, dynamic task planning and coordination issues constitute key enabling technologies for the development of decisional autonomy solutions for robotics in human-robot collaborative scenarios [74]. The integration of such technology with Verification and Validation solutions [75] will foster also robust control solutions for guaranteeing effectiveness and safety of autonomous robots [76].

Multi-robot systems (MRS) under the guidance of a *human orchestrator* can improve the effectiveness of a robotic system both in terms of performance while accomplishing a given task, and of robustness and reliability of the system thanks to modularization. Current multi-robot systems research focuses on the coordination of actions and task execution by groups of robots, which can possibly be relatively large (e.g. swarms) [77, 78]. The main challenge is related to the design of robust and scalable decentralized systems with predictable dynamics, so that tasks can be effectively allocated [79] and coordinated by the *human orchestrator*.

20.2.3.4 Barriers

One of the main barriers to the diffusion of robotic assistants depends on how the presence of robots is acceptable and useful for human purposes in real scenarios (e.g. factories, workplaces, houses, schools, hospitals, museums, shops). Long-term interactions need to be carefully assessed in any robotic application, starting from manufacturing and then moving to other applications like robotic-assisted surgery [80] and healthcare for older or disabled population [81, 82].

Effective human-robot interactions will depend on how people physically and psychologically distance themselves from robots [83]. Fear and suspicion towards robots (so-called *Frankenstein complex*) has been long addressed in several science-fiction novels [84]. Safety has as large impact also on the acceptance of robotic technologies. In several cases, the physical contact can be established mitigating the risks associated with exceeding energy exchanged in the human-robot interaction. Protective safety functions are needed to immediately stop unexpected movements by controlling via software the motion, mechanical parameter limitations, and overall power supply.

The increasing complexity and flexibility of robotic cells, characterized by a large number of sensors and robots working together, require standardization and modularity. The main activities in this field are related to the development of industrial-oriented packages and to guarantee uniform support via industrial platforms (e.g. Robotic Operating System—ROS,[1] and in particular the ROS-Industrial[2] consortium).

---

[1] http://www.ros.org/.

[2] https://rosindustrial.org/.

20.2.4 Factories for Personalised Medicine

20.2.4.1 Definition

The continuous advances in microelectronics and biotechnology open a wide range of innovation opportunities. In particular, microfluidics is an emerging technology dealing with the manipulation of fluids in microchannels for application in biology, chemistry, and other engineering fields to implement detection and separation procedures [85]. A lab-on-chip (LoC) is a microfluidic device integrating functions of a test laboratory (e.g. transfer of samples, use of a precise quantity of a chemical product, titration, mixing with reagents, heating) on a system with a size of a few square centimetres [86].

LoCs for biomedical applications typically include biosensors, i.e. analytical devices that couple *biology* and *human-made artefacts*. Indeed, biological sensing molecules interact with the analyte and are interfaced with a transducer device to convert a biochemical signal into digital signals. The biological material can be composed of enzymes, microorganisms, tissues, cell receptors, organelles, nucleic acids, antibodies or whole cells, whereas the transducer can be electrochemical, thermometric, optical, piezoelectric or magnetic [87]. In addition, more advanced biosensors contain also a series of interconnected zones that enable sophisticated interactions between components [88].

LoCs have the potential to drastically improve people's health through constant fast detection and prevention of diseases, but at the moment they have diffusion only at laboratory scale, since current technologies and factories do not allow sustainable mass production. Herein, the proposed mission aims to *develop factories and technologies for mass production of LoC biodevices that will provide everybody with personalised medicine solutions at affordable costs for better and inclusive healthcare systems.*

Rapid, portable and easy-to-use LoC systems will be used by anyone to provide fast qualitative or quantitative analysis by means of automatic self-testing procedures [89] that will reduce diagnosis time. Individuals will be more responsible for their own health and it will be possible to mitigate treatment delays. Such procedures are named as point-of-care (PoC) testing if they can be performed at the site of patient care [90].

Fully personalised medicine can be boosted by the availability of bioengineered microdevices if the technology of these devices is adequately improved in terms of parallelization, robustness, and throughput. There is still ongoing research on the development of LoCs, but commercial devices using microfluidics and molecular assays already exist [89]. Wider applications will be possible thanks to the identification of disease-specific critical biomarkers and the development of rapid and portable detection schemes [90]. In this direction a developing strategy consists in combining LoC devices with mobile phones [91] to greatly reduce the cost of the system and extend its application. Furthermore, the success of PoC devices largely depends on the development of process technologies for mass production of reliable and accurate LoCs.

20.2.4.2   Societal Impact

The growing demand for personalised care services asks for new technological solutions and business models [92]. PoC devices will provide a cost-effective alternative to time-consuming and expensive laboratory tests, thus speeding up disease diagnosis and treatment decisions. Therefore, the use of PoC devices will potentially improve quality of life and treatment outcomes for all patients [90] (cf. SDG 3.8 [15]); indeed, it must be noted that the unmet need for medical care involved 3.2% of population aged 16 and over in Europe during 2015 [25]. A limited share of hospital budget is earmarked for in vitro diagnostics, but it has an impact on several decisions related to admittance, medication, and discharge. PoC testing enables to decentralize diagnostic testing, thus leading to faster treatment decisions and improved quality of care. Testing can be performed at doctor's office or at home to remotely monitor the progress of patients. Therefore, the number of visits needed to the hospital can be reduced, while personalising and limiting the invasiveness of treatments [94], and minimizing any adverse side effects of drugs [90].

The use of biological material in LoCs [93] may replace complex electronic devices that are characterized by a high cost and environmental impact during their life cycle (i.e. extraction of rare and expansive raw materials, energy and resource consuming production processes, polluting material recovery and disposal processes), thus leading to a positive net environmental contribution (cf. SDGs 12.2 and 12.5 [15]).

The inclusion of mammalian or human cells into biodevices will allow the realization of microfluidic Organ-on-Chips (OoC), i.e. a micro-device that enables to perform in vitro co-culture and perfusion of cells in conditions resembling the physiological environment and to recapitulate cell functions that are not present in conventional culture systems [95]. OoCs will allow scientists to model human physiology so that it is possible to quickly, cheaply and accurately identify chemical hazards, at the cellular and molecular level, as well as potential new medicines [96]. Possible applications with a relevant societal impact include toxicity studies, drug testing, cosmetics industry, cancer research, pharmacology, and testing of materials for implants [97] (cf. SDG 3.3 [15]). OoCs will represent an alternative to animal testing [98] and will enable the development of a personalised human model on chip that can be used to prevent and cure personal diseases [96].

In addition to biological procedures used in clinical and industrial chemistry, biosensor are promising for applications with relevant societal impact in environmental monitoring (cf. SDGs 3.9 and 6.3 [15]), food analysis, and bioterrorism [99]. For instance, applications can be focused on water analysis and air analysis, since contaminants or toxic substances have a strong impact on the environment and on human health. Measurements can be performed in situ and in real time thanks to portable analysis devices, so that it is possible to immediately take action as soon as the problem is detected, thus drastically minimizing the consequences [100].

20.2.4.3 Industrial and Technological Challenges

Even though promising, several industrial and technological challenges must be faced to realize factories that are able to produce large quantities of the most interesting healthcare products, including LoC solutions [101]. The expertise needed for the design and manufacturing of LoC devices includes electronics, advanced materials, photonics, chemistry and microfluidics [90]. LoC devices for PoC applications are needed to meet clinical requirements while being simple to use, reliable, portable, sensitive, self-calibrating, inexpensive, and safe in storage, use and disposal [90].

The manipulation of small quantities of fluids requires a network of microchannels with dimensions ranging from 10 to 100 μm. Depending on the application, a LoC includes other functions such as pumps, valves, sensors, electronics, etc. [86], thus making the fabrication of the device even more challenging [102]. The production of LoCs is more difficult than conventional microelectronic chips because, instead of making only electrical connections, other challenges must be faced [103], e.g.:

- Guaranteeing accurate temperature control (± 1 °C).
- Significantly different pressures (± 1 atm) must be managed across the chip.
- Storage of harsh reagents and solvents.
- Biocompatibility issues (e.g. denaturation, toxicity and adhesion).
- Fluidic manipulations for purification.
- Manipulation of sub-μL volumes of extremely expensive reagents.
- A wide range of electrical signals to be applied and detected.
- Fluidic control by means of valves, pumps, etc.
- Optical probing small quantities of optically-thin materials.

The production of LoCs with biosensors is particularly challenging because the selection of manufacturing technologies needs to take into consideration also the intrinsic molecular properties of the biological material. The stability of biomolecules must be preserved in terms of temperature, with maximum values ranging between 40 and 80 °C. In addition to the thermal stress, also the shear rate and the compression rate must be monitored [88]. Moreover, in case of cells, an adequate perfusion must be guaranteed throughout the production process.

The miniaturization on a single LoC of all steps needed for a portable PoC device can be challenging [102]. The parallelization of testing on a LoC requires the design of separate droplet storage and manipulation sites on the chip [102].

Regardless of the transduction mechanism, the sensing surface of biosensors must be functionalized with selective bio-receptors as biological recognition elements [102]. The functionalized surface is close to the transducer that converts the sensing event into an output signal [90].

A key challenge related to the reuse (or disposal) of biosensors is the biological regeneration (or removal) of the receptors integrated in the biosensors [102].

The enhancement of LoCs with biosensors poses challenges related to the improvement of their sensitivity, small molecule detection, non-toxicity, specificity, and cost-effectiveness [99]. Most of LoCs with biosensors are currently fabricated at laboratory level, therefore new manufacturing techniques have to be developed for large-scale production with affordable costs so that these devices can be readily available for high-throughput biological investigations [95]. Micro/nano replication processes enabled by precision tooling technologies are expected to meet application requirements, such as low cost and high volume production, 3D features/surface properties, high quality, reproducibility and reliability [104]. Recognized as one of the strategic priorities for European industry competitiveness, the scientific and industrial research in the fields of micromanufacturing and microtooling is rapidly growing [105]. Micro- and nano-manufacturing technologies [106, 107] required to produce LoCs are both promising and challenging to enable future batch and mass production [108, 109]. Fabrication methods for LoC devices include soft lithography, hot embossing, (micro) injection moulding, ultrasonic welding, photolithography, three-dimensional (3D) printing techniques, and laser micromachining [85]. Photolithography has high costs and requires cleanrooms that increase the fabrication cost, therefore strong enhancements are needed to make it a viable solution for high volume production. Soft lithography with organic and polymeric materials is a popular method for the fabrication of 2D and 3D structures at high resolution, but is not appropriate for mass production [110]. 3D printing is a promising option and has been increasingly used for microfluidic devices construction thanks to a relatively high resolution, low cost, rapid prototyping, and wide range of materials that can be used [111]. However, the applicability of 3D printing is partially limited because it is not yet possible to reliably print microfluidic channels with dimensions less than several hundred microns [112]. Inkjet printing is a potential technology for mass production of biosensors because it is simple, flexible, rapid, low cost, high resolution, and efficient [88]. Micro injection moulding (μIM) can be employed for the mass production of a polymeric micro-component (e.g. LoC devices) thanks to the high dosing precision and injection speed, while managing a very small amount of material. Recently, higher flexibility has been achieved by introducing tailored changeable inserts in the same master mould plates, so that different part geometries (e.g. removable cavities) can be tested for the injection of specific micro-components [113]. The fabrication of reconfigurable moulds for μIM is still challenging because high accuracy is needed for manufacturing the small features of the whole master mould and tailored inserts.

20.2.4.4   Barriers

Personalised together with *Participatory*, *Preventive* and *Predictive* are the four key characteristics of the future medicine and healthcare. Personalised medicine (PM) can help to radically change patient management and big pharmaceutical companies are already taking actions in this direction. The success of PM will depend on the technological progress and on the market demand for personalisation [90]. Public investments will be fundamental for a democratic development of PM.

LoCs for biomedical applications require a multi-disciplinary approach since biological, engineering and technological research and innovation are needed to create new devices, new materials, new processes, and new production systems that fully exploit the potential of the integration between the biological and artificial world.

LoCs and biosensors will be accepted for routine medical applications only if they are able to meet the requirements of the clinicians [114] and if the devices can be properly calibrated according to standardized procedures as already in place for traditional devices. In addition, it is necessary to address the regulations related to the LoCs market [85].

Diffusion of PoC devices will benefit from cost effective disposable solutions for biosensors that integrate all the needed instrumentation in a portable format [114].

OoCs have a high potential to support drug screen and tissue cure, but effective regulatory mechanisms must be established to deal with ethical issue and security risks [111].

20.2.5 *Internet of Actions*

20.2.5.1 Definition

The fast and the pervasive diffusion of internet, Information and Communication Technologies (ICT), electronics, and Cyber Physical Systems (CPS) paves the way for a new technology shift after *Internet of Things* [115].

The proposed mission is to *develop a novel Internet of Actions that enables everybody to share sensations and actions thanks to the ubiquitous presence of sensors and actuators.*

A specific mission for the manufacturing domain is to *develop by 2030 Internet of Actions solutions to enable a fully remote assistance and maintenance for a class of production plants (e.g. powertrain assembly lines, flexible manufacturing systems).*

Highly controlled and automated factories represent an ideal place where a mission for Internet of Actions (IoA) can be set and later extended to other domains. IoA will enable operators in a distributed set of industrial plants located also at considerable distances from each other to share data, information and actions while guaranteeing that all operators have the same view and perception at the same time. A completely remote assistance and maintenance can be beneficial both if skilled personnel is missing (cf. Sect. 20.2.2) and if intrinsically unhealthy and unsafe environments are involved.

20.2.5.2 Societal Impact

The societal impact of IoA is potentially huge in industry since it can be exploited also to protect labour rights and promote safety in working environments (cf. SDG 8.8 [15]). In 2014 the number of people killed in accidents at work was still 1.83 per 100,000 employees in Europe [25].

Digital technology may become an effective companion for humans to make an impact on the real world with actions generated in a mixed-reality world [116]. However, it will be necessary to realize a shift from supposedly user-centric design of technologies to actually human-centric technologies. A democratic development and spread of IoA will promote the social, economic and political inclusion of everybody (cf. SDG 10.2 [15]). IoA factories will enable people to carry out their job compatibly with the evolution and change of their cognitive and physical abilities [117], for example to cope with the extension of the working life related to ageing population in developed countries [118, 119].

Factories exploiting IoA have the potential of enhancing and better exploiting human skills (cf. SDG 4.4 [15]), thus contributing to higher satisfaction [120] and less work alienation [121]. Factories will be a qualifying learning and training environment [122] to ensure greater usability of work environments and the ability to relate industrial practices to adequate high-level skills.

Factories and hospitals (e.g. remote surgery and rehabilitation) are primary targets for IoA applications because they are characterized by highly structured and controlled environments. However, as soon as the technology is more mature, applications can be foreseen also in many other unstructured environments, such as home, entertainment, inspection and exploration.

20.2.5.3 Industrial and Technological Challenges

The development of IoA devices and systems with processors, sensors, and actuators will require the further development and integration of several enabling technologies, e.g. Augmented and Virtual Reality (AR/VR) [54, 55, 123], High Performance Computing [124], Cloud and Fog Computing [125–127], Cyber-Physical Production Systems [128], Big Data Analytics [129], ultrafast communication infrastructures and standards, artificial intelligence [130], data storage, sensors and monitoring [131], wearable devices [68], and actuator technologies.

Effective IoA systems will need to accurately reproduce sensations [132], so that proper (re)actions by humans can be generated interactively and adaptively. In particular, the development of new sensors and actuators will be fundamental to enable the sense of presence at a distance together with accurate and safe remote actions. The devices employed in IoA architectures will need to properly manage the interaction with the environment and human beings [133].

The extensive use of actuators will need new technological solutions that offer the possibility to build highly miniaturized devices, while keeping low the ratio between system mass and generated power, especially for long-term tasks. Indeed, manufacturing and assembly limitations together with physical scaling laws hinder a further size reduction of traditional actuators (e.g. electromagnetic motors) that are able to provide significant forces and torques. Only piezoelectric motors can be scaled rather efficiently to approach the boundary of the microdomain (below 1 mm) [134]. Bioactuators represent an innovative opportunity to meet the most demanding actuator requirements. Bioactuators exploit the action of live micro-organisms, both as a source of electrical power and a means of actuation (e.g. bioactuators powered by cardiomyocytes, based on bacteria, other motile cells, explanted whole-muscle tissues, engineered skeletal muscle, or insect-derived self-contractile tissues [134]). Bioactuators can reach a volume that is orders of magnitude smaller than the smallest piezoelectric motor. Moreover, the use of living cells offers self-healing capabilities, silent operation, and use of inexpensive and eco-friendly fuel [134]. Bioactuators enable the fabrication of microscale devices and soft robotic artefacts to safely interact with humans [134].

Human Computer Interaction (HCI) technologies will help humans to deal with complex mixed-reality systems full of autonomous agents [116]. Furthermore, HCI and haptic interfaces [64] will help IoA systems to properly interpret the intentions and actions of humans.

In manufacturing applications, various devices connected with the real shop floor can provide detailed data about the status of ongoing processes. Specifically, these devices generate a large amount of intensive and heterogeneous multi-source data (*Factory Telemetry*), which have to be ingested by proper solutions (e.g. simulation tools) capable to process these data streams to extract relevant insights [135]. It will be important to investigate the potential of the new generation of storage and database systems that can also run on distributed cluster systems (e.g. NoSQL databases) [136]. In addition, the need of sharing heterogeneous data demands to develop solutions for interoperability between different systems and between systems and humans. Another important requirement for solutions supporting factory telemetry is a minimal data latency, which can be enabled by the fifth-generation (5G) mobile networks [137]. Moreover, the continuous growth of objects connected to the IoA network requires the identification of valid strategies to distribute intelligence and data among the various components of the infrastructure (sensors, actuators, microcontrollers, services and databases on cloud, etc.). To meet this need, a potential reference model can be Fog Computing that allows transferring part of the computing power and the storage space near the data sources, thus reducing their data transmission time and increasing their availability [127]. Future connected devices will be capable of sophisticated interactions thanks to flexible mechanisms of collaboration and cooperation [138]. Within the IoA network, smart objects can cooperate with humans, with other objects, and with bots (e.g. chatbots, virtual assistants, etc.) or can negotiate offered services with each other.

Workplaces of human-centred and IoA-enabled factories will need to be redesigned on the basis of specific rules of ergonomics and organized according to adaptive work rhythms to provide an environment and working conditions appropriate to the different people, regardless of their age, sex and physiological or pathological status. Indeed, the continuous increase of ageing workers (age: 55+), as well as the need to promote the work (re)integration for impaired individuals, calls for a redesign of workplaces, even going beyond the criteria defined in international standards [139].

In a context characterized by factories where products, processes and technologies evolve through articulated dynamics, a fundamental challenge is represented by the ability to interpret complex production phenomena and identify solutions based on experience. Therefore, it is essential to invest strategically also in enabling technologies (e.g. VR/AR) to support user-centred tasks such as operator training and maintenance support by means of visual, auditory, tactile feedback and interaction, as well as appropriate semantic and ontological representations of information and knowledge to support the formalization and reuse of such experiences.

20.2.5.4 Barriers

The public opinion must be convinced of the centrality of humans in IoA for the realization of the objectives of this mission. An appropriate design of still-new technologies constitutes a fundamental aspect. Technology acceptance depends on two factors that strongly influence the users' attitude toward the employment of a new tool. Firstly, *Perceived usefulness* represents the degree to which a person believes that using a particular system would enhance his or her job performance; secondly, *Perceived ease-of-use* has been defined as "the degree to which a person believes that using a particular system would be free from effort" [140]. Visual technologies like AR/VR will play a key role to involve the population and support the users.

The close collaboration between humans and autonomous systems will pose serious problems of communication and safety. Actuators must be accurately remote controlled to interact with the environment while guaranteeing safety for humans and the environment itself. Companies will strive to exploit the opportunities offered by IoA, but an increasing number of threats could jeopardize the cyber-security of the IoA network. In particular, data confidentiality, integrity, and availability are the main features that could be compromised. For this reason, it is essential to identify new security measures that can contribute to mitigate the risks of attacks against this network, thus guaranteeing the protection of the information exchanged within the IoA network [142].

Algorithms governing IoA systems will have to understand why a user takes specific actions. Moreover, algorithms must be able to explain why specific events happen while facing decisions with life-or-death consequences (e.g. autonomous transportation systems in a factory) [116]. However, there are sectors where algorithms cannot be completely automatized because of the high risks associated with possibly wrong decisions (e.g. medical diagnosis Decision Support System (DSS), or health-related DSSs [141]).

The intensive use of technology in human-centred factories poses relevant ethical issues. Monitoring the activities of human operators and the use of personal data may be needed both for safety reasons and to implement IoA systems, but regulations will be needed to avoid misuses and protect privacy.

Europe has a long tradition in mechatronics, automation and robotics. Therefore, the development of IoA represents an opportunity to exploit a competitive advantage and be a world technology leader in this field, thus generating a relevant economic boost.

20.2.6 Factories Close to the People

20.2.6.1 Definition

Factories and civil population experienced a conflictual relationship since the beginning of the industrial revolution. Positive gains like high employment, mass production of goods, and labour rights have been coupled with industrial pollution, waste of natural resources, and work alienation. In the past, factories were built inside urban areas because of wider availability of workforce and of proximity with working place. However, especially in the case of process manufacturing, this created significant problems of severe environmental impact and disruption of the urban landscape, which is even more negative in the cities with touristic vocation and with relevant historical background, as in the case of many European cities. Therefore, the overall high cost of urban manufacturing led many industrial companies to move their production facilities outside the cities, abandoning urban plants with consequent dismantling problems that are not solved yet in many municipalities. Even if this delocalisation of production had positive impacts on urban environment, it moved environmental problems to previously uncontaminated rural or sub-urban areas. Furthermore, it was the cause of massive traffic flow of workers from cities to production locations, thus increasing the pollution generated by mobility and reducing people's quality of life due to traffic congestion and longer commuting time.

This situation generated over time a general adverse feeling against advanced and intensive manufacturing, which started to be considered as necessary for economic prosperity, but incompatible with green environment, nice landscapes and workers' quality of life. Thus, a paradigm shift is needed to re-think this relationship aiming at factories closer to the people, where closeness must be intended both as proximity and as positive relationship [143, 144]. The proposed mission is to *Design and build symbiotic and sustainable factories that are fully integrated in city districts of large European cities (population higher than one million) by 2030, minimizing their adverse effects at global and local level.*

20.2.6.2 Societal Impact

Factories are a fundamental part of the civil community and their integration in cities and local communities should be based on proper cohesion policies as well as on proper tools to manage risks and challenges, while favouring a positive economic, societal and environmental link between urban, peri-urban and rural areas (cf. SDG 11.a [15]).

Sustainable factories integrated in urban life (i.e. *urban factories*) [143, 145] can help to increase the number of cities adopting and implementing integrated policies and plans towards an efficient use of natural resources (cf. SDG 12.2 [15]), as well as the mitigation and adaptation to climate change (cf. SDG 11.b [15]) [146].

Factories close to the civil population will need to continuously monitor the environmental effects of industrial activities that affect local context, also aiming to reduce the number of illnesses and deaths from air, water and soil pollution and contamination (cf. SDG 3.9 [15]). In 2015 the consumption of chemicals that are toxic to health reached 221 million tonnes in Europe [25]. Monitoring water pollutant in water plants should be clearly addressed both at plant level and at supply chain level (cf. SDG 6.3 [15]).

A symbiotic relationship between factories and cities can help to reduce the overall energy and resource consumption (cf. *urban mining* [147]), secure sustainable energy supply and improve access to affordable energy [145]. In addition, a more homogeneous and distributed presence of factories in the urban areas will help to better integrate production, business, and social activities, thus reducing the phenomenon of *dormitory suburbs*.

20.2.6.3 Industrial and Technological Challenges

Digital technologies for sustainability will enable factories to monitor internal key performance indicators and cooperate with the urban environment and other factories in the supply chain. Indeed, it is fundamental to measure the progress on sustainable development (cf. SDG 17.19 [15]). A close integration of information will be needed between supply chain management, rules and standards suitable for industrial sustainability, internal sensor and monitoring systems, product data management, needs and consumptions of smart cities [148].

A set of analytical methods (Life-cycle Assessment, Life-cycle Costing, Risk assessment, Social Life-cycle Assessment) can be applied by single companies to evaluate effects of industrial activities on product chain, ecosystem, and society [149–153]. Specific assessment methodologies will be needed to identify effective technologies from an environmental and social point of view.

New factory automation and management systems should natively support the participation in the energy market with reference to price and environmental targets. Such approach requires integrating the price of energy and its actual environmental impact within the multi-objective optimization strategy to be pursued. Production units could therefore collaboratively improve their operating margins and environmental benefits by reducing energy costs [154].

The use of CPS may enable a real time quantification of the impact of production activities on two parallel areas: the environmental aspects (e.g. consumption of materials, energy or items as well as solid or fluid emissions) and the related local and global impacts (e.g. contribution to global warming, depletion of local resources and other key action areas).

Further industrial and technological challenges to be addressed include:

- Systems to monitor, manage, and treat water within factories [155, 156].
- Transparent monitoring and reduction of carbon content (e.g. CO_2 pollution) of industrial activities [157] both at plant level and at supply chain level.
- Methodologies and tools to monitor and reduce industrial noise pollution [158] of the factories in the urban tissue.
- Systems for energy management to foster efficient use of energy and to integrate information on energy quality within factories according to a multi-scale approach [159].
- Systems for monitoring and mitigating the effects of industrial activities (e.g. fine and ultrafine particle emissions [160], greenhouse gas emissions [161]) on regional and local areas to intelligently address health and climate effects by ex-ante and real-time analysis (cf. SDG 13.3 [15]).
- Energy-aware real-time control systems through the adoption and extension of predictive and model-based control techniques [162].
- Development of semantic data models for the integrated modeling of sustainability aspects at various factory levels (products, processes, resources and production systems, plant services, industrial building), both in static (e.g. configuration of plant) and dynamic (e.g. evolution of production resources in terms of status and energy consumption profile) terms [163].
- Advanced production scheduling and control to optimize energy and resource consumption profiles at system and machine level [159, 164].
- Integrated management of factory utilities (e.g. optimal real-time cogeneration) considering the thermal/electric consumption profiles of the production processes as well as sales/purchase opportunities in the market [159].

20.2.6.4   Barriers

Symbiotic and sustainable factories require a continuous impact assessment of production operations involving interdisciplinary and data-intensive processes, including data collection systems. Data management systems need to access heterogeneous data sources including sensor networks, factory information systems (e.g. Supply Chain Management tools [165] and Enterprise Resource Planning), and also remote systems (e.g. monitoring of markets for energy trade).

Digitising sustainability requires verifiable and reliable information to be reused in a modular way. Many parameters linked to sustainability are still qualitative or semi-quantitative, therefore tracking systems must be calibrated to provide reliable information. The cost of data tracking is relevant especially in case of real-time sampling systems. Industrial companies avoiding to implement expensive data tracking may benefit from a competitive advantage, therefore the publication and enforcement of regulations together with public funding and support will be fundamental to make a profit-driven business also compatible with sustainability goals. Possible industrial delocalization to countries with less demanding regulations and less efficient production must be regulated to save employment in developed countries [166] and protect the environment in the destination country [167].

The proliferation of different sustainability standards by various public and private bodies can create double counting of specific effects and overlapping in tracking. The overlapping between protocols constitutes a serious limit for implementing accounting and optimization systems. Coordinated standardization initiatives are needed to avoid further computational barriers.

Finally, a cultural barrier should be overcome to establish the common understanding that not only manufacturing is necessary for economic reasons, but that it can successfully co-exist with urban living and can even provide social and cultural advantages to cities. To this aim, future generations should be properly educated to increase the attractiveness of manufacturing and to remove the negative bias that is the heritage of the conflictual relationship between manufacturing and civil society during the last few decades.

20.2.7   Turning Ideas into Products

20.2.7.1   Definition

Personalised production is an evolution of customization [168] that enables companies to differentiate their offer through innovative products for specific needs of a customer or a target group [169] thanks to adaptable and reconfigurable production systems [170] supported by easy-to-use product configuration systems to make processes along the supply chain efficient [171, 172]. In this scope, the proposed mission is to *turn ideas into products by transforming passive consumers into active participants in the production of their own products thanks to new technologies that can enhance and empower their capabilities.* This mission implies a paradigm shift because innovation will not originate anymore from the identification of consumer requirements; indeed, innovation and production will be taken out of the factory boundaries to allow people to be the decision makers during the design and production process in new collaborative supply chain models [173, 174].

Consumer goods (e.g. clothing, footwear, sports items, glasses) [172, 175] but also other kinds of product such as medical products (personalised orthopaedic prosthetics, dental prosthetics, etc.) or durable goods (cars, kitchens, buildings facades, etc.), and even food can be produced based on the ideas of and by customers applying an approach of self-managed personalisation. In this way, they can create products with a unique design and style, along with functional and comfort-related aspects, going beyond the conventional choice dictated by off-the-shelf products.

20.2.7.2 Societal Impact

This mission will lead to a big change for manufacturing heading towards socialization and massive involvement of consumers [176–178]. Many small and medium sized manufacturers (e.g. SMEs, fab-labs and even individuals) will participate in different market segments, while evolving into production service providers to satisfy customers' personalised requirements [28]. These entities will further aggregate into dynamic communities in a decentralized system and win bargaining power and efficiency. Moreover, new companies will basically sell ideas and their integration into new and dynamic value chains and markets (cf. SDG 9.3 [15]).

People empowerment, inclusive society and sustainability are further impacts to be considered. According to a survey by Deloitte, 36% of customers are interested in personalisation and 22% are happy to share personal data in return for a personalised customer services and products [179]. The price is not a barrier, since 20% of consumers who expressed an interest in personalised products or services are willing to pay a 20% premium price [179]. As an example, clothing (19% of customers) and furniture (18%) are two important categories where customers make personalised purchases and an increasing number of manufacturing industries and brands are adopting this paradigm nowadays.

In a broader and inclusive view, it is necessary to ensure that all customers, including people belonging to less-represented communities, are enabled to turn their ideas into design and manufacturing of their own products, according to their specific needs and wishes (cf. SDG 9.2 [15]). It is crucial to promote and increase access to new technologies and the acquisition of knowledge and skills to properly handle them. The direct involvement of consumers in the production processes will increase their awareness of the societal impact of production from economic, environmental, and political perspectives (cf. SDG 12.1 [15]).

The diffusion of technologies for *turning ideas into products* can help the development of an inclusive society (cf. SDG 10.2 [15]) thanks to potential enhancement and better exploitation of personal skills and capabilities (cf. SDG 4.4 [15]) contributing to take people away from alienating working places and facilitating the integration between leisure and work. The creation of new jobs in different sectors exploiting the creativity of people is also enabled. Moreover, consumer entrepreneurship will be promoted through novel network based financing models and means. According to that, new business models to manage such kind of scenario will have to be developed.

The implementation of the proposed mission will have an indirect impact also on sustainability since production will shift from Make-to-Stock to Customise-to-Order, thus avoiding problems related to inventories and, in particular to unsold stocks which are related to economic and environmental costs for society (cf. SDG 12.5 [15]).

20.2.7.3 Industrial and Technological Challenges

Digital technologies play an important role in this mission where the power to design and produce can be at individual or community scale. Customers will produce small and large objects thanks to easy access and integration of technologies and systems suitable for rapid change in their configuration and production. Depending on the type of product, customers may involve people or organizations with various degree of expertise that are able to conceive ideas with different levels of complexity and formalization.

Manufacturing companies need to develop new collaborative systems and easy-to-use manufacturing facilities with flexible automation [180] that are capable of producing relatively small batches of customised products at competitive costs that mimic mass production prices [181] with a human-centred design approach [182]. Not only product design, but also manufacturing operations, services, relationship between people, people and companies, society and companies will need to be changed [183].

Industrial and technological challenges are associated with different processes along the value chain including product development and capability of the customer to use advanced technologies like product configurators, advanced biometric measuring systems, and platforms for production control. Furthermore, there is a need for new flexible and agile supply chain models for decentralized production. One critical aspect is the ability to hide the complexity of technologies and organizations by means of appropriate models, digital twins, artificial intelligence, virtual and augmented reality [55], so that the consumer can concentrate on the idea to be turned into a product but at the same time receives a feedback on the implication of his requirement in terms of time, cost, viability.

New tools for the configuration and design of personalised solutions will enable people to be active (and not passive) actors in the production chain. Sensor models and tools will help to formalize their needs and expectations starting from the design of the product to the innovative services associated with its production. Design and configuration systems need to shift to the mobile economy paradigm. New digital tools will integrate consumer input in terms of requirements and specification into the product design [184]. Moreover, new methods and tools will validate the product design and transform it into manufacturing operations that are taken in consideration also for a dynamic network configuration [185].

Solutions for adaptable and reconfigurable manufacturing will be based on enabling technologies, including not only additive manufacturing [186] but also, for example, laser technologies, hybrid technologies, and other relevant innovative or traditional production technologies. Indeed, the development and production of functional components and parts of the product will be differentiated in response to the needs and demands of the consumer, thus leading to a high product variety [187]. The integration into a single machine tool of different transformation processes is another important challenge. New technologies for innovative production processes and control [164] will allow the development of machines that will hide the complexity of the technologies and through models of the processes, digital twins and artificial intelligence will become easy-to-use for home-based designers as well as for the implementation of the customer-to-customer (C2C) paradigm where each customer can sell to other customers.

Models and tools for the creation of dynamic supply chains for personalised production are needed to implement decentralized production models where manufacturing capabilities are not necessarily concentrated in large plants but under certain conditions may be spread in several locations [188, 189]. This raises also the need to define new logistics systems where digitalization can support the tracking and provision of raw materials and components as well as supporting maintenance and usage of machines with real time remote control. New models and tools should be based on big data analytics to increase the capacity of companies to manage large quantities of data from a variety of sources (client, suppliers, machines, and social media) and support the selection and management of supply and distribution networks, based on real-time exchange of information between the involved actors [190]. Moreover, big data can be used also to activate new blockchain processes ensuring that transferred data are original and to conceive smart contracts for regulating different processes (from design, to production, to logistics).

The production and distribution chain may in certain cases be reorganized in mini-factories, i.e. decentralized and modular production facilities [191] that are managed by consumers willing to develop a new product concept or by technicians (e.g. an orthopaedic technician in a hospital lab producing body prosthesis [92]). The availability of cutting edge technologies and facilities to a wide range of consumers permits the acquisition of digital production knowledge, creativity, and collaboration. The quality and functionality level of one-of-a-kind products will have to be the same as in mass production factories.

New business models need to be developed. For instance, in certain cases the realization of innovative products could be co-financed by groups of customers through micro-sponsoring platforms and peer-to-peer platforms can be adopted for their commercialization. This will enable consumers to learn new skills and create shared job opportunities also promoting consumer entrepreneurship [192]. New business models will have to take in account all these aspects in order to exploit at best the potentialities of this approach.

20.2.7.4   Barriers

The radical innovation of personalised production through dedicated manufacturing technologies in a networked paradigm poses several barriers and critical elements to be faced, both at product and at manufacturing level.

The safe management of all personal information influencing the personalisation process, including biometric ones, is crucial. Distributed manufacturing approaches will require secure sharing of such information along the whole chain, as well as its continuous management and update, in the case of dedicated (web-based) services accompanying the product along its lifecycle, as those intended to monitor health, activity, and performance.

Major barriers must be addressed in terms of liability since one-of-a-kind products must guarantee safety and functionality. Liability therefore will pose limitations to what can be actually decided and what has to be selected among already tested alternatives. This approach will in any case need modifications of regulations and norms, especially in potentially dangerous products as in case of novel products for healthcare. New approaches to liability should consider also the presence of non-stable production networks.

The development of new standards for sharing information on products and processes represents another crucial aspect to fully accomplish this mission.

At manufacturing level, a cultural revolution is needed for the introduction of novel personalised production. Hybrid production technologies will require major transitions both in product design (e.g. design for additive features) and in product realization, where total integration should take place at information level (therefore affecting data formats and interoperability), at technological level (implying the partial disclosure of company mutual technical capabilities) and at organizational level (to become active actors in networked value chain paradigms).

Finally, new regulations for intellectual property rights will be needed to fully take advantage of innovation and knowledge sharing that empowers consumers thanks to the diffusion of ideas and their transformation into products.

20.3   Conclusions

After highlighting the potential of mission-oriented policies to boost research and innovation, this chapter presented seven missions focused on manufacturing industry but with larger societal impact. Some of the enabling technologies needed to accomplish the missions are related to the results of the research projects funded by the Italian Flagship Project *Factories of the Future* [5], as shown in Table 20.1. Further relevant missions can be identified and other specific goals can be associated with the proposed missions.

The accomplishment of a mission for research and innovation requires the clear demonstration of how specific innovation goals have been reached, while supporting the wide uptake of innovation to generate industrial and societal impact. Therefore, effective research infrastructures are needed to improve the exploitation of promising scientific and industrial results. In particular, the following chapter of this book [11] analyses how pilot plants can help to overcome the so-called *Valley of Death*, i.e. the phase between Technology Readiness Level 6–7 and 9. Industrial research and pilot plants can be sustainable only if integrated (public-private) funding mechanisms and innovation partnerships are implemented.

Table 20.1 Mapping of the *Missions* versus *Factories of the Future* research projects

Mission	Related research projects	Topics
Circular Economy (Sect. 20.2.1)	Zero Waste PCBs [33]	Recovery of residual value from post-use products
	WEEE Reflex [36]	Automated technologies for circular factories
Rapid and Sustainable Industrialisation (Sect. 20.2.2)	Pro2Evo [53]	Formalization and sharing of production knowledge using ontologies
	MaCISte [58]	Renewable energy
Robotic Assistant (Sect. 20.2.3)	Xdrone [64]	Multi-sensor, multi-modal map building
	FACTOTHUMS [68]	Safety in human/robot cooperation, wearable devices
Factories for Personalised Medicine (Sect. 20.2.4)	Fab@Hospital [92]	Personalised medicine and business models
	SILK.IT [93]	Biomaterial for LoCs
	PLUS [109]	Fabrication technologies for LoCs
Internet of Actions (Sect. 20.2.5)	Xdrone [64]	Haptic interfaces
	FACTOTHUMS [68]	Wearable devices
Factories close to the People (Sect. 20.2.6)	PROBIOPOL [156]	Waste water treatment
	MECAGEOPOLY [161]	Energy and emission efficient industrial processes
	IMET2AL [162]	Predictive and model-based control techniques
	GECKO [164]	Advanced production scheduling and control
Turning Ideas into Products (Sect. 20.2.7)	Fab@Hospital [92]	Personalised products and mini-factories in hospital labs
	GECKO [164]	Advanced production control
	Made4Foot [189]	Supply chain configuration

Acknowledgements This work has been partially funded by the Italian Ministry of Education, Universities and Research (MIUR) under the Flagship Project "Factories of the Future—Italy" (Progetto Bandiera "La Fabbrica del Futuro") [5].

The authors would like to thank Carlo Brondi, Marcello Colledani, Irene Fassi, Rosanna Fornasiero, Lorenzo Molinari Tosatti, and Marco Sacco for their valuable contributions to this chapter. Special thanks also to Sara Arlati, Andrea Ballarino, Alessandro Brusaferri, Gianfranco Modoni, Loris Roveda, Daniele Spoladore, Gianluca Trotta, and Andrea Zangiacomi for their support during the revision process.

References

1. Haraguchi N, Cheng CFC, Smeets E (2017) The importance of manufacturing in economic development: has this changed? World Dev 93:293–315
2. Marconi N, de Borja Reis CF, de Araújo EC (2016) Manufacturing and economic development: the actuality of Kaldor's first and second laws. Structural Change and Economic Dynamics 37:75–89
3. Su D, Yao Y (2017) Manufacturing as the key engine of economic growth for middle-income economies. J Asia Pac Econ 22(1):47–70
4. Rodrik D (2016) Premature deindustrialization. J Econ Growth 21(1):1–33
5. Terkaj W, Tolio T (2019) The Italian flagship project: factories of the future. In: Tolio T, Copani G, Terkaj W (eds) Factories of the future. Springer
6. Cantner U, Pyka A (2001) Classifying technology policy from an evolutionary perspective. Res Policy 30(5):759–775
7. Edquist C, Zabala-Iturriagagoitia JM (2012) Public procurement for innovation as mission-oriented innovation policy. Res Policy 41(10):1757–1769
8. Mazzucato M (2016) From market fixing to market-creating: a new framework for innovation policy. Ind Innov 23(2):140–156
9. Mazzucato M (2018) Mission-oriented research and innovation in the European Union—a problem-solving approach to fuel innovation-led growth. Publications Office of the European Union. ISBN 978-92-79-79832-0. https://doi.org/10.2777/360325
10. European Commission (2018) Re-finding industry—defining innovation. Publications Office of the European Union. ISBN 978-92-79-85271-8. https://doi.org/10.2777/927953
11. Tolio T, Copani G, Terkaj W (2019) Key research priorities for factories of the future—Part II: Pilot plants and funding mechanisms. In: Tolio T, Copani G, Terkaj W (eds) Factories of the future. Springer
12. Jovane F, Yoshikawa H, Alting L, Boër CR, Westkamper E, Williams D, Tseng M, Seliger G, Paci AM (2008) The incoming global technological and industrial revolution towards competitive sustainable manufacturing. CIRP Ann 57(2):641–659. https://doi.org/10.1016/J.CIRP.2008.09.010
13. Ueda K, Takenaka T, Váncza J, Monostori L (2009) Value creation and decision-making in sustainable society. CIRP Ann 58(2):681–700. https://doi.org/10.1016/J.CIRP.2009.09.010
14. Kaihara T, Nishino N, Ueda K, Tseng M, Váncza J, Schönsleben P, Teti R, Takenaka T (2018) Value creation in production: reconsideration from interdisciplinary approaches. CIRP Ann 67(2):791–813. https://doi.org/10.1016/J.CIRP.2018.05.002
15. United Nations (2015) Transforming our world: the 2030 agenda for sustainable development. Resolution adopted by the General Assembly on 25 September 2015, A/RES/70/1
16. European Commission (2016) Next steps for a sustainable European future—European action for sustainability. COM(2016) 739 final
17. Ghisellini P, Cialani C, Ulgiati S (2016) A review on circular economy: the expected transition to a balanced interplay of environmental and economic systems. J Clean Prod 114:11–32
18. Tolio T, Bernard A, Colledani M, Kara S, Seliger G, Duflou J, Battaia O, Takata S (2017) Design, management and control of demanufacturing and remanufacturing systems. CIRP Ann 66(2):585–609
19. Colledani M, Copani G, Tolio T (2014) De-manufacturing systems. Procedia CIRP 17:14–19
20. Johnson M, McCarthy I (2014) Product recovery decisions within the context of extended producer responsibility. J Eng Tech Manage 34:9–28
21. Copani G, Behnam S (2018) Remanufacturing with upgrade PSS for new sustainable business models. CIRP J Manuf Sci Technol (in press)
22. Knoth R, Kopacek B, Kopacek P (2005) Case study: multi life cycle center for electronic products. In: Proceedings of the 2005 IEEE international symposium on electronics and the environment, New Orleans, LA, USA, 2005, pp 194–198

23. Cerdas F, Kurle D, Andrew S, Thiede S, Herrmann C, Zhiquan Y, Low JSC, Bin S, Kara S (2015) Defining circulation factories—a pathway towards factories of the future. Procedia CIRP 29:627–632. https://doi.org/10.1016/J.PROCIR.2015.02.032
24. Seliger G (2007) Sustainability in manufacturing. Springer, Berlin
25. eurostat (2017) Sustainable development in the European Union—overview of progress towards the SDGs in an EU context. Publications Office of the European Union. ISBN 978-92-79-70973-9. https://doi.org/10.2785/207109
26. World Economic Forum (2014) Towards the circular economy: accelerating the scale-up across global supply chains
27. European Commission (2014) Towards a circular economy: a zero waste programme for Europe. Accessed at http://ec.europa.eu/environment/circular-economy/pdf/circular-economy-communication.pdf. Accessed 07 Aug 2018
28. Meier H, Roy R, Seliger G (2010) Industrial product-service systems—IPS2. CIRP Ann 59(2):607–627. https://doi.org/10.1016/j.cirp.2010.05.004
29. MacArthur Foundation (2015) Towards the circular economy: Economic business rationale for an accelerated transition
30. Bley H, Reinhart G, Seliger G, Bernardi M, Korne T (2004) Appropriate human involvement in assembly and disassembly. CIRP Ann 53(2):487–509
31. Chidi Nnorom I, Osibanjo O (2010) Overview of prospects in adopting remanufacturing of end-of-life electronic products in the developing countries. Int J Innov Manag Technol 1(3):328–338
32. Levänen J, Lindemann S (2016) Frugal innovations in circular economy: exploring possibilities and challenges in emerging markets. In: The international society for ecological economics 2016 conference, transforming the economy: sustaining food, water, energy and justice. Washington D.C., USA
33. Copani G, Colledani M, Brusaferri A, Pievatolo A, Amendola E, Avella M, Fabrizio M (2019) Integrated technological solutions for zero waste recycling of printed circuit boards (PCBs). In: Tolio T, Copani G, Terkaj W (eds) Factories of the future. Springer
34. Westkämper E (2003) Assembly and disassembly processes in product life cycle perspectives. CIRP Ann 52(2):579–588. https://doi.org/10.1016/S0007-8506(07)60205-4
35. Duflou JR, Seliger G, Kara S, Umeda Y, Ometto A, Willems B (2008) Efficiency and feasibility of product disassembly: a case-based study. CIRP Ann 57(2):583–600. https://doi.org/10.1016/J.CIRP.2008.09.009
36. Copani G, Picone N, Colledani M, Pepe M, Tasora A (2019) Highly evolvable e-waste recycling technologies and systems. In: Tolio T, Copani G, Terkaj W (eds) Factories of the future. Springer
37. European Remanufacturing Network (2015) Remanufacturing market study
38. Frosch R, Gallopoulos N (1989) Strategies for manufacturing. Sci Am 261:144–152
39. Guidat T, Seidel J, Kohl H, Seliger G (2017) A comparison of best practices of public and private support incentives for the remanufacturing industry. In: Proceedings of 24th CIRP conference in life cycle engineering, 8–10 Mar, Kamakura, Japan
40. EU H2020 Screen, Synergic circular economy across European Regions, GA 730313
41. Hartwell I, Marco J (2016) Management of intellectual property uncertainty in a remanufacturing strategy for automotive energy storage systems. J Remanuf 6:3
42. Sutherland JW, Richter JS, Hutchins MJ, Dornfeld D, Dzombak R, Mangold J, Robinson S, Hauschild MZ, Bonou A, Schönsleben P, Friemann F (2016) The role of manufacturing in affecting the social dimension of sustainability. CIRP Ann 65(2):689–712. https://doi.org/10.1016/J.CIRP.2016.05.003
43. Ndulu BJ (2014) Human capital flight: stratification, globalization and the challenges to tertiary education in Africa. World Bank Group, Washington, DC, Report n. 90289
44. Koren Y, Heisel U, Jovane F, Moriwaki T, Pritschow G, Ulsoy G, Van Brussel H (1999) Reconfigurable manufacturing systems. CIRP Ann 48(2):527–540. https://doi.org/10.1016/S0007-8506(07)63232-6

45. Wiendahl H-P, ElMaraghy HA, Nyhuis P, Zäh MF, Wiendahl H-H, Duffie N, Brieke M (2007) Changeable manufacturing—classification, design and operation. CIRP Ann 56(2):783–809. https://doi.org/10.1016/J.CIRP.2007.10.003

46. Putnik G, Sluga A, ElMaraghy H, Teti R, Koren Y, Tolio T, Hon B (2013) Scalability in manufacturing systems design and operation: State-of-the-art and future developments roadmap. CIRP Ann 62(2):751–774. https://doi.org/10.1016/J.CIRP.2013.05.002

47. Li Z, Shen GQ, Xue X (2014) Critical review of the research on the management of prefabricated construction. Habitat International 43:240–249

48. Zhong RY, Peng Y, Xue F, Fang J, Zou W, Luo H, Ng ST, Lu W, Shen GQP, Huang GQ (2017) Prefabricated construction enabled by the Internet-of-Things. Autom Constr 76:59–70

49. Hager I, Golonka A, Putanowicz R (2016) 3D printing of buildings and building components as the future of sustainable construction? Procedia Eng 151:292–299

50. Tay YWD, Panda B, Paul SC, Noor Mohame NA, Tan MJ, Leong KF (2017) 3D printing trends in building and construction industry: a review. Virtual Phys Prototyp 12(3):261–276

51. Simeoni F, Cassia F (2017) From vehicle suppliers to value co-creators: the evolving role of Italian motorhome manufacturers. Curr Issues Tour 1–19

52. Thiels CA, Aho JM, Zietlow SP, Jenkins DH (2015) Use of unmanned aerial vehicles for medical product transport. Air Med J 34(2):104–108

53. Urgo M, Terkaj W, Giannini F, Pellegrinelli S, Borgo S (2019) Exploiting modular pallet flexibility for product and process co-evolution through zero-point clamping systems. In: Tolio T, Copani G, Terkaj W (eds) Factories of the future. Springer

54. Alam FD, Katsikas S, Beltramello O, Hadjiefthymiades S (2017) Augmented and virtual reality based monitoring and safety system: a prototype IoT platform. J Netw Comput Appl 89:109–119

55. Leu MC, ElMaraghy HA, Nee AYC, Ong SK, Lanzetta M, Putz M, Zhu W, Bernard A (2013) CAD model based virtual assembly simulation, planning and training. CIRP Ann 62(2):799–822. https://doi.org/10.1016/J.CIRP.2013.05.005

56. Tolio T, Ceglarek D, ElMaraghy HA, Fischer A, Hu SJ, Laperrière L, Newman ST, Váncza J (2010) SPECIES—co-evolution of products, processes and production systems. CIRP Ann 59(2):672–693. https://doi.org/10.1016/J.CIRP.2010.05.008

57. Westkämper E (2014) The objectives of manufacturing development. In: Towards the re-industrialization of Europe: a concept for manufacturing for 2030. Springer, Berlin, Heidelberg, pp 3–6. https://doi.org/10.1007/978-3-642-38502-5_5

58. Gilioli E, Albonetti C, Bissoli F, Bronzoni M, Ciccarelli P, Rampino S, Verucchi R (2019) CIGS-based flexible solar cells. In: Tolio T, Copani G, Terkaj W (eds) Factories of the future. Springer

59. Krüger J, Lien TK, Verl A (2009) Cooperation of human and machines in assembly lines. CIRP Ann 58(2):628–646. https://doi.org/10.1016/J.CIRP.2009.09.009

60. SPARC (2016) Robotics 2020 multi-annual roadmap—for robotics in Europe. https://eu-robotics.net/sparc/upload/Newsroom/Press/2016/files/H2020_Robotics_Multi-Annual_Roadmap_ICT-2017B.pdf

61. Roveda L, Haghshenas S, Prini A, Dinon T, Pedrocchi N, Braghin F, Molinari Tosatti L (2018) Fuzzy impedance control for enhancing capabilities of humans in onerous tasks execution. In: Proceedings of the 15th international conference on ubiquitous robots

62. Sauppé A, Mutlu B (2015) The social impact of a robot co-worker in industrial settings. In: Proceedings of the 33rd annual ACM conference on human factors in computing systems, pp 3613–3622

63. Reina G, Milella A, Galati R (2017) Terrain assessment for precision agriculture using vehicle dynamic modelling. Biosys Eng 162:124–139

64. Aleotti J, Micconi G, Caselli S, Benassi G, Zambelli N, Bettelli M, Calestani D, Zappettini A (2019) Haptic teleoperation of UAV equipped with gamma-ray spectrometer for detection and identification of radio-active materials in industrial plants. In: Tolio T, Copani G, Terkaj W (eds) Factories of the future. Springer

65. Reina G, Milella A, Rouveure R, Nielsen M, Worst R, Blas MR (2016) Ambient awareness for agricultural robotic vehicles. Biosys Eng 146:114–132
66. Petitti A, Di Paola D, Milella A, Lorusso A, Colella R, Attolico G, Caccia M (2016) A network of stationary sensors and mobile robots for distributed ambient intelligence. IEEE Intell Syst 31(6):28–34
67. Scilimati V, Petitti A, Boccadoro P, Colella R, Di Paola D, Milella A, Grieco LA (2017) Industrial Internet of Things at work: a case study analysis in robotic-aided environmental monitoring. IET Wirel Sens Syst 7(5):155–162
68. Pecora A, Maiolo L, Minotti A, Ruggeri M, Dariz L, Giussani M, Iannacci N, Roveda L, Pedrocchi N, Vicentini F (2019) Systemic approach for the definition of a safer human-robot interaction. In: Tolio T, Copani G, Terkaj W (eds) Factories of the future. Springer
69. Bacco FM, Berton A, Ferro E, Gennaro C, Gotta A, Matteoli S, Paonessa F, Ruggeri M, Virone G, Zanella A (2018) Smart farming: opportunities, challenges and technology enablers. In: IEEE IoT vertical and topical summit for agriculture, 8–9 May 2018, Borgo San Luigi in Monteriggioni, Siena, Italy
70. Dariz L, Selvatici M, Ruggeri M, Costantino G, Martinelli F (2017) Trade-off analysis of safety and security in CAN bus communication. In: IEEE International conference on models and technologies for intelligent transportation systems, 26–28 June 2017, Naples, Italy
71. Roveda L, Pedrocchi N, Beschi M, Molinari Tosatti L (2018) High-accuracy robotized industrial assembly task control schema with force overshoots avoidance. Control Eng Pract 71:142–153
72. Roveda L (2018) A user-intention based adaptive manual guidance with force-tracking capabilities applied to walk-through programming for industrial robots. In: Proceedings of the 15th international conference on ubiquitous robots
73. Roveda L, Pallucca G, Pedrocchi N, Braghin F, Molinari Tosatti L (2018) Iterative learning procedure with reinforcement for high-accuracy force tracking in robotized tasks. IEEE Trans Ind Inf 14(4):1753–1763
74. Pellegrinelli S, Orlandini A, Pedrocchi N, Umbrico A, Tolio T (2017) Motion planning and scheduling for human and industrial-robot collaboration. CIRP Ann Manuf Technol 66(1):1–4
75. Bensalem S, Havelund K, Orlandini A (2014) Verification and validation meet planning and scheduling. Int J Softw Tools Technol Transf 16(1):1–12
76. Cesta A, Finzi A, Fratini S, Orlandini A, Tronci E (2010) Analyzing flexible timeline-based plans. In: Proceedings of the 2010 conference on ECAI 2010: 19th European conference on artificial intelligence, pp 471–476
77. Brambilla M, Ferrante E, Birattari M, Dorigo M (2013) Swarm robotics: a review from the swarm engineering perspective. Swarm Intell 7(1):1–41
78. Chen I-M, Yim M (2016) Modular robots. In Siciliano B, Khatib O (eds) Springer handbook of robotics. Springer, Cham, pp 531–542
79. Korsah GA, Stentz A, Dias MB (2013) A comprehensive taxonomy for multi-robot task allocation. Int J Robot Res 32(12):1495–1512
80. BenMessaoud C, Kharrazi H, MacDorman KF (2011) Facilitators and barriers to adopting robotic-assisted surgery: contextualizing the unified theory of acceptance and use of technology. PLoS ONE 6(1):e16395
81. Broadbent E, Stafford R, MacDonald B (2009) Acceptance of healthcare robots for the older population: review and future directions. Int J Soc Robot 1(4):319
82. Mottura S, Fontana L, Arlati S, Redaelli C, Zangiacomi A, Sacco M (2018) Focus on patient in virtual reality-assisted rehabilitation. In: Virtual and augmented reality: concepts, methodologies, tools, and applications. IGI Global, pp 1422–1450
83. Kanda T, Ishiguro H, Ishida T (2001) Psychological analysis on human-robot interaction. In: Proceedings 2001 ICRA. IEEE international conference on robotics and automation, pp 4166–4173
84. Asimov I (1950) I, Robot. Fawcett, Greenwich, CT, 1970. ISBN 0-449-23949-7
85. Khalid N, Kobayashi I, Nakajima M (2017) Recent lab-on-chip developments for novel drug discovery. WIREs Syst Biol Med 9:e1381

86. Abgrall P, Gué A-M (2007) Lab-on-chip technologies: making a microfluidic network and coupling it into a complete microsystem—a review. J Micromech Microeng 17(5):R15–R49

87. Sharma SK, Sehgal N, Kumar A (2003) Biomolecules for development of biosensors and their applications. Curr Appl Phys 3(2–3):307–316

88. Li J, Rossignol F, Macdonald J (2015) Inkjet printing for biosensor fabrication: combining chemistry and technology for advanced manufacturing. Lab Chip 15:2538–2558

89. Romao VC, Martins SAM, Germano J, Cardoso FA, Cardoso S, Freitas PP (2017) Lab-on-chip devices: gaining ground losing size. ACS Nano 11(11):10659–10664

90. Ahmed MU, Saaem I, Wu PC, Brown AS (2014) Personalized diagnostics and biosensors: a review of the biology and technology needed for personalized medicine. Crit Rev Biotechnol 34(2):180–196

91. Martínez Vázquez R, Trotta G, Volpe A, Bernava G, Basile V, Paturzo M, Ferraro P, Ancona A, Fassi I, Osellame R (2017) Rapid prototyping of plastic lab-on-a-chip by femtosecond laser micromachining and removable insert microinjection molding. Micromachines 8(11):328

92. Lanzarone E, Marconi S, Conti M, Auricchio F, Fassi I, Modica F, Pagano C, Pourabdollahian G (2019) Hospital factory for manufacturing customised, patient-specific 3D anatomo-functional models and prostheses. In: Tolio T, Copani G, Terkaj W (eds) Factories of the future. Springer

93. Benfenati V, Toffanin S, Chieco C, Sagnella A, DiVirgilio N, Posati T, Varchi G, Natali M, Ruani G, Muccini M, Zamboni R (2019) Silk Italian technology for industrial biomanufacturing. In: Tolio T, Copani G, Terkaj W (eds) Factories of the future. Springer

94. van Reenen A, de Jong AM, den Toonder JMJ, Prins MWJ (2014) Integrated lab-on-chip biosensing systems based on magnetic particle actuation—a comprehensive review. Lab Chip 14:1966–1986

95. Junaid A, Mashaghi A, Hankemeier T, Vulto P (2017) An end-user perspective on organ-on-a-chip: assays and usability aspects. Curr Opin Biomed Eng 1:15–22

96. Zhang B, Radisic M (2017) Organ-on-a-chip devices advance to market. Lab Chip 17:2395–2420

97. Kodzius R, Schulze F, Gao X, Schneider MR (2017) Organ-on-chip technology: current state and future developments. Genes 8(10):266

98. Shay S (2017) Organs-on-a-chip: a future of rational drug-design. J Biosci Med 5:22–28

99. Mani I, Vasdev K (2018) Current developments and potential applications of biosensor technology. J Biosens Bioelectron 9:253

100. Ríos Ángel, Zougagh Mohammed, Avila Mónica (2012) Miniaturization through lab-on-a-chip: Utopia or reality for routine laboratories? A review. Anal Chim Acta 740:1–11

101. Williams DJ, Ratchev S, Chandra A, Hirani H (2006) The application of assembly and automation technologies to healthcare products. CIRP Ann 55(2):617–642. https://doi.org/10.1016/J.CIRP.2006.10.001

102. Samiei E, Tabrizian M, Hoorfar M (2016) A review of digital microfluidics as portable platforms for lab-on a-chip applications. Lab Chip 16:2376–2396

103. Backhouse CJ (2016) Fabrication challenges of lab-on-chip: an overview of the challenges that have hindered lab-on-chip development, how this may now be dealt with for prototype development and low-to moderate volume manufacture, and finally how this might be dealt with for future high-volume manufacture. In: 2016 IEEE sensors, Orlando, FL, pp 1–3

104. Fassi I, Shipley D (2017) Micro-manufacturing technologies and their applications. Springer, Cham

105. Pourabdollahian G, Copani G (2017) Market analysis, technological foresight, and business models for micro-manufacturing. In: Fassi I, Shipley D (eds) Micro-manufacturing technologies and their applications. Springer, Cham, p 261

106. Van Brussel H, Peirs J, Reynaerts D, Delchambre A, Reinhart G, Roth N, Weck M, Zussman E (2000) Assembly of microsystems. CIRP Ann 49(2):451–472. https://doi.org/10.1016/S0007-8506(07)63450-7

107. Alting L, Kimura F, Hansen HN, Bissacco G (2003) Micro engineering. CIRP Ann 52(2):635–657. https://doi.org/10.1016/S0007-8506(07)60208-X

108. Trotta G, Martínez Vázquez R, Volpe A, Modica F, Ancona A, Fassi I, Osellame R (2018) Disposable optical stretcher fabricated by microinjection moulding. Micromachines 9(8):388

109. Martínez Vázquez R, Trotta G, Volpe A, Paturzo M, Modica F, Bianco V, Coppola S, Ancona A, Ferraro P, Fassi I (2019) Plastic lab-on-chip for the optical manipulation of single cells. In: Tolio T, Copani G, Terkaj W (eds) Factories of the future. Springer

110. Faustino V, Catarino SO, Lima R, Minas G (2016) Biomedical microfluidic devices by using low-cost fabrication techniques: a review. J Biomech 49(11):2280–2292

111. Zhang Y, Ge S, Yu J (2016) Chemical and biochemical analysis on lab-on-a-chip devices fabricated using three-dimensional printing. TrAC Trends Anal Chem 85(part C):166–180

112. Waheed S, Cabot JM, Macdonald NP, Lewis T, Guijt RM, Paull B, Breadmore MC (2016) 3D printed microfluidic devices: enablers and barriers. Lab Chip 16:1993–2013

113. Trotta G, Volpe A, Ancona A, Fassi I (2018) Flexible micro manufacturing platform for the fabrication of PMMA microfluidic devices. J Manuf Process 35:107–117

114. Mohammed M-I, Desmulliez MPY (2011) Lab-on-a-chip based immunosensor principles and technologies for the detection of cardiac biomarkers: a review. Lab Chip 11:569–595

115. Xu LD, He W, Li S (2014) Internet of things in industries: A survey. IEEE Trans Ind Inf 10(4):2233–2243

116. Kite-Powell J (2017) The next technology shift: the internet of actions. Forbes. https://www.forbes.com/sites/jenniferhicks/2017/12/29/the-next-technology-shift-the-internet-of-actions/#5077d11d3227. Accessed 04 Aug 2018

117. Wallen ES, Mulloy KB (2006) Computer-based training for safety: comparing methods with older and younger workers. J Saf Res 37(5):461–467

118. Poscia A, Moscato U, La Milia DI, Milovanovic S, Stojanovic J, Borghini A, Collamati A, Ricciardi W, Magnavita N (2016) Workplace health promotion for older workers: a systematic literature review. BMC Health Serv Res 16(Suppl 5):329

119. Aiyar MS, Ebeke C, Shao X (2016) The impact of workforce aging on European productivity. International Monetary Fund, WP/16/238

120. Fritzsche BA, Parrish TJ (2005) Theories and research on job satisfaction. In: Brown SD, Lent RW (eds) Career development and counselling—putting theory and research to work. Wiley, pp 180–202

121. Fedi A, Pucci L, Tartaglia S, Rollero C (2016) Correlates of work-alienation and positive job attitudes in high- and low-status workers. Career Dev Int 21(7):713–725

122. Abele E, Chryssolouris G, Sihn W, Metternich J, ElMaraghy H, Seliger G, Sivard G, ElMaraghy W, Hummel V, Tisch M, Seifermann S (2017) Learning factories for future oriented research and education in manufacturing. CIRP Ann 66(2):803–826. https://doi.org/10.1016/J.CIRP.2017.05.005

123. Nee AYC, Ong SK, Chryssolouris G, Mourtzis D (2012) Augmented reality applications in design and manufacturing. CIRP Ann 61(2):657–679. https://doi.org/10.1016/J.CIRP.2012.05.010

124. Ahmad A, Paul A, Din S, Rathore MM, Choi GS, Jeon G (2018) Multilevel data processing using parallel algorithms for analyzing big data in high-performance computing. Int J Parallel Prog 46:508–527

125. Stergiou C, Psannis KE, Kim B-G, Gupta B (2018) Secure integration of IoT and cloud computing. Future Gener Comput Syst 78(3):964–975

126. Gao R, Wang L, Teti R, Dornfeld D, Kumara S, Mori M, Helu M (2015) Cloud-enabled prognosis for manufacturing. CIRP Ann 64(2):749–772. https://doi.org/10.1016/J.CIRP.2015.05.011

127. Chiang M, Zhang T (2016) Fog and IoT: an overview of research opportunities. IEEE Internet Things J 3(6):854–864

128. Monostori L, Kádár B, Bauernhansl T, Kondoh S, Kumara S, Reinhart G, Sauer O, Schuh G, Sihn W, Ueda K (2016) Cyber-physical systems in manufacturing. CIRP Ann 65(2):621–641

129. Tsai CW, Lai CF, Chao HC, Vasilakos AV (2016) Big data analytics. In: Furht B, Villanustre F (eds) Big data technologies and applications. Springer, Cham

130. Maamar Z (2017) Ongoing research agenda on the Internet of Things (IoT) in the context of Artificial Intelligence (AI). In: International conference on Infocom technologies and unmanned systems (trends and future directions) (ICTUS), Dubai, 2017
131. Teti R, Jemielniak K, O'Donnell G, Dornfeld D (2010) Advanced monitoring of machining operations. CIRP Ann 59(2):717–739. https://doi.org/10.1016/J.CIRP.2010.05.010
132. Hua Q, Sun J, Liu H, Bao R, Yu R, Zhai J, Pan C, Wang ZL (2018) Skin-inspired highly stretchable and conformable matrix networks for multifunctional sensing. Nat Commun 9(1):244
133. Krüger J, Wang L, Verl A, Bauernhansl T, Carpanzano E, Makris S, Fleischer J, Reinhart G, Franke J, Pellegrinelli S (2017) Innovative control of assembly systems and lines. CIRP Ann 66(2):707–730. https://doi.org/10.1016/J.CIRP.2017.05.010
134. Ricotti L, Trimmer B, Feinberg AW, Raman R, Parker KK, Bashir R, Sitti M, Martel S, Dario P, Menciassi A (2017) Biohybrid actuators for robotics: a review of devices actuated by living cells. Sci Robot 2:eaaq0495
135. Modoni GE, Sacco M, Terkaj W (2016) A telemetry-driven approach to simulate data-intensive manufacturing processes. Procedia CIRP 57(1):281–285
136. Corbellini A, Mateos C, Zunino A, Godoy D, Schiaffino S (2017) Persisting big-data: the NoSQL landscape. Inf Syst 63:1–23
137. Schulz P, Matthe M, Klessig H, Simsek M, Fettweis G, Ansari J, Ashraf SA, Almeroth B, Voigt J, Riedel I, Puschmann A, Mitschele-Thiel A, Muller M, Elste T, Windisch M (2017) Latency critical IoT applications in 5G: perspective on the design of radio interface and network architecture. IEEE Commun Mag 55(2):70–78
138. Modoni GE, Veniero M, Trombetta A, Sacco M, Clemente S (2017) Semantic based events signaling for AAL systems. J Ambient Intell Hum Comput 1–15
139. Arlati S, Spoladore D, Mottura S, Zangiacomi A, Ferrigno G, Sacchetti R, Sacco M (2018) Analysis for the design of a novel integrated framework for the return to work of wheelchair users. Work—A J Prevent Assess Rehabil (in press)
140. Davis FD (1989) Perceived usefulness, perceived ease of use, and user acceptance of information technology. MIS Q 13(3):319–340
141. Spoladore D (2017) Ontology-based decision support systems for health data management to support collaboration in ambient assisted living and work reintegration. In: Camarinha-Matos L, Afsarmanesh H, Fornasiero R (eds) Collaboration in a data-rich world. PRO-VE 2017. IFIP advances in information and communication technology, vol 506. Springer, Cham
142. Conti M, Dehghantanha A, Franke K, Watson S (2018) Internet of Things security and forensics: challenges and opportunities. Future Gener Comput Syst 78(2):544–546
143. Westkämper E (2014) Vision of future manufacturing. Towards the re-industrialization of Europe: a concept for manufacturing for 2030. Springer, Berlin, Heidelberg. https://doi.org/10.1007/978-3-642-38502-5_6
144. Herrmann C, Blume S, Kurle D, Schmidt C, Thiede S (2015) The positive impact factory-transition from eco-efficiency to eco-effectiveness strategies in manufacturing. Procedia CIRP 29:19–27. https://doi.org/10.1016/J.PROCIR.2015.02.066
145. Herrmann C, Schmidt C, Kurle D, Blume S, Thiede S (2014) Sustainability in manufacturing and factories of the future. Int J Precis Eng Manuf-Green Technol 1(4):283–292
146. Juraschek M, Bucherer M, Schnabel F, Hoffschröer H, Vossen B, Kreuz F, Thiede S, Herrmann C (2018) Urban factories and their potential contribution to the sustainable development of cities. Procedia CIRP 69:72–77. https://doi.org/10.1016/J.PROCIR.2017.11.067
147. Brunner PH (2011) Urban mining a contribution to reindustrializing the city. J Ind Ecol 15(3):339–341
148. Kumar M, Graham G, Hennelly P, Srai J (2016) How will smart city production systems transform supply chain design: a product-level investigation. Int J Prod Res 54(23):7181–7192
149. Westkämper E, Alting L, Arndt G (2000) Life cycle management and assessment: approaches and visions towards sustainable manufacturing (keynote paper). CIRP Ann 49(2):501–526. https://doi.org/10.1016/S0007-8506(07)63453-2
150. Hauschild M, Jeswiet J, Alting L (2005) From life cycle assessment to sustainable production: status and perspectives. CIRP Ann 54(2):1–21. https://doi.org/10.1016/S0007-8506(07)60017-1

151. Umeda Y, Takata S, Kimura F, Tomiyama T, Sutherland JW, Kara S, Herrmann C, Duflou JR (2012) Toward integrated product and process life cycle planning—an environmental perspective. CIRP Ann 61(2):681–702. https://doi.org/10.1016/J.CIRP.2012.05.004

152. Brondi C, Carpanzano E (2011) A modular framework for the LCA-based simulation of production systems. CIRP J Manuf Sci Technol 4(3):305–312

153. Ballarino A, Brondi C, Brusaferri A, Chizzoli G (2017) The CPS and LCA modelling: an integrated approach in the environmental sustainability perspective. In: Collaboration in a data-rich world: 18th IFIP WG 5.5 working conference on virtual enterprises PRO-VE 2017, 18–20 Sept 2017

154. Ramin D, Spinelli S, Brusaferri A (2018) Demand-side management via optimal production scheduling in power-intensive industries: The case of metal casting process. Appl Energy 225:622–636

155. Salzman S, Scarborough H, Allinson G (2017) Exploring the economic viability of using constructed wetlands to manage waste-water in the dairy industry. Aust J Environ Manag 24(3):276–288

156. Ortelli S, Costa AL, Torri C, Samorì C, Galletti P, Vineis C, Varesano A, Bonura L, Bianchi G (2019) Innovative and sustainable production of biopolymers. In: Tolio T, Copani G, Terkaj W (eds) Factories of the future. Springer

157. Huisingh D, Zhang Z, Moore JC, Qiao Q, Li Q (2015) Recent advances in carbon emissions reduction: policies, technologies, monitoring, assessment and modeling. J Clean Prod 103:1–12

158. Casas WJP, Cordeiro EP, Mello TC, Zannin PHT (2014) Noise mapping as a tool for controlling industrial noise pollution. J Sci Ind Res 73:262–266

159. Duflou JR, Sutherland JW, Dornfeld D, Herrmann C, Jeswiet J, Kara S, Hauschild M, Kellens K (2012) Towards energy and resource efficient manufacturing: a processes and systems approach. CIRP Ann 61(2):587–609. https://doi.org/10.1016/J.CIRP.2012.05.002

160. Riffault V, Arndt J, Marris H, Mbengue S, Setyan A, Alleman LY, Deboudt K, Flament P, Augustin P, Delbarre H, Wenger J (2015) Fine and ultrafine particles in the vicinity of industrial activities: a review. Crit Rev Environ Sci Technol 45(21):2305–2356

161. Ciccioli P, Capitani D, Gualtieri S, Soragni E, Belardi G, Plescia P, Contini G (2019) Mechano-chemistry of rock materials for the industrial production of new geopolymeric cements. In: Tolio T, Copani G, Terkaj W (eds) Factories of the future. Springer

162. Cataldo A, Cibrario Bertolotti I, Scattolini R (2019) Model predictive control tools for evolutionary plants. In: Tolio T, Copani G, Terkaj W (eds) Factories of the future. Springer

163. Terkaj W, Danza L, Devitofrancesco A, Gagliardo S, Ghellere M, Giannini F, Monti M, Pedrielli G, Sacco M, Salamone F (2014) A semantic framework for sustainable factories. Procedia CIRP 17:547–552

164. Ballarino A, Brusaferri A, Cesta A, Chizzoli G, Cibrario Bertolotti I, Durante L, Orlandini A, Rasconi R, Spinelli S, Valenzano A (2019) Knowledge based modules for adaptive distributed control systems. In: Tolio T, Copani G, Terkaj W (eds) Factories of the future. Springer

165. Fornasiero R, Brondi C, Collatina D (2017) Proposing an integrated LCA-SCM model to evaluate the sustainability of customisation strategies. Int J Comput Integr Manuf 30(7):1–14

166. Bramucci A, Cirillo V, Evangelista R, Guarascio D (2017) Offshoring, industry heterogeneity and employment. In: Structural change and economic dynamics (in press)

167. Forin S, Radebach A, Steckel JC, Ward H (2018) The effect of industry delocalization on global energy use: a global sectoral perspective. Energy Econ 70:233–243

168. Tseng MM, Piller F (2003) The customer centric enterprise—advances in mass customization and personalization. Springer, Berlin, Heidelberg

169. Durá JV, Caprara G, Cavallaro M, Ballarino A, Kaiser C, Stellmach D (2014) New technologies for the flexible and eco-efficient production of customized products for people with special necessities: Results of the FASHION-ABLE project. In: 2014 International conference on engineering, technology and innovation (ICE)

170. Mourtzis D, Doukas M, Foivos P (2015) Design of manufacturing networks for mass customisation using an intelligent search method. Int J Comput Integr Manuf 28(7):679–700

171. Macchion L, Fornasiero R, Vinelli A (2017) Supply chain configurations: a model to evaluate performance in customised productions. Int J Prod Res 55(5):1386–1399
172. Carneiro L, Shamsuzzoha AHM, Almeida R, Azevedo A, Fornasiero R, Ferreira PS (2014) Reference model for collaborative manufacturing of customised products: applications in the fashion industry. Prod Plann Control 25(13–14):1135–1155
173. Wolf M, McQuitty S (2011) Understanding the do-it-yourself consumer: DIY motivations and outcomes. AMS Rev 1:154–170
174. Carelli A, Binachini M, Arquilla V (2014) The 'Makers contradiction'. The shift from a counterculture-driven DIY production to a new form of DIY consumption. In: Proceedings of the 5th STS conference
175. Fornasiero R, Zangiacomi A, Franchini V, Bastos J, Azevedo A, Vinelli A (2016) Implementation of customisation strategies in collaborative networks through an innovative reference framework. Prod Plann Control 27(14):1158–1170
176. Tseng MM, Kjellberg T, Lu SC-Y (2003) Design in the new e-commerce era. CIRP Ann 52(2):509–519. https://doi.org/10.1016/S0007-8506(07)60201-7
177. Jiang P, Ding K, Leng J (2016) Towards a cyber-physical-social-connected and service-oriented manufacturing paradigm: social manufacturing. Manuf Lett 7:15–21
178. Tao F, Cheng Y, Zhang L, Nee AY (2017) Advanced manufacturing systems: socialization characteristics and trends. J Intell Manuf 28(5):1079–1094
179. Deloitte (2015) Making it personal—one in three consumers wants personalised products. https://www2.deloitte.com/uk/en/pages/press-releases/articles/one-in-three-consumers-wants-personalised-products.html. Accessed 04 Sept 2018
180. Jovane F, Koren Y, Boër CR (2003) Present and future of flexible automation: towards new paradigms. CIRP Ann 52(2):543–560. https://doi.org/10.1016/S0007-8506(07)60203-0
181. United Nations Industrial Development Organization (2017) Emerging trends in global advanced manufacturing: challenges, opportunities and policy responses. https://institute.unido.org/wp-content/uploads/2017/06/emerging_trends_global_manufacturing.pdf. Accessed 04 Sept 2018
182. PwC (2017) Creating valuable customer experiences—transforming the customer journey using human-centred design. https://www.pwc.com/it/it/publications/assets/docs/customer-experiences.pdf. Accessed 04 Sept 2018
183. Sana (2017) Manufacturing industry report: mastering industry 4.0 with e-commerce. https://www.hso.com/fileadmin/user_upload/sana-ebook-manufacturing-uk__2_.pdf. Accessed 04 Sept 2018
184. Yin YH, Nee AYC, Ong SK, Zhu JY, Gu PH, Chen LJ (2015) Automating design with intelligent human–machine integration. CIRP Ann 64(2):655–677. https://doi.org/10.1016/J.CIRP.2015.05.008
185. Maropoulos PG, Ceglarek D (2010) Design verification and validation in product lifecycle. CIRP Ann 59(2):740–759. https://doi.org/10.1016/J.CIRP.2010.05.005
186. Thompson MK, Moroni G, Vaneker T, Fadel G, Campbell RI, Gibson I, Bernard A, Schulz J, Graf P, Ahuja B, Martina F (2016) Design for additive manufacturing: trends, opportunities, considerations, and constraints. CIRP Ann 65(2):737–760. https://doi.org/10.1016/J.CIRP.2016.05.004
187. ElMaraghy H, Schuh G, ElMaraghy W, Piller F, Schönsleben P, Tseng M, Bernard A (2013) Product variety management. CIRP Ann 62(2):629–652. https://doi.org/10.1016/J.CIRP.2013.05.007
188. Wiendahl H-P, Lutz S (2002) Production in networks. CIRP Ann 51(2):573–586. https://doi.org/10.1016/S0007-8506(07)61701-6
189. Macchion L, Marchiori I, Vinelli A, Fornasiero R (2019) Proposing a tool for supply chain configuration: an application to customised production. In: Tolio T, Copani G, Terkaj W (eds) Factories of the future. Springer
190. Váncza J, Monostori L, Lutters D, Kumara SR, Tseng M, Valckenaers P, Van Brussel H (2011) Cooperative and responsive manufacturing enterprises. CIRP Ann 60(2):797–820. https://doi.org/10.1016/J.CIRP.2011.05.009

191. Dahlgren E, Göçmen C, Lackner K, Van Ryzin G (2013) Small modular infrastructure. Eng Econ 58(4):231–264
192. Zangiacomi A, Fornasiero R, Franchini V, Vinelli A (2017) Supply chain capabilities for customisation: a case study. Prod Plann Control 28(6–8):587–598

Chapter 21
Key Research Priorities for Factories of the Future—Part II: Pilot Plants and Funding Mechanisms

Tullio Tolio, Giacomo Copani and Walter Terkaj

Abstract Mission-oriented policies have been proposed for research and innovation in the European manufacturing industry to address grand challenges while fostering economic growth and employment. A mission is required to have clear goals that can be demonstrated also to a wide public, therefore research and innovation infrastructures play a key role to create the necessary conditions. Given the fundamental importance of public investment to promote innovation, possible funding mechanisms for industrial research and innovation are discussed. Furthermore, taking advantage of the experience gained during the Italian Flagship Project *Factories of the Future*, this chapter identifies three types of industrial research and innovation infrastructure that can support mission-oriented policies: lab-scale pilot plants, industrial-scale pilot plants, and lighthouse plants.

21.1 Introduction

The adoption of a mission-oriented approach has been proposed to reshape the European research and innovation policy agenda [1, 2]. Mission-oriented policies have the potential to promote continuous innovation while providing solutions for specific problems in the scope of social grand challenges.

The previous chapter of this book [3] adopted a mission-oriented approach to propose seven missions (i.e. circular economy, rapid and sustainable industrialisation, robotic assistant, factories for personalised medicine, internet of actions, factories close to the people, and turning ideas into products) for research and innovation in

T. Tolio
Director of the Italian Flagship Project "Factories of the Future", Direttore del Progetto Bandiera "La Fabbrica del Futuro", CNR - National Research Council of Italy, Rome, Italy

T. Tolio
Dipartimento di Meccanica, Politecnico di Milano, Milan, Italy

G. Copani · W. Terkaj (✉)
CNR-STIIMA, Istituto di Sistemi e Tecnologie Industriali Intelligenti per il Manifatturiero Avanzato, Milan, Italy
e-mail: walter.terkaj@stiima.cnr.it

© The Author(s) 2019
T. Tolio et al. (eds.), *Factories of the Future,*
https://doi.org/10.1007/978-3-319-94358-9_21

manufacturing industry. These missions were designed taking inspiration from the scientific results of the Flagship Project *Factories of the Future* [4].

A mission-oriented approach requires the participation of the civil society both for the identification of the social challenges to be addressed and for the assessment of the results. This chapter deals with the problem of funding mission-related projects and demonstrating their results. Also in this case, the organisation and results of the Flagship Project *Factories of the Future* provided valuable input. Indeed, the flagship project designed open calls for proposals and funded small-sized research projects aimed at realizing hardware and software prototypes demonstrating the key scientific and industrial results [4]. These research projects share common traits with *mission projects* defined in the scope of a mission-oriented approach [1].

Section 21.2 analyses which are the current initiatives and possible funding mechanisms to implement mission-oriented policies. In particular, the need of proving the results of mission-oriented policies leads to design and develop appropriate research and innovation infrastructures that can be accessed by a large set of stakeholders. Therefore, Sect. 21.3 presents three types of industrial pilot plant that can support industrial research and innovation: lab-scale pilot plants, industrial-scale pilot plants, and lighthouse plants. Relevant examples of ongoing initiatives are presented for each type of pilot plant.

21.2 Funding Industrial Research and Innovation

Research and innovation play a relevant role in relation to the prosperity, health and wellbeing of the citizens in Italy and Europe. In this perspective, the public policy to support research and innovation is crucial to fund, activate, and encourage actions and players. After analysing the current research and innovation policy context (Sect. 21.2.1), this section presents a theoretical framework to address the main challenges related to research and innovation funding (Sect. 21.2.2).

21.2.1 Current Research and Innovation Policy Context and Challenges

Due to the fundamental role of manufacturing for guaranteeing sustainable growth and social welfare [4, 5], especially after the recent financial crisis, the European Commission, member states and regions have devoted considerable resources to support manufacturing research and innovation during the last decade, launching a wide number of programs and initiatives.

At European level, the Commission promoted Public-Private Partnerships (PPPs) to strategically address and manage research and innovation programs. In Manufacturing, the PPPs *Factories of the Future* (FoF),[1] *Sustainable Process Industry*

---

[1] http://ec.europa.eu/research/industrial_technologies/factories-of-the-future_en.html.

through Resource and Energy Efficiency (SPIRE),[2] *Robotics*,[3] and *Photonics*[4] were established. In order to address specific innovation and uptake challenges, further initiatives were launched, such as the programs *FTIPilots*,[5] the *SME Instrument*[6] and the *Knowledge and Innovation Communities* (KICs)[7] of the European Institute of Technology (EIT).

Based on the Smart Specialisation Policy, programs for inter-regional cooperation aimed also to industrial technology innovation were funded, such as the INNOSUP[8] and INTERREG[9] programs. The European Commission created the S3 Platform on Industrial Modernisation[10] with the goal of supporting EU Regions in the definition of relevant innovation investment projects based on smart specialization and mobilizing the interest of a high number of stakeholders in Europe. Furthermore, the European Commission invested in creating a better context for technology uptake considering skill, regulation framework, and access to finance for companies, especially SMEs. Examples are Investment Plan for Europe (including the European Fund for Strategic Investments EFSI),[11] the Blueprint for sectoral cooperation on skills,[12] and the Long Life Learning Program.[13] Furthermore, the European Investment Fund (EIF)[14] manages INNOVFIN SME Guarantee Facility, COSME Equity Facility for Growth (EFG) and Loan Guarantee Facility (LGF).

At national level, all most industrialized manufacturing countries have setup research and innovation programs in manufacturing. Some examples are *Platform Industrie 4.0*[15] in Germany, *Catapult network*[16] and its *High Value Manufacturing (HVM) division*[17] in UK, *Usine du future*[18] in France, *Fabbrica del Futuro*[19] and *Piano Industria 4.0*[20] in Italy, *Industrial Conectada 4.0*[21] in Spain, *Made Differ-*

---

[2] https://www.spire2030.eu/.

[3] https://ec.europa.eu/digital-single-market/en/robotics-public-private-partnership-horizon-2020.

[4] https://www.photonics21.org/.

[5] https://ec.europa.eu/programmes/horizon2020/en/h2020-section/fast-track-innovation-pilot.

[6] http://ec.europa.eu/programmes/horizon2020/en/h2020-section/sme-instrument.

[7] https://eit.europa.eu/activities/innovation-communities.

[8] https://ec.europa.eu/easme/en/horizon-2020-innosup.

[9] https://interreg.eu/.

[10] http://s3platform.jrc.ec.europa.eu/industrial-modernisation.

[11] http://www.consilium.europa.eu/en/policies/investment-plan/.

[12] http://ec.europa.eu/social/main.jsp?catId=1415&langId=en.

[13] http://ec.europa.eu/education/lifelong-learning-programme_en.

[14] www.eif.org.

[15] https://www.plattform-i40.de/I40/Navigation/EN/Home/home.html.

[16] https://catapult.org.uk/.

[17] https://hvm.catapult.org.uk/.

[18] http://industriedufutur.fim.net/.

[19] http://www.fabbricadelfuturo-fdf.it/.

[20] http://www.sviluppoeconomico.gov.it/index.php/it/industria40.

[21] http://www.industriaconectada40.gob.es/Paginas/index.aspx.

ent[22] in Belgium, *Smart Industry*[23] in the Netherlands. In general, these programs include actions spanning from industrial research to innovation activities. In some countries, such programs are linked to the setup of National Technology Clusters representing the priorities of the manufacturing community and coordinating manufacturing stakeholders (e.g. the Italian Cluster Intelligent Factories mentioned in Sect. 21.3.3.2).

At regional level, manufacturing policies are implemented in alignment with the concept of Smart Specialisation promoted by the European Commission.[24] The majority of such programs are co-funded through the European Structural and Investment Funds (ESIF).[25] Initiatives at regional level are more bounded towards innovation in specific industrial domains of excellence of local industry. Some examples are:

- innovation vouchers awarded in Baden-Württemberg[26] (Germany), Lombardy[27] (Italy) and Limburg[28] (the Netherlands);
- specific credit and loans schemes for SMEs such as the *Robotic loan* of Pays de la Loire (France) and the financial tools of Finlombarda[29] in Lombardy (Italy);
- the *Innovation Assistant*[30] in Saxony-Anhalt, Brandenburg, North Rhine-Westphalia (Germany), Kärnten and Tyrol (Austria), through which regions co-funds employment of skilled graduates in regional SMEs to boost know-how transfer and innovation;
- measures supporting the development of new manufacturing skills, such as the Industry 4.0 training programme in Navarre[31] (Spain), *Compétences 2020* in Pays de la Loire (France) and the Flemish Cooperative Innovation Networks-VIS in Belgium.

With a focus on innovation infrastructure at regional level, the Vanguard Initiative (see Sect. 21.3.2.2) aims at the synergic cooperation of European Regions to boost innovation through the establishment of a European network of pilot plants based on smart specialisation. The Vanguard network of regions elaborated a specific model to fund the establishment and operation of pilot plants. Such a model implies a decreasing public contribution from the phase of pilot plants implementation, to the phase of

---

[22]http://www.madedifferent.be/.

[23]https://ec.europa.eu/futurium/en/system/files/ged/nl_country_analysis.pdf.

[24]https://ec.europa.eu/jrc/en/research-topic/smart-specialisation.

[25]https://ec.europa.eu/eip/ageing/funding/ESIF_en.

[26]https://www.wirtschaft-digital-bw.de/en/measures/hightech-digital-innovation-voucher/.

[27]http://www.openinnovation.regione.lombardia.it/it/storie-di-innovazione/news/al-via-i-voucher-per-la-digitalizzazione.

[28]https://www.cpb.nl/sites/default/files/publicaties/download/do-innovation-vouchers-help-smes-cross-bridge-towards-science.pdf.

[29]http://www.finlombarda.it/home.

[30]https://ec.europa.eu/growth/tools-databases/regional-innovation-monitor/support-measure/innovation-assistants-0.

[31]http://clusterautomocionnavarra.com/industria-4-0/herramienta-de-auto-formacion-en-industria-4-0/.

operations (service offering) up to the industrial uptake, with the necessary condition that companies complement public intervention through private co-funding.

With the goal of including in virtuous innovation processes not only the most advanced regions but also emerging manufacturing regions, recently new regions from Eastern Europe joined the Vanguard Association and specific initiatives were supported by the European Commission (e.g. the Greenomed INTERREG Mediterranean Project[32]).

Similarly to the national level and stimulated by the European Cluster Excellence Programme,[33] regional Clusters play a central role in coordinating the industrial ecosystems for the definition and exploitation of regional research and innovation policies. As an example, Lombardy Region created Regional Technology Clusters as actors supporting the regional government in the definition and management of research and innovation policies in alignment with National Clusters. Among them, Associazione Fabbrica Intelligente Lombardia (AFIL) is the cluster representing the manufacturing sector.[34]

However, the current manufacturing policy framework presents still challenges for companies and other research and innovation manufacturing stakeholders, which are summarized as follows:

- Even if the coordination among various policies improved in the last few years, the wide number of fragmented initiatives at all geographic levels makes it difficult for companies to address in a synergic and efficient way the different funding opportunities.
- The current research funding approach is mainly technology-based, while a mission-oriented approach would be more effective to address industrial challenges [1–3].
- Existing inter-regional cooperation programs support joint activities based on a geographical proximity base, while common interests might emerge in a wider European supply chain view.
- Apart from digital technologies, funding for innovation and uptake of specific technologies is limited.
- The availability and access to innovation infrastructure to uptake research results is still limited in Europe.
- The access to research and innovation opportunities and services is unbalanced among the most industrialised and less advanced manufacturing regions.
- The role of intermediators for research and innovation, such as Clusters, is very heterogeneous in various regions and countries.

These challenges are acknowledged by European, national and regional institutions that are working cooperatively to improve the current policy context in the scope of the existing governance framework. The next European Research and Innovation programme (2021–2027) is expected to invest about 100 billion euro. Most

---

[32]https://greenomed.interreg-med.eu/.

[33]https://www.clustercollaboration.eu/eu-initiative/cluster-excellence-calls.

[34]http://www.afil.it/.

of the budget will be dedicated to the programme *Horizon Europe*[35] that, building on the previous *Horizon 2020*, will be based on three pillars: Open Science, Global Challenges and Industrial Competitiveness, and Open Innovation.

The main novelties of Horizon Europe include:

- The key role of the European Innovation Council (EIC)[36] to support breakthrough innovation.
- Research & Innovation Missions [1] within the Global Challenges and Industrial Competitiveness pillar.
- Strengthening international cooperation.
- Enhancement of open access dissemination and exploitation.
- Simplified approaches to partnership and funding.

21.2.2 A Framework for Research and Innovation Funding

Research in general, and specifically industrial research, is traditionally organized through a set of sequential steps starting from basic research towards the exploitation of the results in a real industrial environment. The various phases in this chain have different aims and methodologies and, consequently, involve different actors. Basic or fundamental research is aimed at improving the scientific understanding of phenomena in general; it is manly curiosity-driven and is carried out by public bodies like universities and research bodies. On the contrary, the industrial development phase has the main objective of bringing the results of the research (e.g. a prototype or a demonstrated approach) to its industrial maturity. Consequently, the involved actors are those interested in the exploitation of the results.

Funding research has to take this structure into consideration. Basic research always requires a public funding support since it is not explicitly aimed at devising exploitable results within a defined time horizon. When moving to applied research, funding is traditionally public-private, while industrial development is in charge of venture/risk capitals or industrial partners aiming at the exploitation of the research results, although some public support is possible, e.g. in terms of tax credits.

This traditional approach to research funding has a clear and widely-accepted motivation and a fruitful implementation tradition in many countries. Nevertheless, some criticalities arose, in particular because of the increasing requests to rapidly bring the results of the research to the maturity phase and leverage on the consequent high level of innovation and competitiveness of the industry. As stated by Stokes [6], "The belief that the goals of understanding and use are inherently in conflict, and that the categories of basic and applied research are necessarily separate, is itself in tension with the actual experience of science and industry". Indeed, the link between research

---

[35] http://ec.europa.eu/horizon-europe.

[36] https://ec.europa.eu/programmes/horizon2020/en/h2020-section/european-innovation-council-eic-pilot.

and innovation should be strengthened to fast identify research results that have highest potential impact in industry and to set the conditions for successful exploitation. Nevertheless, establishing connections between the results of basic research and its application is a complex process entailing the need of creating the conditions for cross fertilisation, whose success is also sometimes due to serendipitous events. The link between basic and applied research should be strengthened through:

1. The central role of research and innovation infrastructures that enable basic and applied researchers to work together on goal-based research and innovation activities.
2. The possibility to fund joint basic-applied research projects. Funding could proceed through a *stage-gate approach* in which subsequent research and innovation stages are funded based on positive results of previous phases (gates). Funded actors may change in each phase according to the TRL level, required competences and interests.
3. Partnerships between applied and basic research bodies aiming at promoting cross fertilization paths.
4. The role of Clusters and intermediators that mobilise the different research and innovation stakeholders and provide an efficient and coordinated cooperation environment.

In particular, the next section (Sect. 21.3) will delve into the first item of the list, i.e. research and innovation infrastructures.

21.3   Infrastructures for Industrial Research and Innovation

The concept of mission-oriented policy is strictly related to the need of clearly demonstrating how specific innovation goals have been reached, while supporting the uptake of innovation to generate wide industrial and societal impact. Bringing research results to industrial applications is a critical issue for Europe. This is particularly true for the Key Enabling Technologies (KETs) identified by the European Commission, which have the potential to enable disruptive innovation in manufacturing [7]. Innovation infrastructures can play a fundamental role to overcome the *Valley of Death*, i.e. the phase ranging from Technology Readiness Level (TRL) 6-7 to 9 (commercialisation) [8, 9].

Companies naturally tend to stay anchored to technologies and processes that proved to perform well in the past (the path-dependent and *lock-in* effect reported by [10]). The adoption of technologies and solutions implying a change of manufacturing paradigm presents a set of significant concurrent risks (technical, market, organisational, and institutional risks) that companies are not often able to address [11, 12]. Innovation infrastructures can constitute a unique *protected environment* where novel technologies coming from research can be cooperatively further developed and the contextual factors needed for successful technology exploitation can be set-up (such as market existence and acceptance, sustainable network cooperation, institutional and regulatory framework, etc.) [13]. Innovation infrastructures are also useful to set-up and monitor innovation policies, since they can be stimulated by institutional actors to implement industrial policies, or they can be used by the latter to gather trends and assess the performance of various innovations in order to manage policies contents in the long-term [14].

Despite the relevance of innovation infrastructures and the significant investments devoted to them by governments at European, National and Regional level, innovation infrastructures received limited attention from researchers [15], even though multiple definitions and taxonomies were proposed in literature. As an example, Ballon et al. [14] refer to innovation infrastructures as *Test and Experimentation Platforms* (TEPs) and identified five types of them: innovation platforms, living labs, open and closed testbeds, software platforms. Hellsmark et al. [16] call them *Pilot and Demonstration Plants* (PDPs) and classified the following types: high profile pilot and demonstration plants, (lab-scale or industrial-scale) verification pilot and demonstration plants, deployment pilot and demonstration plants, and permanent test centres.

The types of innovation infrastructure can be classified according to several dimensions, such as the maturity of the technologies of the infrastructure (TRL), the focus on technology scale-up versus (market) testing, the degree of openness of the infrastructure, the type of risks they contribute to mitigate and their goal in terms of addressing non-technical challenges (such as the generation of diffused and tacit knowledge on new technologies, the networking dimension and the needed institutional/regulatory framework) [14, 16].

In this chapter, three relevant types of innovation infrastructure are presented: *Lab-scale Pilot Plant* (Sect. 21.3.1), *Industry-scale Pilot Plant* (Sect. 21.3.2) and *Lighthouse Plant* (Sect. 21.3.3). Each type of innovation infrastructure is exemplified with specific reference to the Italian research, industrial and policy context to show the role of innovation infrastructures in a dynamic lifecycle perspective according to the maturity of the technology and of the industrial uptake process, as suggested in [16].

Lab-scale innovation infrastructures are aimed at increasing the TRL of available research results by integrating multiple technologies for the achievement of an industrial objective (*Lab-scale Pilot Plants*). With this goal, a selected community of key-stakeholders that contribute to generate more mature technologies in a cooperative environment should be established. Subsequently, innovation infrastructures for the wide industrial deployment of mature technologies should be built (*Industry-scale Pilot Plants*) to provide innovative solutions solving specific problems of manufacturing companies. Finally, permanent infrastructures are needed to guarantee continuous support for technology uptake and complete the adoption process (*Lighthouse Plants*).

Pilot plants will help the overall innovation system to make cross-fertilization actions in a multi-facetted, highly networked, and dynamic environment where industrial companies, universities, research institutes, policy makers, and civil society can collaborate [17]. Digital technologies dramatically increase the opportunities of vertical and horizontal integration in complex and dynamic eco-systems. The impact

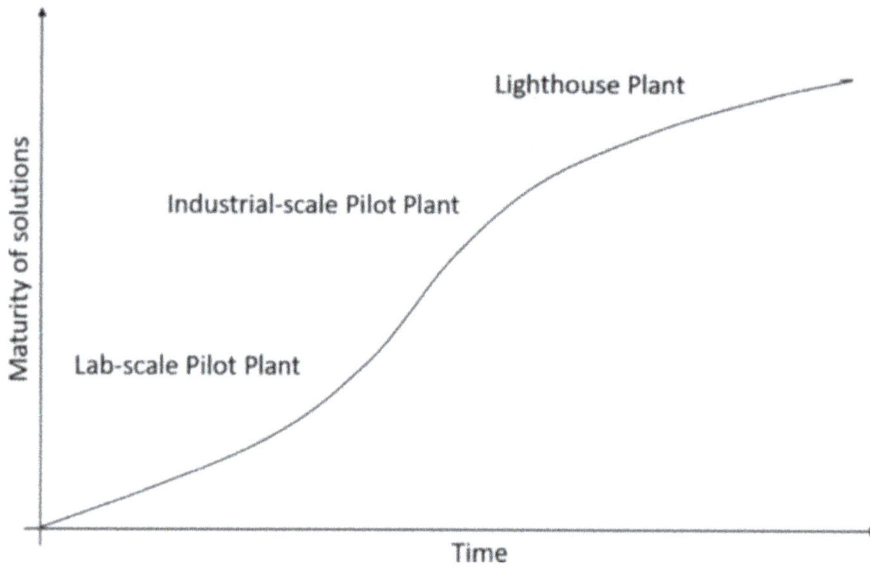

Fig. 21.1 Different type of innovation infrastructure against solutions' maturity and time (adapted from [16])

and strategic importance of pilot plants will be measured in terms of value added, better cohesion, open innovation, and social acceptance of industrial initiatives.

Referring to the framework proposed in [16], Fig. 21.1 shows three different types of infrastructure considering the maturity of solutions (including technologies, organization, business model, supply chain, etc.) and time.

The three types of infrastructure differ for the TRL of their technologies, their main scope, the openness, their funding and business model, as well as for the type of involvement of public authorities.

21.3.1 Lab-Scale Pilot Plants

21.3.1.1 Concept

A Lab-scale Pilot Plant is an innovation infrastructure aimed at supporting research and innovation activities to progress in the TRL scale, making technologies more mature and closer to industrial application (from TRL 4-6 to 5-7). Lab-scale Pilot Plants are focused on the design and finalization of integrated technologies for the solution of specific industrial problems [14], by exploiting solutions that are typically the result of research projects. Therefore, these pilot plants are generally set-up (and owned) by research organisations and universities that define also their strategy and operations rules [16]. Innovative research results are transferred into the pilot plants and the main effort in the setup phase is technology integration under a system engineering perspective. Usually, in fact, research projects generate results in single technology domains, but a systemic perspective to solve specific industrial challenges is missing. Technological equipment of Lab-scale Pilot Plants consists of integrated production systems or lines that can be used in certain industrial domains, but are not customized for specific industrial applications yet. Thus, pilot technologies present a certain degree of flexibility in order to be adapted to different configuration scenarios. The goal of these pilot plants is exactly to define and demonstrate industrial and technological configurations that can represent innovative solutions for the sectors in which they are applied and that can be scaled-up in other pilot plants at industrial level.

Besides the development of new technology setups for specific industrial scenarios, which is a typical innovation activity, Lab-scale Pilot Plants can also perform some research activities that are necessary to integrate the solutions (especially when their nature is highly multi-disciplinary) or to complete the configuration and testing phase. This justifies the central role that research organisations have in this type of innovation infrastructure, in accordance with the main focus of generating scientific and engineering progress. Moreover, a Lab-scale pilot plant has to be in continuous evolution as it tries to integrate and finalize research results as soon as they become available from research projects. To this aim, Lab-scale pilot plants are frequently used as demonstrators in research and innovation projects.

Even though the main goal of Lab-scale Pilot Plants is to lower technology risk and they are generally owned and managed by a single research organization, such facilities constitute an aggregation point for various innovation stakeholders with an important network effect [18]. Besides research organisations, the stakeholders are mainly technology suppliers, that have to cooperate to integrate technologies in manufacturing systems, and manufacturing end-users, that will be the final adopters of technologies. The degree of openness of such facilities will be intermediate: all key-actors at different supply chain and technology levels should be represented, but their number should not be too high in order not to generate competition, conflict of interest and not to reduce the efficiency of cooperation, as also stated by [14]. Usually, stakeholders are highly reputed research and industrial partners in their competence area, that already cooperate in research and innovation activities.

Lab-scale Pilot Plants can be funded and operated exploiting a mix of instruments. Public research and innovation funding (at European, National and Regional levels) supports the development of innovative technology solutions that can be included in the pilot plant. Dedicated funding programs and Regional/National direct funding to research organisations and universities for the setup of infrastructure can support the creation of the pilot plants through the setup of a facility where multiple technologies are integrated. In-kind contribution can be provided by stakeholders, who are interested in the infrastructure because it is a vehicle for the setup of new solutions that later can be sold in the market or can be directly up-taken before competitors. Revenues can be in the form of research and innovation contracts by customers interested in identifying and testing suitable solutions to solve their industrial challenges as well as in the form of incomes from the first sales of such solutions by providers.

21.3.1.2   De- and Remanufacturing Pilot Plant at CNR-STIIMA

A relevant example of Lab-scale Pilot Plant is the "Mechatronics De- and Remanufacturing" pilot plant installed at CNR-STIIMA (ex CNR-ITIA). The pilot plant, in its original configuration, was initially funded by Regione Lombardia with a grant of 1.5 million euro. After several upgrades supported by projects and industrial grants, the pilot plant currently includes innovative technologies and prototypes doubling its initial investment.

The pilot plant goal is to integrate and validate at TRL 5-7 a set of multidisciplinary methodologies, tools and technologies for the smart de- and remanufacturing systems of the future, with specific focus on mechatronic products.

The pilot plant was designed and built according to a precise strategy of CNR-STIIMA (ex CNR-ITIA) that, based on the evidence that End-of-Life (EoL) of mechatronics is addressed in a very fragmented and inefficient way in Europe, decided to invest in the setup of a unique research and innovation facility in terms of process integration and multi-disciplinarity of technological enablers. Single technologies, in fact, are currently available separately as the result of research and innovation projects, but until they were not integrated in a plant that can replicate real industrial processes for the achievement of manufacturing objectives.

The pilot plant includes technologies to support products disassembly, remanufacturing and recycling of materials (addressing mechanical pre-treatments), implementing the most valuable EoL strategy according to the parts to be treated. Innovation is pursued at three levels, as represented in Fig. 21.2: at the level of single process/technologies, of the integrated process chain and of business model.

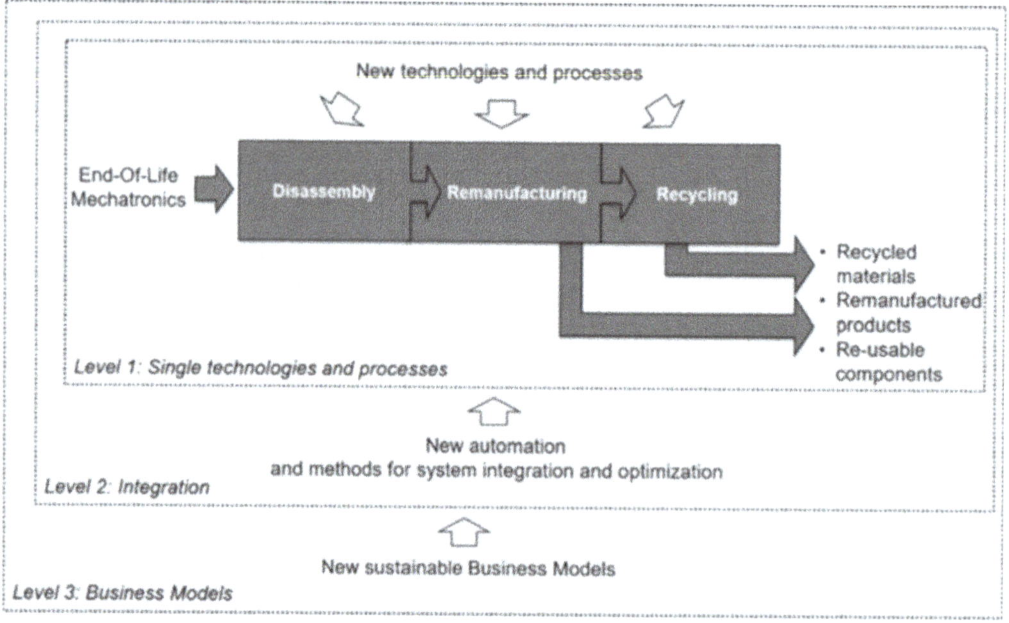

Fig. 21.2 Concept of the lab-scale pilot plant of CNR-STIIMA (ex CNR-ITIA) [19]

The plant consists of three connected cells. The first cell is dedicated to hybrid disassembly of mechatronic components exploiting the human-robot interaction paradigm. The second cell is dedicated to testing and remanufacturing of printed circuit boards (PCBs) and it exploits highly flexible solutions to adapt to the extreme variability of products. The third cell is dedicated to mechanical pre-treatment with low environmental impact (i.e. shredding and materials separation processes) for the recovery of high-value and critical raw materials from PCBs.

A virtual model (digital twin) connected to the real plant was realised for the implementation of the concept of the Digital de-manufacturing factory.

The pilot plant is fully operational and has been employed mainly for research and innovation objectives within several research and innovation projects, for which it is a differential asset. In addition, the plant supports the offering of technology services to companies willing to test the potential of new integrated technological solutions for circular economy, mainly in the automotive, white goods, and telecommunication sectors. These activities allowed building a community of academics, manufacturers, recyclers, remanufacturers and technology providers that constitutes a pool of qualified service offering parties and potential partners for new research and innovation projects. This community is continuously generating new knowledge around the demonstrated technologies and contributes to the consolidation of supply chain relationships that will be necessary when the demonstrated multi-disciplinary solutions will be sold in the market.

Finally, the pilot plant is used also for training and education according to the learning factory paradigm [20–23].

21.3.2　Industrial-Scale Pilot Plants

21.3.2.1　Concept

An Industrial-scale Pilot Plant is an innovation infrastructure aimed at supporting industry in the first uptake of innovative technologies and solutions that have been previously demonstrated in Lab-scale Pilot Plants. Compared to the latter, Industrial-scale Pilot Plants are equipped with technologies at higher TRL level (7-8) which resulted to be successful in precedent innovation phases. The level of flexibility of such technologies is consequently lower, and the pilot plant offers a demonstration facility in real industrial environment to quickly and effectively test the benefits of novel solutions with a setup tailored to specific business applications. Thus, the focus of activity is more on demonstration than on design, which is limited to the final customisation of the solution for the specific users' applications. Limited industrial research activities are carried out.

Demonstration activities are meant to reduce uptake risks. By testing the new technologies on their specific products and processes, companies can better measure expected benefits, thus being able to elaborate robust business plans and to define financial needs. They are also able to anticipate organisational issues linked to new technologies uptake (e.g. production re-organisation and the need of new skills and competences of operators) that the Industrial-scale Pilot Plants might contribute to address through specific industrial-oriented education programs. Finally, technical services received by these pilot plants support industry in the definition of requirements for the technical integration of new technologies in production plants and in the minimization of the inefficiencies during the ramp-up phase.

The main goal of Industrial-scale Pilot Plants is to offer a wide set of technology and business services supporting the uptake. Consequently, these plants are more open that Lab-scale Pilot Plants. Usually, they are not owned by a unique actor, but they present a multi-ownership structure [16]. Public Authorities might also participate or directly influence the governance of such infrastructure, since they are supposed to generate impacts for entire industrial sectors and they need to have a high level of openness to companies, especially to SMEs. This public-private nature of the infrastructure makes the business model challenging, thus becoming a research topic for future research per se.

The networking dimension associated with these pilot plants is very significant, because, besides the goal of supporting technology demonstration and uptake planning, they are supposed to be a meeting point for companies to build new supply chain partnerships that are needed in future operations of novel technologies.

Funding of this type of innovation infrastructure is challenging because required investments are high and a public-private multi-ownership structure may be involved. Public funding plays a major role in triggering the setup of such innovation infrastructures, since the direct benefit for single organisations and private investors is less clear than Lab-scale Pilot Plants [24]. Revenues for the infrastructure derive from direct service contracts with industrial customers, as well as from possible public incentives schemes (such as vouchers) to stimulate the demand of services by industrial companies.

21.3.2.2 Vanguard De- and Remanufacturing Pilot Network for Circular Economy

The "Vanguard Initiative—New Growth Through Smart Specialisation" is a political initiative of more than 30 European Regions aimed at promoting inter-regional cooperation based on smart specialization.[37] The goal of Vanguard is to boost industrial innovation exploiting synergies and complementarities of European Regions. With this goal, regions organize themselves in pan-European partnerships of companies, Research and Technology Organisations (RTOs), Universities and other manufacturing stakeholders that propose and manage strategic projects for the establishment of networks of pilot plants supporting the industrial uptake of innovative technologies and the creation of new European value chains. The network of pilot plants is meant to be a public-private service centre open to companies of all Europe, especially SMEs.

---

[37]https://www.s3vanguardinitiative.eu/.

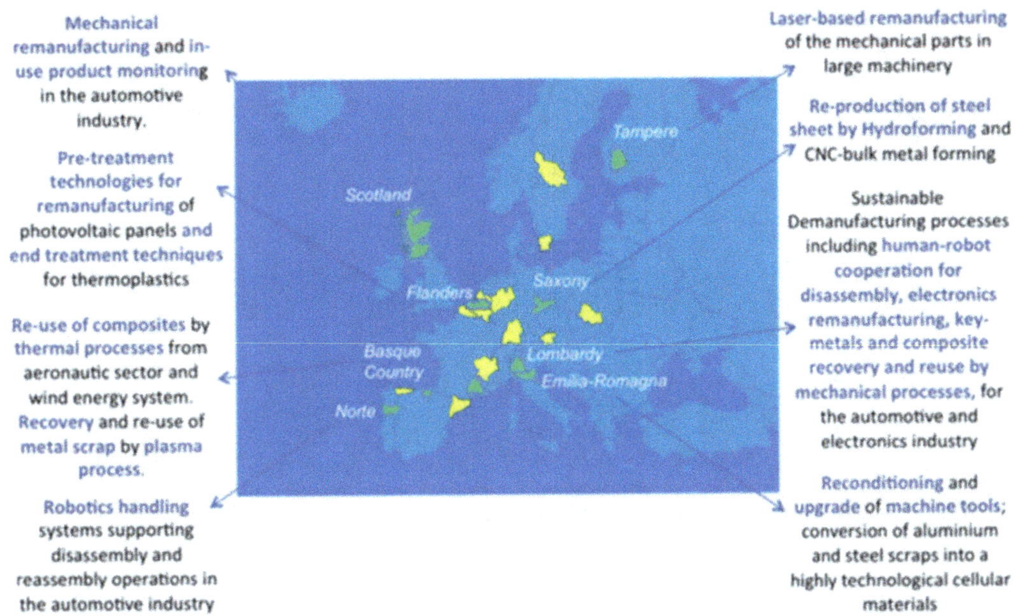

Fig. 21.3 Cross-regional architecture of the De- and Remanufacturing for Circular Economy Pilot Network

Within the Vanguard "Efficient and Sustainable Manufacturing (ESM)" pilot project, the "De- and Remanufacturing for Circular Economy Pilot Plant" was conceived and designed [25]. The cross-regional architecture of the "De- and Remanufacturing" pilot plant currently includes eight Regional Nodes, each of them specialized in a specific testing and demonstration domain (Fig. 21.3).

Each node will be a potential *point of access* for manufacturing end-users in the same or in other regions, depending on the specific capabilities and target sectors. According to regional specialisation, pilot nodes will include a set of advanced technologies to support companies' uptake in specific domains, e.g. the remanufacturing of electronics products, recycling of composites, re-use and recycling of batteries.

The main concept of the De- and Remanufacturing pilot network is represented in Fig. 21.4. For each industrial problem, the most suitable combination of technologies to retrieve the highest residual value from the post-use product will be tested and validated. The output of this process will be a set of demonstrated integrated technological solutions and circular economy business models to support the implementation of the specific business cases at industrial level.

Currently, the definition of the pilot concept is supported by more than 80 private companies (both SMEs and large companies) at European level with a cumulative turnover of 27 billion euro and with some 150,000 employees, and 68 universities and RTOs distributed among the involved regions. These actors also declared their intention to co-fund the development of the pilot network.

It was estimated that this pilot network can lead to about 35 new industrial installations in five years after its setup, bringing a cumulative revenue for the involved companies of about 215 million euro.

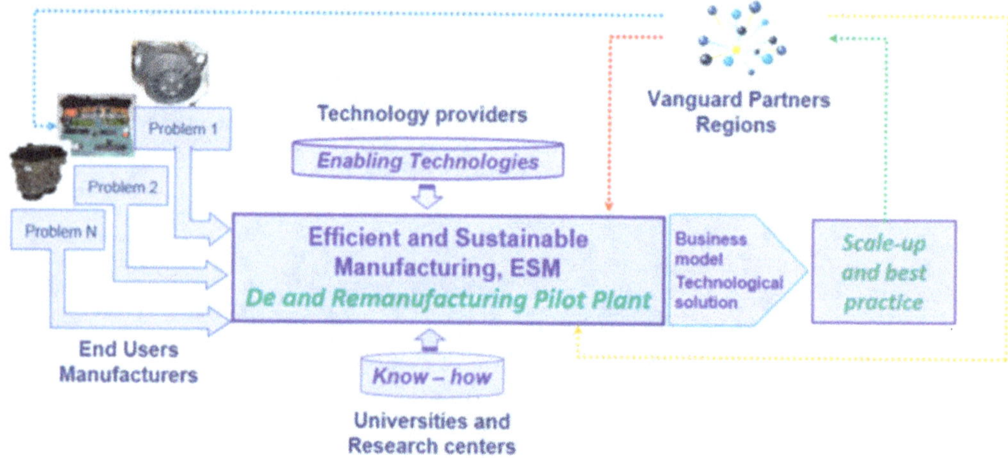

Fig. 21.4 Concept of the De- and Remanufacturing for Circular Economy Pilot Network

The implementation cost of the pilot network is estimated to be 50 million euro. Currently, the Vanguard community and the stakeholders of the pilot plant projects are discussing with European, National and Regional institutions, as well as with banks and other private funding organisations, to define the most appropriate public-private funding mix to establish the infrastructure.

The European Commission has recently selected this partnership to offer support through experts' consulting with the aim of removing the existing implementation bottlenecks in the frame of the S3 Platform on Industrial Modernisation.

21.3.3 Lighthouse Plants

21.3.3.1 Concept

A Lighthouse Plant (LHP) is an infrastructure that aims at creating a reference production plant, owned by a company and operating in a stable industrial environment, based on key enabling technologies whose benefit was previously demonstrated (e.g. in Lab-scale or Industrial-scale pilot plants). The aim of the LHP is twofold: on the one hand, to demonstrate on a long-term basis novel technologies *in operation*, thus supporting the continuous uptake by industry; on the other hand, to trigger the development of industrial research and innovation activities to continuously improve manufacturing solutions according to the progress of technology.

LHPs are conceived as evolving systems and are realized ex-novo or based on an existing plant deeply revisited, where collaborative research and innovation, partially funded by public institutions, is carried out by the owner of the plant together with universities, research centres, and technology providers. The results of research and innovation activities are meant to be readily integrated into the plant. The main difference with respect to the other types of pilot plant presented in Sects. 21.3.1 and

21.3.2 is that LHPs are real plants operated by companies in industrial environments, therefore they prove the sustainability of embedded technologies (TRL 9). A LHP contributes to the generation of new knowledge for the industrial operation of novel technologies. LHPs overcome the purely technology push approach while proposing the use of technologies to solve specific problems, thus creating a link between technologies and a strategy pulled by challenges.

Once realized, a LHP becomes a catalyst for further industrial research and innovation activities, playing the role of test house in subsequent initiatives at regional, national and international level to guarantee that the plant continues supporting over time the uptake of new technologies that are there applied as early user.

Being owned by an industrial company, a LHP will be mainly funded by the company itself, but public authorities can stimulate and co-fund the setup and following research and innovation activities. While guaranteeing intellectual property rights (IPR) and confidentiality of key portions of the plant, the goal is to open as much as possible the LHP to other companies and in general to the industrial system. Educational and training activities must be designed to show how to follow the innovation path, thus increasing the overall culture of the manufacturing network. Various stakeholders can benefit from a LHP:

- *Manufacturers* can set-up innovative plants that are constantly evolving coherently, while receiving a particular and continuous attention on their industrial issues from technology suppliers, universities and research organisations. The manufacturers will have visibility according to the strategic scope of the plant, which is part of a large LHP network.
- *Technology providers* have the opportunity to develop new solutions that can be tested in real production plants and be highly visible to potential buyers. This is particularly strategic for SMEs and startup companies.
- *SMEs* have access to concrete examples of application of new technologies that can inspire several other smaller scale implementations.
- *Universities and Research Organisations* have the opportunity to be involved in research and innovation projects with production plants as a way to enhance the results of their research and to receive new founding for subsequent activities.
- The *supply chain* around the plant is positively influenced by the innovation that often requires commitment from different members of the supply chain upstream and downstream.
- *Local and national governments* have the opportunity to assess concrete results of the implemented innovation actions and to showcase best practices to national and international actors. Based on the results, the local or national governments can also identify strategic initiatives to be funded for basic or applied research in the manufacturing domain.

21.3.3.2   LHP in Italy

The LHPs concept as presented in the previous section has been defined by Italian Cluster Intelligent Factories (CFI)[38] to further boost the National Plan Enterprise 4.0[39] designed by the Ministry of Economic Development in Italy (MISE) in 2017. This plan included incentives for super- and hyper-depreciation as a way to support the implementation of advanced technologies in Italian manufacturing companies.

CFI coordinates the LHP initiative in accordance with the strategic action lines identified in its research and innovation roadmap [26]. A formal procedure has been established for the submission of LHP proposals. Each company interested in the LHP initiative can submit a proposal of research and innovation project linked to a new plant (or a deeply renovated one) to the CFI Technical Scientific Committee that will later express its opinion and possibly admit it to the LHP candidate list. CFI supports the preparation of the LHP proposal till the submission to MISE. If the proposal is approved by MISE, the new LHP will receive funding. After the approval, each LHP proposer is invited to set up a scientific-strategic management board for the project; an expert appointed by CFI Coordination and Management Body is invited to participate at least twice a year in the board meetings to discuss:

- the advancement of the project with respect to the workplan;
- consistency and synergy of the project activities with the CFI activities;
- coordination and planning of joint initiatives with CFI.

Furthermore, each LHP project will participate in the Lighthouse Plant Club[40] managed by CFI to support:

- promotion of the direct interaction with Ministries;
- visibility of the LHP at national and international level;
- access to a set of competences available among CFI members;
- participation in the initiatives promoted by CFI;
- participation in the training and education activities promoted by CFI;
- identification of follow-up research and innovation initiatives.

Currently, four LHP proposals have already been approved by MISE (see Fig. 21.5), each one with a total value for research and innovation activities ranging from 10 to 19 million euro (in addition to the value of the new plant):

- *Ansaldo Energia.*[41] Smart Factory based on the application of Digital Technologies.

---

[38]www.fabbricaintelligente.it.

[39]http://www.sviluppoeconomico.gov.it/index.php/it/industria40.

[40]http://www.fabbricaintelligente.it/english/light-house-club/.

[41]www.ansaldoenergia.com.

Fig. 21.5 Lighthouse plants approved by MISE: **a** Ansaldo Energia, **b** ORI Martin and Tenova, **c** ABB Italy, **d** Hitachi Rail Italy. *Courtesy of Italian Cluster Intelligent Factories (CFI)*

- *ORI Martin*[42] and *Tenova.*[43] Cyber Physical Factory for steel production from scraps.
- *ABB Italy.*[44] Multi-plant factory for the production of the complete Circuit breakers portfolio.
- *Hitachi Rail Italy.*[45] New Products Platforms produced in Digital Factories.

MISE Ministry and the local Regions will contribute to the public funding of these research and innovation projects connected to the plant for 36 months by signing a strategic innovation agreement for each Lighthouse Plant.

The LHPs will enable the CFI community, and in particular SMEs, to have a special access to the technologies of Industry 4.0 in real applications.

Acknowledgements This work has been partially funded by the Italian Ministry of Education, Universities and Research (MIUR) under the Flagship Project "Factories of the Future—Italy" (Progetto Bandiera "La Fabbrica del Futuro") [4].
The authors would like to thank Marcello Colledani, Rosanna Fornasiero, and Marcello Urgo for their valuable contributions to this chapter.

---

[42] www.orimartin.com.

[43] www.tenova.com.

[44] https://new.abb.com.

[45] http://italy.hitachirail.com/en.

References

1. Mazzucato M (2018) Mission-oriented research & innovation in the European Union—a problem-solving approach to fuel innovation-led growth. Publications Office of the European Union. ISBN 978-92-79-79832-0. https://doi.org/10.2777/360325
2. European Commission (2018) Re-finding industry—defining innovation. Publications Office of the European Union. ISBN 978-92-79-85271-8. https://doi.org/10.2777/927953
3. Tolio T, Copani G, Terkaj W (2019) Key research priorities for factories of the future—part I: missions. In: Tolio T, Copani G, Terkaj W (eds) Factories of the future. Springer
4. Terkaj W, Tolio T (2019) The Italian flagship project: factories of the future. In: Tolio T, Copani G, Terkaj W (eds) Factories of the future. Springer
5. Jovane F, Yoshikawa H, Alting L, Boër CR, Westkamper E, Williams D, Tseng M, Seliger G, Paci AM (2008) The incoming global technological and industrial revolution towards competitive sustainable manufacturing. CIRP Ann 57(2):641–659. https://doi.org/10.1016/J.CIRP.2008.09.010
6. Stokes DE (1997) Pasteur's quadrant—basic science and technological innovation. Brookings Institution Press
7. European Commission (2012) A European strategy for key enabling technologies—a bridge to growth and jobs. COM(2012) 341 final
8. European Commission (2016) An analysis of drivers, barriers and readiness factors of EU companies for adopting advanced manufacturing products and technologies. Accessed at https://ec.europa.eu/growth/content/analysis-drivers-barriers-and-readiness-factors-eu-companies-adopting-advanced-1_en. Accessed 01 Aug 2018
9. European Commission (2015) Promoting the access of SMEs to KETs technology infrastructures. Publications Office of the European Union. ISBN 978-92-79-36385-6. https://doi.org/10.2769/92236
10. Schot J, Geels FW (2008) Strategic niche management and sustainable innovation journeys: theory findings, research agenda, and policy. Technol Anal Strateg Manag 20(5):537–554
11. Hekkert M, Suurs R, Negro S, Kuhlmann S, Smits R (2007) Functions of innovation systems: a new approach for analysing technological change. Technol Forecast Soc Change 74(4):413–432
12. Bergek A, Jacobsson S, Carlsson B, Lindmark S, Rickne A (2008) Analyzing the functional dynamics of technological innovation systems: a scheme of analysis. Res Policy 37(3):407–429
13. Smith A, Raven R (2012) What is protective space? Reconsidering niches in transitions to sustainability. Res Policy 41(6):1025–1036
14. Ballon P, Pierson J, Delaere S (2005) Test and experimentation platforms for broadband innovation: examining European practice. http://dx.doi.org/10.2139/ssrn.1331557
15. Frishammar J, Söderholm P, Bäckström K, Hellsmark H, Ylinenpää H (2014) The role of pilot and demonstration plants in technological development: synthesis and directions for future research. Technol Anal Strateg Manag 27(1):1–18
16. Hellsmark H, Frishammar J, Söderholm P, Ylinenpää H (2016) The role of pilot and demonstration plants in technology development and innovation policy. Res Policy 45(9):1743–1761
17. Camarinha-Matos LM, Fornasiero R, Afsarmanesh H (2017) Collaborative networks as a core enabler of Industry 4.0. In: Camarinha-Matos L, Afsarmanesh H, Fornasiero R (eds) Collaboration in a data-rich world. PRO-VE 2017. IFIP advances in information and communication technology, vol 506. Springer, Cham
18. Tichkiewitch S, Shpitalni M, Krause F-L (2006) Virtual research lab: a new way to do research. CIRP Ann 55(2):769–792. https://doi.org/10.1016/J.CIRP.2006.10.007
19. Copani G, Brusaferri A, Colledani M, Pedrocchi N, Sacco M, Tolio T (2012) Integrated De-manufacturing systems as new approach to End-of-Life management of mechatronic devices. In: Proceedings of the 10th global conference on sustainable manufacturing, Istanbul, 31 Oct–2 Nov 2012
20. Mavrikios D, Papakostas N, Mourtzis D, Chryssolouris G (2013) On industrial learning and training for the factories of the future: a conceptual, cognitive and technology framework. J Intell Manuf 24(3):473–485

21. Chryssolouris G, Mavrikios D, Rentzos L (2016) The teaching factory: a manufacturing education paradigm. Procedia CIRP 57:44–48
22. Abele E, Chryssolouris G, Sihn W, Metternich J, ElMaraghy H, Seliger G, Sivard G, ElMaraghy W, Hummel V, Tisch M, Seifermann S (2017) Learning factories for future oriented research and education in manufacturing. CIRP Ann 66(2):803–826. https://doi.org/10.1016/J.CIRP.2017.05.005
23. Caldarola E, Modoni G, Sacco M (2018) Manulearning: a knowledge-based system to enable the continuous training of workers in the manufacturing field. In: Proceedings of 12th International conference on e-Learning
24. Mazzucato M (2014) The entrepreneurial state—debunking public vs private sector myths. Anthem Press, New York, USA
25. Efficient and Sustainable Manufacturing (ESM). https://www.s3vanguardinitiative.eu/cooperations/efficient-and-sustainable-manufacturing-esm. Accessed 31 July 2018
26. Associazione Cluster Fabbrica Intelligente (2015) Roadmap per la ricerca e l'innovazione—research and innovation roadmap. http://www.fabbricaintelligente.it/english/roadmap/

www.ingramcontent.com/pod-product-compliance
Lightning Source LLC
Chambersburg PA
CBHW081300090726
47818CB00079B/207